Springer Series in Optical Sciences

Founding Editor

H. K. V. Lotsch

Volume 248

Springer Series in Optical Sciences is led by Editor-in-Chief William T. Rhodes, Florida Atlantic University, USA, and provides an expanding selection of research monographs in all major areas of optics:

- lasers and quantum optics
- ultrafast phenomena
- optical spectroscopy techniques
- optoelectronics
- information optics
- applied laser technology
- industrial applications and
- other topics of contemporary interest.

With this broad coverage of topics the series is useful to research scientists and engineers who need up-to-date reference books.

Dheeraj Kumar Singh · Ashish Kumar Mishra ·
Arnulf Materny
Editors

Raman Spectroscopy

Advances and Applications

Springer

Editors
Dheeraj Kumar Singh
School of Nano Sciences
Central University of Gujarat
Gandhinagar, Gujarat, India

Ashish Kumar Mishra
School of Materials Science
and Technology
Indian Institute of Technology (BHU)
Varanasi, India

Arnulf Materny
School of Science
Constructor University
Bremen, Bremen, Germany

ISSN 0342-4111 ISSN 1556-1534 (electronic)
Springer Series in Optical Sciences
ISBN 978-981-97-1702-6 ISBN 978-981-97-1703-3 (eBook)
https://doi.org/10.1007/978-981-97-1703-3

This Springer imprint is published by the registered company Springer Nature Singapore Pte Ltd.
The registered company address is: 152 Beach Road, #21-01/04 Gateway East, Singapore 189721, Singapore

Paper in this product is recyclable.

Dedicated to my mother Late Pramila Singh

Preface

Raman spectroscopy is an indispensable analytical technique used for material characterization in modern science. It has witnessed a remarkable journey of discovery since its inception in 1928 by the Indian physicist Sir Chandrasekhara Venkata Raman, also known as Sir C. V. Raman. Over the decades, this technique has experienced a remarkable evolution and has become a crucial tool with a wide range of applications, in many fields like physics, chemistry, biology, medicine, and material science. The various types of Raman spectroscopy including resonance Raman, surface-enhanced Raman spectroscopy (SERS), coherent anti-Stokes Raman scattering (CARS), imaging (Micro-) Raman spectroscopy, and so on have been explored with time and technological advancements. Currently, there are more than 25 variants of Raman spectroscopy techniques available for material characterization, each with a unique range of applications in various research fields. These techniques are constantly being developed. Therefore, through this book, "Raman Spectroscopy: Advances and Applications", the authors aim to provide an overview about the recent development of some of the important advancements in Raman spectroscopic techniques to the researchers and students for enhancing their knowledge and comprehension of the significance of Raman spectroscopy's key concepts.

The present book consists of 14 chapters, which provide information about the development of various fundamental Raman spectroscopy techniques. The book is organized as follows. In Chap. 1, Dr. Dheeraj K. Singh from Central University of Gujarat, Gandhinagar, India, introduces the life journey of Sir. C. V. Raman starting from his birth and education to his research and eventual discovery of the Raman effect. This chapter also sheds some first light on the development of some important Raman spectroscopic techniques such as SERS, TERS, CARS, etc., and their applications. The subsequent 13 chapters are purely dedicated to a comprehensive introduction and discussion of different Raman spectroscopic techniques and their potential applications in a various fields. In Chap. 2, Dr. Pradeep Kumar, IIT Mandi, India, focuses on general resonant and non-resonant Raman spectroscopy techniques aiming at the better understanding of various important properties of the semiconducting transition metal dichalcogenides (TMDs). In Chap. 3, Dr. Surajit Saha IISER Bhopal, India, provides a detailed discussion of magnetic-field- and pressure-dependent Raman spectroscopy and the importance of these techniques for studying fundamental aspects of quantum materials and their potential applications in devices. In Chap. 4, Dr. Ashish Kumar Mishra, IIT (BHU) Varanasi, India, briefly describes optothermal Raman (OTR) spectroscopy to understand phonon anharmonicity, thermal transport properties, and thermal behavior of two-dimensional (2D) materials like graphene and TMDs. In Chap. 5, Dr. Sunil K. Srivastava, MGCU Bihar, India, provides a detailed explanation of polarized Raman spectroscopy and its uses for the characterization of isotropic and anisotropic materials, as well as its practical applications in the field of bioscience. This chapter also covers the integration of polarized Raman spectroscopy with other imaging techniques like scanning electron (SEM) and atomic force microscopy (AFM) for more detailed material characterization. A review of optical tweezers in Raman spectroscopy (OTRS) is given in Chap. 6 by Dr. Saurabh Raj, IIT Delhi, India, where a thorough discussion of the working principle of the OTRS setup along with the various design configurations and applications with enhanced Raman signal is included. In Chap. 7, Dr. Debraj Gangopadhyay, IOCB Prague, Czech Republic provides an overview of the Raman optical activity (ROA) spectroscopy technique beginning with its historical development and theoretical concepts and continuing with the advances of experimental techniques like resonance ROA and surface-enhanced ROA. The applications of ROA to the study of molecular chirality as well as its new possibilities and limitations in analytical chemistry are also discussed. Since Raman scattering only yields relatively weak signals, small concentrations of samples can often not be accessed. Here, the SERS technique is a powerful spectroscopic tool where the Raman signal is enhanced through electromagnetic and chemical enhancement mechanisms when interacting with coin-metal substrates. In Chap. 8, Dr. Arnulf Materny and coworkers from the Constructor University, Bremen, Germany introduce the fundamental study of SERS using the so-called "Kretschmann setup" for examining the interaction between the molecular electronic system and the metal electrons. As an advancement to SERS technique, Dr. P. Tamilarasan from CSIR-CECRI, Tamil Nadu, India gives a comprehensive overview of the early history, fundamentals, significant applications, challenges, and future prospects of the electrochemical SERS (EC-SERS) technique in Chap. 9. In Chap. 10, Dr. Om Prakash, IPCPAS Warsaw, Poland introduces surface-enhanced Raman excitation spectroscopy (SERES) technique with examples for the various applications like plasmon- and wavelength-dependent techniques. In Chap. 11, Dr. Volker Deckert, IPS Jena, Germany briefly describes the tip-enhanced Raman (TERS) technique including the basic working mechanism, development, and potential applications for diverse fields of studies. The nanometer-sized structures can be resolved beyond the diffraction limit of optical microscopes using this TERS technique. In Chap. 12, Dr. Nilesh Kumar Rai, C.M.P. Degree College, Central University of Allahabad, UP, India discusses hyper and stimulated Raman spectroscopy. Theoretical background of these methods is explained and time-resolved studies of the various stimulated Raman scattering processes along with examples and applications are introduced. Chapter 13 focuses on coherent anti-Stokes Raman scattering (CARS) as an important example for Raman spectroscopy based on nonlinear optical processes. In this chapter, Dr. Alexis Bohlin from DUT, Netherlands, and LUT, Kiruna, Sweden outlines the fundamentals of the application of CARS to the investigation of gases and discusses recent advancements and developments. Finally, Chap. 14 demonstrates how the various Raman imaging methods like spontaneous imaging, imaging with SERS, TERS, and CARS can be applied for biomedical research and applications. In this chapter, Dr. S. P. Singh, IIT Dharwad, India introduces the principles and realization of Raman imaging for different applications.

By giving readers a thorough understanding of the various advancements in Raman spectroscopic techniques, this book will hopefully foster the reader's interest for Raman spectroscopy, showing how Sir C.V. Raman's work resulted in the development of a great variety of powerful analytical techniques, which may serve to inspire young researchers to make new discoveries in this area.

Gandhinagar, India — Dheeraj Kumar Singh
Varanasi, India — Ashish Kumar Mishra
Bremen, Germany — Arnulf Materny

Acknowledgements

First of all, I do express my gratitude to Almighty God Prabhu Shree Ram, Bajarang Balee, and Maa Saraswati to give me strength for the successful collections of the quality book chapters on Raman spectroscopy. The book entitled as "Raman Spectroscopy: Advances and Applications" consists of 14 chapters from various reputed institutions across India and abroad. The execution of the book was possible with the unconditional support and help by many individuals, scientists, and academicians. I would like to express my sincere thanks to all of them. At first, I would like to show my heartfelt gratitude and thanks to all contributors for providing a quality chapter in the recent developments of Raman spectroscopy. I extend my heartfelt gratitude to Central University Gujarat, especially Prof. R. S. Dubey, the Vice Chancellor of Central University of Gujarat for a vibrant environment and full support. I also acknowledge Prof. B. N. Singh, Director General, IITRAM for their motivations and good infrastructure.

I owe my deepest gratitude toward Springer publisher team, especially Loyola D'Silva, Prasanth Anandan, Diya, books production team, Editorials team, etc. for their guidance to complete the book. I do take the opportunity to thank my seniors and collaborator Prof. Arnulf Materny, Germany and Dr. Ashish Mishra, IIT BHU for their enormous support and scientific discussions.

I would like to give my sincere regards to the Science and Engineering Research Board (SERB), DST, Government of India, for financial support to conduct research work which makes a basis to the collaboration with the expert researchers and exchange our ideas. My heartfelt and special thanks go to my Ph.D. student Mr. Amit Bhojani for his valuable time to help me during editing the book. Last but not least, I would like to thank my family, my wife Dr. Shweta Singh, and my daughter Drishti Singh, father Jamuna Prasad Singh for their valuable support and sacrifices during the preparation of this book.

Prof. Dr. Dheeraj Kumar Singh
Central University of Gujarat
Gandhinagar, Gujarat, India

Contents

Editors and Contributors

About the Editors

Dr. Dheeraj Kumar Singh (Alexander von Humboldt Fellow, Germany & JSPS Fellow, Japan)
Associate Professor, School of Nano Sciences, Central University of Gujarat, India—382030
Homepage: https://dsinghmscrl.wixsite.com/mysite

He received his Ph.D. degree in Physics from Banaras Hindu University (BHU), Varanasi, India in 2010. He was subsequently a CSIR-Nehru Postdoc at NCL Pune; NRF Postdoc in Seoul, South Korea; and finally was granted the renowned Alexander von Humboldt (AvH) Postdoctoral Fellowship for research in Germany. He was also awarded the prestigious international award of a JSPS Fellowship, Japan, Dr. D. S. Kothari Postdoctoral Fellowship, CSIR-SRF, UGC-Meritorious Fellowship (RFSMS), etc. Currently, he is working as Associate Professor, School of Nano Sciences at Central University of Gujarat. His research group is mainly focused on frequency- and time-domain spectroscopic techniques, their developments and applications on functionalized nanomaterials, ionic liquids, biomolecules, molecular interactions, materials for renewable-energy production, etc. Moreover, he has contributed many book chapters and has a patent granted by the USA and India. He has organized many workshops and a conference on "Spectroscopy and Materials theme". He has edited four books in Springer Nature as a corresponding editor. He has been awarded the prestigious ECR project grant by SERB-DST Government of India and DST-Inspire project for his Ph.D. student. He has published more than 80 research papers in international journals of repute. e-mail: dheeraj.singh@cug.ac.in; dheerajsingh@iitram.ac.in

Dr. Ashish Kumar Mishra Associate Professor
School of Materials Science and Technology
IIT (BHU), Varanasi-221005, India
Homepage: Dr. Ashish Kumar Mishra & Research Group

He is an Associate Professor in School of Materials Science and Technology, IIT(BHU), Varanasi (January 2024–Present). He did his M.Sc. (Physics) from V.B.S. Purvanchal University, India (June 2005) and Ph.D. (*with Best Thesis Award*) from Indian Institute of Technology Madras, India (July 2011). He has worked as an Assistant Professor in School of Materials Science and Technology, IIT(BHU), Varanasi, India (October 2016–January 2024); INSPIRE Faculty at IISER Bhopal, India (February 2015–September 2016); Post Doctoral Fellow at Rensselaer Polytechnic Institute (RPI), Troy (NY), USA (January 2012–December 2014); and Visiting Faculty at Indian Institute of Science Education & Research (IISER) Bhopal, India (July 2011–January 2012). He has published more than 53 papers in peer-reviewed high-impact international journals, 3 books chapters, and has filed 9 national/international patents, out of which 4 are granted. He has been awarded membership of Indian National Young Academy of Science (INYAS-2022), Early Career Research Award (2017), and INSPIRE Faculty Award (2014). e-mail: akmishra.mst@iitbhu.ac.in

Prof. Dr. Arnulf Materny Professor of Chemical Physics
Physics & Earth Sciences
Constructor University, Bremen, Germany
Campus Ring 8, 28759 Bremen, Germany
Homepage: http://amaterny.user.jacobs-university.de/

He received his Ph.D. in 1992 with distinction under the supervision of pioneer Raman spectroscopist Prof. Wolfgang Kiefer at the University of Würzburg, Germany. Further, he was a Postdoc at Caltech, Pasadena, U.S.A. There, he started to work in the field of "femtosecond spectroscopy" in the group of Nobel Laureate Prof. Ahmed H. Zewail. Currently, he is serving as a Full Professor of Chemical Physics at Jacobs University Bremen, Germany. His research interests cover the development and application of frequency and time-domain spectroscopic techniques to various molecules and materials with a special focus on linear and nonlinear Raman spectroscopy. He has received several awards, including a Kekule Fellowship and the Hoechst Prize for his Ph.D. work, a Faculty Award by the JMU Würzburg and the Heisenberg Fellowship by the German Research Foundation (DFG) for his postdoctoral research. More than 20 Ph.D.s and 17 postdoctoral researchers including many Alexander von Humboldt fellows successfully completed their projects under his guidance. His research has been supported by many grants awarded to him, e.g., by DFG and industry. He is the member of several editorial boards including Journal of Raman Spectroscopy. He has published more than 280 articles in international journals of repute. e-mail: a.materny@jacobs-university.de

Contributors

Amit K. Bhojani Department of Basic Sciences, Institute of Infrastructure Technology Research and Management (IITRAM), Ahmedabad, Gujarat, India

Alexis Bohlin Department of Aerospace Engineering, Delft University of Technology, Delft, The Netherlands;
Space Campus, Luleå University of Technology, Kiruna, Sweden;
Inspection Technologies, GKN Aerospace Engine Systems, Trollhättan, Sweden

Petr Bouř Institute of Organic Chemistry and Biochemistry, Czech Academy of Sciences, Prague, Czech Republic;
Department of Analytical Chemistry, University of Chemistry and Technology, Prague, Czech Republic

Leonardo Castellanos Department of Aerospace Engineering, Delft University of Technology, Delft, The Netherlands

Moumita Das Institute of Organic Chemistry and Biochemistry, Czech Academy of Sciences, Prague, Czech Republic;
Department of Analytical Chemistry, University of Chemistry and Technology, Prague, Czech Republic

Volker Deckert Leibniz Institute of Photonic Technology (IPHT), Jena, Germany; Institute of Physical Chemistry, Friedrich Schiller University Jena, Jena, Germany; Abbe Center of Photonics, Friedrich Schiller University Jena, Jena, Germany

Ashish Kumar Dhillon Department of Chemistry, Indian Institute of Technology Delhi, Delhi, India

Debraj Gangopadhyay Institute of Organic Chemistry and Biochemistry, Czech Academy of Sciences, Prague, Czech Republic

Jay Deep Gupta School of Materials Science and Technology, Indian Institute of Technology (Banaras Hindu University), Varanasi, India

Priyanka Jangra School of Materials Science and Technology, Indian Institute of Technology (Banaras Hindu University), Varanasi, India

Aprajita Joshi Department of Physics, Indian Institute of Science Education and Research Bhopal, Bhopal, Madhya Pradesh, India

Jiro Karlo Department of Biosciences and Bioengineering, Indian Institute of Technology Dharwad, Dharwad, India

Dmitrii Kliukin Department of Aerospace Engineering, Delft University of Technology, Delft, The Netherlands

Deepu Kumar School of Physical Sciences, Indian Institute of Technology Mandi, Mandi, India

Pradeep Kumar School of Physical Sciences, Indian Institute of Technology Mandi, Mandi, India

Andrii Kurochka Institute of Organic Chemistry and Biochemistry, Czech Academy of Sciences, Prague, Czech Republic

Pankaj Kumar Kushwaha Department of Physics, Mahatma Gandhi Central University, Motihari, Bihar, India

Arnulf Materny School of Science, Constructor University, Bremen, Germany

Francesco Mazza Department of Aerospace Engineering, Delft University of Technology, Delft, The Netherlands

Ashish Kumar Mishra School of Materials Science and Technology, Indian Institute of Technology (Banaras Hindu University), Varanasi, India

Faezeh Mohaghegh School of Science, Constructor University, Bremen, Germany

N. K. Murugasenapathi Electrodics and Electrocatalysis Division (EEC), CSIR-Central Electrochemical Research Institute (CECRI), Tamil Nadu, Karaikudi, India;
Academy of Scientific and Innovative Research (AcSIR), Ghaziabad, India

Tamilarasan Palanisamy Electrodics and Electrocatalysis Division (EEC), CSIR-Central Electrochemical Research Institute (CECRI), Tamil Nadu, Karaikudi, India;
Academy of Scientific and Innovative Research (AcSIR), Ghaziabad, India

Om Prakash Institute of Physical Chemistry Polish Academy of Sciences, Warsaw, Poland

Nilesh Kumar Rai Department of Physics, C.M.P. Degree College, University of Allahabad, Prayagraj, Uttar Pradesh, India

Saurabh Raj School of Biological Sciences, Indian Institute of Technology Delhi, Delhi, India

Bhumika Ray School of Biological Sciences, Indian Institute of Technology Delhi, Delhi, India;
ASSIST Division, CSIR-Indian Institute of Toxicology Research, Lucknow, Uttar Pradesh, India

Syed S. Razi Department of Chemistry, Gaya College, Bihar, India

Surajit Saha Department of Physics, Indian Institute of Science Education and Research Bhopal, Bhopal, Madhya Pradesh, India

Soumik Siddhanta Department of Chemistry, Indian Institute of Technology Delhi, Delhi, India

Akriti Singh Department of Physics, Indian Institute of Science Education and Research Bhopal, Bhopal, Madhya Pradesh, India

Dheeraj K. Singh School of Nano Sciences, Central University of Gujarat, Gandhinagar, Gujarat, India

S. P. Singh Department of Biosciences and Bioengineering, Indian Institute of Technology Dharwad, Dharwad, India

Shweta Singh Department of Physics, Banaras Hindu University, Varanasi, India

Sunil Kumar Srivastava Department of Physics, Mahatma Gandhi Central University, Motihari, Bihar, India

Alireza Mazaheri Tehrani School of Science, Constructor University, Bremen, Germany

Wei Wang Leibniz Institute of Photonic Technology (IPHT), Jena, Germany;
Institute of Physical Chemistry, Friedrich Schiller University Jena, Jena, Germany

Chapter 1
Saga and Developments of Raman Spectroscopy

Amit K. Bhojani, Shweta Singh, and Dheeraj K. Singh

1.1 From Birth to Discovery

Sir Chandrasekhara Venkata Raman (Fig. 1.1), also known as C. V. Raman was the first Indian Physicist who received a Nobel prize in Physics in 1930 for his path-breaking discovery of the 'Raman effect'. He was born on 7th November 1888, in Tiruchirapalli, present-day Tamil Nadu, India [1]. He came from a family of scholars and academics as his father Chandrasekhara Ramanathan Iyer was a teacher at a local high school and later on, he was appointed as Physics faculty at Mrs. A. V. Narasimha Rao college in Vizagapatam and his mother, Parvati Ammal, was a homemaker belongs to a well-educated family that was renowned for its Sanskrit knowledge. Moreover, his elder brother Chandrasekhara Subrahmanya Iyer was also one of the greatest scientists of the 20th century. Raman has a great interest in Physics and Mathematics as well as like his father, he was excellent in violin and also, deeply interested in music and acoustics [2].

C. V. Raman was an extraordinary child from a young age, he completed his secondary and higher secondary education with the first rank at the age of 11 and 13 from the St Aloysius' Anglo-Indian High School, which was the first English medium school in the state. Raman's father had been transferred to Presidency College, the University of Madras (now Chennai) to teach Physics and Mathematics. From the same university, Raman completed his Bachelor of Arts (B.A.) degree in 1904 with

A. K. Bhojani
Department of Basic Sciences, Institute of Infrastructure Technology Research and Management (IITRAM), Ahmedabad, Gujarat 380026, India

S. Singh
Department of Physics, Banaras Hindu University, Varanasi 221005, India

D. K. Singh (✉)
School of Nano Sciences, Central University of Gujarat, Sector-30, Gandhinagar, Gujarat 382030, India
e-mail: dheerajsingh84@gmail.com; dheeraj.singh@cug.ac.in

D. K. Singh et al. (eds.), *Raman Spectroscopy*, Springer Series in Optical Sciences 248,
https://doi.org/10.1007/978-981-97-1703-3_1

Fig. 1.1 Sir C. V. Raman and his signature. *Source* https://commons.wikimedia.org/wiki/File:Sir_CV_Raman.JPG

first rank in the class and also received gold medals in Physics and English at the age of just 16. After 2 years, he completed his Master of Arts (M.A.) course from the same university with the highest honors. During post-graduation, Raman published his first research work entitled "Unsymmetrical diffraction bands due to a rectangular aperture" in the Philosophical Magazine, a British journal in 1906 [3]. He had also performed many research works in this time interval, however, due to certain circumstances those works have not been published [1]. Prof. Rhishard Llewellyn Jones of Physics was very impressed with Raman's aptitude toward research and also encouraged him to go to England for further research studies. In 1907, he had been appointed as assistant accountant general at the Indian Finance Department, Calcutta by appearing first in the special examination at the age of 19. There he came into contact with the Indian Association for the Cultivation of Science (IACS), which was India's first research institute established by an Indian doctor, Mahendralal Sircar in 1876. This institute allowed Raman to perform his research work independently in their laboratory, where Raman carried out single-handedly his most research works on optics and acoustics till 1917. Raman became a member of life in this institute and in 1919, he became secretary of IACS and took the lead in publishing its Bulletin and Proceedings, the latter in 1926 it was renamed as Indian Journal of Physics. In the same year, 1907, Raman married Lokasundari Ammal via self-arranged marriage at that time his wife was 15 years old.

During studying at Presidency College, Raman performed several Physics experiments and perceived some noteworthy outcomes. Firstly, while performing the standard Melde's experiment on string vibrations, Raman and his colleagues observed a strange effect. For which, on 10th May 1906, Raman wrote a letter to the author of the Theory of sound, i.e., John William, who is also a Nobel laureate and responded to Raman on 15th July 1906. Secondly, Raman also observed another phenomenon based on the diffraction of monochromatic light. Here, Raman identifies that, if we fall monochromatic light on the rectangular aperture of more than 87° then the resulting diffraction bands are not equidistant with the middle band. Further, he also saw that the bandwidth gradually increases while moving from one side to another, as well as each side from the middle band has a limited number of bands. To support this, Raman derived a mathematical expression [3], which was later used to study various phenomena like visual, photography, photometric measurements, etc. [4–7]. Thirdly, Raman observed that the mixing of different colour plates or films results in a transparent media. A similar phenomenon was earlier explained by Charles Fabry (France) and Robert W. Wood (USA) using the standpoint diffraction theory. Here, Raman notices that the resulting features of this phenomenon are different from their expected outcomes through the elementary theory of diffraction. For example, the existing theory was not able to properly explain the gradually increasing widths of dark and bright rings while moving away from the center, the perfect blackness of the dark rings as well as the composite structure of the inner part of the halo. Raman concluded that not the whole surface but only the laminar boundaries between the two media of the film take part in the luminous. Therefore, the optical effects are because of the scattering or radiation of light at laminar diffraction boundaries. Raman further includes that the thickness of the film and its composites are the main factors of a number of rings and derived equations for determining the positions of each dark and bright ring [8]. After that, this effect was successfully reported in several theoretical and experimental studies based on white or monochromatic light diffraction in uniform and non-uniform films as well as also for normal and oblique lights [8].

In 1917, Ashutosh Mukherjee (Judge and Vice-Chancellor) named Raman as the first Palit Professor of Physics at the Rajabazar Science College, University of Calcutta and granted him to access the University laboratories for research work. Nonetheless, Raman continued most of the research works from the IACS. The majority of his work is based on acoustics and optics, which brought him recognition. After four years, in 1921, he was nominated as a Fellow at the Royal Society of London and was selected in 1924.

Prior to 1921, Raman and his students mostly engaged in the fields of acoustic, optical and astronomy branches of Physics. From 1921, Raman diverted his research focus to light diffraction and identified a variety of phenomena. One of these is the discovery of the scattering of water molecules by sunlight. Until 1921, Raman had published approximately 60 research papers on optics and acoustic fields [9]. According to Prof. R. S. Krishnan (a Ph.D. student of Raman), the year 1921 was the watershed moment in Raman's scientific career since he was chosen as a delegate to the Universities congress, which was held in the same year in Oxford, Europe. During his journey, he became stocked when traveling over the Mediterranean Sea due to the deep blue opalescence of the ocean. This phenomenon was already successfully explained by the great scientist Lord Rayleigh. He stated that 'The much admired dark blue colour of the deep sea has nothing to do with the colour of water, it is just simply the reflection of the blue colour of the sky [10]. However, Raman was not satisfied with this explanation, so he began a simple experiment on board the SS Narkunda with optical equipment. He used the suitable Nicol prism to examine the surface of the water at the Brewsterian angle, thereby extinguishing any reflection caused by the sky. The deep blue colour of the Mediterranean and Red Seas is distinct from the suppression of the reflection of the sky, showing that it is a distinct phenomenon rather than just a reflection. Raman successfully explains this phenomenon as a scattering of light by water molecules [11, 12]. After his return to India, Raman immediately started working on the light scattering phenomenon of every transparent media and performed a number of experimental and theoretical studies at IACS. Here, Raman studied how light scatters in crystal and liquid mediums as well as the influence of light frequency in this phenomenon [13]. Furthermore, Raman also studied this phenomenon in dense vapor and gases, where he observed that light was not entirely polarized as expected by theories [14]. Over the next six years, Raman and his students published around 56 research papers in prestigious international journals such as Physical Review, Nature, Proceedings of the Royal Society and many more. These papers contain various laws of molecule scattering in diverse media, state of matter, pressure, phase transitions under pressure and at critical temperature. Dr. Kalpathi Ramakrishna Ramanathan (1893–1985), one of the collaborators of Raman had also conducted a number of experiments on the light scattering intensity of liquids and observed that his results agreed with the theory of moderately anisotropic molecules while diverging from the strong anisotropic molecules theory [15]. These investigations aid Raman in discovering his remarkable work known as the 'Raman Effect'.

Raman was appointed director of the Indian Institute of Science (IIS), Bangalore in 1933. After four years, in 1937, he was forced to resign from the post of director due to some controversies [1, 16]. However, Raman still remained Head of the Department of Physics and served as Professor until his retirement in 1948. Afterward, Raman established a new institute in Bangalore, which is today known as the 'Raman Research Institute'. Here, Raman has worked with dozens of students and given significant contributions to the fields of optics, crystal physics and mineralogy. In the last ten years of his life, Raman worked alone without any students or assistants. Raman served as President of the Indian Academy of Science and Director of the Raman Research Institute in Bangalore until his death in 1970.

1.1.1 Discovery of the Raman Effect

In 1923, K. R. Ramanathan, the oldest and most honored student of Raman, found some unexpected phenomena during his experiment on light scattering in pure water and alcohol. He observed that when sunlight was filtered via violet glass and traveled through the liquid, the resultant scattered rays exhibited some additional rays that were not present in the incident beam. This indicates that the scattering process transforms the light from violet to green rays. Ramanathan referred to this phenomenon as 'weak fluorescence' without understanding the exact significance of this phenomenon [17]. After two years, K. S. Krishnan, another Raman student, noticed a similar phenomenon of light scattering in 65 different purified liquids [18]. The work conducted by K. S. Krishnan was further carried out by S. Venkateswaran (another Raman student) in 1925. He attempted to take a photograph of the spectrum of the scattered light but was unable to report significant results. All these experiments were carried out under the guidance of Raman, but he is still not satisfied with the explanation of the results that the appearance of fluorescence was responsible for the residual scattering. He believed that it was not an unusual fluorescence phenomenon but a new type of secondary radiation. Therefore, Raman carried out further works and also observed a similar phenomenon while performing an experiment on optical glasses and ice crystals [19] towards the end of 1927. At the beginning of 1928, S. Venkateswaran, working under Raman, observed a remarkable result that the pure dry glycerin scatters the colour of sunlight into green colour rather than blue and is also strongly polarized. Following that, they performed a series of experiments with different filters and observed that the resulting scattered light had different colours than the incident one in each case.

Raman wants to find a satisfactory solution and explanation of this phenomenon, therefore, he asked his student K. S. Krishnan to conduct experiments on various vapors and liquids and study their anomalous scattering. Raman was also actively involved in the setup and monitoring of these experiments. In their experiment, they employed an 18 cm telescope objective lens made up of a mixture of short-focus lenses to focus the powerful beam of sunlight. The focused sunlight passed through the blue-violet filter and then through a flask containing double distilled liquid. They observed that when the green glass filter was placed in the incident path, light scattering did not occur in the transverse direction. On the other hand, when the same was placed between the bulb and viewer, the visible opalescent was observed with having less brightness. They repeated the experiment with almost 80 different inorganic, liphatic and aromatic liquids and found the same result, rendering the universality of this phenomenon. Raman recognized that their results were related to the Kramers-Heisenberg processes and called this effect 'modified scattering'. After two days, while performing an experiment on ether and amylene vapours, Raman and Krishnan observed new radiation, a completely new phenomenon of a fundamental property, they communicated these results in the Nature Journal in 1928, entitled 'A New Type of Secondary Radiation' [20].

On 28th February 1928, Raman and Krishnan investigated the effect of incident light wavelength on this phenomenon while performing an experiment in which they employed a blue-violet filter in conjunction with a uranium glass and observed the scattered trace with a direct vision spectroscope. From their observations, they found a great surprise with the results of the experiment that the modified scattering and unmodified light were separated through dark space between them. The detailed description of this work is provided by Raman in an extract on 16th March 1928,

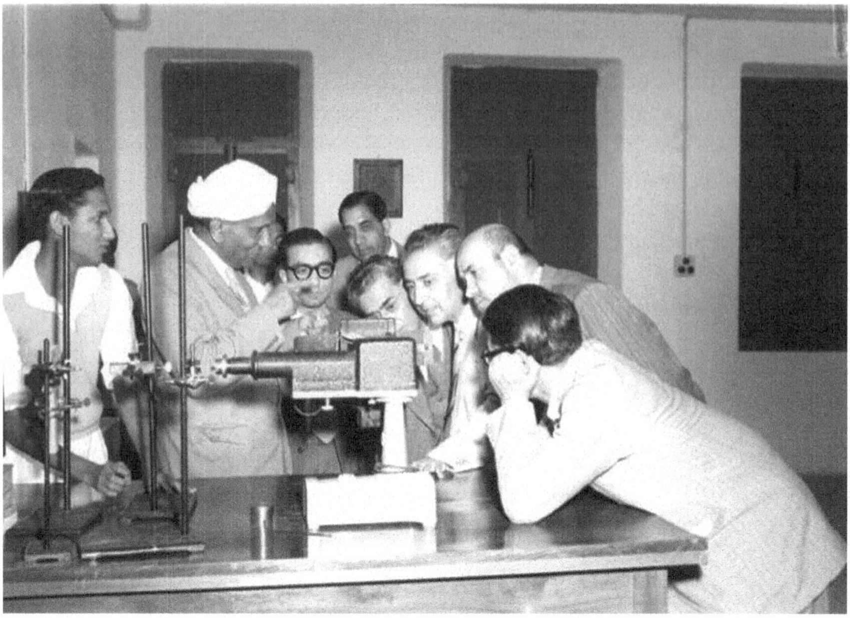

Fig. 1.2 Sir C. V. Raman explaining the Raman effect. Image Courtesy https://imgur.com

entitled 'New Radiation' to the South Indian Science Association [21]. At the time of discovery, Krishnan and Venkateswaran were both with Raman, hence, they were the first candidates to have knowledge of this phenomenon, afterwards, other scholars subsequently got an opportunity to see this phenomenon (see Fig. 1.2).

Through this experiment and observations, the discovery of a new phenomenon was completed, later called as 'Raman Effect' on the name of great Physicist Sir C. V. Raman. On the following day, this first-ever recorded line spectrum was published in the Associated Press of India and Raman communicated the above results in a letter to Nature in 1928. The first photograph of the scattered spectrum of ether vapour was taken by Ramdas [22], who was working in Raman's laboratory and he also referred to the effect as the 'Raman effect' when communicating the results to Nature. Unfortunately, this note was rejected by the referees, however, the Editor of Nature, Sir Richard Gregory took responsibility for the publication. Raman had already announced this discovery with the title of 'A New Radiation' during the inauguration of the South Indian Science Association, Bangalore. On 31st March 1928, the full text of this lecture was published in the Indian Journal of Physics together with the scattered spectrum photograph of the Benzene [21]. Figure 1.3 depicts Raman and Krishnan's obtained scattering spectrum of liquid benzene in the presence of a mercury arc lamp source, published in Springer Nature Journal [23].

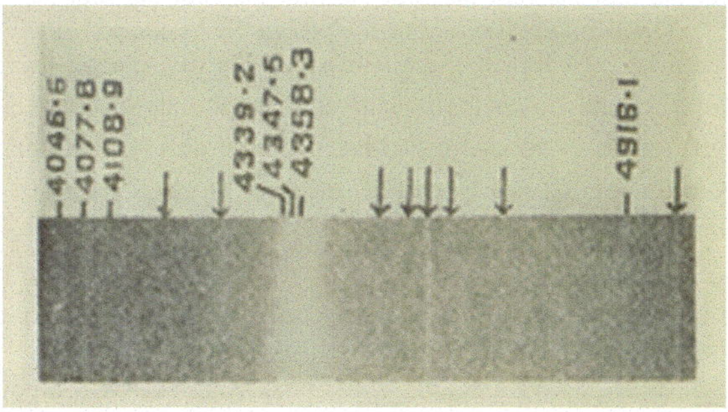

Fig. 1.3 Benzene Raman spectrum published by Raman and Krishnan. Adapted with permission from [23]. Copyright (1928) (Springer Nature)

This historic paper contains the following characteristics of the new effect.

- Each incident radiation will produce a set of modified polarized lines.
- The resultant modified lines have lower frequencies than the incident radiation.
- Some materials possess an increased frequency of modified lines with less intensity.
- The resultant absorption frequency of scattered radiation usually lies in the IR region.
- This effect is universal as it was observed on a variety of materials including liquids, gases and vapours as well as also in ice crystal and optical glasses.

According to quantum theory, among the entire incident energy only some portion of the radiation was absorbed by the molecule while the remaining one is scattered. The quantum theory of dispersion given by Kramer and Heisenberg also shows a possible relationship with the new phenomenon of Raman. Therefore, the observations made by Raman provide the fundamental character of the new effect.

As his historic paper was published in the Indian Journal of Physics in 1928 [21], Raman immediately reprinted around 2000 copies of his paper and posted it to various physicists working on the light scattering phenomenon all over the world, including the USA, Germany, France, Russia and Canada, as well as other scientific institutes. Raman's decision also helped to raise awareness of the Indian Journal of Physics, which he founded in 1927. After receiving the reprint of Raman's historic paper, many research scientists from around the world, including physicists and chemists began to work on this phenomenon.

The authorities in optics and scattering immediately recognized the importance of Raman's discovery and confirmed it. Pringsheim et al. [24] in their review paper titled 'The Raman Effect, a New Radiation Effect Discovered by C. V. Raman' provides a detailed analysis of this new scattering effect and also mentions that Raman discovered a totally new phenomenon which should be designed as the 'Raman Effect' as well as the resultant new line spectrum as the Raman Spectrum. In the paper's conclusion, he also stated that Raman has opened a vast and completely new field of spectroscopy for research. Furthermore, a large number of physicists accepted Raman's discovery and nominated him for the Nobel Prize in Physics in 1930. The Raman technique provides more information with better results in a single observation, whereas IR gives a series of observations, making it more difficult for scientists. In the decade following its discovery, the Raman technique became so popular that it was utilized to investigate approximately 2500 chemical compounds. Furthermore, according to Krishnan et al. [18] around 23,600 research papers using Raman spectroscopy were published up to the end of 1978, rendering the potential of this technique in all fields.

It is also worth noting that while performing the experiments, the total cost of the apparatus used by Raman was not exceeding the amount of $25. At the time of discovery, they used a mirror for sunlight deflection, a complimentary glass filter pair, benzene in a flask and a spectroscope of pocket size. Further, this discovery was not made by accidentally or as a result of only a few efforts, but it took almost seven years of hard work and devotion from Raman and his students. It is also more creditable when we consider the period of discovery, scientific research did not have as much freedom and support of the Government of India as it has today. For his path-breaking discovery of the 'Raman effect', Sir. C. V. Raman received the Nobel Prize in Physics in the year of 1930. Figure 1.4 depicts Sir C. V. Raman with fellow Nobel Prize winners Hans Fischer, Karl Landsteiner and Sinclair Lewis at the Nobel Prize ceremony. In the same year, Professor Raman was also awarded the Hughes Medal of the Royal Society of England (see Fig. 1.5). Apart from these, Sir. C. V. Raman has also received many prestigious awards, prizes and medals throughout his research career, including the Bharat Ratna in 1954 and the Lenin Peace Prize in 1957, as summarized in Fig. 1.5.

Since the introduction of laser [25], Raman spectroscopy has become a more powerful technique for the characterization of all types of materials. With time, several Raman techniques have been developed with their potential applications in

Fig. 1.4 Sir C. V. Raman with other Nobel Prize Award winners in 1930, from left C. V. Raman (physics), Hans Fischer (chemistry), Karl Landsteiner (medicine) and Sinclair Lewis (literature). *Source* https://commons.wikimedia.org/wiki/File:Nobel_ceremony_1930.jpg

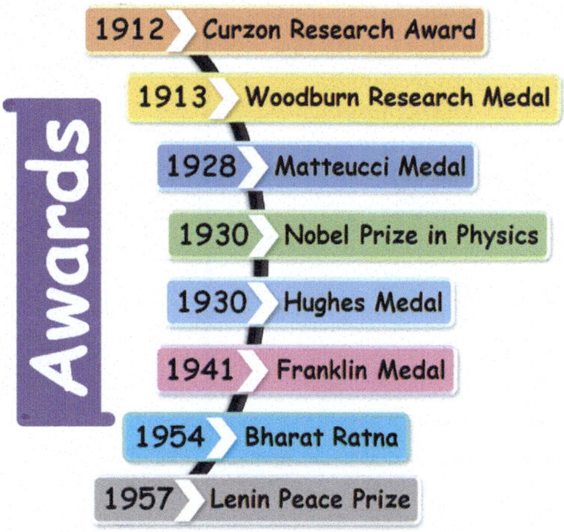

Fig. 1.5 The awards, medals and prizes received by Sir C. V. Raman during his research journey

physics, chemistry and medical fields and currently, there are almost different 25 Raman techniques available. Some of the well-known Raman spectroscopy techniques are Surface Enhanced Raman Spectroscopy (SERS), Surface enhanced hyper Raman Scattering (SEHRS), Tip-enhanced Raman Spectroscopy (TERS), Coherent anti-Stokes Raman Spectroscopy (CARS), Stimulated Raman Scattering (SRS) and Resonance Raman Spectroscopy (RRS) are briefly describing in the following section.

1.2 Development of Raman Spectroscopy

Raman spectroscopy has been a very expensive technique in the earlier decades due to the high cost of the spectrometer as well as the poor efficiency rate of normal Raman scattering, which limits its application range [26]. The significant development in technology boots spectroscopic capability and overcomes the limitations of Raman spectroscopy [27]. As a result, very low-cost and portable devices that can easily characterize the different compounds and treat the Raman spectrum as a fingerprint tool are now accessible. Thus, through this section, the authors will make an attempt to cover the development of some of the most utilized Raman spectroscopy techniques.

1.2.1 Surface Enhanced Raman Spectroscopy (SERS)

SERS is one of the most powerful and sensitive technique used for the detection of highly sensitive structures and very low concentrations of molecules by amplifying electromagnetic fields [28]. In 1974, Fleischmann and co-workers reported the first

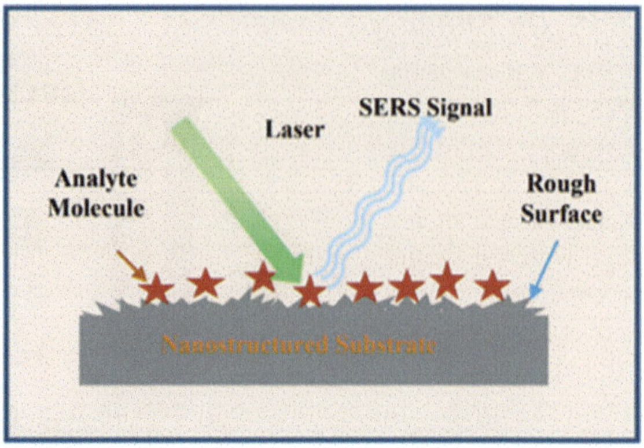

Fig. 1.6 Schematic diagram of Surface Enhanced Raman Spectroscopy technique. Adapted with permission from [34]. Copyright (2021) (Springer Nature)

work based on SERS and claimed that the high intensity of the Raman signal on the roughened surface is due to the huge number of adsorption sites [29]. In contrast, two other groups in 1977 revealed that the enhancement in the Raman cross-section of the adsorbed molecules results in a high intensity of Raman signal, which is known as surface enhanced Raman spectroscopy [30, 31]. However, some additional investigations proved that SERS is the consequence of a combination of two different effects that are electromagnetic effect and a charge transfer effect similar to resonance [30]. SERS enhances the magnitude of weak Raman signals up to 15 times [26, 32], making it a widely utilized technique in the last two decades for detecting industrial waste, food industries, organic pollutants, impurities, chemicals and medical sciences [33]. Figure 1.6 depicted the schematic diagram of SERS technique.

1.2.2 Surface Enhanced Hyper Raman Scattering (SEHRS)

SEHRS technique is an inelastic sum of two photons scattered frequencies, whereas the normal Raman scattering is caused by a sole photon. Here, the enhancement in the factor of SERS results in a molecular electronic resonance in which the molecular transition corresponds to the frequency sum of the incident photons [35]. In order to achieve a higher enhancement for this technique, both the surface and resonance effects are need to be applied to each other [36]. The first report on SEHRS including the experiment and theory results was presented by Van Duyne's group in 1988 [35]. SERHS techniques have been employed in biological applications such as recording bond stretches, detecting bonds in DNA like amide I and III and C–O, studying the redox reaction at the electrode surface [37, 38]. The reports show that SEHRS is being an ultrafast potential technique in order to study the dynamics of complex reactions with much higher sensitivity than the SERS as well as also to detect the excited states [35].

1.2.3 Tip Enhanced Raman Spectroscopy (TERS)

TERS is a combined approach of Raman spectroscopy and scanning probe microscopy (SPM) that is used for the investigation of various materials at the nanoscale [39, 40]. TERS is the only technique that confines the light and measures the scattered light of the nanostructured sample and provides a Raman image with the spatial resolution of the nano sample (typically 10–20 nm). The first concept regarding TERS was proposed by Synge in the same year of Raman's discovery (i.e., 1928) [41], however, it was not accepted at that time and considered inappropriate until the first theoretical work reported on TERS measurement by Wessel in 1985 [42]. Later on, Stöckle et al. [43], Anderson [44] and Kawata et al. [45] reported the experimental evidence of TERS in the year 2000. TERS has the ability to work in water and humidity atmospheres and also it increases the reproducibility problem of SERS [46]. Thus, in the past few years, TERS has become a powerful technique with a wide range of nano-structural investigations of organic, inorganic and bioorganic compounds [47]. Such as TERS is used to detect single-walled carbon nanotubes, graphene, fluctuation in spectra due to structural change, provide chemistry and structural information of nanomaterials, identification of nucleic acid, DNA, RNA, as well as for analyzing the cells, lipids and virus particles [48–50]. The schematic diagram of the TERS technique along with the obtained carbon nanotube image and its Raman spectrum are illustrated in Fig. 1.7.

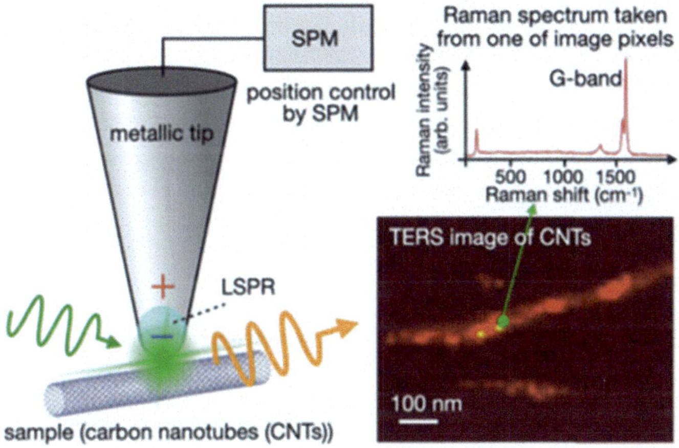

Fig. 1.7 Schematic diagram of Tip Enhanced Raman Spectroscopy technique, spectrum and an image of carbon nanotubes. Adapted with permission from [51]. Copyright (2021) (Springer Nature)

1.2.4 Coherent Anti-Stokes Raman Spectroscopy (CARS)

CARS is a highly efficient nonlinear Raman technique that uses coherent laser beams in order to enhance the normal Raman signal [52]. Thus, it is referred to as an anti-stoke frequency technique. In 1965, P.D. maker and R.W. Terhune gave an initial idea of CARS while working on "Raman laser" and described it as a three wave mixing process [31], which was later named "coherent anti-Stokes Raman spectroscopy" by R. F. Begley in 1974 [53]. In 1982, the first CARS was constructed [54] and with the involvement of near infrared laser beams instead of a visible dye layer source in 1999 [55], CARS became one of the most popular technique with a wide range of applications in physics, chemistry and biology fields. CARS offers excellent spatial and spectral resolution with high sensitivity and chemical specificity. Therefore, CARS is utilized to investigate the fluorescing samples, discharge gases, plasmas and also in combustion and atmospheric chemistry, imaging biological samples, living cells, etc. [56, 57]. Nonetheless, CARS has also shown some promising applications in material science [58], catalysis [59] and kinetic reactions [60].

1.2.5 Stimulated Raman Scattering (SRS)

SRS is a nonlinear simulated Raman technique, which was accidentally discovered by Woodbury and Ng in 1962 during an experiment [61, 62]. They saw an increased intensity in radiation without the knowledge of the source, later, Eckhardt explained it as two photon process and SRS [63]. SRS delivers real time images, high signal intensity of spontaneous Raman technique with improved single-noise ratio, precise determination of bond length and to analyze the molecular and atomic gases, solids and liquids [64, 65]. Furthermore, this technique has biological applications such as lipid components measurement, skin tissue analysis, liver tissue analysis and mapping, as well as agrochemical and plant studies [66, 67].

1.2.6 Resonance Raman Spectroscopy (RRS)

RRS is an instrument that measures the shift in photon frequency during the electronic transition process [31]. Due to resonance excitation, the oscillation of the electron charge displacement increases and introduces the dipole moment, which as a result directly enhances the Raman scattering efficiency [68]. The use of UV resonance eliminates the fluorescence effect and enhances the sensitivity of this technique with potential applications in materials science and biological fields [69, 70]. RRS is used to explore the structure of food and hemi protein, cell function of the human body, shape of potential surfaces and molecular entities during the excited state, analysis of various biological samples like enzymes and pigment [31, 71, 72]. RRS is also utilized to generate vibrational spectra for specific sites in macromolecular systems [73].

1.2.7 Confocal Raman Spectroscopy (CRS)

CRS is the most popular and widely used Raman spectroscopy technique for determining the chemical composition as well as the internal structure of the material. This approach can produce images of interior structure along the x, y and z Cartesian coordinate axes with lateral resolution at the micrometer level [74]. Marvin Minsky invented the first ever confocal Raman microscopy in 1955 [74]. In CRS, the probe head uses the microscope objective to focus the laser light on the sample and then the resulting signal is recorded on the camera, which creates a spectrum [31]. The non-destructive nature of the CRS demonstrates its applications in a wide range of fields, including in situ chemical analysis of plants, provides tissue information without harming the plants, direct visualization of the polymer composition of the cell wall, in-depth analysis of the various samples, molecular composition and many more [74–77]. In addition, this technique is effective for generating high spatial resolution images without the need for sample preparation [31, 75].

Despite these techniques, there are various other Raman spectroscopy tools accessible, including resonance, polarization, optical, imaging and so on, with potential applications in the diverse fields of science and technology. In the following chapters of this book, the authors from various institutes have attempted to provide a detailed analysis of these Raman spectroscopy techniques along with their working mechanism, development, advances and range of applications.

1.3 Summary

Sir. C. V Raman, the man behind the discovery of the 'Raman effect' was the first Indian scientist to receive the prestigious Noble Prize in 1930 for his outstanding research on the light scattering phenomenon. Raman was a brilliant student from childhood and possessed very much interest in the acoustic, optical and astronomy branches of Physics and persuaded numerous research works in these fields. However, the journey of Europe in 1921, was a turning point of Raman's scientific career since it redirected his attention towards the light scattering phenomenon. Raman and his students' dedication to research and years of hard work help them to achieve the millstone of their research careers. Nowadays, Raman spectroscopy has become a powerful tool for material characterization with the scientific advancement in technology and possesses applications in diverse fields of study. Currently, there are more than 25 variants of Raman spectroscopic techniques accessible including enhanced, coherent, resonance, etc., demonstrating their potential in almost all sectors of physics, chemistry and biological fields.

References

1. R. Singh, Phys. Perspect. **4**, 399 (2002)
2. S. Ramaseshan, Indian Inst. Sci. Bangalore (1978)
3. C.V. Raman, London, Edinburgh, Dublin Philos. Mag. J. Sci. **12**, 494 (1906)
4. C.V.RM.A., London, Edinburgh, Dublin Philos. Mag. J. Sci. **21**, 618 (1911)
5. C.V. Raman, Nature **82**, 69 (1909)
6. C.V. Raman, Nature **78**, 55 (1908)
7. C.V.R.M.A., London, Edinburgh, Dublin Philos. Mag. J. Sci. **17**, 204 (1909)
8. C.V. Raman, Part III Philos. Mag. **42**, 679 (1921)
9. R. Singh, F. Riess, Notes Rec. R. Soc. Lond. **58**, 47 (2004)
10. J.W.S. Rayleigh, *Scientific Papers by Lord Rayleigh (John William Strutt). Vol. V, 1902–1910 (1912); Vol. VI, 1911–1919 (1920)* (Dover Publications, 1964)
11. C.V. Raman, Nature **108**, 367 (1921)
12. C.V. Raman, G.T. Walker, Proc. R. Soc. London. Ser. A, Contain. Pap. A Math. Phys. Character **101**, 64 (1922)
13. C.V. Raman, Nature **109**, 75 (1922)
14. C.V.R.M.A. Hon.D.Sc., K.R.R.M.A., London, Edinburgh, Dublin Philos. Mag. J. Sci. **45**, 113 (1923)
15. K.R. Ramanathan, Proc. Indian Assoc. Cultiv. Sci. **8**, 1 (1923)
16. S. Chatterjee, Indian Phys. Soc. Diam. Jubil. Number **43** (1995)
17. C.V. Raman, K.R. Ramanathan, (1923)
18. R.S. Krishnan, R.K. Shankar, J. Raman Spectrosc. **10**, 1 (1981)
19. C.V. Raman, J. Opt. Soc. Am. **15**, 185 (1927)
20. C.V. Raman, Nat. **121**, 501 (1928)
21. C.V. Raman, Indian J. Phys. **2**, 387 (1928)
22. L.A. Ramdas, Nature **122**, 57 (1928)
23. C.V. Raman, K.S. Krishnan, Nature **122**, 12 (1928)
24. P. Pringsheim, Naturwissenschaften **16**, 597 (1928)
25. S.P.S. Porto, D.L. Wood, J. Opt. Soc. Am. **52**, 251 (1962)
26. A. Kudelski, Talanta **76**, 1 (2008)
27. X. Zhang, M.A. Young, O. Lyandres, R.P. Van Duyne, J. Am. Chem. Soc. **127**, 4484 (2005)
28. B. Sharma, R.R. Frontiera, A.-I. Henry, E. Ringe, R.P. Van Duyne, Mater. Today **15**, 16 (2012)
29. G.C. Schatz, J. Mol. Struct. Theochem **573**, 73 (2001)
30. C.L. Haynes, A.D. McFarland, R.P. Van Duyne, Analytic Chemistry, **77**, 338 (2005)
31. R.S. Das, Y.K. Agrawal, Vib. Spectrosc. **57**, 163 (2011)
32. D. Zeisel, V. Deckert, R. Zenobi, T. Vo-Dinh, Chem. Phys. Lett. **283**, 381 (1998)
33. F.-G. Banica, *Chemical Sensors and Biosensors: fundamentals and Applications* (Wiley, 2012)
34. B.P. Majee, A.K. Mishra, in *Modern Techniques of Spectroscopy: basics, Instrumentation, and Applications*, edited by D.K. Singh, M. Pradhan, A. Materny (Springer Singapore, Singapore, 2021), pp. 185–208
35. N.L. Gruenke, M.F. Cardinal, M.O. McAnally, R.R. Frontiera, G.C. Schatz, R.P. Van Duyne, Chem. Soc. Rev. **45**, 2263 (2016)
36. W.-H. Li, X.-Y. Li, N.-T. Yu, Chem. Phys. Lett. **312**, 28 (1999)
37. K. Kneipp, Y. Wang, H. Kneipp, L.T. Perelman, I. Itzkan, R.R. Dasari, M.S. Feld, Phys. Rev. Lett. **78**, 1667 (1997)
38. W.-H. Li, X.-Y. Li, N.-T. Yu, Chem. Phys. Lett. **327**, 153 (2000)
39. P. Verma, Chem. Rev. **117**, 6447 (2017)
40. S. Kawata, T. Ichimura, A. Taguchi, Y. Kumamoto, Chem. Rev. **117**, 4983 (2017)
41. E. Synge, London, Edinburgh, Dublin Philos. Mag. J. Sci. **6**, 356 (1928)
42. J. Wessel, J. Opt. Soc. Am. B **2**, 1538 (1985)
43. R.M. Stöckle, Y.D. Suh, V. Deckert, R. Zenobi, Chem. Phys. Lett. **318**, 131 (2000)
44. M.S. Anderson, Appl. Phys. Lett. **76**, 3130 (2000)

45. N. Hayazawa, Y. Inouye, Z. Sekkat, S. Kawata, Opt. Commun. **183**, 333 (2000)
46. B.-S. Yeo, J. Stadler, T. Schmid, R. Zenobi, W. Zhang, Chem. Phys. Lett. **472**, 1 (2009)
47. A.J. Driscoll, M.H. Harpster, P.A. Johnson, Phys. Chem. Chem. Phys. **15**, 20415 (2013)
48. Y. Okuno, Y. Saito, S. Kawata, P. Verma, Phys. Rev. Lett. **111**, 216101 (2013)
49. W. Su, D. Roy, J. Vac. Sci. Technol. B **31** (2013)
50. N. Kumar, S. Mignuzzi, W. Su, D. Roy, EPJ Tech. Instrum. **2**, 1 (2015)
51. T. Umakoshi, P. Verma, in *Modern Techniques of Spectroscopy: basics, Instrumentation, and Applications*, edited by D.K. Singh, M. Pradhan, A. Materny (Springer Singapore, Singapore, 2021), pp. 209–234
52. A. Zumbusch, G.R. Holtom, X.S. Xie, Phys. Rev. Lett. **82**, 4142 (1999)
53. R.F. Begley, A.B. Harvey, R.L. Byer, Appl. Phys. Lett. **25**, 387 (1974)
54. J.-X. Cheng, Y.K. Jia, G. Zheng, X.S. Xie, Biophys. J. **83**, 502 (2002)
55. C. Heinrich, S. Bernet, M. Ritsch-Marte, Appl. Phys. Lett. **84**, 816 (2004)
56. S. Roy, J.R. Gord, A.K. Patnaik, Prog. Energy Combust. Sci. **36**, 280 (2010)
57. W.M. Tolles, J.W. Nibler, J.R. McDonald, A.B. Harvey, Appl. Spectrosc. **31**, 253 (1977)
58. M. Müller, A. Zumbusch, ChemPhysChem **8**, 2156 (2007)
59. M.H.F. Kox, K.F. Domke, J.P.R. Day, G. Rago, E. Stavitski, M. Bonn, B.M. Weckhuysen, Angew. Chemie Int. Ed. **48**, 8990 (2009)
60. D. Schafer, J.A. Squier, J. van Maarseveen, D. Bonn, M. Bonn, M. Müller, J. Am. Chem. Soc. **130**, 11592 (2008)
61. T.T. Basiev, A.A. Sobol, P.G. Zverev, L.I. Ivleva, V.V. Osiko, R.C. Powell, Opt. Mater. (Amst). **11**, 307 (1999)
62. E.J. Woodbury, W.K. Ng, Proc. IRE **50**, 2347 (1962)
63. P. Rostron, S. Gaber, D. Gaber, Laser **21**, 24 (2016)
64. P. Nandakumar, A. Kovalev, A. Volkmer, New J. Phys. **11**, 33026 (2009)
65. V. Boudon, D. Bermejo, R.Z. Martínez, J. Raman Spectrosc. **44**, 731 (2013)
66. D. Zhang, P. Wang, M.N. Slipchenko, J.-X. Cheng, Acc. Chem. Res. **47**, 2282 (2014)
67. D. Fu, Y. Yu, A. Folick, E. Currie, R.V.J. Farese, T.-H. Tsai, X.S. Xie, M.C. Wang, J. Am. Chem. Soc. **136**, 8820 (2014)
68. S.A. Asher, Anal. Chem. **65**, 201A (1993)
69. E. Smith, G. Dent, Mod. Raman Spectrosc. Pract. Approach 1 (2005)
70. C. Li, J. Catal. **216**, 203 (2003)
71. E.C.Y. Li-Chan, Trends Food Sci. Technol. **7**, 361 (1996)
72. B.R. Wood, P. Caspers, G.J. Puppels, S. Pandiancherri, D. McNaughton, Anal. Bioanal. Chem. **387**, 1691 (2007)
73. M.A. Marcus, A. Lewis, Science **195**, 1328 (1977)
74. N. Gierlinger, T. Keplinger, M. Harrington, Nat. Protoc. **7**, 1694 (2012)
75. N.J. Everall, Spectroscopy **19**, 22 (2004)
76. L. Cabrales, N. Abidi, F. Manciu, Fibers **2**, 285 (2014)
77. N. Gierlinger, M. Schwanninger, Plant Physiol. **140**, 1246 (2006)

Chapter 2
Resonant and Non-resonant Raman Spectroscopy

Deepu Kumar and Pradeep Kumar

2.1 Introduction

Light has played a very important role in rearing all forms of life on our planet earth. Without light, we cannot even imagine the existence of life. When light interacts with materials, it gives birth to many crucial phenomena which play a significant role to understand the fundamental properties associated with the materials and giving rise to important applications. For example, lasers, solar cells and photodiodes are the results of light-matter interaction. Furthermore, light-matter interaction leads to the formation of various quasiparticles such as excitons, polariton and many more as well as their hybrid bound structure [1–8]. Generally, light-matter interaction leads to fundamental phenomena such as absorption, reflection, transmission, photoluminescence and scattering. The different phenomenon provides different information about the intriguing properties associated with the materials. Light scattering is one of the important phenomena and plays an important role in understanding many intriguing and exotic properties of various kinds of materials. Generally, light scattering may be classified into two categories (i) elastic light scattering (ii) inelastic light scattering. In the case of elastic scattering, also known as Rayleigh scattering, the frequency of scattered light (ω_s) is equal to the frequency of incident light (ω_i), and is the most scattered part of the light [9].

On the other hand, inelastic scattering is the scattering where the frequency of the scattered light is different compared to the frequency of the incident light. In this chapter, we focus on the interaction between light and matter, in particularly inelastic light scattering. Foundation of inelastic light scattering was laid around the third decade of the 20th century. Brillouin, in 1922, proposed the inelastic light scattering by acoustic waves. Adolf Smekal, in 1923, theoretically predicted this scattering by a system with two quantized energy levels [10], while it was experimentally discovered

D. Kumar (✉) · P. Kumar
School of Physical Sciences, Indian Institute of Technology Mandi, Mandi 175005, India
e-mail: deepu7727@gmail.com

D. K. Singh et al. (eds.), *Raman Spectroscopy*, Springer Series in Optical Sciences 248,
https://doi.org/10.1007/978-981-97-1703-3_2

by Sir C. V. Raman and K. S. Krishnan in 1928 [11, 12]. Inelastic scattering is also known as the Raman scattering/Raman effect after the name of C. V. Raman. He was awarded the Physics Nobel prize in 1930 for the discovery of the Raman effect. Classically, Raman scattering originates due to change in the electronic polarizability of the system by the quasiparticle excitations which leads to inelastic scattering of the incident light. Raman scattering is classified into two parts (i) Stokes Raman scattering, where the frequency of the scattered light is less than the incident light, and (ii) Anti-Stokes Raman scattering, where frequency of the scattered light is larger than that of the incident light, see Fig. 2.1 for schematic representation of the Stokes and Anti-Stokes Raman process. During Stokes Raman scattering process, elementary excitations such as phonons, magnons etc., are created in the material, on the other hand in the Anti-Stokes Raman scattering process, these are annihilated for every scattered photon. In this chapter, we have adopted particularly the phonons (lattice vibrations) to understand the various aspects of Raman scattering. For excitations beyond Raman scattering from phonons, we recommend the following ref [13–20].

If we consider that a phonon with frequency (ω_{ph}) and wave vector ($\vec{q}$) is either created or annihilated during the Raman scattering process. The energy and momentum remain conserved in both the processes (i.e. Stokes and Anti-Stokes), and is expressed as $\hbar\omega_s = \hbar\omega_i - \hbar\omega_{ph}$; $\vec{k}_s = \vec{k}_i - \vec{q}$ for Stokes and $\hbar\omega_s = \hbar\omega_i + \hbar\omega_{ph}$; $\vec{k}_s = \vec{k}_i + \vec{q}$ for Anti-Stokes process, where ω_s and ω_i are the frequency of scattered and incident light, respectively; while $\vec{k}_s$ and $\vec{k}_i$ are the wave vector of scattered and incident light, respectively.

The energy of created or annihilated phonons is equal to the energy difference between the incident and scattered light, and is generally referred to as Raman shift. In the Raman scattering experiment, the Raman spectrum consists of the scattering cross section/intensity, generally known as Raman scattering intensity of the phonons, of

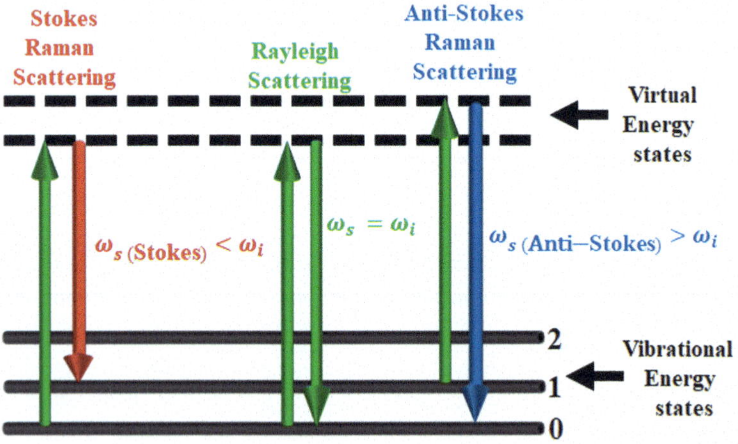

Fig. 2.1 Schematic representation of the Raman scattering (Stokes and Anti-Stokes) and Rayleigh scattering process. ω_s and ω_i are frequency of the scattered and the incident light, respectively

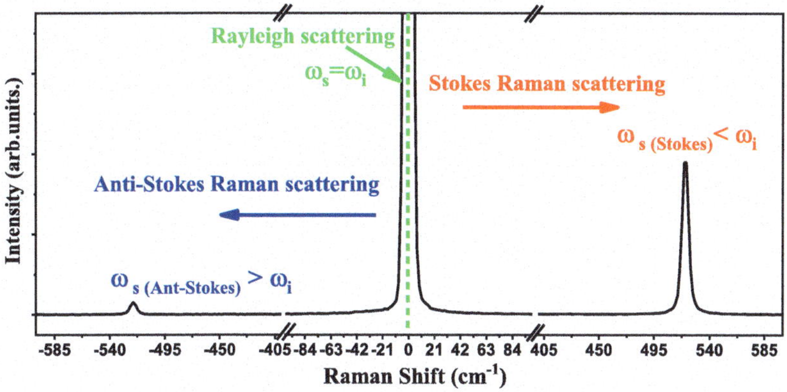

Fig. 2.2 Raman spectrum for silicon showing the Anti-Stokes, Stokes and Rayleigh scattering lines. Adapted from Deepu Kumar (Author of the Chapter) Ph.D. Thesis work (2022), Indian Institute of Technology Mandi

the scattered light as a function of frequency of elementary excitations or vibrational modes. The probability to create or annihilate the phonon may be given by the Bose-Einstein distribution function (n). At finite temperature, n can be written as $n = [\exp\,(\hbar\omega_{ph}/k_bT) - 1]^{-1}$, where k_b is the Boltzmann constant, T is the temperature and $\hbar = h/2\pi$ where h is the Planck's constant. During the Stokes process, a phonon is created which increases the population of phonons by 1, therefore the number of phonons changes from n to $n + 1$. Intensity for the Stokes process is proportional to $n + 1$. While the intensity for the Anti-Stokes process is proportional to n. The intensity ratio of Stokes and Anti-Stokes process may be written as $\frac{I_S}{I_{AS}} = \frac{n+1}{n} = \exp\,(\hbar\omega_{ph}/k_bT)$ and it may be used to determine the local temperature of the system for the given laser excitation energy. Figure 2.2 illustrates the Raman spectrum for the case of silicon showing the Stokes, Anti-Stokes and Rayleigh scattering lines. The spectrum consists a very intense peak centered at 0 cm^{-1} corresponding to the elastic or Rayleigh scattering ($\omega_s = \omega_i$) peak. On the other hand, Raman spectrum towards the right and left side of the elastic peak corresponds to the Stokes and Anti-Stokes Raman scattering part, respectively.

2.2 Raman Scattering: Classical Picture

In this section we focus on macroscopic (classical) theory of the Raman scattering. When light interacts with the molecules or atoms, which is in their equilibrium position, the molecules or atoms undergo some dynamic fluctuation such as vibration as shown in Fig. 2.3. This may be because of the incident light on the sample induces the electric dipole moment which can be expressed as (within the linear approximation)

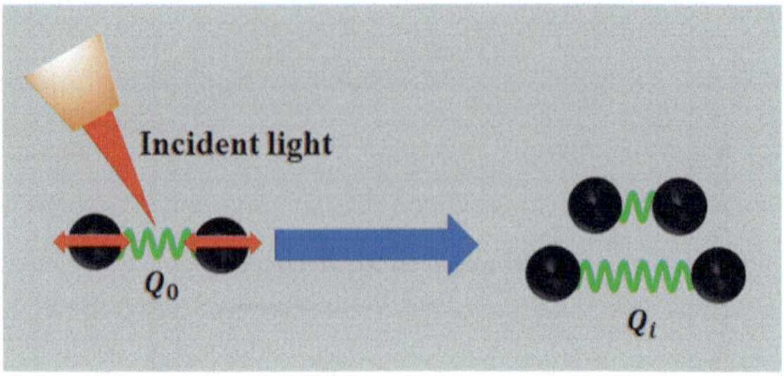

Fig. 2.3 Schematic diagram showing the creation of dynamic fluctuation upon incident light

$$\vec{\mu}_{ind} = \alpha \vec{E} \tag{2.1}$$

where $\vec{E}_i$ is the electric field vector of the incident light and is given a $\vec{E}_i = \vec{E}_0(\vec{k}_i, \omega_i)\cos\left(\vec{k}_i \cdot \vec{r} - \omega_i t\right)$, and α stands for the induced polarizability tensor which is a measure of how easy is to polarize the atoms.

Generally, α is the function of atomic coordinates and hence dependence on the atomic vibrational mode frequencies. The atomic vibrations modify the induced polarizability, around the equilibrium position (Q_0), which can be expanded using Taylor expansion, and is given as

$$\alpha = \alpha_0 + \sum_i \frac{\partial \alpha}{\partial Q_i}|_{Q_0} Q_i + \frac{1}{2}\sum_{ij} \frac{\partial^2 \alpha}{\partial Q_i \partial Q_j}|_{Q_0} Q_i Q_j + \cdots \tag{2.2}$$

where α_0 is the polarizability at equilibrium configuration without fluctuations, and Q_p is the pth, where $p = i, j$, normal mode coordinate of atomic vibrations associated with vibrational frequency ω_p. The second term in (2.2) gives rise to first-order scattering, and the higher-order terms are associated with higher-order scattering which may be ignored generally because of the low intensity of the light used in the Raman scattering experiments. Therefore, (2.2) can be reduced as

$$\alpha = \alpha_0 + \sum_i \frac{\partial \alpha}{\partial Q_i}|_{Q_0} Q_i \tag{2.3}$$

For the case of a dynamic harmonic fluctuation, the normal mode coordinate associated with atomic vibrations can be given as

$$Q_i = Q_0(\vec{q}, \omega_{vib})\cos(\vec{q} \cdot \vec{r} - \omega_{vib} t) \tag{2.4}$$

where ω_{vib} corresponding to the vibrational frequency of the atomic vibrations. By putting (2.3) and (2.4) into (2.1), and using $\vec{E}_i = \vec{E}_0(\vec{k}_i, \omega_i)\cos\left(\vec{k}_i \cdot \vec{r} - \omega_i t\right)$, the induced electric dipole moment may be written as

$$\vec{\mu}_{ind} = \left(\alpha_0 + \sum_i \frac{\partial \alpha}{\partial Q_i}|_{Q_0} Q_0(\vec{q}, \omega_{vib}) \cos(\vec{q} \cdot \vec{r} - \omega_{vib} t) \right) \times \vec{E}_0(\vec{k}_i, \omega_i) \cos\left(\vec{k}_i \cdot \vec{r} - \omega_i t\right) \tag{2.5}$$

$$\vec{\mu}_{ind} = \alpha_0 \vec{E}_0(\vec{k}_i, \omega_i) \cos\left(\vec{k}_i \cdot \vec{r} - \omega_i t\right) + \sum_i \frac{\partial \alpha}{\partial Q_i}|_{Q_0} Q_0(\vec{q}, \omega_{vib}) \cos(\vec{q} \cdot \vec{r} - \omega_{vib} t) \times \vec{E}_0(\vec{k}_i, \omega_i) \cos\left(\vec{k}_i \cdot \vec{r} - \omega_i t\right) \tag{2.6}$$

Using the trigonometric relations, (2.6) can be written as

$$\vec{\mu}_{ind} = \alpha_0 \vec{E}_0(\vec{k}_i, \omega_i) \cos\left(\vec{k}_i \cdot \vec{r} - \omega_i t\right) + \frac{1}{2} \sum_i \frac{\partial \alpha}{\partial Q_i}|_{Q_0} Q_0(\vec{q}, \omega_{vib}) \vec{E}_0(\vec{k}_i, \omega_i) \times \begin{pmatrix} \cos\left((\vec{k}_i + \vec{q}) \cdot \vec{r} - (\omega_i + \omega_{vib})t\right) \\ + \cos\left((\vec{k}_i - \vec{q}) \cdot \vec{r} - (\omega_i - \omega_{vib})t\right) \end{pmatrix} \tag{2.7}$$

The first term in (2.7) corresponds to Rayleigh scattering. On the other hand, the second term represents the Raman scattering. The second term is divided into two sinusoidal parts, the first part $(\vec{k}_i + \vec{q}, \omega_i + \omega_{vib})$ represents the Anti-Stokes Raman scattering while the second part $(\vec{k}_i - \vec{q}, \omega_i - \omega_{vib})$ corresponds to the Stokes Raman scattering. Further, second term of (2.7) indicates that for Raman scattering, first-order derivative of the polarizability $\left(\frac{\partial \alpha}{\partial Q_i}\right)$ must be non-zero.

2.3 Raman Scattering: Quantum Picture

The classical approach as discussed in Sect. 2.2 has limitations such as it does not explain the real picture associated with the Raman scattering process. For example, (i) Raman scattering, in classical treatment, is associated with phonons, quanta of atomic vibrations and no atomic vibrations are expected at 0 K suggesting that no Raman scattering at 0 K. (ii) How does the Raman scattering cross section changes if one changes the laser excitation energy? (iii) How the electrons interact with other elementary quasiparticles? (iv) Multi-phonon Raman scattering etc. Limitations of classical treatment may arise because the light is considered as a wave not explicitly quantized particle. Therefore, the microscopic treatment (quantum mechanical picture), where the light is considered as both wave and particle, may be invoked to overcome above mentioned limitations in classical treatment. In quantum picture, first-order Raman scattering, which will be discussed later in more details, involves three step processes: (i) photon absorption; electron-photon interaction (ii) creation (stokes) or annihilation (Anti-Stokes) of quanta of excitation such as phonon; electron-phonon interaction (iii) emission of photon; electron-photon interaction.

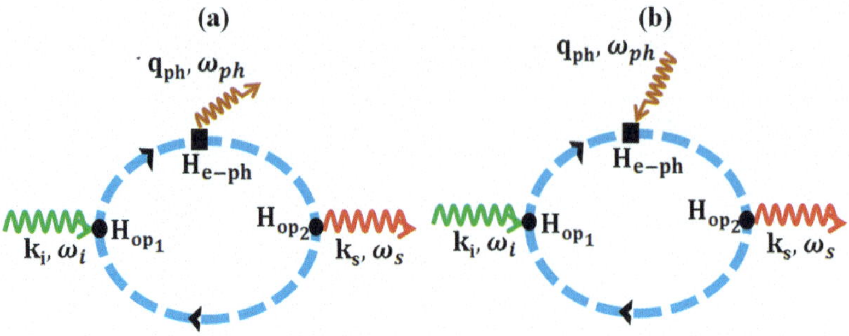

Fig. 2.4 **a, b** Feynman diagram for the Stokes and Anti-Stokes Raman scattering process, respectively. H_{op_1} and H_{op_2} are the Hamiltonians corresponding to the electron-photon interaction during photon absorption and photon emission, respectively; and H_{e-ph} is the Hamiltonian corresponding to electron-phonon interaction during phonon creation/annihilation

Figure 2.4a, b demonstrate the Feynman diagram corresponding to the Stokes and Anti-Stokes first-order Raman scattering process, respectively.

Considering we have N number of electrons in a system. In the presence of electromagnetic radiation, the electron momentum $\vec{P}$ becomes $\vec{P} - e\vec{A}$, where e is charge of the electron and $\vec{A}$ is the vector potential of electromagnetic radiation. The Hamiltonian resulting from the electron-radiation $\left(H_{op}\right)$ interaction is written as [21, 22]

$$H_{op} = H'_{op} + H''_{op} \tag{2.8}$$

$$H_{op} = \frac{e^2}{2m}\sum_j \vec{A}(\vec{r}_j)\cdot\vec{A}(\vec{r}_j) - \frac{e}{m}\sum_j \vec{A}(\vec{r}_j)\cdot\vec{P}_j \tag{2.9}$$

where, m is mass of the electron. $\vec{r}_j$ and $\vec{P}_j$ are the position and momentum of jth electron. Both terms $\vec{A}(r_j)\cdot\vec{A}(r_j)$ and $\vec{A}(r_j)\cdot\vec{P}_j$ contributes to the Raman scattering. However, the contribution from $\vec{A}(r_j)\cdot\vec{P}_j$ is significant and contributes to the scattering from almost all kinds of excitations. While contribution from $\vec{A}(r_j)\cdot\vec{A}(r_j)$ in Raman scattering is significant only for some special kinds of excitation such as scattering by free electrons, plasmons and conduction electrons in medium [21]. Raman scattering in quantum mechanical picture has been proved to be a very crucial for understanding the fundamental properties associated with various materials.

2.4 Resonant and Non-resonant Effect

The advent and development in laser technology plays a key role in advancement of Raman spectroscopic technique. In early days, the available light sources allowed Raman studies only for transparent materials due to their larger scattering length. However, the advancement in laser technology makes feasible Raman scattering studies of opaque materials. Furthermore, due to the small Raman scattering cross section, usually defined as Raman scattering intensity or intensity of the phonon modes in the Raman experiment, it was difficult to observe weak signal. The discovery of lasers with different wavelengths plays a very important role in Raman studies particularly for semiconducting materials and lead to the observation of scattering cross section as a function of laser excitation energy. In addition, if the laser excitation energy is close to the band gap or available excitonic energy gap of materials, results in the resonance effect which enhances the Raman scattering intensity manifolds and leads to the observation of even extremely weak peaks in the Raman scattering experiments. Raman spectroscopic studies as a function of laser excitation energy give rise to cross over from non- resonance to resonance conditions and results in a significant effect of resonance and non-resonance conditions on the Raman spectrum.

2.4.1 *Resonant Effect and First-Order Raman Scattering*

If only one phonon is created or annihilated during the Raman scattering process-it is known as the first-order Raman scattering, and generally the Brillouin zone center ($\vec{q} \sim 0$) optical phonon participates in the first-order Raman scattering process. This could be understood as: upon incident of the light of wave vector $\vec{k}_i$, the momentum is transferred to the electron and the electron is excited to some intermediate state and at the same time a hole is created in the ground state. Excited electron creates a phonon, Stokes Raman process, with a wave vector $\vec{q}$ and the wave vector of the excited electron becomes $\vec{k}_i - \vec{q}$. Therefore, for excited electron to recombine with hole in ground state, the $\vec{q}$ should be equal to zero and it is well known that the zone centered phonons satisfy the $\vec{q} \sim 0$ condition suggesting that the only zone centered phonons participate in the first-order Raman scattering. For convenience, we restrict our discussion to the Stokes Raman scattering process because formalism of Stokes and Anti-Stokes Raman process is similar.

First-order Raman scattering process in quantum mechanical picture, may be understood via Feynman diagram and schematic representation as shown in Figs. 2.4a and 2.5a for the Stokes Raman scattering process, respectively; which involve three elementary processes: (i) When a laser light impinges on the system, the electronic system absorbs a photon and an electron-hole pair is created that results in a finite electron-photon/radiation interaction (H_{op_1}). (ii) The electron goes to the excited intermediate state (real/virtual) $|i>$ from the ground state $|g>$, and creates a phonon (Stokes process) and results in an electron-phonon interaction (H_{e-ph}) and then the

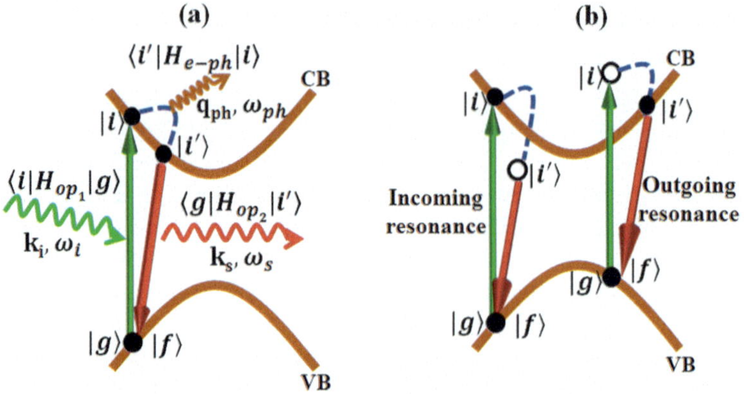

Fig. 2.5 **a** Schematic representation for the Stokes first-order Raman scattering process. **b** illustrating the resonance conditions for incident and outgoing photons. Solid and open circles correspond to the real and virtual energy states, respectively

electron undergo transition from intermediate state $|i>$ to another intermediate state $|i'>$ with the transfer of energy and momentum. (iii) Finally, the electron comes back to the initial/ground state ($|g>/|f>$) and recombines with the hole via emitting the photon/radiation results in electron-photon/radiation interaction (H_{op_2}). If during step (ii) a phonon is annihilated -it is referred as the Anti-Stokes process.

Raman scattering intensity of the phonon, in quantum mechanical picture, for the first-order stokes Raman scattering process as shown in Fig. 2.5a can be written as [20–26]

$$I_{Stokes} \propto \left| \sum_{g,i,i'} \frac{<g|H_{op_2}|i'> <i'|H_{e-ph}|i> <i|H_{op_1}|g>}{(E_L - \Delta E_{ig})\,(E_S - \Delta E_{i'g})} \right|^2 \tag{2.10}$$

where $|i>$, $|i'>$ represent the intermediate states, whereas $|g>$ stands for the ground state. H_{op_1} and H_{op_2} are the Hamiltonians representing the electron-photon (absorption) and electron-photon (emission) process, respectively; and H_{e-ph} is the electron-phonon interaction Hamiltonian. Matrix elements $<i|H_{op_1}|g>$, and $<g|H_{op_2}|i'>$ correspond to the electron-photon (absorption) and the electron-photon (emission), respectively; while the matrix element $<i'|H_{e-ph}|i>$ correspond to the electron-phonon interaction. $\Delta E_{ig} = (E_i - E_g + i\gamma_1)$ represents the energy difference between the intermediate state E_i and the final/ground state E_g, and $\Delta E_{i'g} = (E_{i'} - E_g + i\gamma_2)$ corresponds to the energy separation between the intermediate state $E_{i'}$ and the final/ground state E_g. γ_1 and γ_2 are the linewidth/full width at half maximum (FWHM) or damping constants and are inversely proportional to the lifetime of the intermediate states. γ_1 and γ_2 are introduced phenomenologically due to finite lifetime of these excitations. The energy $E_L = \hbar\omega_L$, and $E_S = \hbar\omega_s = \hbar\omega_L - \hbar\omega_{ph}$ are energies of the incoming and outgoing photon, respectively. In the above equation, there are two terms in the denominator which give rise to two resonance conditions. The incidence resonance/incoming resonance condition is given as

$$E_L = \Delta E_{ig} \tag{2.11}$$

While the scattered resonance/outgoing resonance condition is given as

$$\left\{ \begin{array}{c} E_S = \Delta E_{i'g} \\ \text{or} \\ E_L = \hbar\omega_{ph} + \Delta E_{i'g} \end{array} \right\} \tag{2.12}$$

When one of the two denominators' terms becomes very small it gives rise to the resonance effects which lead to the tremendous enhancement in the Raman scattering intensity. On the other hand, in case when the denominator terms are very large, there will be a reduction in intensity such a condition generally known as non-resonance condition. Under the non-resonance, the Raman scattering intensity of the phonon does not depend on the incident laser excitation energy. Here it is also important to note that the intensity never approaches infinity because the denominator terms will never be equal to zero because of the finite lifetime of these intermediate states. The other important point to be noted is that the intermediate states could be both real excitonic energy states as well as virtual energy states. For the case of incoming resonance, the intermediate state $|i>$ behave as a real excitonic energy state while $|i'>$ behaves as a virtual energy state. On the other hand, for outgoing resonance, $|i'>$ behave as real excitonic energy state while $|i>$ behaves as a virtual energy state.

If the intermediate states are real excitonic energy states rather than virtual states, the $E_i - E_g$ and $E_{i'} - E_g$ could correspond to the bandgap or different available excitonic energy states for the semiconducting materials. By considering the excitonic energy states, ΔE_{ig} and $\Delta E_{i'g}$ could be written as [24, 26–28]

$$\left\{ \begin{array}{l} \Delta E_{ig} = (E_i - E_g + i\gamma_1) = E_x + i\gamma_x \\ \Delta E_{i'g} = (E_{i'} - E_g + i\gamma_2) = E_x + i\gamma_x \end{array} \right\} \tag{2.13}$$

where E_x and γ_x is the energy corresponding to the energy and lifetime of the different available excitonic energy transition states. In the vicinity of resonance conditions, the role of the numerator term (matrix elements) becomes very small. However, here it should be noted that the numerator term could not be zero by the symmetry. Therefore, the numerator term may be considered as a constant and the (2.10) could be written as

$$I_{Stokes} \propto \left| \frac{A}{(E_L - E_x - i\gamma_x)\,(E_L - \hbar\omega_{ph} - E_x - i\gamma_x)} \right|^2 \tag{2.14}$$

The (2.14) suggests that if the laser excitation energy (E_L) is very close to the E_x or $E_x + \hbar\omega_{ph}$ it results into the resonance effect which leads to the tremendous enhancement in intensity of the phonon modes.

2.4.2 Resonant/Double Resonant Effect and Second/ Higher-Order Raman Scattering

In the first-order Raman scattering, only one phonon participates in the Raman scattering process. However, when two phonons participate in the Raman scattering it is called the second-order Raman scattering. In the second-order Raman scattering both phonons may be created (Stokes phonons) or one may be created and the other annihilated (Stokes and Anti-Stokes phonons) or both phonons may be annihilated (Anti-stokes phonons). Second-order Raman scattering may be classified into two categories: (i) Second-order line Raman scattering and (ii) Second-order continuum Raman scattering [20]. In the second-order line Raman scattering process, two individual successive first-order Raman scattering occurs. On the other hand, in the second-order continuum Raman scattering, two phonons participate in Raman scattering at a single event. It should be noted that the conservation of wave vector during the scattering process is the most important selection rule. Since in the case of second-order line Raman scattering, the two successive first-order Raman scattering processes occur, therefore, the wave vector should be conserved in each individual first-order Raman scattering. This may lead to the participation of zone center phonon in second-order line Raman scattering suggesting that the second-order line Raman scattering provides no additional information about the materials in addition to already provided by first-order Raman scattering. On the other hand, for the case of second-order continuum Raman scattering, since the two phonons participate in a single event during the scattering process, the sum of wave vectors of two phonons should balance for the change in the wave vector of the scattered photon or after recombination process. Therefore, the second-order continuum scattering process allows the phonons from the entire Brillouin zone to participate in the Raman scattering process. Figure 2.6a, b illustrate the Feynman diagram and schematic representation for the continuum second-order Stokes Raman scattering process and showing the creation of two phonons. The continuum second-order Raman scattering as shown in Fig. 2.6a, b involve four processes: first two process, i.e. excitation of an electron from the ground state $|g>$ to the intermediate state (real/ virtual) $|i>$ and creation of a phonon takes place similar to the first-order process as shown in the Fig. 2.5a. The electron undergo transition from intermediate state $|i>$ to another intermediate state $|i'>$. If the created phonon does not conserve the wave vector, a second phonon is created to conserve the wave vector and now electron undergo transition from intermediate state $|i'>$ to $|i''>$. Once the wave vector is conserved, the excited electron comes back to the ground state and recombines with the hole via emitting the photon.

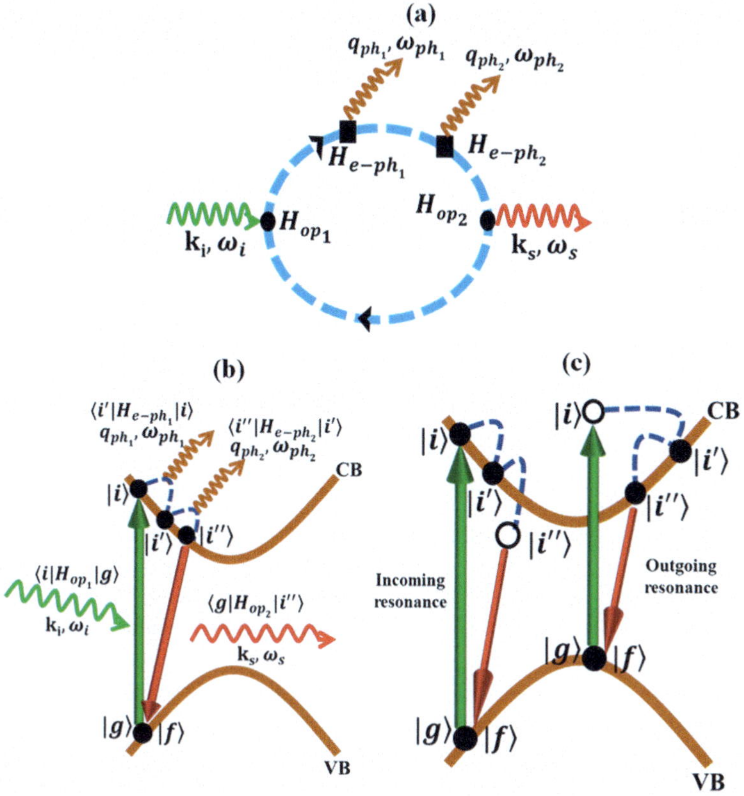

Fig. 2.6 **a**, **b** Feynman diagram and schematic representation for the second-order Stokes Raman scattering process, respectively. **c** Illustrating the resonance conditions for incident and outgoing photons. Solid and open circles are corresponding to the real and virtual energy states, respectively

The conservation of wave vector for the entire process may be understood as: after creation of first phonon with wave vector $\vec{q}$, the wave vector of excited electron becomes $\vec{k}_i - \vec{q}$. If the created phonon wave vector is not equal to zero i.e. $\vec{q} \neq 0$, the excited electron cannot recombine with hole, then the excited electron stays in an excited state and waits for further relaxation by creating the second phonon. As soon as second phonon with wave vector $-\vec{q}$ is created, the excited electron recombines with the hole. Here, it should be noted that the created two phonons have equal but opposite wave vectors in order to conserve the wave vector. However, the frequency of the created phonons could be the same or different. If the created phonons have equal frequency, called overtone. On the other hand, if the frequency is not equal, is called combination mode. Furthermore, if the created two phonons have no wave vectors in pair, then the excited electron waits to create a third phonon before recombination which gives rise to higher-order Raman scattering phenomenon. Thus, the second/higher-order continuum Raman scattering process has freedom for selecting any values of $\vec{q}$, suggesting that the phonons from the entire Brillouin zone may participate.

Intensity of the second-order Raman scattering considering the quantum mechanical picture can be given as [21–26]

$$I_{Stokes} \propto \left| \sum_{g,i,i'i''} \frac{< g|H_{op_2}|i'' >< i''|H_{el-ph_2}|i' >< i'|H_{el-ph_1}|i >< i|H_{op_1}|g >}{(E_L - \Delta E_{ig})\,(E_L - \hbar\omega_{ph_1} - \Delta E_{gi'})(E_L - \hbar\omega_{ph_1} - \hbar\omega_{ph_2} - \Delta E_{i''g})} \right|^2 \tag{2.15}$$

where, $\hbar\omega_{ph_1}$ and $\hbar\omega_{ph_2}$ are the energy of the first and second created phonon, respectively. Matrix elements $< i|H_{op_1}|g >$ and $< g|H_{op_2}|i'' >$ correspond to the electron-photon (absorption), the electron-photon (emission), respectively. Matrix element $< i'|H_{el-ph_1}|i >$ correspond to the electron–phonon interaction of the first created phonon. These three matrix elements in the second-order Raman scattering process are similar to what we have seen in (2.10) for the first-order Raman scattering process. However, in contrast to (2.10), there is an additional matrix element $< i''|H_{el-ph_2}|i' >$ and energy term $E_L - \hbar\omega_{ph_1} - \hbar\omega_{ph_2} - \Delta E_{i''g}$ in (2.15) which appears due to the presence of second electron-phonon interaction as a result of creation of a second phonon. The (2.15) gives rise to three resonance conditions i.e. $E_L = \Delta E_{ig}$, $E_L = \hbar\omega_{ph_1} + \Delta E_{gi'}$ and $E_L = \hbar\omega_{ph_1} + \hbar\omega_{ph_2} + \Delta E_{i''g}$. Generally, the second-order Raman scattering is much weaker compared to the first-order Raman scattering but it could be significantly enhanced under the resonance effect. Furthermore, in case when two of the three denominator terms in (2.15) show resonance simultaneously then the intensity of second-order Raman scattering phonon becomes comparable or even larger than that of the first-order Raman scattered phonon and such a condition is known as double resonance condition [26–28]. During the incidence double resonance condition, $E_L = \Delta E_{ig}$ and $E_L = \hbar\omega_{ph_1} + \Delta E_{gi'}$ is expected to resonate simultaneously and the intermediate states $|i >$ and $|i' >$ behave as a real excitonic energy state while $|i'' >$ behaves as virtual energy state. On the other hand, for outgoing double resonance $E_L = \hbar\omega_{ph_1} + \Delta E_{gi'}$ and $E_L = \hbar\omega_{ph_1} + \hbar\omega_{ph_2} + \Delta E_{i''g}$ is expected to resonate simultaneously and the intermediate states $|i' >$ and $|i'' >$ behave as a real excitonic energy state while $|i >$ behaves as a virtual energy state. The double resonance condition has been realized in various materials such as graphite, graphene, group VI-TMDCs materials etc. [27–35]. Similar to the first-order Raman scattering, if one consider the intermediate states as the real excitonic energy states rather than the virtual states, then $E_i - E_g$ and $E_{i'} - E_g$ could correspond to the bandgap or different excitonic energy states of the semiconducting materials. By considering the excitonic energy states, the (2.15) could be written as

$$I_{Stokes} \propto \left| \frac{B}{(E_L - E_x - i\gamma_x)\,(E_L - \hbar\omega_{ph_1} - E_x - i\gamma_x)(E_L - \hbar\omega_{ph_1} - \hbar\omega_{ph_2} - E_x - i\gamma_x)} \right|^2 \tag{2.16}$$

The (2.14) and (2.16) may be used to understand the non-resonant and resonant Raman studies for the case of semiconducting or two-dimensional (2D) layered semiconducting materials such as group VI-TMDCs which will be discussed in the next section.

2.5 Resonance Effect in Semiconducting Transition Metal Di Chalcogenides

Discovery of graphene in 2004 [36], a first member of the 2D family, led to the discovery of a large variety of 2D materials with a wide range of properties including semiconducting, metallic, insulating, superconducting and so on. Recently, the group VI-TMDCs with common atomic formula MX_2, where M = Mo, W and X = Se, S, Te, have gained the considerable attention owing to their intriguing properties. These are the layered materials and the two adjacent layers are coupled with weak Van der Waals forces which lead to the isolation of a single layer from the bulk crystal. Monolayer (1L) of these materials have emerged as a very promising class of 2D materials, and has attained a special interest by the scientific community due its possibilities and promising advantages in optical, electronic, optoelectronic device applications due to their finite direct bandgap [37–42]. Moreover, the properties of these materials, especially optical, electronic and vibrational, are modulated by the layer thickness which leads these materials to the thickness controlled potential applications. For example, Fig. 2.7a shows transition of the indirect bandgap in bulk to the direct bandgap in monolayer for MoS_2. The change in the bandgap from indirect to direct bandgap enhances the photoluminescence of the materials, see Fig. 2.7b. Additionally, these materials also offer an astounding and interesting physics attributed to the strong spin-orbit coupling and splitting of energy bandgap (see Fig. 2.7c) [43–47], valley nature of the conduction and valence band (see Fig. 2.7d) and coupled spin-valley physics [48–51]. Furthermore, in atomically thin group VI-TMDCs, the reduced dielectric screening effect due the reduction in the dielectric constant led to the enhancement in Coulombic interaction between electrons and holes because of the increase in binding energy, (see Fig. 2.7e, f) which gives rise to the formation of many excitonic quasiparticles transition states like excitons, trions, and biexcitons etc. (see Fig. 2.7g) [52–58]. These excitonic quasiparticle remains stable even at room temperature and play a vital role in controlling the optical properties of group VI-TMDCs. However, stability of the exciton is more than that of trion or biexciton due to the large binding energy of the excitons which is in the range of hundreds of meV to 1 eV for the atomically thin group VI-TMDCs. While trions and biexcitons show binding energy in the range of tens of meV. Therefore, particularly, the excitons play more important role in the optical properties of these materials compared to the trions or biexcitons.

Excitons are the bound states of one electron and one hole and are the electrically neutral excitonic quasiparticles. In the monolayer group VI-TMDCs, the main excitons are A, B and C. A and B excitons correspond to the direct excitonic transitions from the minima of the conduction band to the maxima of split valence band at the

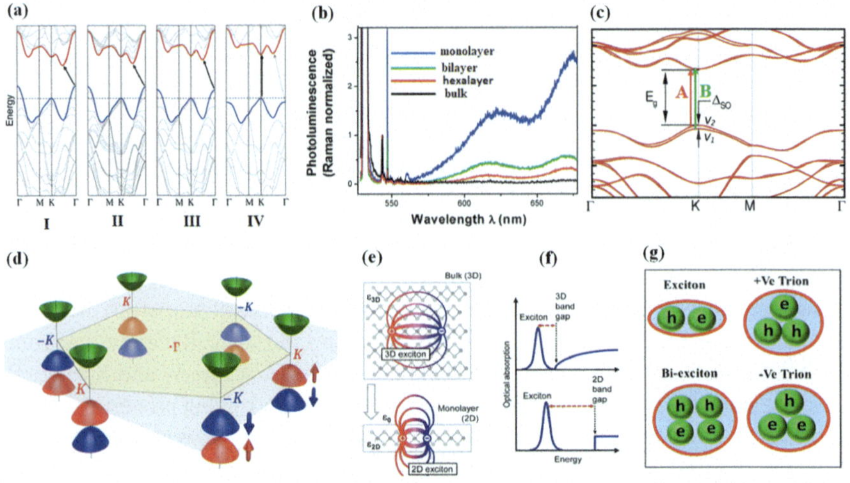

Fig. 2.7 **a** Calculated band structures of (I) bulk, (II) quadrilayer, (III) bilayer, and (IV) monolayer MoS_2; **b** Layer dependence of the photoluminescence emission for MoS_2. Adapted with permission from [38]. Copyright (2010) (American Chemical Society). **c** Typical band structure for an MX_2 monolayer. The valence-band maxima is split due to spin-orbit coupling. Direct excitonic transitions from minima of the conduction band to maxima of the split valence band v_1 and v_2 at the K point of the Brillouin zone correspond to the A and B excitons, respectively. Adapted with permission from [45]. Copyright (2012) (American Physical Society). **d** Schematic of the band structure at the band edges located at K points and showing the Valley nature of the band structure. Adapted with permission from [48]. Copyright (2012) (American Physical Society). **e** Representation of holes and electrons bound states (excitons) in bulk (3D) and monolayer (2D). ε_{3D}, ε_{2D} and ε_0 are the dielectric constants for the bulk, monolayer and free space; **f** schematic representation of the optical absorption indicating the effects of dimensionality on the electronic and excitonic properties. Bandgap and exciton binding energy (dashed red line) is expected to increase in monolayer compared to that of bulk. Adapted with permission from [57]. Copyright (2014) (American Physical Society). **g** Schematic representation of the excitonic quasiparticle states

K point of the Brillouin zone, (see Fig. 2.7c). A exciton corresponds to the fundamental bandgap of the material. On the other hand, B exciton appears due to the splitting of valence band because of the lift in degeneracy of the spin owing to strong spin-orbit coupling. The origin of the C exciton in literature is still under debate as few studies have also attributed the origin of C exciton to the Van Hove singularity [59]. In addition to the excitons, other excitonic quasiparticles such as trions (charged excitons) and biexcitons (exciton molecule) are also a very important excitonic energy transition states and have a fundamental importance in optical devices based on these materials. The formation of the trions or biexcitons may occur due to the presence of excess number of electrons or holes in the materials. Trions are the three-particle bound states of either two electrons and one hole (negative trion) or one electron and two holes (positive trion) while biexcitons are the electrical neutral excitonic quasiparticle states consist of four particle bound states (i.e. consisting of the two excitons) as shown in Fig. 2.7g. Due to the existence of various excitonic transition energy states, group VI-TMDCs materials could be considered as ideal candidates to understand the resonance versus non-resonance effect. Whenever the laser excitation energy is close to one of these excitonic energy states, it results into the resonance effect and may be called as excitonic resonance conditions which lead to the observation of various intriguing properties as well as fundamental physics associated with these materials. For example, observation of multi-phonon Raman scattering, backscattering forbidden and infrared active modes, splitting of some of the first-order Raman active phonon modes, electron-phonon interactions etc. While the observations of these various aspects would have not been possible under the non-resonance effect. In subsections of this section, we will focus on the resonant and non-resonant Raman scattering as well as various consequences of resonance effect in group VI-TMDCs.

2.5.1 Multi-phonon Raman Scattering

Resonance effect in group VI-TMDCs not only enhances Raman scattering intensity of the phonons but also led to the observation of multi-phonon Raman scattering and has been proved to be an important tool to explore the phonons from the entire Brillouin zone. Under the resonance/double resonance effect first, second or higher-order phonon modes, in addition to the zone centered first-order Raman active phonon modes, from the high symmetry such as M and K points of the Brillouin zone have been observed in these materials which are normally absent or extremely weak in non-resonance Raman scattering measurements [60–72]. Figure 2.8a shows the room temperature Raman spectrum for bulk MoS_2 collected for four different laser excitations. Photoluminescence or absorption measurement for few layers of MoS_2 shows two excitonic features centered at ~670 nm (~1.85 eV) and ~620 nm (~2.0 eV) corresponding to the A and B excitonic features, see Fig. 2.8b, c, respectively. It should be noted that the absorption spectrum is observed even for the bulk MoS_2 while the photoluminescence has been found to be absent for the bulk MoS_2 and other group VI-TMDCs because of the indirect band gap in bulk group VI-TMDCs [39]. For the Raman spectrum collected with laser 488 nm (2.4 eV) and 532 nm (2.33 eV), one can see only two prominent features near ~380 and ~410 cm^{-1} corresponding to the first-order Raman active mode E_{2g}^1 (in plane vibration of both Mo and S atoms in opposite direction) and A_{1g} (out of plane vibration of only S atoms in opposite direction) from Γ point of the Brillouin zone, respectively. We note that the phonon mode notation changes for the even/odd layers as well as for the bulk. The energy of 488 and 532 nm laser excitation is far away from the A and B excitonic features and as a result it gives rise to the non-resonance condition. Therefore, only the first-order Raman active E_{2g}^1 and A_{1g} modes are predominantly observed under the excitations of 488 and 532 nm.

On the other hand, when the spectra is excited with 594 nm (2.09 eV) and 633 nm (1.96 eV) laser, one could see many additional prominent features apart from the first-order Raman active E_{2g}^1 and A_{1g} phonon modes. These additional features may

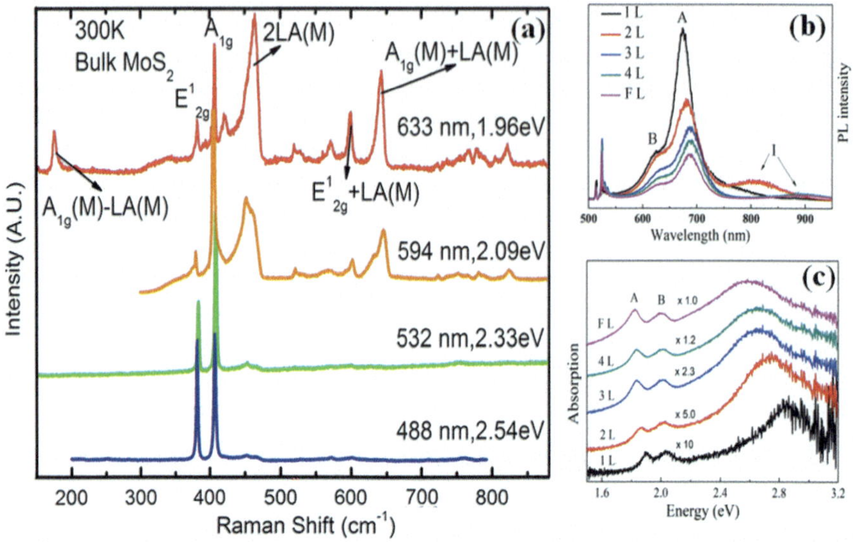

Fig. 2.8 **a** Room temperature Raman spectrum of bulk MoS_2 crystal collected for different laser excitation. Adapted with permission from [64]. Copyright (2014) (American Institute of Physics). **b**, **c** Photoluminescence and absorption spectrum for 1L, 2L, 3L, 4L and few layers (FL) MoS_2, respectively. Adapted with permission from [73]. Copyright (2014) (Royal Society of Chemistry) under a Creative Commons Attribution 3.0 Unported License

appear due to the multi-phonon Raman scattering which may be assigned as overtone, combination and difference of phonon modes and also appears from the high symmetry such as M/K points of the Brillouin zone. For example, a broad and strong asymmetric feature, centered at ~460 cm^{-1} is attributed to the second-order longitudinal acoustic ($2LA(M)$) phonon mode which is the overtone of the $LA(M)$ phonon. Further, the observed phonon modes centered near ~180 cm^{-1} and ~640 cm^{-1} are attributed to the difference and combination of A_{1g} and LA from M point of the Brillouin zone, respectively. The observed phonon mode near ~600 cm^{-1} could be assigned as combination of E^1_{2g} and LA from M point of the Brillouin zone. Moreover, a large number of other phonon modes could also be seen in the Fig. 2.8a. The appearance of these multi-phonon Raman scattering modes, in addition to the first-order Raman active phonon, may be understood via resonance effect. 594 nm (2.09 eV) and 633 nm (1.96 eV) laser excitation energy lie in the vicinity of A and B excitonic features and results in the resonance effects, and it led to the appearance of these multi-phonon Raman scattering modes. The observation of these multi-phonon Raman scattering is very important from the perspective of fundamental research as well as applications. For example, the anomalous behavior of the self-energy parameters such as frequency and FWHM as function of temperature lead to the gauging of electron-phonon coupling in these materials which will be discussed later in more detail. Understanding the electron-phonon coupling in these materials is important for better performance of the electrical devices based on these materials. Temperature dependent intensity of the phonon modes in the vicinity of resonance effect could be used to extract the excitonic features in these materials.

2.5.2 *Back Scattering Forbidden and Infrared Active Phonons*

In addition to the multi-phonon Raman scattering modes, resonance effects could lead to the observation of generally forbidden phonons as well as infrared active phonon modes in group VI-TMDCs. These phonon modes are generally absent in normal (non-resonance) Raman scattering measurements. Earlier studies have reported the forbidden phonons in back scattering geometry (i.e. E_{1g}) and infrared active A_{2u} in group VI-TMDCs under the resonance effect [71, 72, 74–78]. The observation of back scattering forbidden/infrared active phonons under the resonance condition may be understood by considering the Fröhlich mechanism [79]. The Fröhlich mechanism emerges due to the coupling between electrons/excitons and longitudinal optical (LO) phonons. Understanding of the coupling of excitons with phonons in these group VI-TMDCs and their heterostructure is the central focus of research in the area of quasiparticle excitations because it governs the many aspects of the optical properties of these materials. A lot of studies have been devoted to understand the Fröhlich interaction in these 2D materials and their heterostructure [80–87]. The Fröhlich mechanism of exciton-phonon coupling also play an important role in Raman scattering and give rise to the finite intensity of the back scattering forbidden/infrared phonons proportional to the $(a\ q)^2$ where q is the wave vector of the phonon and a is the Bohr radius of the exciton [79]. The large Bohr radius compared to the lattice parameter could lead to the breakdown in the momentum selection rule under the resonance condition which may give rise to the observation of back scattering forbidden/infrared phonons in the Raman spectra. For the case of group VI-TMDCs, large Bohr radius of the excitons than that of the lattice parameter has been reported [88–94], and it results into strong exciton-phonon coupling giving rise to a finite intensity even for the back scattering forbidden/infrared phonons.

Figure 2.9a shows the room temperature Raman spectrum for 1L, 2L and 3L MoS_2 collected using 2.7 eV laser excitation energy. The observed phonon modes for 2L and 3L centered near ~290 cm^{-1} and ~470 cm^{-1} corresponds to the backscattering forbidden E_{1g} and infrared active A_{2u} (in figure it has been denoted by Raman active E'/E_g and A'_1/A_{1g}, respectively) phonons, respectively. Here, we have considered the peak at ~470 cm^{-1} to be infrared active A_{2u} mode because many previous reports confirmed its symmetry to be of infrared active mode. The laser excitation energy 2.7 eV is very close to energy of the C exciton suggesting the resonance effect with C exciton [95]. Therefore, appearance of the E_{1g} and A_{2u} phonons may be attributed to the resonance effect with the C exciton. Quite surprisingly, these E_{1g} and A_{2u} phonon modes even becomes stronger as compared to the first-order Raman active

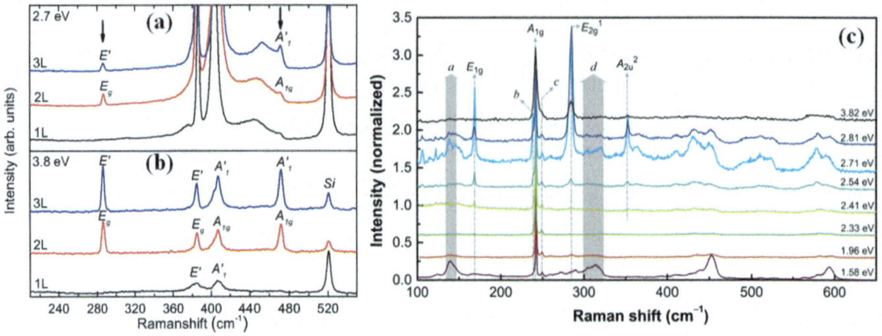

Fig. 2.9 Room temperature Raman spectrum for 1L, 2L, 3L MoS_2 on Si/SiO_2 excited with **a** 2.7 eV and **b** 3.8 eV. Adapted with permission from [78]. Copyright (2015) (American Physical Society). **c** Raman spectra of bulk $MoSe_2$ as a function of laser excitation energies. Adapted with permission from [76]. Copyright (2015) (Springer Nature), under Creative Commons CC BY license

E_g/E_{2g}^1 and A_1'/A_{1g} phonons in the frequency range of ~380–390 cm^{-1} and ~402–410 cm^{-1}, respectively; when the spectra is excited with 3.8 eV laser which lie above the energy of the C exciton, see Fig. 2.9b. Generally, the peak attributed to the C exciton is very broad, in contrast to A and B excitons, which show narrow peaks, and looks like contribution from multiple peaks [52]. The appearance of E_{1g} and A_{2u} phonons above 2.7 eV may be attributed to the broadness of the C exciton. Lee et al. also observed E_{1g} and A_{2u} phonon modes in few layer MoS_2 for the excitation energy above 2.4 eV and both become intense close to the C exciton energy or above it [75]. Similar to the case of MoS_2, E_{1g} (~170 cm^{-1}) and A_{2u} (~350 cm^{-1}) phonons are also observed for $MoSe_2$ for laser excitation energies above 2.2 eV and becomes intense for the laser excitation energy close to the C exciton, see Fig. 2.9c [65, 67, 72, 76]. Further, it should be noted that the E_{1g} and A_{2u} phonons are found to be either absent or extremely weak in monolayer even under the resonance conditions [65, 67, 72, 75]. The observation of E_{1g} and A_{2u} phonons in few layers compared to the monolayer group VI-TMDCs may be attributed to the change in band gap, however full understanding is still lacking and further studies are required. Nevertheless, from the above discussion it is clear that the appearance of the back scattering forbidden/infrared active phonons in Raman spectrum suggests that the C exciton plays a prominent role in activation of these forbidden phonon modes. In other word, the appearance of the E_{1g} and A_{2u} phonons clearly reflects the resonance effect with the C exciton and these could also be used as a fingerprint method to differentiate the monolayer and few layers of group VI-TMDCs.

2.5.3 *Davydov Splitting*

In group VI-TMDCs, in addition to the observation of multi-phonon Raman scattering under the resonance effect, it also led to the observation of splitting of some of the first- order Raman active phonon modes and such kind of splitting in these group VI-TMDCs materials is known as Davydov splitting [65, 68, 96–100]. Davydov splitting is a very special kind of splitting and occurs in electronic and vibrational spectra due to the presence of more than one equivalent interacting entity in the unit cell [101]. In few layers group VI-TMDCs, Davydov splitting occurs due to the presence of more than one equivalent monolayer MX_2. For example, in N layer MX_2 ($N \geq 2$), the A_{1g} mode will split into N Davydov components which may be expressed as $\frac{N}{2}A_{1g} + \frac{N}{2}A_{2u}$ and $\frac{N+1}{2}A_{1g} + \frac{N-1}{2}A_{2u}$ for even and odd numbers of layers, respectively. These suggesting that for even number of layers, A_{1g} will split into $\frac{N}{2}A_{1g}$ Raman active and $\frac{N}{2}A_{2u}$ infrared active modes. On the other hand, for the case of odd number of layer, A_{1g} will split into $\frac{N+1}{2}A_{1g}$ Raman active and $\frac{N-1}{2}A_{2u}$ infrared active modes. For the case of bilayer (2L), A_{1g} mode will split into one Raman active and one infrared active mode while for trilayer (3L), it will split into two Raman active and one infrared active mode indicating that the total number of Davydov components are equal to the number of layers. Therefore, Davydov splitting of A_{1g} mode may also be used to identify the number of layers in these materials. Furthermore, each monolayer MX_2 is coupled via interlayer interaction. Therefore, appearance of the Davydov splitting may also be inferred as a consequence of the interlayer interaction. By carefully analysing the Davydov splitting one could understand the crucial role of interlayer interaction in these TMDCs materials.

Moreover, this splitting occurs only under the resonance effect. In other words, the splitting is more prominent and visible for a specific wavelength which is close to the excitonic energy state, and hence the occurrence of such type of splitting reflects the resonance effect. However, it is still not understood completely, why such kind of splitting occurs in these materials only under the resonance effect. Figure 2.10a shows the Raman spectrum near the A_{1g} mode for 1–6 layers of $MoSe_2$ using different laser excitation energies. Figure 2.10b illustrates schematic vibration of A_{1g} mode for 2L, 3L, and 4L. For the case of 1L, only one peak could be seen for all excitation energy. If one goes from 1 to 2L, the A_{1g} peak split into two phonon modes for excitation energy 1.58 eV which is close to the energy of A exciton indicating resonance effect with A exciton [72]. However, only one peak is clearly observed for the other laser exciton energies. On the other hand, the splitting increases from 3 to 6L for the laser exaction energies above 2.41 eV and becomes prominent for 2.71 and 2.81 eV which is close to the C exciton indicating resonance effect with C exciton. Further, one could clearly see two resolved phonons for 3L and 4L and three well resolved phonons for 5L and 6L. These well resolved Davydov components may be assigned as Raman active A_1'/A_{1g} mode. However, it should be noted that the one, two and three infrared active phonons could also be expected for 2L and 3L, 4L and 5L and 6L, respectively. But similar to $MoSe_2$, only Raman active A_1'/A_{1g} mode is clearly observed in other group VI-TMDCs and generally the infrared active modes are not clearly resolved even under the resonance effect may be due to the very weak intensity as compared to the A_1'/A_{1g} phonon modes. In addition to the A_{1g} mode, Davydov splitting could also be seen for the in-plane E_{2g}^1 mode. However, Davydov splitting is more prominent for the case of A_{1g} compared to E_{2g}^1 phonon mode. This may be understood as:

the A_{1g} mode appear due to out-of-plane vibration of atoms while E_{2g}^1 appears due to in-plane vibration of atoms. The phonon modes associated with the out-of-plane vibration of atoms are directly affected by the interlayer interaction compared to the phonon mode associated with in-plane vibration of atoms in these kinds of materials. Therefore A_{1g} mode is more affected by interlayer interaction compared to the E_{2g}^1 mode and hence the Davydov splitting. Figure 2.10c shows the frequency of the split component as a function of layer which is in good agreement with the characteristics of A_{1g} mode i.e., frequency increases with increasing number of layers [102]. Frequency position of these Davydov components as a function of layers is widely used to estimate the mechanical properties such as spring/force constant [65, 97]. Further, people have also reported the Davydov splitting components as a function of external perturbation such as temperature. Tan et al. for WS_2, [68] and Kumar et al. for MoS_2 [100] conducted the temperature dependent Raman study in order to understand the temperature dependence of the Davydov splitting. They found that the Davydov splitting components are more pronounced and well resolved at the lowest recorded temperature (i.e. close to 4 K) compared to the room temperature and this was attributed to the increase in interlayer strength because of the reduction in the interlayer spacing with decreasing temperature. Figure 2.10d, e show the 2D color contour maps demonstrating the Davydov splitting of the modes (A_{1g} ~ 400–412 cm^{-1} and E_{2g}^1 ~ 370–388 cm^{-1}) as a function of temperature for few layers (FL) and three layer (3L) MoS_2, respectively and clearly reflecting the Davydov splitting is pronounced and well resolved at low temperature. Temperature dependent Davydov splitting in these group VI-TMDCs layered materials also reflects that the Davydov splitting indeed originates because of the interlayer interaction. Therefore, the observation of Davydov splitting gives rise to the following important consequences (i) Identification of number of layers (ii) Signature of resonance effect (iii) Estimation of the strength of interlayer coupling as well as mechanical properties.

2.5.4 Electron-Phonon Coupling

Interaction of electrons with the phonons known as the electron-phonon coupling (EPC), play a crucial role in wide range of phenomena. For example, the strong EPC reduces the charge carrier mobility which leads to the reduction in the electrical conductivity and as a result the reduction in the performance of thermoelectric as well as electrical devices. Additionally, the coupling also plays an important role in heat generation and dissipation, superconductivity, phonon transport and phonon heat conduction, etc. [103–105]. Dimensionality of the materials has drastic effect on the EPC. The reduced dimensionality of the materials may decrease the EPC. Therefore, EPC could be an important parameter in 2D materials and give rise to the modulation in the electrical and thermal properties. For example, high quality monolayer graphene exhibits high carrier mobility compared to bulk/thick graphene even at room temperature may be due to decrease in EPC. This allows the charge carriers, in monolayer graphene, to propagate over a large distance without any

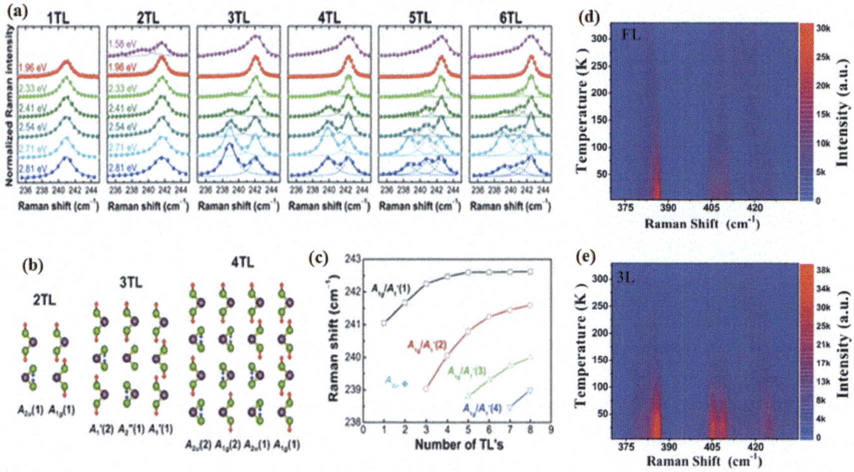

Fig. 2.10 **a** Raman spectra near the A_{1g} mode showing Davydov splitting of the A_{1g} modes for 1–6 layers $MoSe_2$ for various excitation energies. **b** Schematic representations of out-of-plane vibrational A_{1g} mode from 2 to 4L. **c** Frequency of split Raman modes as a function of layers. Adapted with permission from [65]. Copyright (2016) (American Chemical Society). **d, e** 2D color contour maps showing Davydov splitting of the modes (A_{1g} ~ 400–412 cm^{-1} and E^1_{2g} ~ 370–388 cm^{-1}) as a function of temperature for few layers (FL) and three-layer (3L) MoS_2, respectively. Adapted with permission from [100]. Copyright (2021) (IOP Publishing)

kind of scattering, and makes it an ideal as well as promising candidate for ballistic electronic device applications [106, 107]. Similar to graphene, high carrier mobility has also been reported for monolayer of group VI-TMDCs compared to the bulk/thick group VI-TMDCs at room temperature [108, 109]. The increase in carrier mobility in these atomically thin/monolayers materials compared to bulk may be due to decrease in the EPC with decreasing number of layers. Furthermore, EPC also put significant effect on the optical properties via controlling the rate of decay/dynamics of excitonic quasiparticles. Since, the group VI-TMDCs offer spin and valley degrees of freedom which make these materials an ideal candidate for the spintronics, valleytronics, and quantum phenomenon such as spin and valley quantum Hall effect. The EPC may also play an important role in these above-mentioned phenomena [110, 111].

The interactions between the various quasi and elementary particles in these 2D materials may be classified into two categories (i) intralayer interactions; the interaction between the particles within the same layer and (ii) interlayer interactions; the interaction between the particles from the different layers. For example, for the intralayer excitons; electron and hole reside within the same layer. On the other hand, for interlayer excitons; the electron resides in one layer while hole in another layer. Similarly, EPC in 2D layered materials may also be classified into two categories (i) intralayer EPC; coupling between electrons and phonons within the same layer and (ii) interlayer EPC; coupling between electrons and phonons in the different layers. Similar to the intralayer EPC, interlayer EPC may also play a very important role to control the fundamental properties of these materials. For example, one of the important consequences of interlayer EPC, the high temperature superconductivity in heterostructure [112]. Recently, the interlayer EPC has been realized in various heterostructures of 2D materials [113–118]. EPC is also affected by the symmetry of the phonons which led to the phonon symmetry-dependence of the EPC [27, 117, 119]. Therefore, understanding the EPC in group VI-TMDCs and other such layered materials become pertinent for fundamental physics as well as device applications. The resonance Raman spectroscopy has been proven to be a powerful and effective tool to investigate and explore the EPC and different aspects such as phonon symmetry dependent EPC in these 2D materials. Under the resonance effect, the EPC in these materials may be gauged via understanding the self-energy parameter i.e. frequency and linewidth of the phonon modes. Recently, the EPC has been investigated in group VI-TMDCs for example in mono and bilayer $MoSe_2$ [72], monolayer WS_2 [120], in bulk and nano flakes of MoS_2 [121] and in layered MoS_2 and $MoTe_2$ [122] via understanding the frequency and linewidth/full width at half maximum (FWHM) of the phonon modes under the resonance effect. However, linewidth of the phonon modes, in contrast to frequency, could be the best parameter to extract the EPC [123–127]. Therefore, we have considered here only the linewidth to discuss the role of EPC in group VI-TMDCs materials. Remarkably, in a high quality group VI-TMDCs, without any kind of impurities, the linewidth at a finite temperature may be affected by the phonon–phonon coupling (PPC) and EPC. The spin-phonon coupling, a very important coupling in magnetic materials, may be another interesting parameter which may affect the linewidth of the phonon modes but their role in linewidth may be excluded from the fact that the generally group VI-TMDCs are not magnetic materials. Therefore, considering the contributions from the EPC and PPC, the temperature dependent linewidth of phonon modes in high quality group VI-TMDCs may be written as

$$\gamma(T) = \gamma_{ph-ph}(T) + \gamma_{e-ph}(T) \tag{2.17}$$

where, the first term $\gamma_{ph-ph}(T)$ is the temperature dependent linewidth of the phonon and result from PPC. It is mainly governed by the three-phonon anharmonic effect, i.e. an optical phonon decay into two acoustic phonons with equal frequency but opposite momentum, as suggested by Klemens and is given as $\gamma_{ph-ph}(T) = \gamma_{ph-ph}(0) + D\ (1 + \frac{2}{e^x-1})$, where $x = \hbar\omega_0/2k_BT$, $\gamma_{ph-ph}(0)$ is the linewidth at 0 K and D is the self-energy parameter representing the contributions from the three-phonon anharmonic effect [128]. Since $\gamma_{ph-ph}(T)$ is generally governed by the phonon's population which increases with the rise in temperature and therefore increase in $\gamma_{ph-ph}(T)$ is expected with the rise in temperature. It may be understood as: the phonon population is governed by the Bose-Einstein distribution function, which increases with increasing temperature which lead to the increase in PPC resulting in the decrease in phonon's lifetime. We note that the linewidth of the phonons is inversely proportional to lifetime of the phonons (linewidth $\propto\ 1/\tau$, where τ is the lifetime of phonons) and therefore one could expect the increase in the linewidth with increase in temperature. The second term $\gamma_{e-ph}(T)$ in (2.17) is the temperature dependent linewidth of phonon and appear due to contributions from the EPC. It may be written as $\gamma_{e-ph}(T) = \gamma_{e-ph}(0)[(\frac{1}{e^{-y}+1}) - (\frac{1}{e^{y}+1})]$, where $y = \hbar\omega_0/2k_BT$ and $\gamma_{e-ph}(0)$ is the linewidth at 0 K [123]. The term $\gamma_{e-ph}(T)$ is expected to decrease with increasing temperature [123]. However, it should be noted that the second term in (2.17), generally, has been used to extract the EPC for the case of metallic systems such as graphene but this may also be used to extract EPC in narrow band gap semiconductor. Figure 2.11a shows the temperature dependent linewidth of the E_{2g} mode of graphene commonly known as G mode. Dashed (dotted) line represents individual contributions from the EPC (PPC) to temperature dependent linewidth and one could see a clear decrease (increase) in linewidth with rise in temperature. Figure 2.11b, c shows the temperature dependent linewidth of the E_{1g} and A_{1g} mode for 1L $MoSe_2$, respectively. One could see that both modes showing normal temperature dependence i.e. linewidth decreases with decreasing temperature and may be understood considering only PPC contributions.

Figure 2.11d–f show the temperature dependent linewidth of the E_{2g}^2 (interlayer interaction mode), E_{1g} and A_{1g} modes for 2L $MoSe_2$, respectively. The linewidth of the both E_{1g} and A_{1g} mode shows normal temperature dependence from 330 to ~100 K and interestingly below 100 K, an increase in linewidth could be seen with further lowering the temperature till 4 K. The linewidth of the E_{2g}^2 mode show normal temperature dependence from 300 to ~120 K. While one could see a sharp

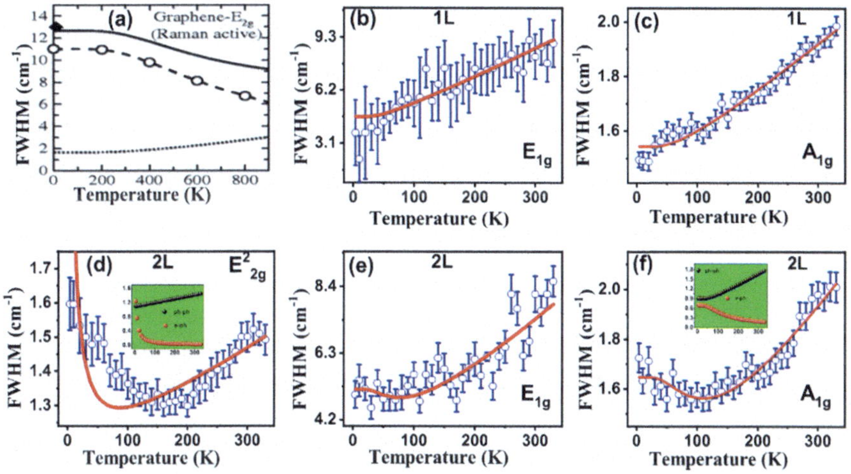

Fig. 2.11 **a** Temperature dependent total linewidth (solid line) for the E_{2g} mode of graphene. Dashed and doted lines represent the EPC and PPC contributions to the linewidth, respectively. Adapted with permission from [123]. Copyright (2007) (American Physical Society). **b, c** Temperature dependent linewidth of the E_{1g} and A_{1g} mode for 1L $MoSe_2$, respectively. **d–f** Temperature dependent linewidth of the E_{2g}^2, E_{1g} and A_{1g} modes for 2L $MoSe_2$, respectively. Empty circle are the experimental data and solid red lines are the fitted curves using (2.17). The insets in **d**, **f** are the individual contributions from the EPC (red) and PPC (black) for E_{2g}^2 and A_{1g} mode for the 2L $MoSe_2$, respectively. Adapted with permission from [72]. Copyright (2022) (American Physical Society)

increase with further lowering the temperature below ~120 K. The observed anomalous increase in linewidth in low temperature regime is attributed to the contributions from the EPC. It should be noted that the increase in linewidth is also observed in interlayer interaction mode for the case of graphene and has been attributed to the contributions from EPC [124]. The insets in (d) and (f) are the individual contributions from the EPC (red) and PPC (black) for the E_{2g}^2 and A_{1g} mode for the 2L $MoSe_2$, respectively; suggesting significant role of EPC (PPC) to linewidth broadening in the low temperature (high temperature) regime. From Fig. 2.11b–f one can see the significant role of EPC is observed only for the case of 2L $MoSe_2$ which is in line with previous studies which reported that the EPC increases with increase in thickness of the materials.

2.6 Crossover Between Resonant and Non-resonant Conditions

From the above discussions, it is clear that the resonance effect not only enhance the Raman scattering intensity and led to the observation of multi-phonon Raman scattering from the entire Brillouin zone in group VI-TMDCs but it also led to the activation of back scattering forbidden/infrared active optical phonon modes. From (2.14) and (2.16) it is clear that the crossover between the resonant and non-resonant conditions may be easily achieved by either changing laser excitation or band gap energy as well as energy of different available excitonic energy states in the semiconducting materials. The crossover between the resonant and non-resonant conditions may lead to the modulation in the Raman spectrum. In this section we will focus on the crossover of resonant and non-resonant conditions for the case of group VI-TMDCs as a function of different external perturbations particularly incident laser excitation energy and temperature.

2.6.1 Laser Excitation Energy Induced Crossover of Resonant and Non-resonant Conditions

Energy of the incident photons plays a very important role in the Raman scattering for semiconducting materials. The variation of laser excitation energy has been proved to be an important tool to achieve the crossover between resonant and non-resonant conditions. It is clear from (2.14) to (2.16) that one could modulate the Raman scattering intensity as a function of laser excitation energy by keeping fixed the energy of the band gap or different available excitonic energy states. Figure 2.9c shows the Raman spectra of bulk $MoSe_2$ as a function of laser excitation energies and one could see the Raman scattering intensity is strongly dependent on the excitation energy. The modulation in the Raman scattering intensity with laser excitation energy may occurred due to the crossover between resonant and non-resonant conditions. Laser excitation energy induced crossover between resonant and non-resonant conditions has been reported in earlier studies for the case of group VI-TMDCs as well as for other semiconducting materials. In these group VI-TMDCs, some of the phonons become intense for a particular excitonic energy state i.e. the phonons with different symmetry show different resonance characteristics (intensity become maximum) with changing the laser excitation energies. For example, the back scattering Raman active (E_{1g}) and infrared active (A_{2u}) phonons resonate with the C exciton, see Fig. 2.9. Therefore, one can selectively modulate the intensity of particular phonons by selectively choosing the laser excitation energy which lead to the observation of the phonon symmetry induced exciton-phonon coupling. Figure 2.12a shows the Raman excitation profile of the modes A_{1g} (solid triangles) and E^1_{2g} (open squares) for (a_1) monolayer, (a_2) bilayer (a_3), trilayer and (a_4) for bulk MoS_2. The red curves are fitting of the experimental data as described in the [129]. The Raman excitation profile cover the energy range of the main excitonic transition states i.e. A, B and C excitons for the case of MoS_2. Following observation could be made: (i) E^1_{2g} particularly resonate for laser excitation energy close to the C exciton. (ii) A_{1g} resonant for laser excitation energy close to the all A, B and C excitons. (iv) Intensity of A_{1g} is larger for excitation energy close to the C exciton than that of A and B excitons for the 1L,2L and 3L MoS_2. On the other hand, the intensity of A_{1g} is smaller for excitation energy close to the C exciton than that of A and B excitons for the case of bulk MoS_2.

Figure 2.12b shows the Raman excitation profiles of the modes (b_1) E''/E_{1g}, (b_2) A'_1/A_{1g} , (b_3) $E'/E_g/E^1_{2g}$,(b_4) $A''_2/A_{1g}/A_{2u}$ and (b_5) $3LA$ for the case of monolayer (black), bilayer (red) trilayer (green) and quadrilayer (blue) $MoSe_2$. Following observation could be made: (i) Both E''/E_{1g} and $A''_2/A_{1g}/A_{2u}$ resonate close to C exciton. (ii) A'_1/A_{1g} show resonance close to A, B and C excitons and the intensity is larger for excitation energy close to C exciton than that of A and B excitons similar to the MoS_2. (ii) $E'/E_g/E^1_{2g}$ shows the resonance nature close to the A, B and C excitons. Generally, this mode is divided into two components. The lower frequency component of $E'/E_g/E^1_{2g}$ ~289 cm^{-1} resonate with the C exciton while the higher frequency component of $E'/E_g/E^1_{2g}$ ~292 cm^{-1} resonant with the A and B excitons. These two components have been attributed to the longitudinal optical (LO) and transverse acoustic (TA) phonons corresponding to the higher and lower frequency, respectively. However, these two components have not been observed simultaneously, see Ref [67] for more details. The intensity profile of TA $E'/E_g/E^1_{2g}$ as a function of laser excitation energy is the continuation of the $LOE'/E_g/E^1_{2g}$, see Fig. 2.12b. (iv) $3LA$ mode, the third-order overtone of the LA shows the resonance characteristics close to the A and B excitons. This selective enhancement in intensity of the phonons for a specific excitonic state may be attributed to the different exciton-phonon coupling. This different resonance characteristic of the phonons with A, B and C excitons may originate due to different symmetry of the phonons as well as different electronic nature of wave function, effective mass and Bohr radius of the A, B and C excitons [27, 93] and lead to the observation of symmetry induced exciton-phonon coupling. Furthermore, from the Fig. 2.12a, b, one could also see

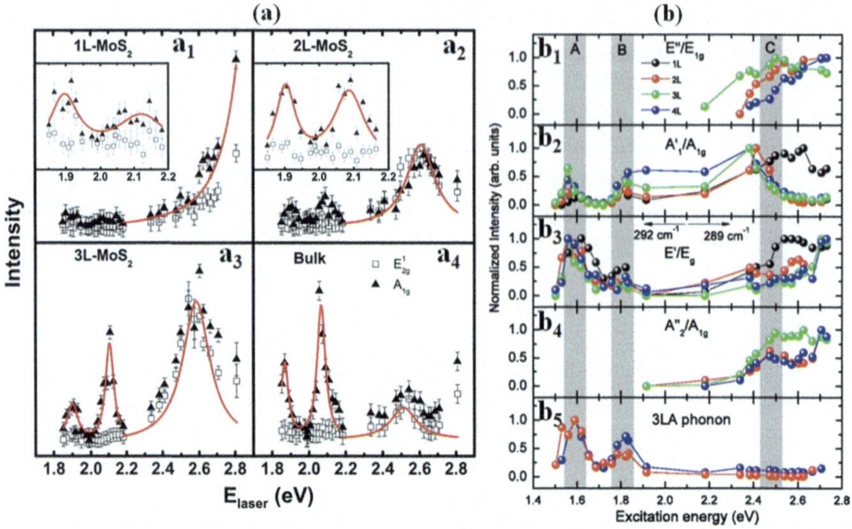

Fig. 2.12 **a** Raman excitation profiles of the modes A_{1g} (solid triangles) and E^1_{2g} (open squares)) for (a_1) monolayer, (a_2) bilayer (a_3), trilayer and (a_4) for bulk MoS_2. The insets in a_1 and a_2 show the amplified data in the energy range of 1.8–2.2 eV. Adapted with permission from [129]. Copyright (2016) (American Physical Society). **b** Raman excitation profiles of the modes (b_1) E''/E_{1g}, (b_2) A'_1/A_{1g}, (b_3) $E'/E_g/E^1_{2g}$, (b_4) $A''_2/A_{1g}/A_{2u}$ and (b_5) $3LA$ for the case of monolayer (black), bilayer (red) trilayer (green) and quadrilayer (blue) $MoSe_2$. Adapted with permission from [67]. Copyright (2016) (American Physical Society)

that the intensity maxima of the phonons for both MoS_2 and $MoSe_2$ slightly changes with changing in the thickness of the system. This thickness dependence resonance behavior may occur due to the variation in the energy of the excitonic states as a function of thickness of the materials. Therefore, the laser excitation energy induced crossover of resonant and non-resonant conditions not only modulated the Raman spectrum but also provide the information about the exciton-phonon coupling in group VI-TMDCs.

2.6.2 Temperature Induced Crossover of Resonant and Non-resonant Conditions

In the above section, we discussed the laser excitation energy induced crossover of resonant and non-resonant conditions. In addition to the laser excitation energy, the crossover between the resonant and non-resonant conditions could also be achieved by the variation of temperature, pressure, stress and so on. Because, the variation in these parameters lead to the change in the band structure which results in modulation in resonant and non-resonant conditions. Moreover, the group VI-TMDCs show the thickness dependent modulation of the band structure which may also give rise to the thickness induced crossover between the resonant and non-resonant conditions. One can modulate the resonant and non-resonant conditions by changing one of these above parameters even by keeping fixed the laser excitation energy. Here we limit our discussion only to the temperature induced crossover of resonant and non-resonant conditions. Under the non-resonant condition, Bose-Einstein distribution function, which is associated with the thermal population of the phonons, plays an important role in temperature dependence of intensity of the phonons. For the case of first-order Raman scattering process, temperature dependent intensity of the phonon mode is given as

$$\begin{aligned} I(T)_{\text{Stokes}} &\propto [n(\omega, T) + 1] \\ I(T)_{\text{Anti - Stokes}} &\propto n(\omega, T) \end{aligned} \tag{2.18}$$

However, For the case of second-order Raman scattering process, temperature dependent intensity of the phonon mode is given as [23]

$$\begin{aligned} &\text{Stokes}\begin{cases} I(T)_{combination} \propto [n_1(\omega, T) + 1][n_2(\omega, T) + 1] \\ I(T)_{Overtone} \quad \propto [n(\omega, T) + 1][n(\omega, T) + 2] \end{cases} \\ &\text{Anti - Stokes}\begin{cases} I(T)_{combination} \propto [n_1(\omega, T)][n_2(\omega, T)] \\ I(T)_{Overtone} \quad \propto [n(\omega, T)][n(\omega, T) - 1] \end{cases} \\ &\text{Difference}\{I(T)_{difference}\,[n_1(\omega, T) + 1][n_2(\omega, T)] \end{aligned} \tag{2.19}$$

The expressions given by (2.18) and (2.19) may be used to understand the temperature dependent intensity of the phonons under the non-resonance conditions. While these expressions fail to explain the Raman scattering intensity of the phonon modes in the vicinity of resonance conditions because the resonance effect play prominent role compared to the thermal population of the phonons. It should be noted that generally band structure is affected by the temperature which may lead to the modulation in the resonant and non-resonant conditions. Temperature dependent Raman scattering intensity, in quantum mechanical picture, for the first and second-order Stoke Raman scattering intensity may be written as [24]

$$I_{Stokes}(T) \propto \left| \frac{A}{[E_L - E_x(T) - i\gamma_x(T))\ (E_L - \hbar\omega_{ph} - E_x(T) - i\gamma_x(T)]} \right|^2 \tag{2.20}$$

$$I_{Stokes}(T) \propto \left| \frac{B}{[E_L - E_x(T) - i\gamma_x(T)]\ [E_L - \hbar\omega_{ph_1} - E_x(T) - i\gamma_x(T)]\ [E_L - \hbar\omega_{ph_1} - \hbar\omega_{ph_2} - E_x(T) - i\gamma_x(T)]} \right|^2 \tag{2.21}$$

The (2.20) and (2.21) may be used to understand the temperature dependent intensity of the phonons in the vicinity of resonance conditions. In (2.20) and (2.21), $E_x(T)$ and $\gamma_x(T)$ represent the temperature dependent transition energies and linewidth corresponding to the excitonic energy states such as $x = A$, B, C excitons for the case of group VI-TMDCs, respectively. Temperature dependent transition energy is understood using two empirical relations in which the first one was proposed by Varshni [130] and second one by Donnell and Chen [131]. Considering the Varshni relation, the temperature dependent $E_x(T)$ is given as:

$$E_x(T) = E_0 - \frac{\alpha T^2}{(T + \beta\ )} \tag{2.22}$$

where E_0 is the energy of excitonic quasiparticle at 0 K. α and β are the fitting parameters associated with the electron (exciton)-phonon interaction and Debye temperature, respectively.

On the other hand, considering the Donnell and Chen empirical relation, the temperature dependent $E_x(T)$ is given as:

$$E_x(T) = E_0 - S^* E_p \left[\cot h\left(\frac{E_p}{2k_b T}\right) - 1\right] \tag{2.23}$$

where E_0 is energy of the excitonic quasiparticle at 0 K. S is a dimensionless Huang-Rhys factor describing the strength of electron (exciton)-phonon interactions. $E_p = < \hbar\omega >$ is average energy of the contributing phonons to the shift in energy bands with temperature.

The temperature dependent linewidth of the excitonic quasiparticle transition states is given as [132]

$$\Gamma(T) = \Gamma_0 + \Gamma_{AC}(T) + \Gamma_{LO}(T) = \Gamma_0 + \lambda_{AC} T + \lambda_{LO} N_{LO}(T) \tag{2.24}$$

where Γ_0 is the linewidth at 0 K and is attributed to the scattering of excitons with impurities and imperfections. $\Gamma_{AC}(T) = \lambda_{AC} T$ and $\Gamma_{LO}(T) = \lambda_{LO} N_{LO}(T)$ correspond to the contributions from the acoustic and LO phonons, respectively. λ_{AC} (λ_{LO}) describing the coupling strength between exciton and acoustic (LO) phonons. $N_{LO}(T)$ is the Bose-Einstein distribution function which is given as $N_{LO}(T) = 1/[e^{(E_{LO}/k_b T)} - 1]$, where E_{LO} is energy of the participating LO phonons. As we have already mentioned that the exciton and LO phonon interaction is described by the Fröhlich mechanism [79], on the other hand interaction between the exciton and acoustic phonon may be described by deformation-potential [132].

Therefore, the temperature is an additional perturbation, in addition to the laser excitation energy, which may play a very important role in resonant Raman studies for group VI-TMDCs as well as other such kinds of materials. Equations (2.20) and (2.21) suggests that the modulation in resonant and non-resonant condition is not only achieved by changing the incident laser excitation energy but could also be achieved by changing the sample temperature. Furthermore, the group VI-TMDCs exhibit various excitonic energy states such as A , B and C excitons. For a fixed laser excitation energy, with variation in temperature, in the vicinity of resonance condition, laser excitation energy move away from one excitonic energy state while at the same time laser excitation energy approaching to the another excitonic energy

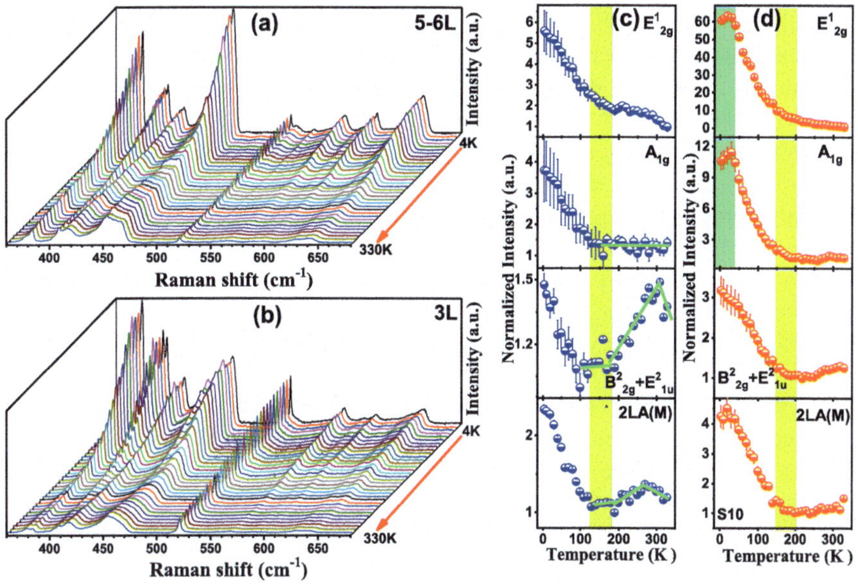

Fig. 2.13 **a, b** show the Raman spectrum in 3D wall view in the temperature range of 4–330 K, in a spectral range from 330 to 700 cm^{-1} for a few layers (5–6 layer) and 3L MoS_2, respectively. **c, d** show the temperature dependence of normalized intensity of the phonon modes E^1_{2g}, A_{1g}, $B^2_{2g} + E^2_{1u}$ and $2LA(M)$ phonon mode for a few layer and 3L MoS_2, respectively. The yellow shaded part depicts the region where resonance effect is tuned from B exciton to A exciton. Adapted with permission from [100]. Copyright (2021) (IOP Publishing)

state due to change in the energy of excitonic states with temperature resulting in the tuning of resonant condition from one excitonic energy state to another. Figure 2.13a, b show the Raman spectra in 3D wall view, in the temperature range of 4–330 K for few layers (5–6 layers) and 3L MoS_2, respectively. The Raman spectrum was recorded using 633 nm (1.96 eV) laser which is very close to the B exciton at room temperature. One can see a clear and sharp variation in the Raman scattering intensity with temperature.

Figure 2.13c, d show the temperature dependence of the normalized intensity of the E^1_{2g}, A_{1g}, $B^2_{2g} + E^2_{1u}$ and $2LA(M)$ phonon modes for a few layers and 3L MoS_2, respectively. The following observation can be made: (i) For both the systems, E^1_{2g} mode show only mild increase in intensity with lowering temperature from 330 to 160 K. However below, ~160 K a sharp increase could be seen on further lowering the temperature. (ii) Intensity of A_{1g} remain nearly constant in temperature range of 330 to ~160 K for the case of a few layer MoS_2. On the other hand, for the case of 3L MoS_2, intensity of A_{1g} first decrease slightly from 300 K to ~250 K and then increase slightly from 250 to ~160 K. Below 160 K, one could see a sharp increase in the intensity with further lowering the temperature for both systems. (iii) For both systems, one could also see the sharp enhancement in intensity for the $B^2_{2g} + E^2_{1u}$ and $2LA(M)$ modes with lowering temperature below ~140–160 K, while only slight change could be observed above this temperature. Therefore, from Fig. 2.13 it is clear that such temperature dependence of the intensity could not be explained using the Bose-Einstein distribution function because $n(\omega, T)$ factor is expected to increase with increase in temperature and hence the intensity of the phonon. However, this observed trend of intensity may be understood easily using the quantum mechanical relations for intensity of the phonons as well as considering variations in energy of the excitonic transition states with temperature. Since the authors used the 633 nm (1.96 eV), which is close to the energy of B (~1.96 to 197 eV) exciton at room temperature for present case which give rises to the resonance effect with B exciton. With lowering temperature, the energy of B exciton moves away from the laser excitation energy because of increase in the energy of B exciton, which may explain the decrease in intensity from 330 to ~200 −160 K. On the other hand, energy of the A exciton (~1.82–1.83 eV at room temperature for the present case) approaches towards the incident laser excitation energy and it may resonate with the laser excitation energy which may explain the sharp enhancement in intensity with lowering the temperature below 160 K, reflecting a prominent role of A exciton in the low temperature range. Furthermore, some of the phonons, also show anomalous change at low temperature, i.e. below ~40 K, as well as above room temperature, this trend of intensity may be understood using the thermal expansion effect and lattice mismatch between MoS_2 flakes and the substrate [133]. The temperature dependent intensity given by (2.20) and (2.21) is not only used to understand the temperature induced crossover of resonant and non-resonant conditions but can also be used to extract the electronic information such as excitonic energy states using either Varshni or Donnell and Chen empirical relation for the case of $E_x(T)$ and the linewidth using relation as given by (2.24). Fan [64] and Zobeiri et al. [133] reported the temperature dependent resonant Raman studies on group VI-TMDCs materials and extracted the energy of the excitonic states and their dependence of the temperature by using the temperature dependent intensity of the Raman active A_{1g} phonon.

2.7 Conclusion and Future Perspective

In this chapter, we discussed the non-resonance versus resonance Raman scattering and their significant in understanding the underlying physics associated with the group VI-TMDCs materials. We highlighted that resonance Raman scattering allows the observation of various intriguing phenomena such as multi-phonon Raman scattering from the entire Brillouin zone and backscattering forbidden/infrared active modes, Davydov splitting and investigation of the electron-phonon coupling. We also discussed the cross over from resonance to non-resonance conditions and vice versa as a function of two very important external perturbations i.e. laser excitation energy and the temperature. Hitherto, a lot of efforts have been made by the scientific community to study the group VI-TMDCs materials using the resonance Raman scattering. However, to the best of our knowledge, the detailed analysis and full understating of resonance Raman effect in group VI-TMDC materials is still lacking. For example, previous studies reported that the resonance effect led to the splitting of the first-order modes (Davydov splitting) but still it is not clear why the phonon mode splitting occurs only under the resonance effect. In addition to the resonance Raman effect, the contribution of double resonance effect has also been seen in group VI-TMDCs and it led to the observation of intense overtones of the longitudinal acoustic phonon from the M point of the Brillouin zone as well as intravalley and intervalley scattering phonon modes. However only few studies are available which focus on the double resonance effect in these group VI-TMDCs. Therefore, we emphasize here that the understanding the various phenomena in group VI-TMDCs and other such kind of low dimensional materials under the resonance/double resonance effect is still not fully explored and further theoretical and experimental studies are required to explore the resonance/double resonance Raman scattering in these materials.

References

1. N. Rivera, I. Kaminer, Nat. Rev. Phys. **2**, 538 (2020)
2. P. Forn-Díaz, L. Lamata, E. Rico, J. Kono, E. Solano, Rev. Mod. Phys. **91**, 25005 (2019)
3. C. Weisbuch, M. Nishioka, A. Ishikawa, Y. Arakawa, Phys. Rev. Lett. **69**, 3314 (1992)
4. C. Noh, D.G. Angelakis, Rep. Prog. Phys. **80** (2017)
5. J. Kwon, Y. Kim, A. Lanuza, D. Schneble, Nat. Phys. **18**, 657 (2022)
6. T. Low, A. Chaves, J.D. Caldwell, A. Kumar, N.X. Fang, P. Avouris, T.F. Heinz, F. Guinea, L. Martin-Moreno, F. Koppens, Nat. Mater. **16**, 182 (2017)
7. F.H.L. Koppens, D.E. Chang, F.J.G.D. Abajo, Nano Lett. **11**, 3370 (2011)
8. C. Möhl, A. Graf, F.J. Berger, J. Lüttgens, Y. Zakharko, V. Lumsargis, M.C. Gather, J. Zaumseil, ACS Photonics **5**, 2074 (2018)
9. L. Rayleigh, Phil. Mag. **41**, 107 (1899)
10. A. Smekal, Naturwissenschaften **11**, 873 (1923)
11. C.V. Raman, K.S. Krishnan, Ind. J. Phys. **2**, 387 (1928)
12. C.V. Raman, K.S. Krishnan, Nature **121**, 501 (1928)
13. P.A. Fleury, R. Loudon, Phys. Rev. **166**, 514 (1968)
14. M. Cardona, G. Guntherodt, *Light Scattering in Solids VII* (Springer, Berlin Heidelberg, 2000)
15. C.C. Chen, C.J. Jia, A.F. Kemper, R.R.P. Singh, T.P. Devereaux, Phys. Rev. Lett. **106**, 067002 (2011)
16. C.J.C. Filho, P.F. Gomes, A.F. García-Flores, G.E. Barberis, E. Granado, J. Magn. Magn. Mater. **377**, 430 (2015)
17. C.H. Henry, J.J. Hopfield, Phys. Rev. Lett. **15**, 964 (1965)
18. A.S. Barker, R. Loudon, Rev. Mod. Phys. **44**, 18 (1972)
19. F. Cerdeira, N. Mestres, M. Cardona, Phys. Rev. B **29**, 3737 (1984)
20. R. Loudon, Adv. Phys. **50**, 813 (2001)
21. W. Hayes, R. Loudon, *Scattering of Light by Crystals* (Wiley, New York, 1978)
22. M. Cardona, *Light Scattering in Solid* (Springer-Verlag, Berlin Heidelberg GmbH, 1975)
23. M. Cardona, G. Guntherodt, *Light Scattering in Solid II* (Springer, Berlin, Heidelberg, New York, 1982)
24. P.Y. Yu, M. Cardona, *Fundamentals of Semiconductors* (Springer, Berlin Heidelberg, 2010)
25. A. Jorio, R. Saito, G. Dresselhaus, M.S. Dresselhaus, *Raman Spectroscopy in Graphene Related Systems* (Wiley-VCH Verlag GmbH & Co. KGaA, Weinheim, 2011)
26. R. Saito, Y. Tatsumi, T. Yang, H. Guo, S. Huang, L. Zhou, M.S. Dresselhaus, in *Raman Spectroscopy of Two-Dimensional Materials*, ed. by P.-H. Tan (Springer Series in Materials Science, 2019), p. 131

27. B.R. Carvalho, L.M. Malard, J.M. Alves, C. Fantini, M.A. Pimenta, Phys. Rev. Lett. **114**, 136403 (2015)
28. R.N. Gontijo, G.C. Resende, C. Fantini, B.R. Carvalho, J. Mater. Res. **34**, 1976 (2019)
29. C. Thomsen, S. Reich, Phys. Rev. Lett. **85**, 5214 (2000)
30. E.H.M. Ferreira, M.V.O. Moutinho, F. Stavale, M.M. Lucchese, R.B. Capaz, C.A. Achete, A. Jorio, Phys. Rev. B **82**, 125429 (2010)
31. L.M. Malard, M.A. Pimenta, G. Dresselhaus, M.S. Dresselhaus, Phys. Rep. **473**, 51 (2009)
32. H.L. Liu, S. Siregar, E.H. Hasdeo, Y. Kumamoto, C.C. Shen, C.C. Cheng, L.J. Li, R. Saito, S. Kawata, Carbon **81**, 807 (2015)
33. H. Guo, T. Yang, M. Yamamoto, L. Zhou, R. Ishikawa, K. Ueno, K. Tsukagoshi, Z. Zhang, M.S. Dresselhaus, R. Saito, Phys. Rev. B **91**, 205415 (2015)
34. A.C. Ferrari, J.C. Meyer, V. Scardaci, C. Casiraghi, M. Lazzeri, F. Mauri, S. Piscanec, D. Jiang, K.S. Novoselov, S. Roth, A.K. Geim, Phys. Rev. Lett. **97**, 187401 (2006)
35. A. Berkdemir, H.R. Gutiérrez, A.R.B. Méndez, N.P. López, A.L. Elías, C.I. Chia, B. Wang, V.H. Crespi, F.L. Urías, J.C. Charlier, H. Terrones, M. Terrones, Sci. Rep. **3**, 1755 (2013)
36. K.S. Novoselov, A.K. Geim, S.V. Morozov, D. Jiang, Y. Zhang, S.V. Dubonos, I.V. Grigorieva, A.A. Firsov, Science **306**, 666 (2004)
37. K. Albe, A. Klein, Phys. Rev. B **66**, 073413 (2002)
38. A. Splendiani, L. Sun, Y. Zhang, T. Li, J. Kim, C.-Y. Chim, G. Galli, F. Wang, Nano Lett. **10**, 1271 (2010)
39. K.F. Mak, C. Lee, J. Hone, J. Shan, T.F. Heinz, Phys. Rev. Lett. **105**, 136805 (2010)
40. I.G. Lezama, A. Arora, A. Ubaldini, C. Barreteau, E. Giannini, M. Potemski, A.F. Morpurgo, Nano Lett. **15**, 2336 (2015)
41. Q.H. Wang, K. Kalantar-Zadeh, A. Kis, J.N. Coleman, M.S. Strano, Nat. Nanotechnol. **7**, 699 (2012)
42. W. Zhao, Z. Ghorannevis, L. Chu, M. Toh, C. Kloc, P.-H. Tan, G. Eda, ACS Nano **7**, 791 (2013)
43. Z.Y. Zhu, Y.C. Cheng, U. Schwingenschlögl, Phys. Rev. B **84**, 153402 (2011)
44. G. Wang, C. Robert, A. Suslu, B. Chen, S. Yang, S. Alamdari, I.C. Gerber, T. Amand, X. Marie, S. Tongay, B. Urbaszek, Nat. Commun. **6**, 10110 (2015)
45. A. Ramasubramaniam, Phys. Rev. B **86**, 115409 (2012)
46. G.B. Liu, W.Y. Shan, Y. Yao, W. Yao, D. Xiao, Phys. Rev. B **88**, 085433 (2013)
47. K. Komider, J.W. González, J. Fernández-Rossier, Phys. Rev. B **88**, 245436 (2013)
48. D. Xiao, G.B. Liu, W. Feng, X. Xu, W. Yao, Phys. Rev. Lett. **108**, 196802 (2012)
49. R. Suzuki, M. Sakano, Y.J. Zhang, R. Akashi, D. Morikawa, A. Harasawa, K. Yaji, K. Kuroda, Nat. Nanotechnol. **9**, 611 (2014)
50. Y. J. Zhang, M. Yoshida, R. Suzuki, Y. Iwasa, 2D Mater. **2**, 044004 (2015)
51. L. Du, T. Hasan, A. Castellanos-gomez, G.B. Liu, Y. Yao, C.N. Lau, Z. Sun, Nat. Rev. Phys. **3**, 193 (2021)
52. D.Y. Qiu, F.H. da Jornada, S.G. Louie, Phys. Rev. Lett. **111**, 216805 (2013)
53. G. Wang, A. Chernikov, M.M. Glazov, T.F. Heinz, X. Marie, T. Amand, B. Urbaszek, Rev. Mod. Phys. **90**, 021001 (2018)
54. J. Pei, J. Yang, T. Yildirim, H. Zhang, Y. Lu, Adv. Mater. **31**, 1706945 (2019)
55. J.W. Christopher, B.B. Goldberg, A.K. Swan, Sci. Rep. **7**, 14062 (2017)
56. K.F. Mak, K. He, C. Lee, G.H. Lee, J. Hone, T.F. Heinz, J. Shan, Nat. Mat. **12**, 207 (2013)
57. A. Chernikov, T.C. Berkelbach, H.M. Hill, A. Rigosi, Y. Li, O.B. Aslan, D.R. Reichman, M.S. Hybertsen, T.F. Heinz, Phys. Rev. Lett. **113**, 076802 (2014)
58. Z. He, W. Xu, Y. Zhou, X. Wang, Y. Sheng, Y. Rong, S. Guo, J. Zhang, J.M. Smith, J.H. Warner, ACS Nano **10**, 2176 (2016)
59. A.R. Klots, A.K.M. Newaz, B. Wang, D. Prasai, H. Krzyzanowska, J. Lin, D. Caudel, N.J. Ghimire, J. Yan, B.L. Ivanov, K.A. Velizhanin, A. Burger, D.G. Mandrus, N.H. Tolk, S.T. Pantelides, K.I. Bolotin, Sci. Rep. **4**, 6608 (2014)
60. T. Sekine, K. Uchinokura, T. Nakashizu, E. Matsuura, R. Yoshizaki, J. Phys. Soc. Jpn. **53**, 811 (1984)

61. G.L. Frey, R. Tenne, M.J. Matthews, M.S. Dresselhaus, G. Dresselhaus, Phys. Rev. B **60**, 2883 (1999)
62. B.R. Carvalho, Y. Wang, S. Mignuzzi, D. Roy, M. Terrones, C. Fantini, V.H. Crespi, L.M. Malard, M.A. Pimenta, Nat. Commun. **8**, 14670 (2017)
63. H. Terrones, E. Del Corro, S. Feng, J.M. Poumirol, D. Rhodes, D. Smirnov, N.R. Pradhan, Z. Lin, M.A.T. Nguyen, A.L. Elías, T.E. Mallouk, L. Balicas, M.A. Pimenta, M. Terrones Sci. Rep. **4**, 4215 (2014)
64. J.H. Fan, P. Gao, A.M. Zhang, B.R. Zhu, H.L. Zeng, X.D. Cui, R. He, Q.M. Zhang, J. Appl. Phys. **115**, 053527 (2014)
65. K. Kim, J.U. Lee, D. Nam, H. Cheong, ACS Nano **10**, 8113 (2016)
66. X. Huang, Y. Gao, T. Yang, W. Ren, H.M. Cheng, T. Lai, Sci. Rep. **6**, 32236 (2016)
67. P. Soubelet, A.E. Bruchhausen, A. Fainstein, K. Nogajewski, C. Faugeras, Phys. Rev. B **93**, 155407 (2016)
68. Q.H. Tan, Y.J. Sun, X.L. Liu, Y. Zhao, Q. Xiong, P.H. Tan, J. Zhang, 2D Mater. **4**, 031007 (2017)
69. I. Bilgin, A.S. Raeliarijaona, M.C. Lucking, S.C. Hodge, A.D. Mohite, A. De Luna Bugallo, H. Terrones, S. Kar, ACS Nano **12**, 740 (2018)
70. M. Grzeszczyk, K. Gołasa, M.R. Molas, K. Nogajewski, M. Zinkiewicz, M. Potemski, A. Wysmołek, A. Babiński, Sci. Rep. **8**, 17745 (2018)
71. D. Kumar, B. Singh, P. Kumar, V. Balakrishnan, P. Kumar, J. Phys. Condens. Matter **31**, 505403 (2019)
72. D. Kumar, V. Kumar, R. Kumar, M. Kumar, P. Kumar, Phys. Rev. B **105**, 085419 (2022)
73. K.P. Dhakal, D.L. Duong, J. Lee, H. Nam, M. Kim, M. Kan, Y.H. Lee, J. Kim, Nanoscale **6**, 13028 (2014)
74. C. Sourisseau, F. Cruege, M. Fouassier, M. Alba, Chem. Phys. **150**, 281 (1991)
75. J.U. Lee, J. Park, Y.W. Son, H. Cheong, Nanoscale **7**, 3229 (2015)
76. D. Nam, J.U. Lee, H. Cheong, Sci. Rep. **5**, 17113 (2015)
77. M.R. Molas, K. Nogajewski, M. Potemski, A. Babiński, Sci. Rep. **7**, 5036 (2017)
78. N. Scheuschner, R. Gillen, M. Staiger, J. Maultzsch, Phys. Rev. B **91**, 235409 (2015)
79. R.M. Martin, T.C. Damen, Phys. Rev. Lett. **26**, 86 (1971)
80. T. Sohier, M. Calandra, F. Mauri, Phys. Rev. B **94**, 085415 (2016)
81. L. Du, M. Liao, J. Tang, Q. Zhang, H. Yu, R. Yang, K. Watanabe, T. Taniguchi, D. Shi, Q. Zhang, G. Zhang, Phys. Rev. B **97**, 235145 (2018)
82. S. Shree, M. Semina, C. Robert, B. Han, T. Amand, A. Balocchi, M. Manca, E. Courtade, X. Marie, T. Taniguchi, K. Watanabe, M.M. Glazov, B. Urbaszek, Phys. Rev. B **98**, 035302 (2018)
83. Z.W. Wang, X.Y. Dong, R.Z. Li, Y. Xiao, Z.Q. Li, Phys. Status Solidi—Rapid Res. Lett. **12**, 1800306 (2018)
84. B. Miller, J. Lindlau, M. Bommert, A. Neumann, H. Yamaguchi, A. Holleitner, A. Högele, U. Wurstbauer, Nat. Commun. **10**, 807 (2019)
85. Y. Jiang, S. Chen, W. Zheng, B. Zheng, A. Pan, Light Sci. Appl. **10**, 72 (2021)
86. P. Merkl, C.K. Yong, M. Liebich, I. Hofmeister, G. Berghäuser, E. Malic, R. Huber, Nat. Commun. **12**, 1719 (2021)
87. E.C. Castro, D.S. Brandão, H. Bragança, A.S. Martins, F. Riche, A.C. Dias, J.H. Zhao, A.L.A. Fonseca, F. Qu, Phys. Rev. B **107**, 035439 (2023)
88. C. Ataca, H. Şahin, S. Ciraci, J. Phys. Chem. C **116**, 8983 (2012)
89. B. Rajbanshi, S. Sarkar, P. Sarkar, Phys. Chem. Chem. Phys. **17**, 26166 (2015)
90. T.C. Berkelbach, M.S. Hybertsen, D.R. Reichman, Phys. Rev. B **88**, 045318 (2013)
91. C. Yelgel, O.G, J. Appl. Phys. **122**, 065303 (2017)
92. A.V. Stier, N.P. Wilson, G. Clark, X. Xu, S.A. Crooker, Insight from high magnetic fields. Nano Lett. **16**, 7054 (2016)
93. G.Y. Jia, Y. Liu, J.Y. Gong, D.Y. Lei, D.L. Wang, Z.X. Huang, J. Mater. Chem. C **4**, 8822 (2016)
94. R.Z. Li, X.Y. Dong, Z.Q. Li, Z.W. Wang, Solid State Commun. **275**, 53 (2018)

95. A. Castellanos-Gomez, J. Quereda, H.P. Van Der Meulen, N. Agraït, G. Rubio-Bollinger, Nanotechnology **27**, 115705 (2016)
96. M. Staiger, R. Gillen, N. Scheuschner, O. Ochedowski, F. Kampmann, M. Schleberger, C. Thomsen, J. Maultzsch, Phys. Rev. B **91**, 195419 (2015)
97. G. Froehlicher, E. Lorchat, F. Fernique, C. Joshi, A. Molina-Sánchez, L. Wirtz, S. Berciaud, Nano Lett. **15**, 6481 (2015)
98. Q.J. Song, Q.H. Tan, X. Zhang, J.B. Wu, B.W. Sheng, Y. Wan, X.Q. Wang, L. Dai, P.H. Tan, Phys. Rev. B **93**, 115409 (2016)
99. W. Na, K. Kim, J.U. Lee, H. Cheong, 2D Mater. **6**, 015004 (2019)
100. D. Kumar, B. Singh, R. Kumar, M. Kumar, P. Kumar, Nanotechnology **32**, 285705 (2021)
101. A.S. Davydov, Sov. Phys. Usp. **7**, 145 (1964)
102. C. Lee, H. Yan, L.E. Brus, T.F. Heinz, J. Hone, S. Ryu, ACS Nano **4**, 2695 (2010)
103. J. Bardeen, L.N. Cooper, J.R. Schrieffer, Phys. Rev. **108**, 1175 (1957)
104. E. Pop, Nano Res. **3**, 147 (2010)
105. J. Zhou, H.D. Shin, K. Chen, B. Song, R.A. Duncan, Q. Xu, A.A. Maznev, K.A. Nelson, G. Chen, Nat. Commun. **11**, 6040 (2020)
106. Y. W. Tan, Y. Zhang, K. Bolotin, Y. Zhao, S. Adam, E. H. Hwang, S. Das Sarma, H. L. Stormer, P. Kim, Phys. Rev. Lett. **99**, 246803 (2007)
107. K.I. Bolotin, K.J. Sikes, Z. Jiang, M. Klima, G. Fudenberg, J. Hone, P. Kim, H.L. Stormer, Solid State Commun. **146**, 351 (2008)
108. B. Radisavljevic, A. Radenovic, J. Brivio, V. Giacometti, A. Kis, **6**, 147 (2011)
109. D. Ovchinnikov, A. Allain, Y.S. Huang, D. Dumcenco, A. Kis, ACS Nano **8**, 8174 (2014)
110. N.F. Hinsche, A.S. Ngankeu, K. Guilloy, S.K. Mahatha, A.G. Čabo, M. Bianchi, M. Dendzik, C.E. Sanders, J.A. Miwa, H. Bana, E. Travaglia, P. Lacovig, L. Bignardi, R. Larciprete, A. Baraldi, S. Lizzit, K.S. Thygesen, P. Hofmann, Phys. Rev. B **96**, 121402 (2017)
111. W.Y. Shan, Phys. Rev. B **102**, 241301 (2020)
112. J.F. Ge, Z.L. Liu, C. Liu, C.L. Gao, D. Qian, Q.K. Xue, Y. Liu, J.F. Jia, Nat. Mater. **14**, 285 (2015)
113. C. Jin, J. Kim, J. Suh, Z. Shi, B. Chen, X. Fan, M. Kam, K. Watanabe, T. Taniguchi, S. Tongay, A. Zettl, J. Wu, F. Wang, Nat. Phys. **13**, 127 (2017)
114. C.M. Chow, H. Yu, A.M. Jones, J. Yan, D.G. Mandrus, T. Taniguchi, K. Watanabe, W. Yao, X. Xu, Nano Lett. **17**, 1194 (2017)
115. G.S.N. Eliel, M.V.O. Moutinho, A.C. Gadelha, A. Righi, L.C. Campos, H.B. Ribeiro, Po-Wen Chiu, K. Watanabe, T. Taniguchi, P. Puech, M. Paillet, T. Michel, P. Venezuela, M.A. Pimenta, Nat. Commun. **9**, 1221 (2018)
116. M.L. Lin, Y. Zhou, J. Bin Wu, X. Cong, X.L. Liu, J. Zhang, H. Li, W. Yao, P.H. Tan, Nat. Commun. **10**, 2419 (2019)
117. C. Chen, X. Chen, H. Yu, Y. Shao, Q. Guo, B. Deng, S. Lee, C. Ma, K. Watanabe, T. Taniguchi, J.-G. Park, Sh. Huang, W. Yao, F. Xia, ACS Nano 13, 552 (2019)
118. Y. Li, X. Zhang, J. Wang, X. Ma, J.-A. Shi, X. Guo, Y. Zuo, R. Li, H. Hong, N. Li, K. Xu, X. Huang, H. Tian, Y. Yang, Z. Yao, P. Liao, X. Li, J. Guo, Y. Huang, P. Gao, L. Wang, X. Yang, Q. Dai, E. Wang, K. Liu, W. Zhou, X. Yu, L. Liang, Y. Jiang, X.-Z. Li, L. Liu, Nano Lett. **22**, 2725 (2022)
119. N. Mao, X. Wang, Y. Lin, B.G. Sumpter, Q. Ji, T. Palacios, S. Huang, V. Meunier, M.S. Dresselhaus, W.A. Tisdale, L. Liang, X. Ling, J. Kong, J. Am. Chem. Soc. **141**, 18994 (2019)
120. A.P.S. Gaur, S. Sahoo, J.F. Scott, R.S. Katiyar, J. Phys. Chem. C **119**, 5146 (2015)
121. R. Rao, R. A. Yadav, N. Padma, Jagannath, A. Arvind, J. Appl. Phys. **128**, 165703 (2020)
122. S. Paul, S. Karak, A. Mathew, A. Ram, S. Saha, Phys. Rev. B **104**, 075418 (2021)
123. N. Bonini, M. Lazzeri, N. Marzari, F. Mauri, Phys. Rev. Lett. **99**, 176802 (2007)
124. C. Cong, T. Yu, Nat. Commun. **5**, 4709 (2014)
125. H.N. Liu, X. Cong, M.L. Lin, P.H. Tan, Carbon **152**, 451 (2019)
126. G.B. Osterhoudt, Y. Wang, C.A.C. Garcia, V.M. Plisson, J. Gooth, C. Felser, P. Narang, K.S. Burch, Phys. Rev. X **11**, 011017 (2021)

127. H.-Y. Yang, X. Yao, V. Plisson, S. Mozaffari, J.P. Scheifers, A.F. Savvidou, E.S. Choi, G.T. McCandless, M.F. Padlewski, C. Putzke, P.J.W. Moll, J.Y. Chan, L. Balicas, K.S. Burch, F. Tafti, Nat. Commun. **12**, 5292 (2021)
128. P.G. Klemens, Phys. Rev. **148**, 845 (1966)
129. B.R. Carvalho, L.M. Malard, J.M. Alves, C. Fantini, M.A. Pimenta, Phys. Rev. Lett. **116**, 089904 (2016)
130. Y.P. Varshni, Temperature dependence of the energy gap in semiconductors. Physica **34**, 149 (1967)
131. K.P. O'Donnell, X. Chen, Appl. Phys. Lett. **58**, 2924 (1991)
132. S. Rudin, T.L. Reinecke, B. Segall, Phys. Rev. B **42**, 11218 (1990)
133. H. Zobeiri, S. Xu, Y. Yue, Q. Zhang, Y. Xie, X. Wang, Nanoscale **12**, 6064 (2020)

Chapter 3
Magnetic Field and Pressure-Dependent Raman Spectroscopy

Akriti Singh, Aprajita Joshi, and Surajit Saha

3.1 Basic Principle

Quantum materials are a class of materials where the quantum interactions among the various degrees of freedom (DOF) i.e., charge, spin, lattice, and orbitals, lead to the emergence of exotic phenomena thus dictating their properties. Quantum phases like superconductivity, charge density waves, quantum hall effect, and spin-liquids are some consequences of the interactions between the different DOF [1–4]. Often, low-energy phenomena are represented as bosonic particles such as magnons, skyrmions, polarons and so on. There are many reports on the interaction of the low-energy excitations or quasiparticles, for example, magnon and lattice coupling leads to exciting magnetic properties in the material [5]. Investigation on the various DOF as well as their interactions with the quasiparticles allow us to unravel the rich physics behind the emergent phases.

Raman spectroscopy is a non-destructive technique that is used for studying the crystal structure of materials and their associated properties. The effect of the interactions or any changes in the local crystal structure or symmetries is reflected in the Raman signatures as a change in the phonon parameters such as, the phonon frequency, linewidth, and the intensity. Notably, the phonon frequency gives a measure of the phonon energy, linewidth relates to the phonon lifetime/relaxation, and intensity represents the phonon density of states. In addition, the low-energy excitations can also be detected in Raman spectroscopy, if they fulfill the criteria of selection rules [6].

Since Raman spectroscopy is sensitive to the changes in the structure or local symmetry or interactions, external perturbations, for instance, charge or chemical

Akriti Singh and Aprajita Joshi—Equal contributors.

A. Singh · A. Joshi · S. Saha (✉)
Department of Physics, Indian Institute of Science Education and Research Bhopal, Bhopal, Madhya Pradesh 462066, India
e-mail: surajit@iiserb.ac.in

D. K. Singh et al. (eds.), *Raman Spectroscopy*, Springer Series in Optical Sciences 248,
https://doi.org/10.1007/978-981-97-1703-3_3

doping, temperature, pressure, strain, and electric or magnetic field, are great ways to influence and tailor the symmetries/interactions that can be probed by Raman measurements [6–9]. In this chapter, we primarily focus on the basic instrumentation and application of pressure and magnetic field to perform Raman investigations unraveling some new domains of physics having potential applications.

3.2 Instrumentation

In this section, we will discuss how magnetic field and high pressure can be coupled to perform Raman spectroscopic studies.

3.2.1 Magnetic Field Dependent Raman Spectroscopy

Condensed matter system exhibits unique signatures arising due to quantum interactions among various degrees of freedom as discussed above, especially, spin. Application of external magnetic field opens several new possibilities to not only better understand the quantum interactions but also to explore new ground states and properties. Raman scattering under magnetic field can probe such ground states, magnetic excitations, spin-lattice interactions, and many other phenomena.

A magnetic field can be applied to a material in various ways. Electromagnets in the form of long solenoids or Helmholtz geometry are some of the commonly used techniques that give the freedom to generate uniform magnetic field over a sizeable length scale. They work on the principle that a magnetic field is produced on passing an electric current through the coil(s) attaining fields up to a few Tesla. Figure 3.1 shows the types of arrangements that can be employed as a magnet where (a) is a solenoid coil and (b) represents two coils (Helmholtz coil) arranged in a parallel configuration. In solenoids, we can achieve a uniform field at the center inside the long solenoid coil while in the case of Helmholtz arrangement of the coils, uniform magnetic field is produced in the region between them. Though electromagnets are good sources of magnetic field, the field produced by these magnets is of the order of a few Tesla. Many experiments require quite a high magnetic field (~10 T) to probe their properties. This requirement along with the freedom of variable field and efficacy can be fulfilled by superconducting magnets. Nonetheless, specialized requirements of very high fields (~ a few tens of Tesla) may be met by using pulsed magnets which is beyond the scope of our discussion.

Figure 3.2a shows the schematic of a setup used for magnetic field-dependent Raman scattering measurements. The setup consists of a superconducting coil wound around the sample chamber. The coil is placed in a liquid helium-filled chamber to cool down the magnet to its superconducting state. The sample chamber is composed of a sample holder equipped with x-, y-, and z-translation stages and a set of mirrors, filters, and an objective to guide a laser beam to the sample. Notably, all these

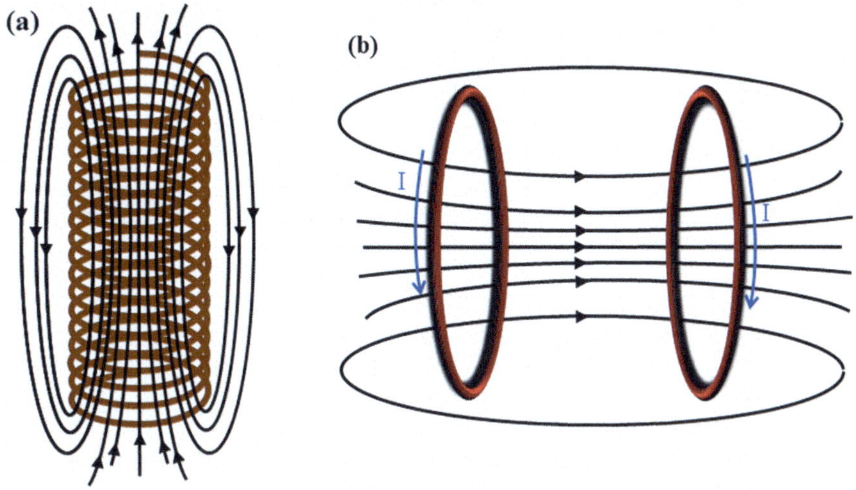

Fig. 3.1 Magnetic field lines produced by **a** solenoid, and **b** Helmholtz coil. Field is nearly uniform at the center of the solenoid and at the middle of the Helmholtz coil

components are made of non-magnetic materials. As shown in Fig. 3.2b, one can perform experiments with the simultaneous application of pressure and magnetic field by appropriately introducing a diamond anvil cell (DAC) on the translation stages. The instrumentation related to DAC are discussed later. After backscattering of the incident light from the sample, the scattered beam is collected using the required optics and is guided to a Raman spectrometer. Coupling of the spectrometer with the magnet can be done as follows: one of the ways is through an optical fiber where the laser beam is incident on the sample through a fiber and the backscattered beam is guided to the spectrometer through another fiber. The other way is the application of free beam method where the laser beam is directly guided to the sample with the help of mirrors in place of fibers and backscattered light travels the same path through optical components to the spectrometer. Both the ways are equally useful for collecting the spectra having respective drawbacks that may be appropriately taken care by using specific optics and relevant components.

In the case of anisotropy in a crystal under investigation, the polarization of the incident laser light and magnetic field greatly influence the properties of the material and the corresponding Raman signatures. Therefore, it is important to consider the geometry of placing the crystal with respect to the direction of field and light. So, here we will briefly introduce two such geometries, the Faraday and Voigt geometries. In the Faraday configuration, the direction of light (propagation vector) is parallel to the field direction, so the induced magnetization is also parallel to the direction of propagation of the light, as shown in Fig. 3.3a, while in the Voigt configuration the field direction is perpendicular to the direction of the light and thus the induced magnetization direction is perpendicular to the propagation vector of the light, as depicted in Fig. 3.3b [10].

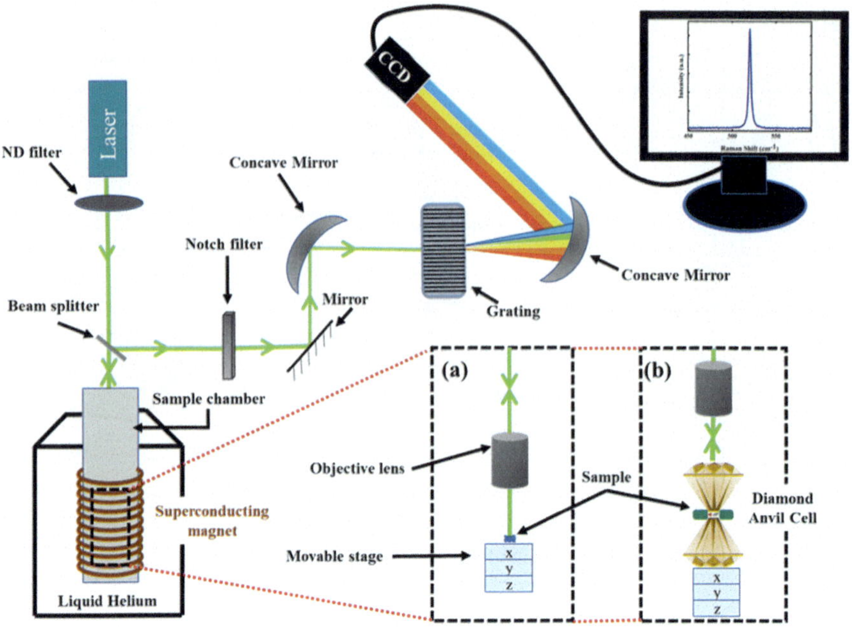

Fig. 3.2 Schematic for Magneto-Raman setup configured for **a** measurement in presence of external magnetic field, and **b** measurement in-situ magnetic field and pressure

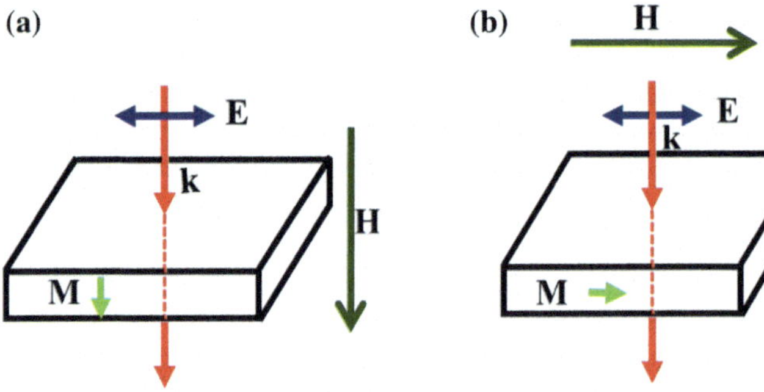

Fig. 3.3 Arrangements showing the application of magnetic field with respect to the incident laser light in **a** Faraday geometry and **b** Voigt geometry

3.2.2 Pressure Dependent Raman Spectroscopy

The application of high pressure is an important tool to study the properties of matter. However, despite success over more than 50 years, the generation of high pressure has remained a challenging task to accomplish. In fact, it was first time in the year 1950 when diamond, due to its extreme hardness and transparency to x-rays (light), was exploited for the generation and containment of high pressure [11]. Later in

1959, two groups independently came up with different designs of DACs. Jamieson et al. (University of Chicago) used a clamped DAC with x-ray beam perpendicular to the stress direction to study x-ray powder diffraction [12] while Weir et al. (National Bureau of Standards) used a DAC where light was allowed to incident along the stress direction to perform infrared absorption measurements [13].

Since the discovery of the latter version of DAC, several changes have been incorporated to make DAC a worthy tool to carry out the modern research. Some remarkable developments include introduction of metal gaskets to generate hydrostatic pressure, addition of new pressure transmitting media, usage of ruby fluorescence for pressure calibration, etc. [14, 15]. A typical DAC, as depicted in Fig. 3.4a, comprises of diamond anvils, a metal gasket, the specimen under investigation (sample), and a ruby crystal. The details of these components are discussed below.

Diamond. A brilliant cut gemstone is preferred for constructing the anvils of a DAC. Usually, the anvil flats are placed parallel to the (100) or (110) plane of the diamond [11]. The choice of diamond depends on the nature of study, for instance,

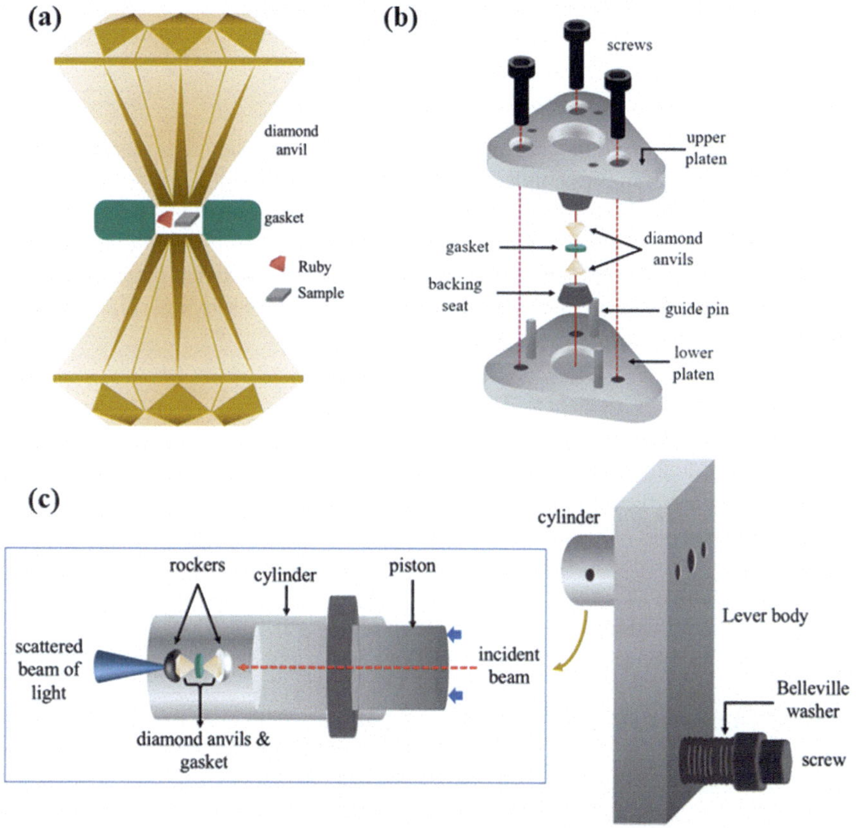

Fig. 3.4 Schematic of **a** diamond anvil cell (DAC), **b** Merrill-Bassett cell, and **c** Mao-Bell cell with the enlarged view of the cylinder assembly

optical studies require diamonds of low luminescence [16], while x-ray diffraction is independent of the luminescent properties of diamonds [11].

Gasket. It was for the first time in the year 1965, when van Valkenburg showed that a gasket can be employed to contain a hydrostatic medium [17]. A gasket is made by digging a hole of required diameter, say 200 μm, at the center of the mark (indent) made by the two anvil faces on a metal foil. The gasket hole contains the sample, ruby chip, and a hydrostatic medium, and is placed on the flat surface (culet) of the diamond anvil. Some of the common materials used to make gasket are tempered 301 stainless steel, Inconel X750, Waspalloy, rhenium, Kapton, tungsten, etc. [18, 19].

Ruby. Forman et al. [20] in 1972, proposed the ruby fluorescence technique to calibrate pressure while performing high-pressure studies. He put forward that R-lines of Cr^{3+}-doped Al_2O_3 shifts linearly for hydrostatic pressures up to 22 kbar. In the following year, Barnett, Block, and Piermarini, placed a ruby chip of dimension ~10 μm in the hydrostatic medium and showed that the ruby doublet lines, R_1 and R_2 (wavelength 6927 Å and 6942 Å, respectively at 1 atm), exhibited a linear shift to higher wavelengths under pressure [21]. In 1975, Piermarini et al. [22] suggested the following relation for pressures up to 190 kbar (19 GPa),

$$P(\text{kbar}) = 2.746\ \Delta\lambda(\text{Å}) \tag{3.1}$$

where $\Delta\lambda$ is the wavelength shift of R_1 line between 1 bar and high pressure. Some worthy remarks from studies conducted in successive years includes that the linearity is good up to a higher pressure of 300 kbar (30 GPa) and temperatures down to 4.2 K and allowed value of ruby lineshift being 0.365 Å kbar^{-1} or $-0.753\ \text{cm}^{-1}\ \text{kbar}^{-1}$ [22]. Later, Mao et al. [23] performed experiments and concluded that there exists some positive non-linearity in the pressure dependence of the ruby lineshift and came up with the modified form of equation (3.1) given by,

$$P(\text{Mbar}) = 3.808[(\Delta\lambda/6942 + 1)^5 - 1] \tag{3.2}$$

The above equation holds good for non-linear pressure dependence above 300 kbar where the reported error at 1 Mbar pressure was around 3%.

The basic principle underlying the working of a DAC is that when exposed to an external force, it creates a very high pressure on a sample placed between the two parallel faces (culets) of opposed anvils [24]. Based on the anvil alignment and force generating assembly, DACs can be classified into Piermarini-Block cell, Mao-Bell cell, Bassett-Merrill cell, and Syassen-Holzapfel cell. Here, we describe the two most popular DACs, namely, Merrill-Bassett and Mao-Bell cells.

Merrill-Bassett Cell. Designed by Leo Merrill and William A. Bassett in the year 1974, this is a compact cell with major application in single crystal x-ray diffraction studies. As shown in Fig. 3.4b, the cell comprises of two steel platens, which are pulled together with the help of three screws. A beryllium backing seat is present on each platen which serves four main purposes: (a) provides support to the diamond anvils, (b) presence of confocal holes on the seat enables optical alignment of anvils,

(c) function as an x-ray window, and (d) allows to view the loaded sample from outside. A metal gasket confining the sample, ruby and a hydrostatic medium is placed between the diamond faces (culets) [25].

Mao-Bell Cell. This cell was designed by H. K. Mao and P. M. Bell in the year 1978, comprising of a long piston-cylinder assembly that is embedded in a pressure cell holder (refer to Fig. 3.4c). The diamond anvils are placed on the half-cylinders (rockers) that are fixed in the fissure made in the respective piston and cylinder. A metal gasket of desired dimensions containing the sample and other important accessories is fitted between the anvils. The translation and tilting of the rockers (usually made from steel RC60 or tungsten carbide) enables the centering and alignment of the diamond anvils, and the pressure is generated by the Belleville spring-loaded lever-arm mechanism or by tightening the driving screw [11, 26].

Yet another important component of DAC while performing the high-pressure studies is the pressure medium that should reduce the pressure gradient and maintain a uniform pressure inside the sample chamber [27]. The medium should exhibit properties like, being hydrostatic for the pressure range under study, should be stable and inert so that it does not undergo any reaction with the sample while performing the experiment. Some of the media that have been used and reported are listed in Table 3.1 along with their tolerable pressure ranges.

Common signatures to look upon for hydrostaticity of a medium includes R-linewidths as well as the R_1–R_2 peak separation of Ruby fluorescence [30]. Quasi-hydrostatic conditions can be regained by heating the cell and slight release of force during cooling.

To perform the optical studies using high-pressure cells, require a few changes in the conventional cells that were used in the x-ray diffraction studies. Some changes include (i) higher aperture and (ii) improvement in the signal to noise ratio, which can be accomplished by using diamonds of low luminescence within the spectral range under study [30]. Qualitative test for diamonds for performing Raman measurements is the color classification, wherein, light yellow and commercial white diamonds are favorable. Since the Raman spectra are sensitive to non-hydrostatic stresses, the hydrostaticity is indicated by the quality of Raman spectra and Ruby luminescence lineshape [30].

Table 3.1 Pressure media and their (quasi)hydrostatic pressure range [11, 28, 29]

Medium	Pressure tolerance (kbar)
Methanol:ethanol (4:1)	~200
Methanol:ethanol:water (16:3:1)	~200
Helium (He)	>600
Argon (Ar)	~100
Nitrogen (N_2)	130

3.3 Application of the Techniques for Fundamental Science

3.3.1 Application of Magneto-Raman

Quantum many-body interactions play a vital role in defining the materials properties and their emergent phenomena. Often, external perturbations are the key to tailor these interactions opening up the possibilities to explore new emergent phenomena and ground states which are otherwise absent in the parent systems. Magnetic field is one of such tuning knobs that can engineer the quantum interactions. Discussed below are some of the examples that are of fundamental and contemporary interests.

Induction of electric dipole from magnetic monopoles. The existence of an electric monopole or a charge is a well-known concept where the basic definition of the monopole is that a particle possesses a single negative or positive charge, but its magnetic equivalent is fundamentally prohibited. Theoretically, magnetic monopole was predicted by Dirac in 1931, but the first experimental signature was reported in 2009 by Morris et al. [31] in a condensed matter system called spin-ice. These magnetic systems are magnetically frustrated and have corner-shared tetrahedra with spins at the vertices of the tetrahedra. The spins are arranged in a 2-in-2-out configuration to minimize the energy of the frustrated system. If by means of an external perturbation (magnetic field or thermal energy) one of the spins of the tetrahedra is flipped, we get a 3-in-1-out configuration in one tetrahedron while 3-out-1-in configuration in the other tetrahedron and thus an equivalence of a pair of magnetic monopoles is formed [32]. An electric dipole is induced in a spin-ice system by tailoring the magnetic monopole using an external magnetic field. Spin-ice $Tb_2Ti_2O_7$ is a fine example of inducing electric dipole moment by magnetic monopoles upon applying an external magnetic field, as was reported by Jin et al. [33], by performing magneto-Raman studies. Figure 3.5a shows the monopole-antimonopole state of the spin-ice with an oxygen atom at the center of the tetrahedron as all four Tb-O bonds are equivalent. The orbital shape of Tb and oxygen ions are shown in Fig. 3.5b where the green arrow represents the spin of Tb. When a magnetic field is applied parallel to [111] direction, the Tb2 magnetic moment remains in the same direction as the field while the moment associated with Tb1 starts to orient in the direction of the field.

Due to strong spin-orbit coupling in rare-earth elements, the orbital of Tb1 also rotates such that there is an overlap between the oxygen and Tb1 orbitals. Due to this, the coulomb repulsion of oxygen and Tb electron clouds increases. To decrease the energy of the system, the oxygen atom displaces from its original position (represented by dashed circle in Fig. 3.5a) which creates a displacement of the charges thus creating an electric dipole. The signature of the dipole formation with a magnetic field is reportedly observed in the magnetic field-dependent Raman studies of $Tb_2Ti_2O_7$. The Raman spectrum at 10 K in the absence of magnetic field shows three Raman modes assigned as F_{2g} (289 cm^{-1}), E_g (320 cm^{-1}), and A_{1g} (511 cm^{-1}) along with three crystal-field split modes (CFE) observed at ~2117 (CFE1), 2072 (CFE2), and 2060 (CFE3) cm^{-1}. Magnetic field-dependent Raman spectrum at 10 K revealed that

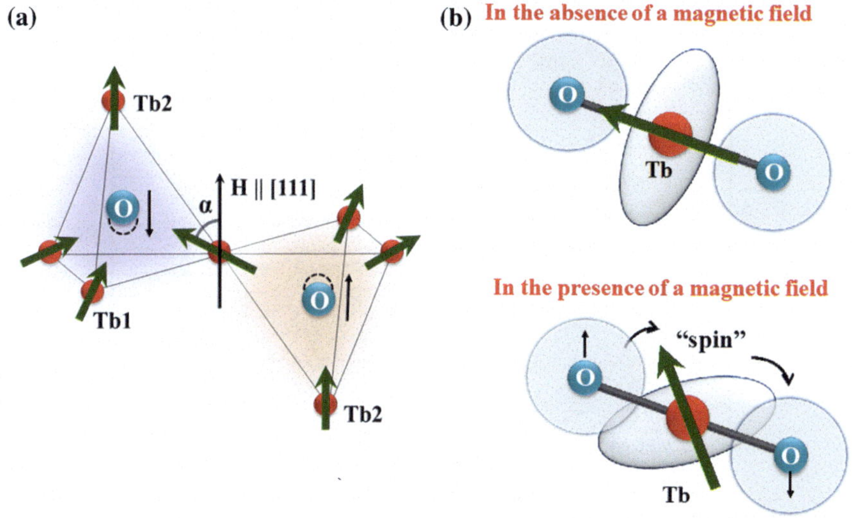

Fig. 3.5 Representation of **a** electric dipole and magnetic monopoles and magnetic field applied || [111] (magnetic field making an angle α with Tb1 magnetic moments). Oxygen atom gets displaced along the direction of the applied magnetic field ($\overrightarrow{H}$) and **b** orbital angular momentum picture of Tb1 atom and oxygen atoms (top panel) and its rotation in presence of external magnetic field (bottom panel). Adapted with permission from [33]. Copyright, 2020, American Physical Society

the mode A_{1g} is independent of the field while the F_{2g} and E_g modes are field dependent, as shown in Fig. 3.6a. On increasing the field, the F_{2g} mode gradually splits into two components P1 and P2 at 9 T. Splitting of the mode implies a structural change on applying field. For a better understanding, the CFE modes were also studied as a function of field. It can be inferred from the Fig. 3.6b that at around 2.5 T, new CFE modes appear which is an indication of a change in the local symmetry or the environment around Tb ion. The field dependent studies conclude that there occurs a structural change at around H_C (~2.3 T).

This also supports the discussion on formation of electric dipole due to the displacement of oxygen ion thus changing the crystal field environment under external magnetic field.

Magnon. In an ideal ferro- and anti-ferromagnets, spins are aligned or ordered in their ground state and interact with an exchange force. When the system is excited either due to thermal energy or laser, the spin gets deviated and due to the exchange force among them, this deviation of the spins is propagated to the neighboring spins. These disturbances produce a wave-like character known as spin-wave and the quasi-particle of the spin-wave is called a magnon [34]. In literature there are several reports on the presence of magnon in the magnetic system such as, ferromagnetic CrI_3 and antiferromagnetic $MnPSe_3$, etc. [35, 36]. Let us consider a case of magnon in an antiferromagnetic system. The ground state of antiferromagnets can be considered as two sublattices with magnetization in opposite directions (+z and −z), as shown in Fig. 3.7a. The magnon in an antiferromagnet is seen as the quasiparticle of the

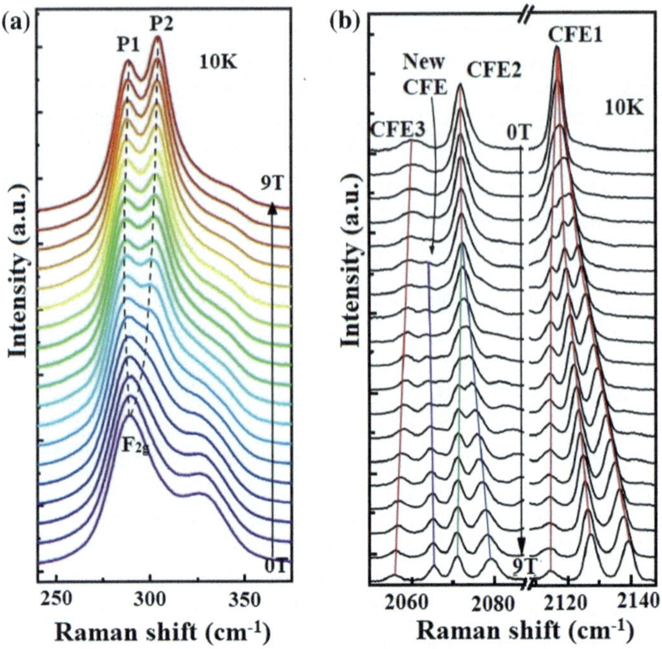

Fig. 3.6 Raman spectra of $Tb_2Ti_2O_7$ from 0 to 9 T at 10 K, **a** showing splitting of F_{2g} mode, and **b** crystal field splitting mode. Adapted with permission from [33]. Copyright, 2020, American Physical Society

spin-wave produced by each sublattices. Thus, two types of magnons are formed in the system. If an external magnetic field (along the easy axis i.e., z direction) is applied, the magnetization (M1 and M2) of the sublattices precess about the field. The frequency of the two non-degenerate magnon modes is given by [37]:

$$\omega_{k=0} = \gamma\{(2H_E + H_A)H_A\}^{1/2} \pm \gamma H_0 \quad (3.3)$$

where H_E is the exchange field, H_A is the anisotropy field, H_0 is the applied field, and γ is the gyromagnetic ratio.

In the absence of magnetic field i.e., $H_0 = 0$ the magnon modes are degenerate. The degeneracy of the magnon is lifted due to Zeeman effect when an external field is applied, thus creating two components of the magnon as shown in Fig. 3.7b, c, with the precession of both in opposite directions [38]. Importantly, if magnons are Raman active, the effect of magnetic field on magnons can be probed using magneto-Raman spectroscopy. Here, we refer to a report on the magneto-Raman study performed on $FePS_3$ (T_N ~ 120 K) by McCreary et al. [38]. Group theory predicts 15 Raman active phonon modes for $FePS_3$. From the temperature-dependent Raman measurements of $FePS_3$, the mode ψ_4 is attributed to be a magnon. To support that the mode ψ_4 is a magnon, Raman spectra were recorded at 5 K in the presence of a magnetic field

(0–9 T). As the magnetic field is increased, ψ_4 splits into two components as shown in Fig. 3.8.

Magneto-optic effect. Interaction of light with a material in the presence of a magnetic field brings changes in the characteristics of the light such as polarization, intensity etc. after propagating through the material. This phenomenon is known as magneto-optical effect. One such example of the magneto-optical effect is the Faraday effect where the polarization of the light gets rotated after passing through the material kept in an external magnetic field. Since Raman spectroscopy is sensitive to the state of polarization of light scattered by a material, similar changes can be measured using Raman scattering in the presence of a magnetic field [39].

Here, we have taken the example of $CrBr_3$ [40] to explain how magneto-Raman spectroscopy can be used as a tool to probe and investigate the magneto-optical property of the system. Two-dimensional (2D) ferromagnet $CrBr_3$ with $T_C \sim 50$ K has a rhombohedral structure. Raman measurements were performed by Yin et al. [40] on the bulk flakes, as the Raman spectra were independent of layer thickness. Figure 3.9a shows the Raman spectra at 10 K in both parallel (XX: polarization axis for incident and scattered light are parallel) and cross-(XY: polarization axis for incident and scattered light are perpendicular) polarization configurations.

The figure depicts the behavior of Raman modes at 106 cm^{-1} (A_g^1), 143 cm^{-1} (E_g^1), 153 cm^{-1} (E_g^2), and 184 cm^{-1} (A_g^2) with the change in polarization axis, where A_g^2 mode disappears in XY configuration. Figure 3.9b shows the Raman spectra when a magnetic field (0–5 T) is applied in the out-of-plane direction at 8 K. The intensity of the mode A_g in XX configuration gradually decreases while it increases in XY configuration with the increase in magnetic field. The results suggest the rotation of polarization axis with increase in magnetic field. A similar signature of the magneto-optical effect was obtained for a non-magnetic 2D system, MoS_2, where A_{1g} peak in XY polarization gets enhanced with the magnetic field and gets suppressed in the XX polarization [41].

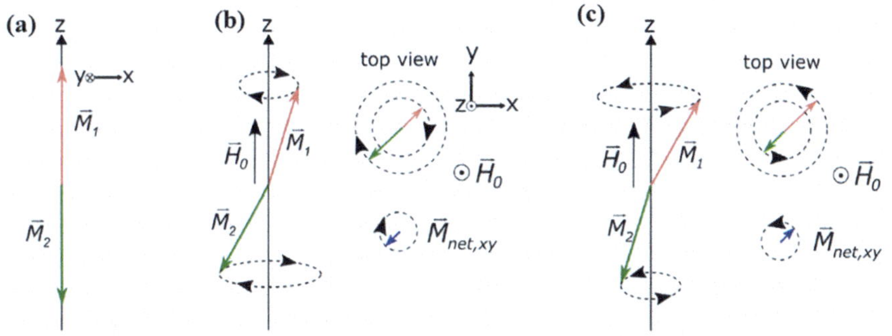

Fig. 3.7 Schematic depicting direction as well as the precession of the magnetization M1 and M2. **a** Showing ground state of $FePS_3$ in the absence of magnetic field having degenerate energy of the magnon mode. **b** and **c** represents the non-degenerate magnon modes which splits in the presence of magnetic field showing side and top view. Adapted with permission from [38]. Copyright, 2020, American Physical Society

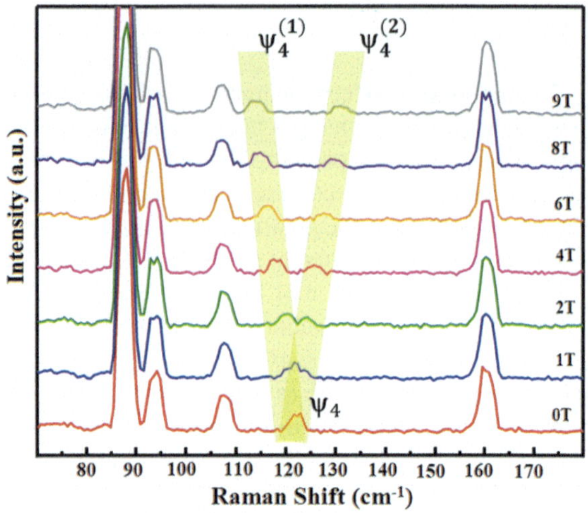

Fig. 3.8 Magnetic field dependent Raman spectrum of $FePS_3$ at 5 K showing splitting of ψ_4 (magnon mode). Adapted with permission from [38]. Copyright, 2020, American Physical Society

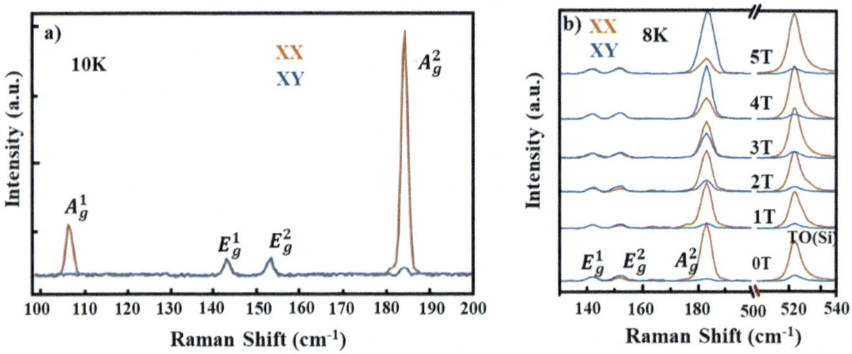

Fig. 3.9 **a** Raman spectrum of $CrBr_3$ in parallel and cross-polarization configuration at 10 K, and **b** Magnetic field-dependent Raman for fields 0–5 T at 8 K in parallel and cross-polarization. Adapted with permission from [40]. Copyright, 2021, John Wiley and Sons

Magnetostriction. Magnetic materials that are susceptible to changes in their volume i.e., the unit cell dimensions in the presence of an external magnetic field, are known as magnetostrictive materials and the property to change their shape is called magnetostriction [42]. Poojitha et al. [43], reported the magnetostriction in 15R-$BaMnO_3$ with the help of magneto-Raman spectroscopy and x-ray diffraction. Upon cooling 15R-$BaMnO_3$ to 4 K i.e., way below the magnetic transition temperature (T_N ~ 230 K), the phonon frequency exhibited a redshift with the field, as shown in Fig. 3.10, indicative of the magnetostrictive property of the material and was further supported with x-ray diffraction results [43].

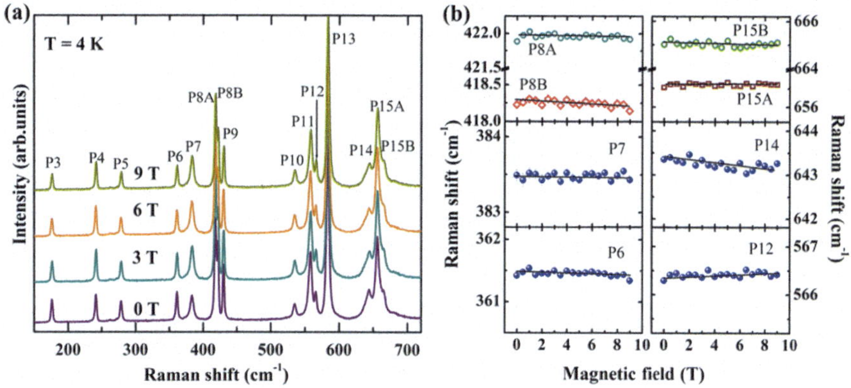

Fig. 3.10 **a** Raman spectra of 15R-$BaMnO_3$ at various magnetic fields at 4 K, and **b** magnetic field dependence of phonon frequency for selected modes. Adapted with permission from [43]. Copyright, 2020, American Physical Society

3.3.2 Application of High-Pressure Raman

Application of pressure induces changes in the crystallographic parameters which in turn can impart changes in the interactions across various degrees of freedom, thus tuning the properties of the material under study. In this section, we will look at some of the properties in different classes of materials that can be tuned with pressure.

Structural Phase Transitions. Pressure can modulate the interatomic distances and their interactions, which in turn may cause structural transition(s), that may/may not be reversible when pressure is released. In certain systems, the structural transitions can influence the magnetic states and electronic properties and even lead to development of superconductivity [44–46]. To probe these transitions, one can employ techniques like x-ray diffraction, transport measurements, Raman spectroscopy, etc. Some of the signatures for pressure-induced structural phase transitions in Raman spectra includes disappearance of phonon modes, emergence of new modes, and a decrease in Raman intensity beyond the transition pressure. For instance, let us consider Cs_2CuCl_4, a frustrated antiferromagnet with distorted triangular lattice [47]. Xue et al. studied the structural changes in Cs_2CuCl_4 by Raman spectroscopy and made the following observations: (a) at low pressure, two new modes appeared (at 200 and 302 cm^{-1}), (b) for pressure above 4.5 GPa, further new modes emerged (at 250 and 305 cm^{-1}), and (c) some modes disappeared beyond 6.2 GPa. Moreover, there was a change in slope in the frequency shift with pressure at around 5 GPa (Fig. 3.11a) that was consistent with the phase transition (~4 GPa) reported by x-ray diffraction studies [47, 48].

On a different system, Wang et al. [49] reported phase transitions due to compression in $TmVO_4$ using Raman spectroscopy with a maximum pressure of 55 GPa. For pressure of 8.8 GPa, several new peaks (141, 219, 258, 343, 374, 397, 450, 470, 792, 834, and 861 cm^{-1}) emerged that marked the transition to a new crystallographic

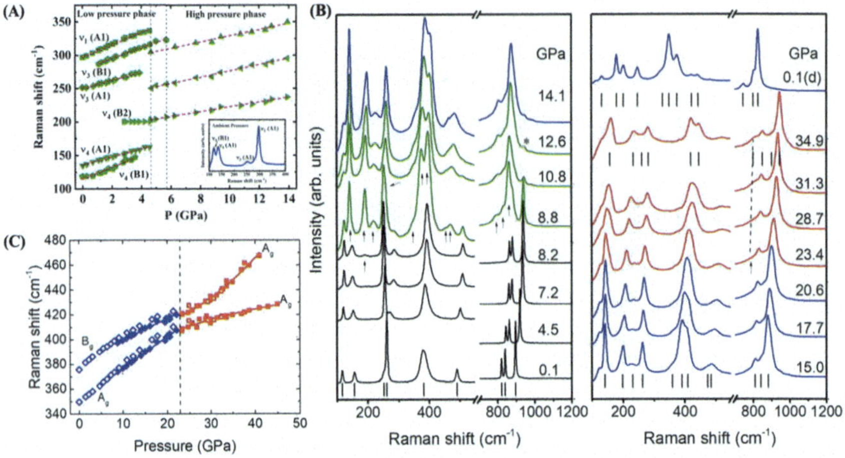

Fig. 3.11 (clockwise) **a** Plot for frequency as a function of pressure in Cs_2CuCl_4 (the inset shows the spectrum at ambient pressure), adapted from [47]. **b** Raman spectra for $TmVO_4$ from ambient pressure to 34.9 GPa, the arrows show the emergence of new modes at 8.2, 8.8, and 23.4 GPa, **c** B_g and A_g mode move closer to each other and transform into A_g mode having non-linear behavior (Fermi Resonance). Adapted with permission from [49]. Copyright, 2023, Elsevier

phase (scheelite phase). The appearance of another new peak at 23 GPa was considered as the second phase transition to fergusonite phase (Fig. 3.11b). In addition, at a pressure of 37 GPa, the A_g mode at 420 and 446 cm^{-1} exhibited a non-linear behavior, that was attributed to Fermi resonance (Fig. 3.11c).

In yet another system, Cu_2OSeO_3 [50], a skyrmion lattice system, when exposed to pressure showed a monotonous decrease (with different rates) in the intra- and inter-tetrahedral Cu–Cu distances, using pressure-dependent x-ray technique. The variation of lattice parameters marked the presence of multiple phase transitions in the system. These transitions were quite easily visible from the complementary Raman studies as well. Figure 3.12a shows the Raman spectra for pressures up to 16.5 GPa along the spectra collected after releasing the pressure. The disappearance and emergence of new modes, around 5.5 and 11 GPa, led to the conclusion that the system is driven into a lower symmetry structure at both the stated pressures. This lowering is due to pronounced deformation in the CuO_5 polyhedra, distortion in Cu–O bond lengths and Cu–O–Cu angles. Raman studies performed on pyrochlore systems, like $Tb_2Ti_2O_7$ [51], reflect a phase transition through some changes in modes at around 200, 550, and 680 cm^{-1}, under hydrostatic pressure upto 9 GPa (Fig. 3.12b).

Sometimes the pressure transmitting media (PTM) also influence the phase transitions. It is noted that the non-hydrostatic stresses can shift the phase transition to lower pressure as compared to (quasi)hydrostatic ones [52]. Pressure studies on nickel titanate ($NiTiO_3$) using Argon and silicone oil as the PTM manifested 31.4 GPa and 17.7 GPa as the phase transition pressure, respectively. The former was due to disappearance of E_g mode, while the latter involved splitting of E_g mode. Similar

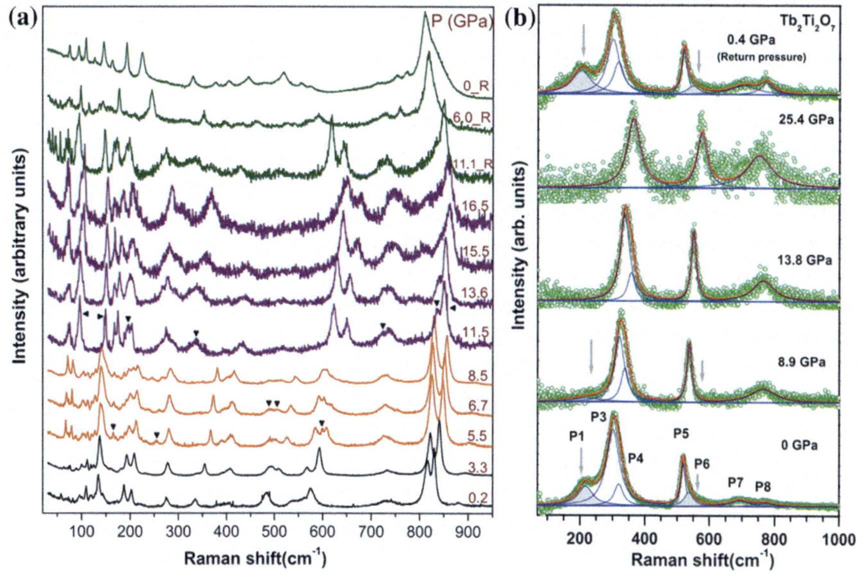

Fig. 3.12 Plot for frequency vs. pressure for **a** Cu_2OSeO_3, where the arrows mark the appearance of new modes at 5.5 and 11 GPa. Adapted with permission from [50]. Copyright, 2020, American Physical Society. **b** $Tb_2Ti_2O_7$, a pyrochlore system. Adapted with permission from [51]. Copyright, 2009, American Physical Society

observations are also reported in RVO_4 type materials, where non-hydrostaticity of PTM impact the onset and pathway of phase transitions [53].

Investigation of Magnetic Ground States. The application of a magnetic field is the most common way to instigate quantum criticality in a magnetic system. However, one can tune the strength of magnetic interactions by applying external pressure as well. For instance, pressure can reduce the interatomic distance in a Kagome plane, which would increase the superexchange interactions and lead to a quantum spin-liquid ground state [54]. We will bring out a few recent studies that highlight the importance of high pressure in understanding and exploring novel magnetic phases, investigated by Raman spectroscopy.

Pressure-dependent quantum phase transition in magnetic systems (e.g., $KCuCl_3$, $TlCuCl_3$, piperazinium hexachlorodicuprate—PHCC) has been studied through techniques like inelastic neutron scattering, muon spin relaxation (μSR), etc. [55–57]. However, to gain an insight on whether such transitions are accompanied by any structural phase transition or to acquire knowledge on to what extent the phonons are involved in driving the magnetic transition, one can employ laboratory-based pressure-dependent Raman spectroscopic measurements for an easy, in-depth, and quick investigation.

In 2020, Malavi et al. [54] reported pressure induced effect on an S = 1/2 antiferromagnetic quantum spin-liquid candidate, Herbertsmithite ($ZnCu_3(OH)_6Cl_2$), wherein Cu^{2+} ions occupy a perfect kagome lattice and are separated by non-magnetic Zn^{2+} ions that rest on a triangular lattice (refer to Fig. 3.13a).

The pressure-dependent x-ray diffraction studies showed a structural transition at around 8 GPa from rhombohedral to monoclinic structure with an increase in pressure. In the range where this system exhibits rhombohedral structure, the hydrostatic pressure reduces the intra-kagome and inter-kagome distance by nearly 4% and 2%, respectively. Figure 3.13b illustrates the pressure evolution of the seven observed Raman modes. A quasi-elastic scattering (QES) Lorentzian peak ($\omega < 300$ cm^{-1}) associated with spin density fluctuations, along with an asymmetric Fano-like feature near 230 cm^{-1} is also observed in the Raman spectra. The latter can be induced either by coupling to continuum of magnetic DOF, or by structural disorder. New modes appeared in the Raman spectrum beyond 8 GPa signifying a phase transition within the system. However, Vries et al. [58] attributed the Fano resonance to coupling of phonon mode to continuum of magnetic fluctuations, that was observed through Neutron spectroscopy.

For pressure upto 8 GPa, a decrease of Cu–Cu spin separation in the kagome plane leads to increased magnetic frustration in the system. While, in the high-pressure regime (beyond 8 GPa), the inter-kagome separation is heavily reduced, and the kagome lattice undergoes large distortion. As observed in Fig. 3.13c, the QES intensity reduces, or in other words, there is suppression of spin density fluctuations,

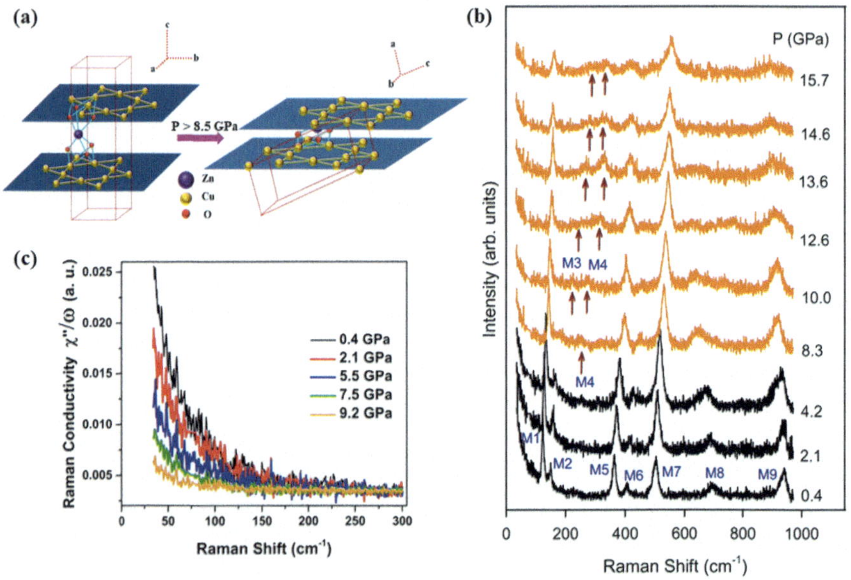

Fig. 3.13 (clockwise) **a** Pressure-induced distortion in unit cell. The Kagome planes in low pressure cell (rhombohedral) are stacked along the c-axis while high-pressure (monoclinic) cell have stacking along [101], **b** pressure evolution of Raman Spectra, **c** low frequency Raman response at a few pressures, after removing phonon contribution. Adapted with permission from [54]. Copyright, 2020, American Physical Society

monotonically up to 8 GPa. The lattice distortion and reduction in the spin frustration, directs towards the surfacing of a weak magnetic ordered state in the monoclinic phase. Thus, magnetic ordering in Kagome lattice systems can be tuned via pressure.

Electronic Properties. The application of high pressure induces changes in the lattice (bond distances/angles) which in turn can change the electronic band structure leading to intriguing phases. In certain materials it can lead to surfacing or intensification of superconductivity. This phenomenon is strongly dependent on the material properties and can originate via different mechanisms. Consider the example of FeSe, an iron-based superconductor, that shows an increase in its superconductivity transition temperature to 36.7 K, when subjected to hydrostatic pressure of about 6 GPa [59]. In this system, pressure (i) shifts the Se Wyckoff positions, which alters the electronic states near the Fermi level and (ii) increase the phonon frequencies (blueshift). Both these factors together increase the electron-phonon coupling in the system, which may be responsible for superconductivity in FeSe [60].

2D systems. 2D materials and their heterostructures have the interlayer interaction as an additional degree of freedom, which when modulated by external perturbation can lead to changes in the material properties. Applying pressure can influence the inter-layer distances (weak vdW interaction) more effectively as compared to the intralayer distances (covalent bonds), and thus, can modulate their structural as well as electronic properties. In some materials, like WS_2 and WSe_2, pressure decrease the interlayer spacing to an extent that it increases the interactions between the chalcogen anions tremendously. As a result, these systems transform from 2 to 3D, at 22 GPa and 51.7 GPa, respectively [61, 62]. The pressure-dependent Raman study on ReS_2, indicates phase transition in the material through the emergence of new modes. In addition, for pressure range 15–25 GPa, Raman modes show anomalous softening. The possible reason for such behavior was accounted to a decrease in the bond strength due to an increase in volume because of interlayer sliding [63]. In some of the 2D materials, the Raman study also enables to investigate which vibrations are the most affected with pressure. In the WSe_2–$MoSe_2$ heterostructure, the greater Raman shift ratio of the A_{1g} mode than E_g mode is indicative that the out-of-plane vibrations have a greater impact of the applied pressure [64].

3.3.3 *Magneto-Raman at High Pressure*

The above sections dealt with the application of pressure and magnetic field to explore exotic phenomena in different classes of materials. However, the entanglement between various degrees of freedom and its complexity can be further investigated with the simultaneous application of magnetic field and pressure to gain an insight into interactions and their rich phase diagram. On applying pressure, one can alter the lattice, while magnetic field can change the magnetic structure, thus, the simultaneous application enables to study magnetostructural properties within the system. Apart from this, novel magnetic ground states can also be explored by in-situ field and pressure studies. One of the recent in-situ studies is on strontium ruthenate that exhibits a metamagnetic phase [65].

$Sr_4Ru_3O_{10}$ (SRO) is a triple-layer canted ferromagnet below $T_c \sim 105$ K. It consists of RuO_6 octahedra in such a manner that below T_c, two of the octahedra are rotated by 5.25° to the c-axis and the intermediate octahedra are rotated by an angle of 10.6° to the c-axis in the opposite direction. B_{1g} (~380 cm^{-1}) phonon mode of SRO shows signatures of spin-lattice excitation below T_c in temperature-dependent Raman measurements performed on the sample. On applying the external magnetic field parallel to the c-axis, the frequency of B_{1g} is reported to increase with the field. This was explained as, due to the external magnetic field the canted moment of Ru gets reduced such that the RuO_6 is elongated along the c-axis and in turn Ru–O bond is contracted in the ab-plane resulting in an increase in the frequency of the in-plane B_{1g} mode.

The metamagnetic transition (occurs at $H > H_C$ (~2 T) and $T < 50$ K) was also realized by applying a magnetic field along the ab-plane. In the presence of a field, the mode frequency doesn't show any significant change above $T \sim 50$ K. In this case, the Ru moments are canted ferromagnetically along the c-axis while along the ab-plane, the moments lack ordering. However, when the temperature is lowered below 50 K, the Ru moments are canted antiferromagnetically or paramagnetically, when $H < 2$ T and with the increment in the field the moments are ferromagnetically canted thus attaining metamagnetic transition from AFM or PM canted to ferromagnet canted.

The report also discussed the influence of pressure on the magnetic property (i.e., magnetoelastic effect) of SRO. The pressure was applied in-situ with temperature and an anomalous decrease in frequency was observed on increasing pressure below $T < 70$ K. The decrease in the frequency of B_{1g} mode with pressure is because of the canting of Ru moments along the ab-plane in contrast with the likelihood of moments to align along the c-axis.

Apart from the appearance of novel magnetic states, strong coupling between the different degrees of freedom manifests itself with the emergence of coupled excitations when exposed to external perturbations like magnetic field and pressure. In rare-earth materials, the crystal electric field (CEF) states of localized f-orbitals which are energetically resonant with the phonon bands, lead to strong vibronic coupling between the CEF and phonons, and offers novel spin-lattice coupling phenomena. For example, in Pr_2O_3, Raman active CEF excitations were observed at low temperatures, and their coupling with phonons were studied as a function of temperature, magnetic field, and pressure [66]. At low temperatures, four phonon modes were present at 105 (P1), 189 (P2), 396 (P3), and 414 (P4) cm^{-1} and CEF excitations at 124 and 527 cm^{-1}. The temperature-dependent study shows that, P1 and P3 modes exhibit deviation from the two- phonon decay model, pointing towards some additional interactions with other degrees of freedom, shown in Fig. 3.14a, b.

The application of magnetic field results into splitting of the 124 cm^{-1} CEF level into two excitations, with their energy gap increases linearly with the magnetic field. Further, pressure-dependent study reflected hardening of all the phonons and CEF excitations with increasing pressure, along with splitting of CEF excitation above 10 kbar. This pressure-induced splitting becomes unclear at elevated temperatures. The

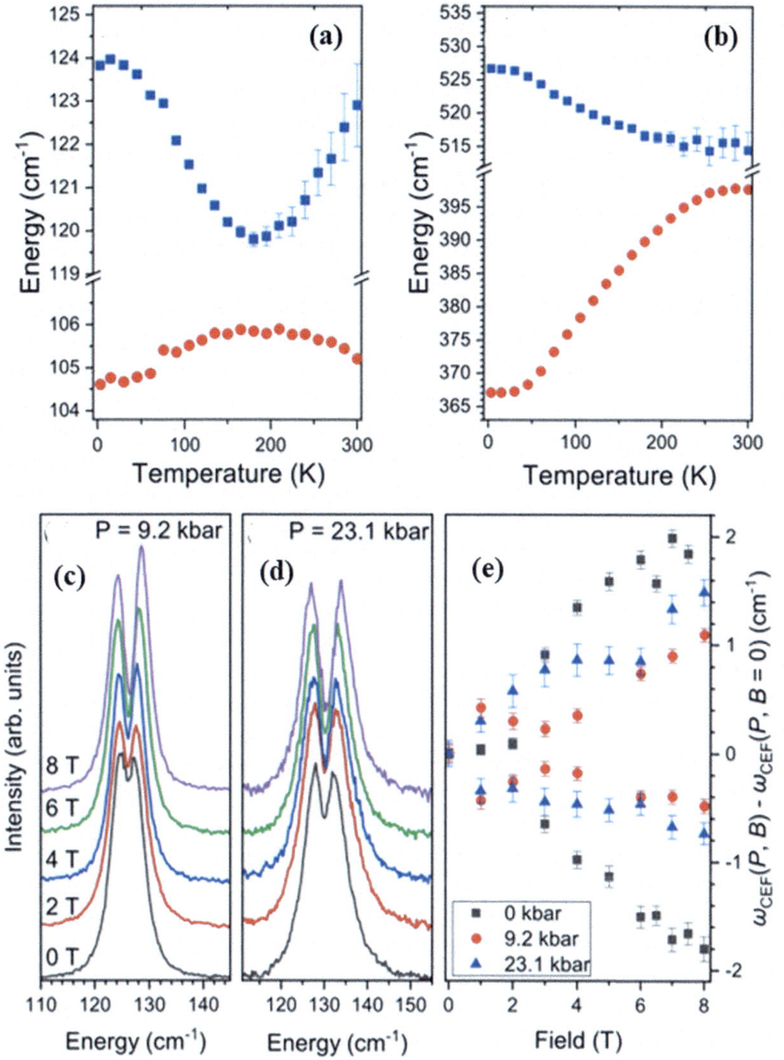

Fig. 3.14 Temperature dependence of **a** P1 mode (red) and 124 cm^{-1} CEF excitation (blue), **b** P3 mode (red) and 527 cm^{-1} CEF excitation (blue), **c**, **d** are the field dependence plots of 124 cm^{-1} excitation at P = 9.2 kbar, and 23.1 kbar, respectively, and **e** Difference in energies of CEF excitations at zero and higher field at a few pressures. Adapted from [66]

simultaneous application of magnetic field and pressure shows that pressure-induced splitting of the above-mentioned CEF level at zero field increases symmetrically with increasing field, depicted in Fig. 3.14c–e. The similar behavior of pressure and magnetic field induced splitting corroborates pressure-dependent magnetostructural effects in Pr_2O_3.

3.4 Conclusion and Future Prospects

In this chapter, we have highlighted the basics behind the technique of magnetic field and pressure-dependent Raman spectroscopy and discussed their importance to elucidate fundamental aspects and manifestation of quantum many-body interactions in the light of some recent studies on a few classes of materials. Quantum materials offer a plethora of possibilities to tailor their quantum interactions and consequently their properties. High pressure serves as a thermodynamic tuning parameter to explore the emergent properties of quantum materials while magnetic field allows to explore novel magnetic ground states. Topological materials, superconductivity, quantum spin liquids, etc. are some examples of such manifestations. Raman scattering is sensitive to changes in the local environment or symmetries induced by various means. The above-mentioned external perturbations, such as pressure and magnetic field (in-situ) prove to be useful in studying magnetic ground states, phase transitions, low energy excitations, and unravel various other exotic phenomena in condensed matter systems. In combination, Raman spectroscopy at high pressure and magnetic field offers an extremely powerful technique that can be vital not only to elucidate the fundamental effects of quantum interactions in complex systems but also to engineer their properties for potential device applications.

References

1. J. Orenstein, A.J. Millis, Science **288**, 468 (2000)
2. J.C. Tsang, J.E. Smith Jr., M.W. Shafer, Phys. Rev. Lett. **37**, 1407 (1976)
3. D.S. Lee, C. Riedl, T. Beringer, A.C. Neto, K. von Klitzing, U. Starke, J.H. Smet, Phys. Rev. Lett. **107**, 216602 (2011)
4. Y. Okamoto, M. Nohara, H. Aruga-Katori, H. Takagi, Phys. Rev. Lett. **99**, 137207 (2007)
5. S. Park, N. Nagaosa, B.J. Yang, Nano Lett. **20**, 2741–2746 (2020)
6. S. Paul, S. Karak, M. Mandal, A. Ram, S. Marik, R.P. Singh, S. Saha, Phys. Rev. B **102**, 054103 (2020)
7. M.W. Iqbal, K. Shahzad, R. Akbar, G. Hussain, Microelectron. Eng. **219**, 111152 (2020)
8. M. Kim, X.M. Chen, X. Wang, C.S. Nelson, R. Budakian, P. Abbamonte, S.L. Cooper, Phys. Rev. B **84**, 174424 (2011)
9. A.K. Sood, G. Gruner, Phys. Rev. B **32**, 2711 (1985)
10. A. Sethi, T. Byrum, R.D. McAuliffe, S.L. Gleason, J.E. Slimak, D.P. Shoemaker, S.L. Cooper, Phys. Rev. B **95**, 174413 (2017)
11. A. Jayaraman, Rev. Mod. Phys. **55**, 65 (1983)
12. J.C. Jamieson, A.W. Lawson, N.D. Nachtrieb, Rev. Sci. Instrum. **30**, 1016 (1959)
13. C.E. Weir, E.R. Lippincott, A.V. Valkenburg, E.N. Bunting, J. Res. Natl. Bur. Stand. A Phys. Chem. **63A**, 55–62 (1959)
14. G.J. Piermarini, S. Block, J.D. Barnett, R.A. Forman, J. Appl. Phys. **44**, 5377 (1973)
15. A. Jayaraman, Rev. Sci. Instrum. **57**, 1013–1031 (1986)
16. R.J. Wijngaarden, I.F. Silvera, in *High Pressure Science and Technology*. ed. by B. Vodar, P.H. Martean (Pergamon, Oxford, 1980), p.157
17. G.J. Piermarini, High Press. Res. **11**, 279–284 (1993)
18. A.S. Pereira, C.A. Perottoni, J.A.H. Jornada, J. Raman Spectrosc. **34**, 578–586 (2003)
19. G. Shen, H.K. Mao, Rep. Prog. Phys. **80**, 016101 (2016)

20. R.A. Forman, G.J. Piermarini, J.D. Barnett, S. Block, Science **176**, 284–285 (1972)
21. J.D. Barnett, S. Block, G.J. Piermarini, Rev. Sci. Instrum. **44**, 1–9 (1973)
22. G.J. Piermarini, S. Block, J.D. Barnett, R.A. Forman, J. Appl. Phys. **46**, 2774–2780 (1975)
23. H.K. Mao, Science **200**, 1145–1147 (1978)
24. E. Soignard, P.F. McMillan, *High-Pressure Crystallography* (2004), pp. 81–100
25. L. Merrill, W.A. Bassett, Rev. Sci. Instrum. **45**, 290–294 (1974)
26. H.K. Mao, Carnegie Inst. Washington Yearb. **77**, 904–908 (1978)
27. N. Tateiwa, Y. Haga, J. Phys. Conf. Ser. **215**, 012178 (2010)
28. X.B. Wang, Z.X. Shen, S.H. Tang, M.H. Kuok, J. Appl. Phys. **12**, 8011–8017 (1999)
29. S. Klotz, J.-C. Chervin, P. Munsch, G.L. Marchand, J. Phys. D Appl. Phys. **42**, 075413 (2009)
30. K.R. Hirsch, W.B. Holzapfel, Rev. Sci. Instrum. **52**, 52–55 (1981)
31. D.J.P. Morris, D.A. Tennant, S.A. Grigera, B. Klemke, C. Castelnovo, R. Moessner, C. Czternasty, M. Meissner, K.C. Rule, J.-U. Hoffmann, K. Kiefer, S. Gerischer, D. Slobinsky, R.S. Perry, Science **326**, 411–414 (2009)
32. C. Castelnovo, R. Moessner, S.L. Sondhi, Nature **451**, 42–45 (2008)
33. F. Jin, C. Liu, Y. Chang, A. Zhang, Y. Wang, W. Liu, X. Wang, Y. Sun, G. Chen, X. Sun, Q. Zhang, Phys. Rev. Lett. **124**, 087601 (2020)
34. P.A. Fleury, R. Loudon, Phys. Rev. **166**, 514 (1968)
35. L. Chen, J.-H. Chung, B. Gao, T. Chen, M.B. Stone, A.I. Kolesnikov, Q. Huang, P. Dai, Phys. Rev. X **8**, 041028 (2018)
36. T.T. Mai, K.F. Garrity, A. McCreary, J. Argo, J.R. Simpson, V. Doan-Nguyen, R.V. Aguilar, A.R.H. Walker, Sci. Adv. **7**, eabj3106 (2021).
37. F. Keffer, C. Kittel, Phys. Rev. **85**, 329 (1952)
38. A. McCreary, J.R. Simpson, T.T. Mai, R.D. McMichael, J.E. Douglas, N. Butch, C. Dennis, R. Valdés Aguilar, A.R. Hight walker. Phys. Rev. B **101**, 064416 (2020)
39. T. Haider, Int. J. Elec. Appl. **7**, 17–24 (2017)
40. T. Yin, K.A. Ulman, S. Liu, A.G. del Águila, Y. Huang, L. Zhang, M. Serra, D. Sedmidubsky, Z. Sofer, S.Y. Quek, Q. Xiong, Adv. Mater. **33**, 2101618 (2021)
41. J. Ji, A. Zhang, J. Fan, Y. Li, X. Wang, J. Zhang, E.W. Plummer, Q. Zhang, Proc. Natl. Acad. Sci. U.S.A. **113**, 2349 (2016)
42. A.H. Reid, X. Shen, P. Maldonado, T. Chase, E. Jal, P.W. Granitzka, K. Carva, R.K. Li, J. Li, L. Wu, T. Vecchione, T. Liu, Z. Chen, D.J. Higley, N. Hartmann, R. Coffee, J. Wu, G.L. Dakovski, W.F. Schlotter, H. Ohldag, Y.K. Takahashi, V. Mehta, O. Hellwig, J. Cao, E.E. Fullerton, J. Stöhr, P.M. Oppeneer, X.J. Wang, H.A. Dürr, Nat. Commun. **9**, 388 (2018)
43. B. Poojitha, A. Rathore, A. Kumar, S. Saha, Phys. Rev. B **102**, 134436 (2020)
44. Y. Wang, Z. Zhou, T. Wen, Y. Zhou, N. Li, F. Han, Y. Xiao, P. Chow, J. Sun, M. Pravica, A.L. Cornelius, W. Yang, Y. Zhao, J. Am. Chem. Soc. **138**, 15751–15757 (2016)
45. M.P. Pasternak, G.K. Rozenberg, G.Y. Machavariani, O. Naaman, R.D. Taylor, R. Jeanloz, Phys. Rev. Lett. **82**, 4663 (1999)
46. Y. Wang, J. Ying, Z. Zhou, J. Sun, T. Wen, Y. Zhou, N. Li, Q. Zhang, F. Han, Y. Xiao, P. Chow, W. Yang, V.V. Struzhkin, Y. Zhao, H. Mao, Nat. Commun. **9**, 1914 (2018)
47. K. Xue, X. Chen, J. Zhang, Z. Mao, L. Tang, I.O.P. Conf, Ser. Mater. Sci. Eng. **612**, 022083 (2019)
48. Y. Xu, S. Carlson, K. Söderberg, R. Norrestam, J. Solid State Chem. **153**, 212–217 (2000)
49. B. Wang, K. Sun, G. Chu, Spectrochim. Acta A Mol. Biomol. Spectrosc. **286**, 121945 (2023)
50. S. Pal, P. Malavi, S. Chaturvedi, S. Das, S. Karmakar, D.V.S. Muthu, U.V. Waghmere, A.K. Sood, Phys. Rev. B **102**, 214107 (2020)
51. S. Saha, D.V.S. Muthu, S. Singh, B. Dkhil, R. Suryanarayanan, G. Dhalenne, H.K. Poswal, S. Karmakar, S.M. Sharma, A. Revcolevschi, A.K. Sood, Phys. Rev. B **79**, 134112 (2009)
52. W. Qi, A. Mattursun, M. Gao, A. Hushur, H. Zhang, Results Phys. **43**, 106114 (2022)
53. V. Panchal, S. López-Moreno, D. Santamaría-Pérez, D. Errandonea, F.J. Manjón, P. Rodríguez-Hernandez, A. Muñoz, S.N. Achary, A.K. Tyagi, Phys. Rev. B **84**, 024111 (2011)
54. P. Malavi, S. Pal, D.V.S. Muthu, S. Sahoo, S. Karmakar, A.K. Sood, Phys. Rev. B **101**, 214402 (2020)

55. K. Kakurai, T. Osakabe, K. Goto, A. Oosawa, M. Fujisawa, H. Tanaka, Phys. B: Condens. Matter. **385**, 450–452 (2006)
56. A. Oosawa, M. Fuzisawa, T. Osakabe, K. Kakurai, H. Tanaka, J. Phys. Soc. Jpn. **72**, 1026–1029 (2003)
57. M. Thede, A. Mannig, M. Mansson, D. Hüvonon, R. Khasanov, E. Morenzoni, A. Zheludev, Phys. Rev. Lett. **112**, 087204 (2014)
58. M.A. Vries, D. Wulferding, P. Lemmens, J.S. Lord, A. Harrison, P. Bonville, F. Bert, P. Mendels, Phys. Rev. B **85**, 014422 (2012)
59. A.E. Böhmer, A. Kreisel, J. Phys. Condens. Matter **30**, 023001 (2018)
60. T. Li, X. Zhang, Z. Zeng, Phys. Chem. Chem. Phys. **23**, 25107 (2021)
61. A.P. Nayak, Z. Yuan, B. Cao, J. Liu, J. Wu, S.T. Moran, T. Li, D. Akinwande, C. Jin, J. Lin, ACS Nano **9**, 9117–9123 (2015)
62. P. Shen, X. Ma, Z. Guan, Q. Li, H. Zhang, R. Liu, B. Liu, X. Yang, Q. Dong, T. Cui, B. Liu, J. Phys. Chem. C **121**, 26019–26062 (2017)
63. Y. Yan, C. Jin, J. Wang, T. Qin, F. Li, K. Wang, Y. Han, C. Gao, J. Phys. Chem. Lett. **8**, 3648–3655 (2017)
64. J. Xia, J. Yan, Z. Wang, Y. He, Y. Gong, W. Chen, T.C. Sum, Z. Liu, P.M. Ajayan, Z. Chen, Nat. Phys. **17**, 92–98 (2021)
65. R. Gupta, M. Kim, H. Barath, S.L. Cooper, G. Cao, Phys. Rev. Lett. **96**, 067004 (2006)
66. J.E. Slimak, A. Sethi, T. Kolodiazhnyi, S.L. Cooper, Phys. Rev. Res. **2**, 043169 (2020)

Chapter 4
Optothermal Raman Spectroscopy for Thermal Transport Study

Jay Deep Gupta, Priyanka Jangra, and Ashish Kumar Mishra

4.1 Introduction

Effective heat management has become increasingly crucial in various fields of materials research and development. Achieving optimal performance of devices often necessitates the presence of extremely high or low thermal conductivities. For example, in the microelectronics industry, efficient cooling of integrated circuits is needed to prevent diffusion or electromigration, demanding high thermal conductance at the sub-micrometer level [1]. Recently, there has been a rising interest in exploring the properties and applications of 2D materials like graphene, TMDCs, black phosphorene, and boron nitride. It has been proposed that high thermal conductivity (k) materials, including 2D materials, are suitable for various applications [2]. However, there is also a recognition that materials with low thermal conductivity, yet capable of enduring temperature fluctuations, offer advantages in thermoelectric devices or sensors.

Raman spectroscopy is non-destructive and powerful technique that is used to investigate the thermal transport properties of 2D materials. Raman spectroscopy indicates the inelastic scattering of photons, providing valuable information about the vibrational modes and lattice dynamics of materials. In the context of 2D materials, like graphene and other single-layer materials, Raman spectroscopy is particularly useful for probing their thermal properties. The Raman peaks because inplane and out-of-plane atomic vibrations can shift with temperature, providing the perception of thermal expansion, phonon–phonon interactions, and thermal conductivity. Raman spectroscopy also allows the investigation of phonon modes influenced by variations in temperature [3]. For instance, the frequency shift and linewidth

J. D. Gupta · P. Jangra · A. K. Mishra (✉)
School of Materials Science and Technology, Indian Institute of Technology (Banaras Hindu University), Varanasi 221005, India
e-mail: akmishra.mst@iitbhu.ac.in

D. K. Singh et al. (eds.), *Raman Spectroscopy*, Springer Series in Optical Sciences 248,
https://doi.org/10.1007/978-981-97-1703-3_4

broadening of Raman peaks with temperature can reveal the anharmonic phonon–phonon interactions and thermal expansion coefficient. These parameters are crucial for understanding the heat transport mechanisms and thermal conductivity of 2D materials.

4.2 Recent Advancements in Raman Technique

Raman-based thermometry is widely used as a reliable micro- and nanoscale temperature-probing technique, which is developed to fulfil the demands to measure the temperature of various nanomaterials and structures. In order to investigate the heat transport process in the suspended film of 2D materials and measure their thermal properties of micro/nanowires, a novel concept of laser flash-Raman spectroscopy approach was initially introduced in July 2014 [5]. For the purpose of thermal characterization of 1D structures, the frequency-resolved Raman method and time-domain differential Raman approach were invented in 2015 [6]. These two techniques eliminate the necessity for temperature-based calibration for optical and Raman properties during data processing. Also, a supporting 2D materials measurement technique using laser flash-Raman spectroscopy was invented at the same time. A significant change in Raman thermometry in 2017 was energy transport state-resolved Raman (ET-Raman), which is focused on the interface between energy transfer and measurement of heat carrier diffusion [7]. Raman spectroscopy was used to study hot carriers for the first time, and the time resolution was increased to picoseconds. Based on ET-Raman, frequency-domain energy transport state-resolved Raman (FET-Raman) to improve measurement accuracy was developed in 2018 [8]. In the same year, a variable-spot-size laser flash Raman spectroscopy approach was developed for thermal and optical observation of supported 2D van der Waals heterostructures. In order to achieve high temporal resolution for transient thermal measurement and resolve the modulation limit of the heating laser, a dual-wavelength laser flash Raman spectroscopy approach incorporating a picosecond laser and the heating laser was developed in 2018 [9]. This method was improved in 2019 to map the anisotropic thermal properties of materials [10].

4.3 Thermal Transport Properties Using Different Techniques

Phonons are quantized vibrations of the crystal lattice that carry heat in materials. In 2D materials like Graphene, transition metal dichalcogenides, boron nitride, black phosphorous, etc. phonon transport is affected by the reduced dimensionality, phonon confinement, and scattering processes. The scattering mechanisms can include phonon–phonon interactions, phonon-electron interactions, phonon-defect interactions, and boundary scattering at edges and interfaces. Thermal conductivity of 2D materials is influenced by the nature and strength of these scattering mechanisms. Thermal conductivity (k_s) of a solid is a measure of its ability to conduct heat and is described by Fourier's law of heat conduction $\mathbf{J} = -k\,\nabla T$, where it relates the heat flux (**J**) to the temperature gradient (∇T). The kinetic theory predicts that the group velocity (v), mode-dependent heat capacity (C), relaxation time (τ), and mean free path (Λ) all influence the thermal conductivity of crystalline materials.

$$\mathrm{ks} = \sum_{qs} C_{qs} v_{qs}^2 \tau_{qs} = \sum_{qs} C_{qs} v_{qs} \Lambda_{qs} \tag{4.1}$$

Subscript qs represents a phonon mode with momentum $\hbar q$ in the sth phonon branch, where $\hbar$ is the Planck constant. The heat capacity of a specific mode, denoted as C_{qs}, is express as $\hbar\,\omega_{qs}\partial f_{qs}^0/\partial T$ where, $\partial f_{qs}^0 = \frac{1}{[\exp\left(\frac{\hbar\omega_{qs}}{k_B T}\right)-1]}$ is the Bose-Einstein statistics-obeying equilibrium phonon population function, where ω_{qs} is the phonon frequency and $k_B T$ is the product of the Boltzmann constant and the absolute temperature. $\Lambda_{qs} = (v_{qs}\tau_{qs})$ represents the mean free path i.e., phonons move through the sample without encountering significant scattering [11–14]. This behavior is observed in thin films, where the thermal conductivity decreases as the films become thinner due to an increase in the scattering of phonons at the film boundaries.

Figure 4.1 shows the thermal conductivity of 2D materials like graphene, hBN, silicene, TMDCs (MoS_2, WS_2 etc.) [15]. The determination of temperature-dependent thermal conductivity values for 2D materials involves both first-principles calculations and experimental measurements, whenever possible. However, it is still challenging to carry out experimental measurements to determine the thermal conductivity of 2D materials. Understanding the thermal properties of 2D materials is crucial, but conventional methods used to examine the thermal properties of thin films cannot be directly applied to 2D materials because of the challenges posed by their atomic-layer thickness. As a result, in the last twenty years, considerable research efforts have been directed toward measuring the thermal transport properties of 2D materials.

Experimental techniques have played a vital role in validating theoretical models, and numerous methods have been proposed over the past five decades to investigate thermal transport properties. Depending on whether the heat flow is constant or vary over time, this experimental characterization technique can generally be divided into two categories: transient methods and steady-state methods [16, 17]. Transient methods involve measuring thermal conductivity by observing the thermal gradient as a function of time during the heating process. One advantage of these methods is that they can be performed more rapidly since there is no requirement to wait for a steady-state condition. The 3ω method, time-domain thermoreflectance method (TDTR), and laser flash technique (LFT) are example of the transient method, employed to measure the thermal diffusivity of various materials. Cahill et al. developed the 3ω technique to investigate the thermal conductivity of bulk materials. The technique involves applying a periodic heating source, such as a metal heater, to the sample

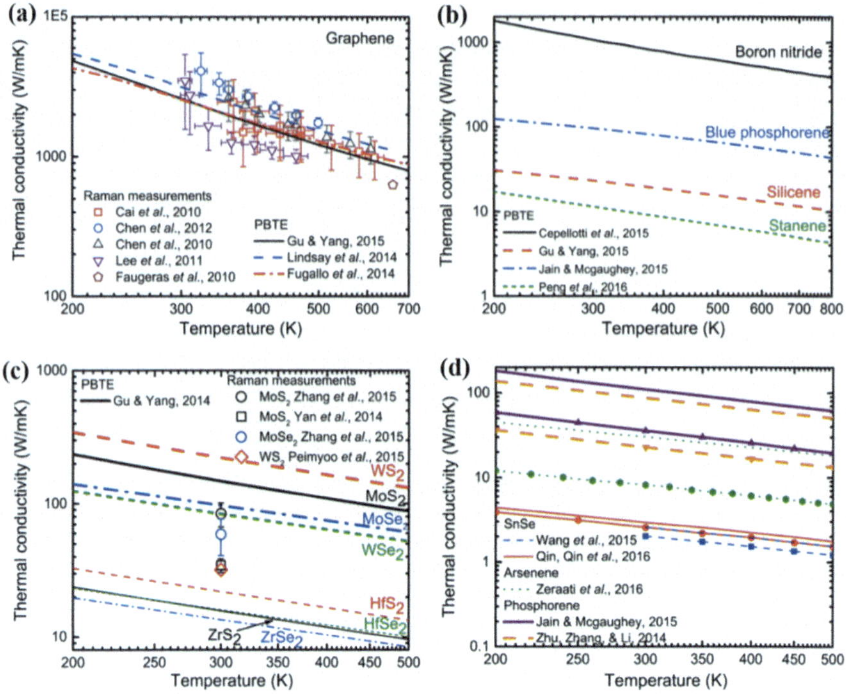

Fig. 4.1 Thermal conductivities of: **a** Graphene. **b** Hexagonal crystal structures: boron nitride, stanine, blue phosphorene, and silicene. **c** TMDCs and **d** Puckered structures: a few 2D crystals; In panel, **d** lines with marks show thermal conductivity along the armchair direction, the lines represent thermal conductivity along the zigzag direction. Adapted with permission from [15]. Copyright 2018 APS publishing

[18]. The resulting temperature oscillations are then measured, and the frequency dependence of these oscillations can be used to determine the thermal conductivity and thermal diffusivity of the sample. However, when applying the 3ω technique to 2D materials, the sample preparation process becomes complex and expensive. It requires microfabrication techniques such as e-beam evaporation and lithography to pattern electrodes. As a result, the 3ω method is not well-suited for 2D thin film. The TDTR technique is used for investigating the thermal conductivity and thermal conductance of micro/nanomaterials. TDTR technique is based on the observation that changes in surface reflectance can be utilized to determine thermal properties when a material is heated. The initial development of the transient thermo-reflectance (TR) method by Paddock et al. paved the way for TDTR, enabling the measurement of thermal diffusivity in metal thin films [19]. One advantage of noncontact TDTR method involves a small surface area for heating and probing. However, it should be noted that the initial experimental setup costs for TDTR can be high, and there may be uncertainties associated with the thickness of the metal films, which could pose challenges during measurements.

In LFT, it should be noted that if the laser beam with uneven energy distribution or if the sample unevenly absorbs thermal energy, it can lead to inaccurate diffusivity measurements of the sample. This issue becomes more pronounced in thin films, exacerbating the impact of uneven energy distribution or absorption. Therefore, careful attention must be given to ensure accurate measurements, particularly in the case of thin films. Transient methods are extremely challenging and involve complex sample preparation procedures, including the fabrication of electrodes for measurements. During this sample preparation process, there is a risk of damaging the sample, which can significantly impact the reliability of the obtained thermal property values. Steady-state techniques are considered relatively straightforward and dependable for characterizing thermal properties and it involve conducting measurements when the temperature of the material remains constant over time. This characteristic simplifies signal analysis since steady-state implies consistent signals. However, these techniques often require a well-designed experimental setup. An example of a steady-state technique widely employed in this chapter is the Optothermal Raman (OTR) measurement. The subsequent section will provide a detailed discussion of the OTR technique.

4.3.1 Optothermal Raman Technique

The optothermal Raman (OTR) technique is a direct steady-state measurement method used to determine the thermal conductivity (k) of materials without the need to calculate it from thermal diffusivity data. Initially applied for measuring the thermal properties of graphene, this technique utilizes a micro-Raman spectrometer as a thermometer to measure the local temperature rise. The temperature dependence of Raman peak positions enables the measurement of temperature rise through Raman spectroscopy. In the OTR technique, the same Raman excitation laser is utilized both as a spectroscopic tool and as a heating source. The measurement procedure involves two main steps: calibration measurements and power-dependent Raman measurements. The details of both procedures will be discussed in subsequent sections. The pioneering work by Balandin et al. introduced the application of OTR technique to investigate in-plane thermal conductivity of mono-layer graphene (suspended) [20]. The OTR technique offers simplicity as it does not require complex sample preparation procedures. Furthermore, it allows non-contact probing, precisely measuring thermal properties without damaging the sample. The laser beam used in OTR acts as a heat source as well as a tool for probing to obtain the Raman spectrum. Figure 4.2a shows the experimental setup for this method. A monolayer graphene is suspended between heat sinks situated around edges. Once an incident laser beam is directed on suspended graphene, the absorbed laser energy generates heat, which then dissipates radially, leading to a temperature increase within the graphene. As shown in Fig. 4.2b, the laser beam shape can be determined using knife edge method, although it is commonly assumed to have a Gaussian beam profile. Temperature distribution within the graphene follows contour of the laser beam, by the local heating of the

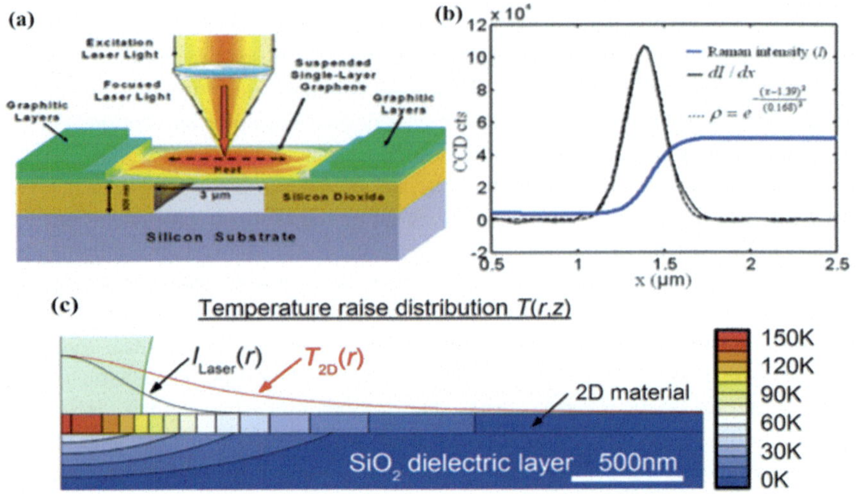

Fig. 4.2 **a** Experimental setup for OTR technique. Adapted with permission from [20]. Copyright 2008 ACS publishing. **b** The Raman laser beam distribution is measured and shown to have a Gaussian profile Adapted with permission from [21]. Copyright 2010 ACS publishing. **c**. Correspondingly, the temperature distribution within the graphene sample also follows a Gaussian profile reflecting the distribution of laser beam. Adapted with permission from [22]. Copyright 2021 scientific reports

sample within a small radius beam (approximately 1 μm) over radiation (Fig. 4.2c) [21].

To investigate the k of 2D materials such as Graphene, TMDCs, boron nitride, black phosphorous, etc., the OTR technique is employed. This technique involves conducting temperature and power dependent Raman studies. The temperature distribution within these materials can be determined by solving the heat diffusion equation in cylindrical coordinates, which takes the following form [22, 23]:

$$\frac{1}{r}\frac{d}{dr}\left(r\frac{dT(r)}{dr}\right)-\frac{g}{kt}(T(r)-T_{amb})=\frac{Q}{k} \tag{4.2}$$

where T(r) and r denote the temperature distribution and radial position, respectively, as measured from the centre of the laser beam. Here, T_{amb} is the ambient temperature, t is the thickness of the film, k is the thermal conductivity of supported film and g is the total interfacial thermal conductance per unit area. In this case, T_{amb} denotes the ambient temperature, t is the thickness of film, k is the thermal conductivity of the supported film, and g is the total interfacial thermal conductance per unit area. Q is representing as volumetric optical heating, which is given by

$$Q = \frac{q_0 \alpha}{t} \exp\left(-\frac{r^2}{r_0^2}\right) \tag{4.3}$$

where q_0, α and r_0 are the maximum absorbed laser power per unit area, absorption coefficient of material and radius of laser beam spot, respectively. Total absorbed laser power (P_{abs}) is calculated as follows

$$P_{abs} = \int_0^\infty q_0 \exp\left(-\frac{r^2}{r_0^2}\right) 2\pi r \, dr = q_0 \pi r_0^2$$

By considering $\theta(r) = T(r) - T_{amb}$ and $z = \left(\frac{g}{k_s t}\right)^{1/2} r$, (4.3) becomes

$$\frac{d^2\theta}{dz^2} + \frac{1}{z}\frac{d\theta}{dz} - \theta = -\frac{q_0}{g}\exp(-\frac{z^2}{z_0^2}) \tag{4.4}$$

The analytical solution for the supported film can be written as

$$\theta(z) = C_1 I_0(z) + C_1 K_0(z) + \theta_P(z) \tag{4.5}$$

where I_0 and K_0 are zero-order of first and second-kind of modified Bessel functions, respectively. In the (4.5), $\theta_P(z)$ is denoted as the solution of nonhomogeneous differential equation and can be written as follows

$$\begin{aligned} \theta_P(z) = & I_0(z) \int K_0(z) \frac{\frac{q_0}{g}\exp\left(-\frac{z^2}{z_0^2}\right)}{-I_0(z)K_1(z) - K_0(z)I_1(z)} dz \\ & - K_0(z) \int I_0(z) \frac{\frac{q_0}{g}\exp\left(-\frac{z^2}{z_0^2}\right)}{-I_0(z)K_1(z) - K_0(z)I_1(z)} dz \end{aligned} \tag{4.6}$$

where $I_1(z)$ and $K_1(z)$ are the first order of first and second-kind modified Bessel functions, respectively. To obtained the value of C_1 and C_2 in (4.5), we have incorporated the following boundary conditions

$$\frac{d\theta}{dz} = 0 \text{ at } z = 0,$$

$$\theta(z) = 0 \text{ at } z \to \infty$$

After using these boundary conditions, we get $C_2 = 0$ and $C_1 = -\lim_{z \to \infty} \frac{\theta_P(z)}{I_0(z)}$, which approaches a constant value for large z. Now,

$$\theta(z) = -I_0(z)\lim_{z\to\infty}\frac{\theta_P(z)}{I_0(z)} + \theta_P(z)$$

The temperature rises in 2D film, measured by the Raman laser beam can be given by

$$\theta_m = \frac{\int_0^\infty \theta(r)\exp\left(-\frac{r^2}{r_0^2}\right)rdr}{\int_0^\infty \exp\left(-\frac{r^2}{r_0^2}\right)r\,dr} \tag{4.7}$$

Then, the thermal resistance of the supported film is defined as R_m and it is the ratio of θ_m and P_{abs}. It can be expressed as follows

$$R_m = \frac{\theta_m}{P_{abs}} \tag{4.8}$$

where, R_m is the function of $r_{0,}$ k_s and g. To avoid the artificial shift in mode frequency

$$R_m = \frac{\partial\theta_m}{\partial P_{abs}}$$

At the same time, R_m can be measured experimentally using the relation below

$$R_m = \chi_P/\alpha\chi_T \tag{4.9}$$

where χ_P, χ_T are first order power and temperature coefficients of the material and α is the absorption coefficient. In order to get the value of R_m we need the value of χ_p and χ_T, which can be determine by power and temperature dependent Raman studies, respectively [24]. The temperature and power dependent Raman studies of 2D materials have been discussed below.

4.3.2 Temperature Dependent Raman Study

To analyze the thermal transport properties and understand the effect of phonon confinement in 2D films, temperature-dependent Raman spectroscopy is commonly employed. This technique allows for the investigation of various properties, including lattice vibrations, potential energy, thermal conductivity, and thermal expansion of 2D materials. These observed thermal properties greatly influence the efficiency of electrical and optoelectronic devices made from these materials. The temperature-dependent Raman shift, represented by the peak position ω (in cm^{-1}), provides valuable information about different interactions such as electron-phonon and phonon-phonon interactions. The Raman shift exhibits a linear dependence on temperature, and it can be expressed as follows

$$\omega(T_2) - \omega(T_1) = \chi_T(T_2 - T_1) \tag{4.10}$$

where ω represents the peak position at a reference temperature T and, χ_T is the temperature coefficient. This equation describes the linear relationship between the Raman shift and temperature, allowing for the determination of the temperature coefficient (χ_T) through experimental measurements. By analyzing the temperature-dependent Raman spectra of the 2D film, researchers can gain insights into the behavior of phonons and their interactions within the material. These studies provide crucial information for understanding the thermal response, lattice dynamics, and thermal transport properties of 2D materials, aiding in the design and optimization of electronic and optoelectronic devices based on these materials. The linear variation of phonon frequency in 2D materials over any substrate can be attributed to the anharmonic vibrations of the lattice. When the temperature changes, the lattice of the material undergoes expansion or contraction. This change in lattice spacing affects the equilibrium positions of the atoms and consequently alters the interatomic forces within the material. These variations in the interatomic forces lead to shifts in the energies of phonons. The anharmonic vibrations of the lattice involve deviations from the idealized harmonic oscillations. In a perfect harmonic lattice, the phonon frequencies would remain constant regardless of temperature. However anharmonic interactions between atoms introduce non-linear effects, causing phonon frequencies to shift with temperature. As the lattice expands or contracts with increasing or decreasing temperature, the interatomic distances and bond strengths change. This alteration in the lattice potential energy affects the vibrational modes and phonon frequencies. Consequently, the Raman-active phonon modes exhibit shifts in their energies, which can be observed through temperature-dependent Raman spectroscopy. By studying the linear variation of phonon frequencies through temperature-dependent Raman measurements, researchers can gain insights into the anharmonic behavior of the lattice in 2D materials. Understanding these thermal effects and the resulting changes in phonon energies is crucial for characterizing the lattice dynamics and thermal properties of 2D materials, which, in turn, influences the performance of optoelectronic and electronic devices.

4.3.3 Power Dependent Raman Study

To calculate thermal conductivity of a supported 2D film and investigate the local temperature rise caused by phonon confinement, it is necessary to analyze the shifting of Raman modes by heating the film with different laser powers. In this experimental setup, a laser beam is directed onto the top surface of the sample, leading to local heating and resulting changes in the Raman spectra. When the laser power is increased, the Raman peak corresponding to the 2D film exhibits several noticeable changes. Firstly, the intensity of the Raman peak increases, indicating a higher number of scattered photons due to enhanced phonon-phonon scattering caused by the local temperature rise. Additionally, a red shift in the Raman peak is observed.

This red shift occurs because the bonds in the 2D film soften as a result of the increased local temperature. The softening of bonds leads to a decrease in the phonon frequency, resulting in a lower Raman shift. The relationship between the Raman peak position and the laser power can be described as linear. This linear change in the peak position with respect to the laser power can be expressed by the following equation

$$\omega(P_2) - \omega(P_1) = \chi_P(P_2 - P_1) \tag{4.11}$$

where ω is the shifted peak position, χ_P is the power coefficient, and P is the laser power. This equation demonstrates the linear dependence between the Raman peak shift and the laser power, providing a means to quantify the relationship between local temperature rise and phonon confinement. By analyzing the Raman spectra under different laser powers and examining the linear change in peak position, researchers can gain insights into the thermal conductivity of the supported 2D film. The understanding of local temperature rise and its effects on phonon behavior is essential for characterizing the thermal properties of 2D materials and optimizing their performance in electronic and optoelectronic devices.

4.3.4 Example of OTR Study in 2D Materials

Thermal conductivity (k) of a vertical few-layer (VFL) 2D film can be determined using the OTR technique. In the case of VFL, there is minimal contact area between the 2D film and the substrate, allowing it to be treated as a free-standing or suspended film. This implies that the interfacial thermal conductance between the VFL and the substrate is considered to be zero [25]. To calculate the thermal conductivity in this scenario, different models are employed. Because of the low strain in VFL-2D film, greater thermal transport might be possible. The following equation can be used to explain the conduction of heat (Q) via a planar laminar surface with cross sectional area A over time (t)

$$\frac{\partial Q}{\partial t} = -k \oint \nabla T \cdot dA \tag{4.12}$$

where T is the absolute temperature and k is thermal conductivity of film. The following equation can be used to consider a plane wave heat flow for the thermal conductivity of a suspended film

$$k = \chi_T \left(\frac{L}{2hW} \right) \left(\frac{\delta \omega}{\delta P} \right)^{-1} \tag{4.13}$$

where the lateral dimension of 2L suspended flake, the first order temperature coefficient is χ_T, and W is the size of individual flakes and h is their thickness, $\frac{\delta \omega}{\delta P}$ is shift in the Raman peak caused by a variation in incident laser intensity that heats the sample surface. The first order power coefficient, χ_P, is the same as this $\delta\omega/\delta P$. It is believed that heat spreads from the centre in two opposing directions for a suspended 2D film cross trench with width larger than the size of the laser spot. Consequently, the following statement can be used to determine thermal conductivity.

$$k = \left(\frac{L}{2hW}\right)\left(\frac{\chi_T}{\chi_P}\right) \tag{4.14}$$

where W is the size of individual flakes and h is their thickness. The optothermal Raman technique was used to examine the thermal behaviour of VFL-2D film, as shown schematically in Fig. 4.3. Majee et al. investigated the in-plane thermal conductivity of chemical vapor deposition (CVD) grown vertical few-layer (VFL) film of MoS_2 on a silicon (Si) substrate [25]. They estimated the thermal conductivity using (4.13). To understand the thermal conductivity of the MoS_2 VFL film, they performed temperature-dependent Raman spectroscopy in a range of 83–453 K as shown in Fig. 4.4a, b. They utilized a long working distance (LWD) 50X objective lens and applied a laser power of 0.44 mW at the sample surface. By measuring the Raman spectra of the VFL-MoS_2 film at different temperatures, they observed shifts in the peak positions, indicating a temperature-dependent response. The excitation power of the laser and the temperature dependency of vibrational modes play a crucial role in interpreting the anharmonicity among phonons and electron-phonon interactions, providing insights into the thermal properties of materials.

In this study, the researchers examined the Raman spectra of the material under different incident laser powers. Figure 4.4c demonstrates the increase in the intensity of the Raman peaks (E^1_{2g} and A_{1g}) with increasing incident laser power. This increase in intensity is attributed to enhanced phonon-phonon scattering resulting from the local temperature rise induced by the laser. Additionally, both Raman modes show red shifts in their frequencies due to the softening of bonds caused by the elevated local temperature, as depicted in Fig. 4.4d. This linear behavior further confirms the connection between the local temperature rise and the phonon dynamics. By analyzing these experimental results, the researchers were able to determine the in-plane thermal conductivity of the material at room temperature. The obtained thermal conductivity was found to be relatively high, with a value of 100 ± 14 W m^{-1} K^{-1}. Thermal behavior of 2D TMDCs, such as MoS_2, is important as it can impact the operation and performance of optoelectronic devices.

It is crucial to carefully consider thermal transport in the cross-plane direction, especially in experiments conducted in ambient condition or when materials is supported on substrate. Here, it should be noted that the heat transport from materials to its substrate can vary reliant over various experimental conditions, including the type of substrate used and interface of material with substrate. To study the thermal transport behaviour through the interface of materials, it is essential to investigate and characterize the interfacial thermal conductance. This involves studying the thermal transport behavior at the interface, including effects of substrate properties, interfacial bonding, and other factors that influence the thermal transfer between materials and

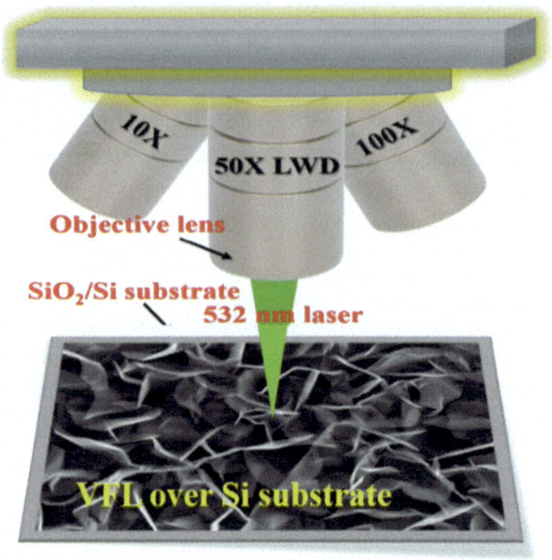

Fig. 4.3 Schematic diagram of OTR setup for measuring thermal conductivity of VFL film. Adapted with permission from [25]. Copyright 2020 ACS publishing

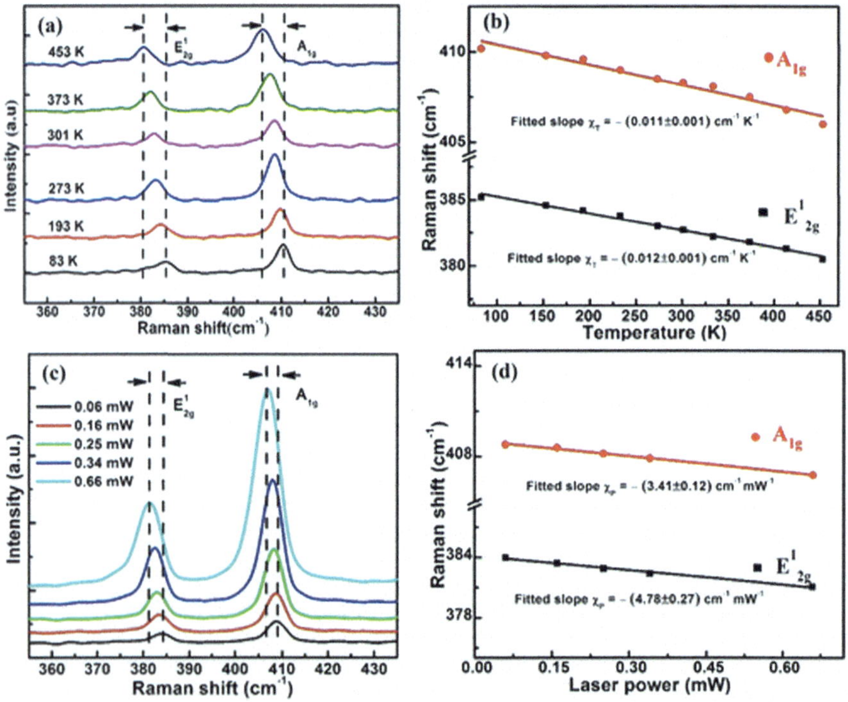

Fig. 4.4 **a** Temperature dependent Raman spectra and **b** corresponding Raman shift with temperature; **c** power dependent Raman spectra and **b** corresponding Raman shift with temperature of VFL-MoS_2 film over Si film. Adapted with permission from [25]. Copyright 2020 ACS publishing

its substrate. By conducting experimental investigations and studying the interfacial thermal conductance, researchers can enhance their understanding of thermal transport across materials-substrate interface. This knowledge is crucial for accurately characterizing and modeling the thermal properties of graphene-based systems and optimizing their performance in various applications.

Cai et al. employed two different objective lenses (50X and 100X) in their experiment to perform OTR technique [21]. They conducted simultaneous measurements of thermal conductivity (k) and interfacial thermal conductance (g) between graphene supported by an Au/SiN substrate. In the same way, Judek et al. measured g and k_s of graphene/SiO_2 and MoS_2/SiO_2 using the OTR technique. They employed a motorized XYZ stage to introduce variations in size of laser beam [22]. By moving stage in the Z-direction, they were able to alter the beam spot size, by using the same lens magnification (as depicted in Fig. 4.5a). This technique allows for the establishment of an additional governing equation for the substrate, which is provided by

$$k_{sub}\frac{1}{r}\frac{d}{dr}\left(r\frac{dT_{sub}(r,z)}{dr}\right)+k_{sub}\frac{d^2T_{sub}(r,z)}{dz^2}=0 \tag{4.15}$$

where k_{sub} and, $T_{sub}(r, z)$ are thermal conductivity and temperature dispersion of substrate respectively, while z is its distance from the upper surface. The effective Raman temperature of 2D film fluctuates depending on size r_0 of the laser beam which is a significant parameter in (4.9). As shown in Fig. 4.5b solution, pairs of (g and k) of (4.2) and (4.4) can be calculated by MATLAB software. Another infinite set of (k and g) solution pairs, represented as the temperature ratio depicted in Fig. 4.5c, is obtained by combining various effective Raman temperatures with various laser beam sizes.

Similarly, with separate objective lenses (50X and 100X), Taube et al. measured k and g at the MoS_2/SiO_2 contact using the OTR technique [26]. The beam radius r_0 is determined by the lens magnification, which is written as $r_0 = \lambda/\pi{\cdot}NA$ where λ is laser wavelength and NA is the numerical aperture of lens. In this study, r_0 was regulated by utilising two different objective lenses, which allowed for the precise determination of k and g for the substrate and the supported 2D material. By using OTR method, Bishnu et al. demonstrated the thermal transport properties for the CVD grown triangular shaped MoS_2 film over SiO_2/Si substrate [23]. In this work, they perform power and temperature-dependent Raman measurement using two different objective lenses LWD 50X and 100X having NA 0.6 and 0.9, respectively as shown in Fig. 4.6. By performing the temperature and power dependent Raman measurement, they calculated χ_T and χ_P. With the help of χ_T and χ_P thermal conductivity (42 ± 8 Wm^{-1} K^{-1}) and interfacial thermal conductance (1.264 ± 0.128 MW m^{-2} K^{-1}) at room temperature were calculated for supported bilayer MoS_2, suggesting their suitability for thermal management in devices. In our recent study, to determine the interfacial thermal conductance and thermal conductivity of snow-like and hexagonal $MoSe_2$ thin film on a SiO_2/Si substrate, OTR technique is performed [27]. This study includes the phonon response to the environment temperature by performing temperature dependent Raman study and phonon response to the local rise in temperature

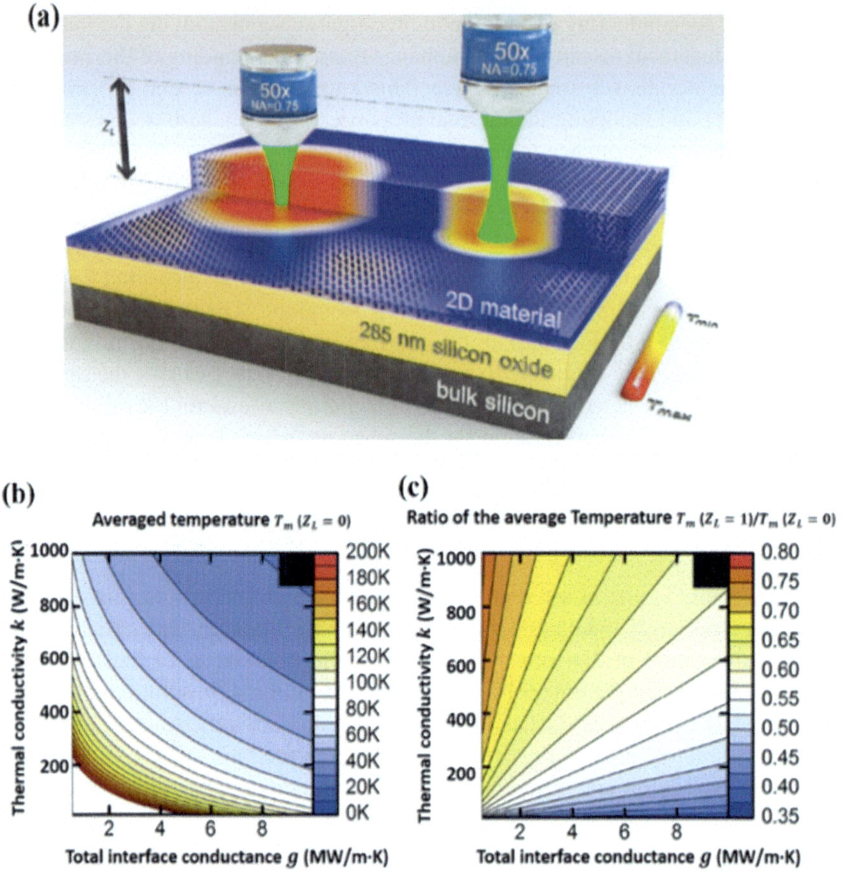

Fig. 4.5 For optothermal Raman thermometry, there are two additional objective lenses (beam diameters). **a** Experimental schematic illustration employing different OTR technique beam widths. **b** Depending on the effective Raman temperature infinite solution pairings for (g and k). **c** The ratio of the effective Raman temperature from the various beam widths determines the infinite number of solution pairings of (k and g). Adapted with permission from [22]. Copyright 2015 scientific reports publishing

due to laser heating by performing power dependent Raman study. For the snow-like $MoSe_2$ film, we obtained experimental values of thermal resistance, 3.8521×10^5 and 3.2886×10^5 KW^{-1} for spot sizes of 0.19 μm (using a 100 × objective lens) and 0.28 μm (using a 50X LWD objective lens), respectively. Similarly, for the hexagonal $MoSe_2$ film, we measured experimental values of thermal resistance, 6.6086×10^5 and 4.4871×10^5 KW^{-1} for spot sizes of 0.19 and 0.28 μm, respectively. By solving MATLAB simulations, we determined the values of g ~ 6.0788 ± 1.3707 MW m^{-2} K^{-1} and k ~ 36 ± 6 Wm^{-1} K^{-1} for the snow-like $MoSe_2$ film, and g ~ 3.2278 ± 0.58565 MW m^{-2} K^{-1} and k ~ 41 ± 7 Wm^{-1} K^{-1} for the hexagonal $MoSe_2$ film.

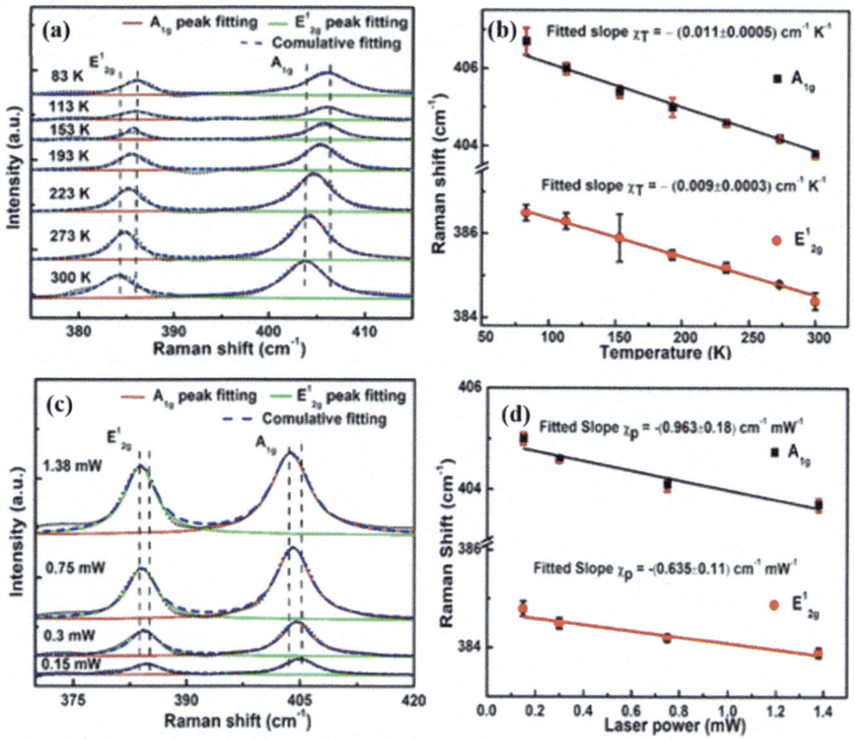

Fig. 4.6 **a** Temperature dependent Raman spectra, **b** corresponding Raman shift with temperature. **c** Power dependent Raman spectra, **b** corresponding Raman shift with temperature of triangular shaped MoS_2 film over SiO_2/Si film. Adapted with permission from [23]. Copyright 2023 ACS publishing

These findings highlight the thermal transport characteristics of the respective morphologies and provide valuable insights for thermal management applications in $MoSe_2$-based systems. Yue et al. performed Raman measurement with an electrical heating technique to study the thermal contact resistance between SiC and graphene [28]. This method involves using a heat source called Joule heating, which required an electrical connection to heat the graphene (middle in Fig. 4.7a). By applying uniform steady-state Joule heating, the temperature of the graphene increased uniformly. The heat was subsequently transferred into the substrate, causing a temperature difference between the graphene and its substrate.

Raman spectrum was measured for both Raman peak of graphene and SiC substrate simultaneously (upper in Fig. 4.7a) due to the transparency of graphene. The Raman spectra of the substrate had to be sensitive to temperature fluctuations in order to determine the temperature of substrate. Through each Raman thermometry experiment, the correlations between temperature and power have been identified (Fig. 4.7b). Raman thermometry allowed for the independent determination of the

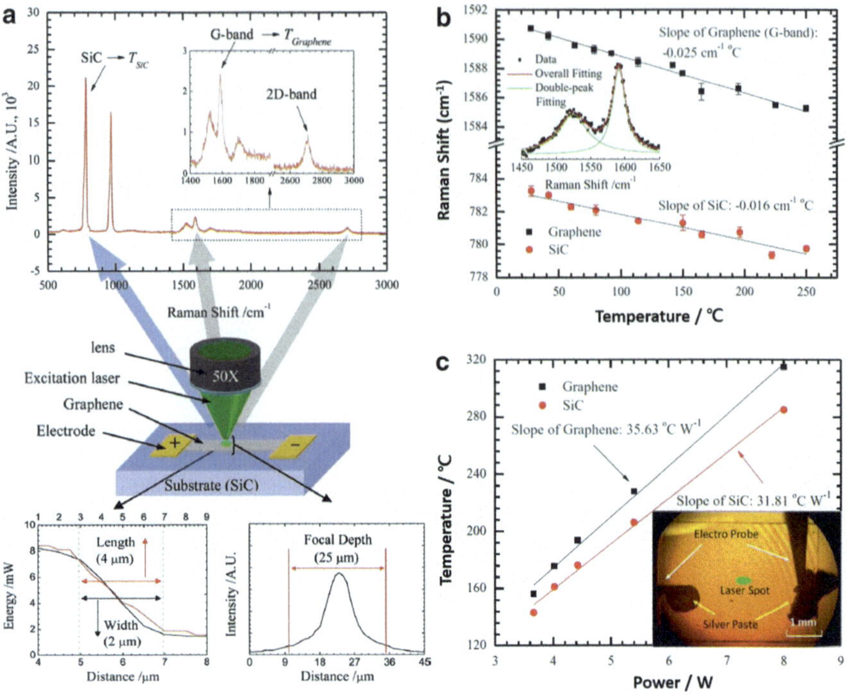

Fig. 4.7 Electrical heating and Raman thermometry. **a** Experimental schematic showing the Raman thermometry with electrical heating (middle) and the Raman spectrum of graphene and its substrate (upper). **b** Linear temperature dependent Raman peak for graphene and SiC substrate. **c** The relationship between Raman power and temperature of the graphene and SiC substrate. Adapted with permission from [28]. Copyright 2011 Small publishing

temperatures of the graphene (T_g) and the SiC substrate (T_{sic}) (Fig. 4.7c). The electrical resistance (R) and controlled applied current (I) could be used to compute the heat produced by the electrical circuit. As a result, it was possible to obtain the thermal contact resistance (R_{tc}), which is provided by a specific formula.

$$R_{tc} = \frac{T_g - T_{sic}}{I^2R/A} \tag{4.16}$$

where A represents the interfacing area. In contrast to the OTR technique, the utilization of Joule heating offers certain advantages. With Joule heating, samples can be uniformly heated, ensuring a more controlled and consistent temperature profile throughout the material. By controlling the electrical current, the power input can be precisely adjusted, allowing for accurate control of the heating power. OTR with electrical heating is a technique commonly employed to obtain thermal contact resistance in 2D materials like MoS_2 and h-BN.

4.3.5 Phonon Anharmonicity by Optothermal Raman Technique

Understanding the thermal expansion coefficient is essential for designing optoelectronic devices based on 2D transition metal dichalcogenides (TMDCs) in order to prevent problems with heat dissipation, internal stress/strain, thermal mismatch, and self-heating of the devices. Raman spectroscopy is most useful technique for characterizing 2D nanostructured materials, can provide valuable insights in this regard. By analyzing the first-order temperature coefficients of Raman modes, the thermal conductivities of TMDCs can be derived as discussed in previous sections. These coefficients offer information about how the lattice constant changes as a function of temperature, allowing for a deeper understanding of thermal expansion. Additionally, Raman spectroscopy can provide insights into phonon-phonon coupling, including three- and four-phonon anharmonic coupling, which further contribute to the understanding of thermal transport properties. Temperature-dependent Raman spectroscopy is particularly useful in investigating the vibrational properties and electron-phonon coupling in TMDCs. Thermal conductivity of TMDCs based nanostructured materials is denoted as k, is a fundamental physical parameter that can be expressed as the sum of thermal conductivity (k_l) due to the lattice and thermal conductivity (k_e) due to the electron i.e. $k = k_l + k_e$. In TMDCs, the contribution of ke to k is often negligible due to the low carrier concentration. Therefore, k_l dominant in thermal transport. It is measured by the harmonic vibration of the lattice and anharmonic phonon scattering processes. The OTR technique provides direct insights into the optical phonons near the center of the Brillouin zone and allows the observation of lattice anharmonicity. The frequency shift and variation in the full width at half-maximum (FWHM) of Raman peaks can reveal the presence of lattice anharmonicity. This technique has been widely employed to examine the phonon anharmonicity in various materials, including silicon, graphene, black phosphorus, MoS_2, $MoSe_2$, WS_2, and WSe_2. In summary, Raman spectroscopy, particularly temperature-dependent Raman scattering, is a powerful technique for characterizing the thermal properties of TMDCs. It provides information on the thermal expansion coefficient, lattice anharmonicity, and phonon-phonon interactions, all of which are crucial for understanding and addressing heat dissipation challenges in next generation electronic devices based on TMDCs.

For better understanding of nonlinear temperature dependence observed in Raman modes frequencies, a semiquantitative model is employed. This model considers both pure volume effect and temperature effect, which arise from thermal expansion and phonon-phonon coupling (specifically, three- and four-phonon scattering processes). Mathematically, this model can be represented as [29–31]

$$\omega(T) = \omega_0 + \Delta\omega_A + \Delta\omega_E \tag{4.17}$$

The model combines the contributions from phonon-phonon coupling and thermal expansion to explain the observed temperature dependence of the Raman mode. By incorporating these effects, it aims to provide a comprehensive understanding of the nonlinear behavior observed in Raman spectra.

where, $\Delta\omega_E$ and $\Delta\omega_A$ stand for pure volume and temperature effect produced by Raman shifts respectively. Additionally, Grüneisen constant model is used to show how volume expansion affects the resulting Raman shift [29–31]

$$\Delta\omega_E(T) = \omega_0 \exp\left(-n\gamma \int_0^T \alpha_T dT\right) - \omega_0 \tag{4.18}$$

where n is the degeneracy Raman mode. Here, α_T is the thermal expansion coefficient, ω_0 is the extrapolated frequency at zero kelvin, and γ is the Grüneisen parameter. The negative sign in the previous equation shows how the volume expansion decrease vibration frequency.

The product of thermal expansion coefficient and Grüneisen parameters expressed as in terms of polynomial function of temperature [29–31]

$$\gamma\alpha_T = a + bT + cT^2 \tag{4.19}$$

with a, b, and c as fitting constants. The pure temperature effect, primarily arising from the anharmonic behaviour due to the three- and four-phonon scattering processes, contributes to the observed Raman shift. These scattering processes involve the emission absorption and scattering of photons, leading to the generation of two or three phonons. The pure temperature effect can be expressed using a semiquantitative model, which considers the nonharmonic behavior induced by these scattering processes. By incorporating this semiquantitative model, the contribution of the nonharmonic behavior arising from three- and four-phonon scattering phenomena to the temperature-dependent Raman shift can be better understood. This provides basic idea of temperature-dependent vibrational properties of the material and allows for a more comprehensive analysis of the observed Raman spectra [31].

$$\Delta\omega_A(T) = A\left(1 + \frac{2}{e^x - 1}\right) + B\left(1 + \frac{3}{e^y - 1} + \frac{3}{(e^y - 1)^2}\right) \tag{4.20}$$

where, $x = \frac{\hbar\omega}{2k_B T}$ and $y = \frac{\hbar\omega}{3k_B T}$, T is the temperature (in Kelvin), k_B is the Boltzmann constant, and coefficients A and B represent for the individual contributions of the three- and four-phonon processes, respectively. These equations are further used to fit the nonlinear temperature dependency of Raman shifts. Figure 4.8 is the demonstration for the thermal expansion coefficient of MoS_2 is sturdily correlated with the number of layers, which can only be clearly exhibited after isolating it the from the substrate effect. Then, the peak positions for Raman modes (E_{2g} and A_{1g}) of MoS_2 were fitted as a function of temperature to obtain the regularities of the temperature dependence of the peak positions. The temperature evolution of Raman peaks for the supported MoS_2 shows a nearly linear behavior as a function of temperature. On the other hand, the temperature evolution for the suspended MoS_2 cannot be well fitted using a linear function, in which large discrepancies between the linearly fitted curves and experimental results are observed.

In our recent work, we conducted a temperature-dependent Raman analysis to examine various properties such as phonon-phonon interactions and electron-phonon. The measurements were performed using a long working distance 50X LWD objective lens by varying the temperature from 93 to 300 K [27]. The temperature-dependent Raman spectra were obtained for both snow-like and hexagonal $MoSe_2$ nanostructured films on SiO_2/Si substrates at different temperatures. Figure 4.9a, c illustrate the Raman spectra of the snow-like and hexagonal $MoSe_2$ nanostructured films, respectively. To analyze the behavior of the Raman peaks with temperature, we examined the shifting of peak position of the A_{1g} modes for both morphologies of $MoSe_2$ films. Figure 4.9b, d display the associated shifts in the peak position with temperature. We analysed the linear and nonlinear responses of the Raman shifts using the following equations: $\omega(T) = \omega_0 + \chi T$ for linear response, and $\omega(T) = \omega_0 + \chi_1 T + \chi_2 T^2$ for nonlinear response. Here, ω_0 and $\omega(T)$ represent the Raman mode frequency at absolute zero temperature (0 K) and at any temperature (T), respectively.

The χ represents the slope of first-order thermal coefficient. By applying these equations, we can quantitatively analyze the linear and nonlinear responses of the Raman shifts observed in the snowlike and hexagonal $MoSe_2$ films, providing insights into their thermal properties and behavior at different temperatures. Through our analysis, we have determined that a polynomial fit provides a better representation for the nonlinearity in thermal response of the Raman mode in snow-like $MoSe_2$. This implies that the anharmonic effect produced by the phonon-phonon interaction is the major contributing to the observed nonlinear behavior. To quantitatively evaluate the nonlinear thermal response of $MoSe_2$, we employ a semi-quantitative model and fit it to (4.17)–(4.20). By fitting these equations, we can obtain a comprehensive understanding of the relationship between temperature and the Raman mode in snowlike $MoSe_2$. This approach allows us to accurately assess the influence of anharmonic effects (Fig. 4.10) and phonon-phonon coupling on the thermal response observed in $MoSe_2$.

Singh et al. measure layer-dependent thermal transport properties i.e., thermal conductivity, interfacial conductance, and phonon anharmonicity for CVD-grown MoS_2 thin film using OTR method. They showed that the thermal conductivity decreases while interfacial thermal conductance increases with increasing layer number. They also discussed that the higher-order scattering decreases as the layer number increases. Thermal conductivities of CVD-grown (1L,3L, and 5L) MoS_2 are found to be 50 ± 2, 34 ± 2, and 28 ± 2 W m^{-1} K^{-1}, respectively [33]. The shifting in Raman peaks with temperature variation corresponding to E^1_{2g} and A_{1g} modes for 1L, 3L and 5L MoS_2 are shown in Fig. 4.11. The recent published work based on optothermal Raman spectroscopy by Li et al. used a physical model with involving temperature effect and volume effect for understanding the nonlinear dependence on temperature of E^1_{2g} Raman mode in case of 1L–5L WSe_2 supported with SiO_2/

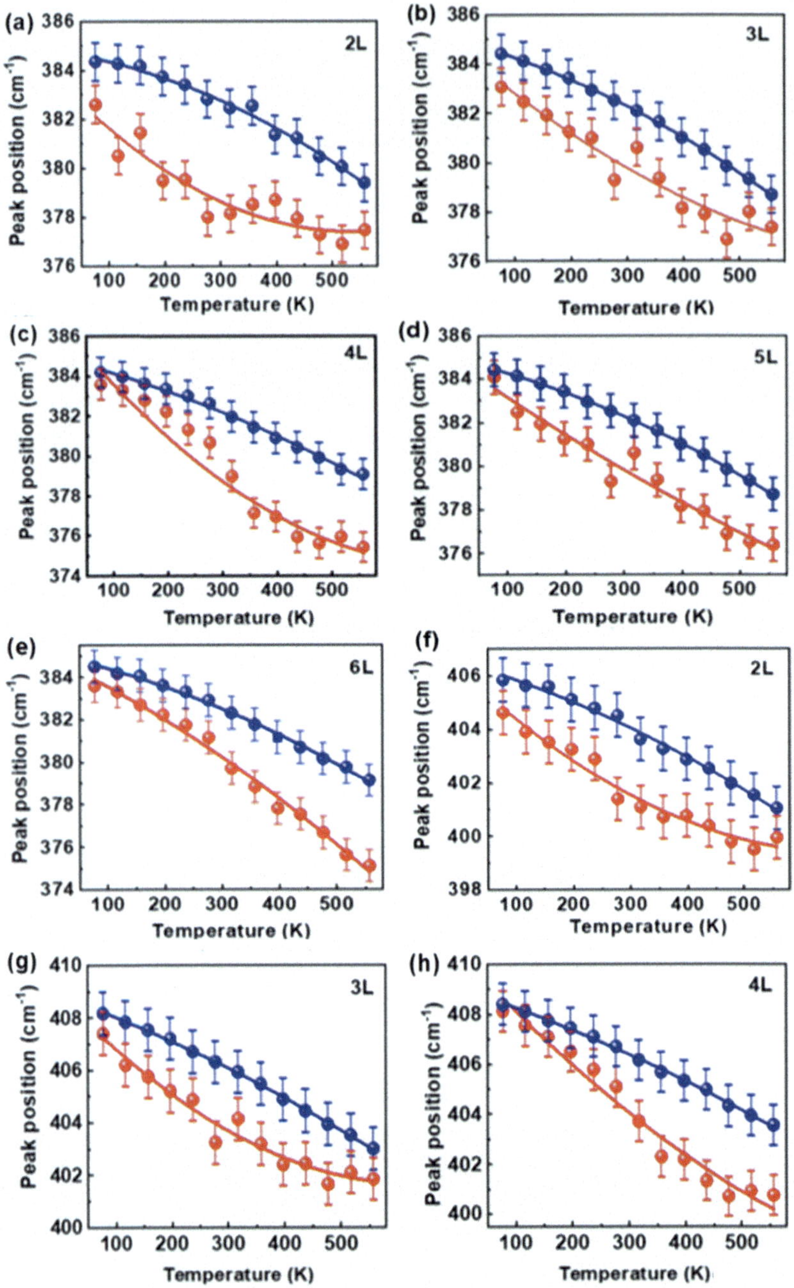

Fig. 4.8 Positions of the Raman peaks. **a–e** E_{2g} and, **f–j** A_{1g} modes for supported and suspended MoS_2 with various numbers of layers as a function of temperature. The experimental findings of suspended and supported MoS_2 are shown by the red and blue spheres, respectively. This curve is fitting via a polynomial function (second-order) of temperature are represented by the blue and red lines. Adapted with permission from [33]. Copyright 2021 Scientific reports

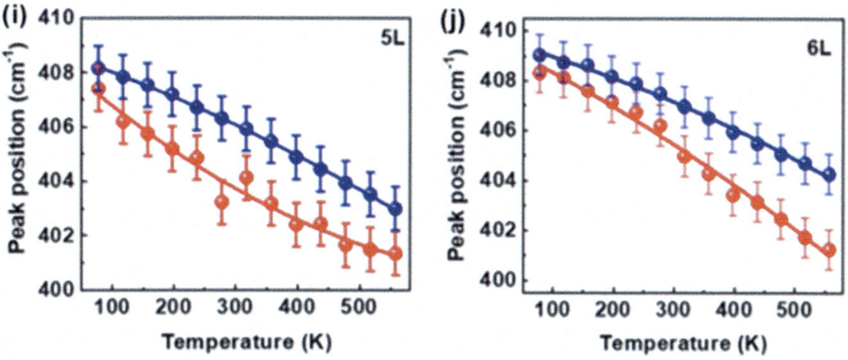

Fig. 4.8 (continued)

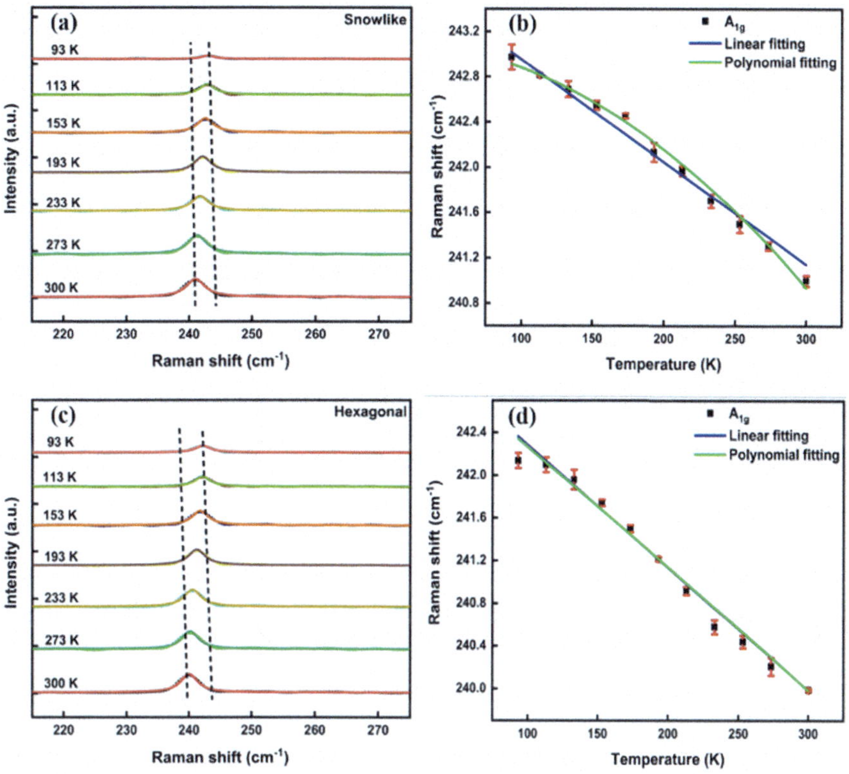

Fig. 4.9 Raman spectra of **a** snow-like and **c** hexagonal $MoSe_2$ with variation of temperature; corresponding Raman shifts in **b** snow-like and **d** hexagonal $MoSe_2$ nanostructured film with variation of temperature using LWD 50X objective lens. Adapted with permission from [27]. Copyright 2023 Royal Society of Chemistry

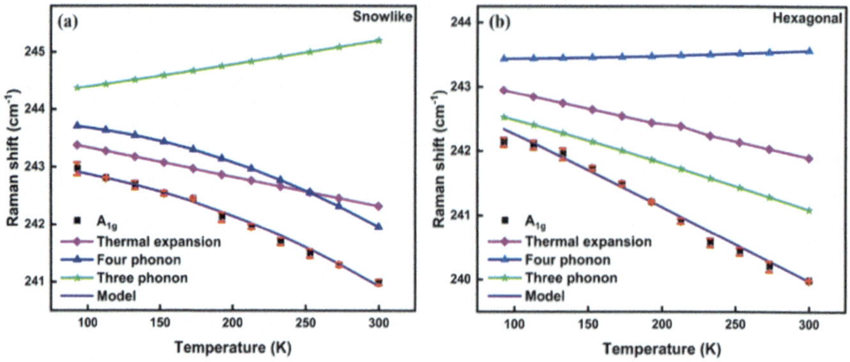

Fig. 4.10 To pretend the temperature dependence of the Raman shift in both snowlike and hexagonal $MoSe_2$ films, we consider the specific contributions from the procedures of four-phonon, three-phonon interactions, and thermal expansion. **a** For the snow-like $MoSe_2$ nanostructured film **b** for the hexagonal $MoSe_2$ nanostructured film. Adapted with permission from [27]. Copyright 2023 Royal Society of Chemistry

Si substrate, as shown in Fig. 4.12 [34]. They observed that this nonlinear dependency of Raman modes basically generates from three phonon scattering and thermal expansion effect.

As the number of layers increases the phonon scattering effect decreases but thermal expansion effect increases. The temperature dependent Raman study from 133 to 413 K and full width at half maxima was studied systematically corresponding to E^1_{2g} mode. It is found that with decaying E^1_{2g} phonon, the FWHM decreases. Here the anharmonic interaction between phonons was promoted in WSe_2 flakes with different thickness. The physical quantitative model was used to explain the three phonon-four phonon scattering and nonlinear dependency on temperature. The contribution of anharmonicity and pure volume effect in 1L-5L WSe_2 is shown in Fig. 4.12.

4.4 Conclusions and Future Perspective

In this chapter, we discussed different methods for examining the thermal properties of 2D materials mainly focussing on optothermal Raman (OTR) technique. Here, we discussed the measuring process in OTR technique and explained how to use different thermodynamical models to calculate thermal conductivity and interfacial thermal conductance of 2D materials using OTR technique. We have discussed examples on thermal transport studies in graphene and TMDCs. Additionally, the phonon anharmonicity has also been explained using OTR technique and appropriate models for better understanding of nonlinear thermal transport behavior (higher order phonon scattering) of 2D materials. In the end this OTR study provides understanding for

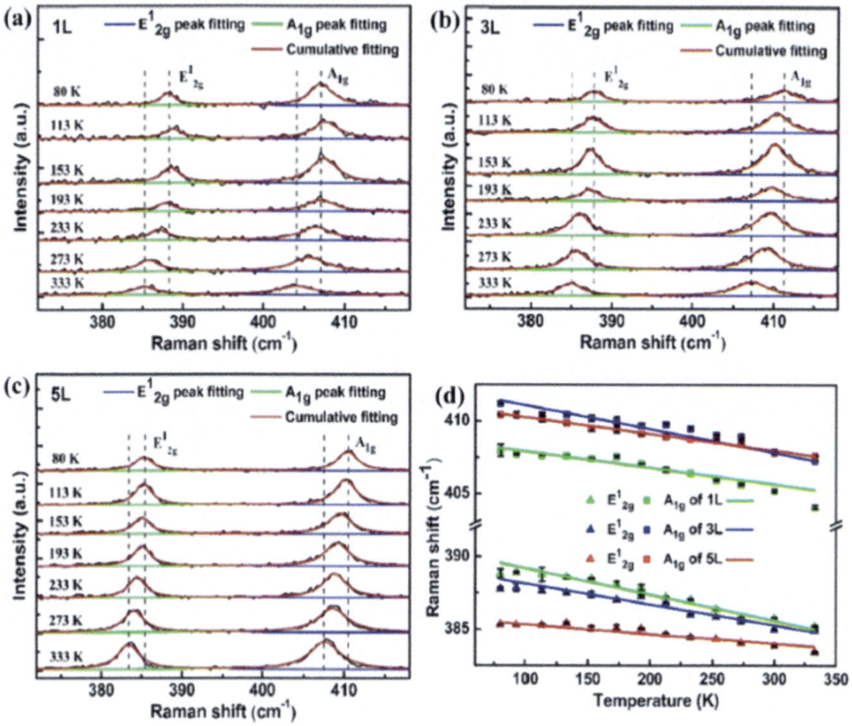

Fig. 4.11 Temperature-dependent phonon response using 50X LWD lens for triangular, **a** 1L, **b** 3L, and **c** 5L CVD-grown MoS_2 on the SiO_2/Si substrate. **d** Variations of the peak position for E^1_{2g} and A_{1g} Raman modes with temperature for 1L, 3L, and 5L triangular MoS_2. Adapted with permission from [33]. Copyright 2023 ACS publishing

thermal behavior of 2D materials in different electronic and optoelectronic devices, which can be crucial for thermal management in those devices.

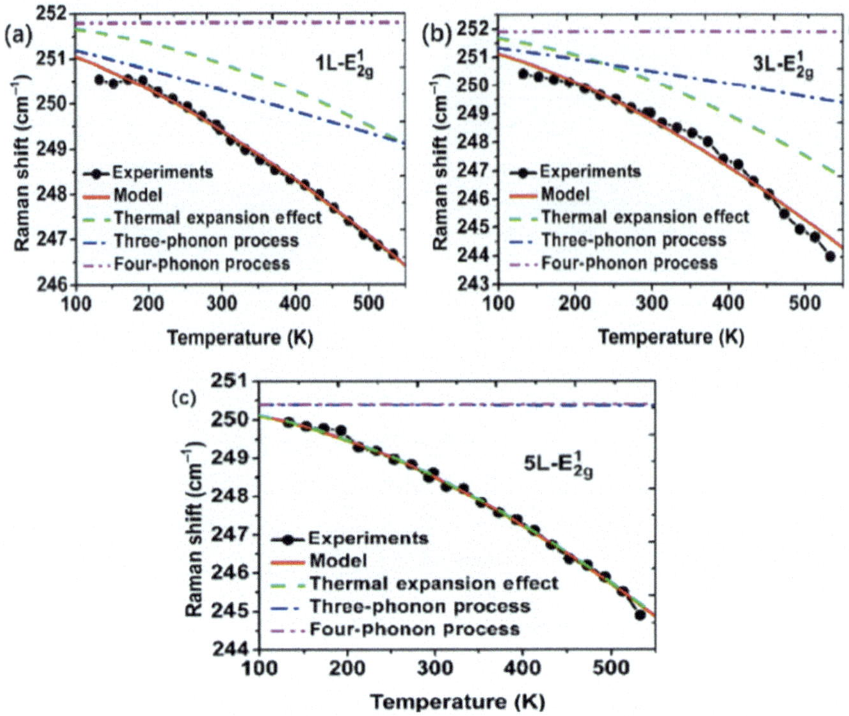

Fig. 4.12 a–c The experimental results for temperature dependent Lorenzen fitting of E^1_{2g} Raman mode shifts in 1L, 3L and 5L WSe_2 (scattered filled black circles). The solid red lines represent the theoretical results for both temperature effect and volume effect. Adapted with permission from [34]. Copyright 2020 springer publishing

References

1. W.Y. Chen, X.L. Shi, J. Zou, Z.G. Chen, Mater. Sci. Eng. R Rep. **151** (2022)
2. D. Li, Y. Gong, Y. Chen, J. Lin, Q. Khan, Y. Zhang et al., J. Mater. Chem. A **12**, 1–40 (2020)
3. Z. Dong, H. Xu, F. Liang, C. Luo, C. Wang, Z.Y. Cao et al., J. Mater. Chem. A **24**(1), 88 (2018)
4. S. Mosca, C. Conti, N. Stone, P. Matousek, Nat. Rev. Methods Primers **1**(1), 21 (2021)
5. Q.Y. Li, X. Zhang, Y.D. Hu, Thermochim. Acta **592**, 67–72 (2014)
6. T. Wang, S. Xu, D.H. Hurley, Y. Yue, X. Wang, Opt. Lett. **41**, 80–83 (2016)
7. P. Yuan, R. Wang, H. Tan, T. Wang, X. Wang, ACS Photon. **4**, 3115–3129 (2017)
8. H. Zobeiri, R. Wang, T. Wang, H. Lin, C. Deng, X. Wang, Int. J. Heat Mass Transfer **133**, 1074–1085 (2019)
9. Q.Y. Li, X. Zhang, K. Takahashi, Int. J. Heat Mass Transfer **125**, 1230–1239 (2018)
10. Y. Zhang, A. Fan, S. Luo, H. Wang, W. Ma, X. Zhang, Int. J. Heat Mass Transfer **145**, 118795 (2019)
11. J.M. Ziman, *Electrons and Phonons: The Theory of Transport Phenomena in Solids, Oxford Classic Texts in the Physical Sciences*, online edn. (Oxford Academic, Oxford, 2001, 2007), pp. 45–56
12. N.W. Ashcroft, N.D. Mermin, *Solid State Physics* (Holt, Rinehart and Winston, New York, 1978), pp.45–56

13. K. Esfarjani, G. Chen, H.T. Stokes, Phys. Rev. B **84**, 085204 (2011)
14. J. Garg, N. Bonini, B. Kozinsky, N. Marzari, Phys. Rev. Lett. **106**, 045901 (2011)
15. X. Gu, Y. Wei, X. Yin, B. Li, R. Yang, Rev. Mod. Phys. **90**, 041002 (2018)
16. H. Dai, R. Wang, Nanomaterials **12**, 589 (2022)
17. D. Zhao, X. Qian, X. Gu, S.A. Jajja, R. Yang, J. Electron. Packag. **138**, 040802 (2016)
18. D.G. Cahill, R.O. Pohl, Phys. Rev. B **35**, 4067 (1987)
19. C.A. Paddock, G.L. Eesley, J. Appl. Phys. **60**, 285–290 (1986)
20. A.A. Balandin, S. Ghosh, W. Bao, I. Calizo, D. Teweldebrhan, F. Miao, C.N. Lau, Nano Lett. **8**, 902–907 (2008)
21. W. Cai, A.L. Moore, Y. Zhu, X. Li, S. Chen, L. Shi, R.S. Ruoff, Nano Lett. **10**(5), 1645–1651 (2010)
22. J. Judek, A.P. Gertych, M. Świniarski, A. Łapińska, A. Dużyńska, M. Zdrojek, Sci. Rep. **5**, 12422 (2015)
23. B.P. Majee, J.D. Gupta, A. Sanskrityayn, A.K. Mishra, J. Phys. Chem. C **125**, 14865–14873 (2021)
24. R. Yan, J.R. Simpson, S. Bertolazzi, J. Brivio, M. Watson, X. Wu et al., ACS Nano **8**, 986–993 (2014)
25. B.P. Majee, Bhawna, A. Singh, R. Prakash, A.K. Mishra, J. Phys. Chem. Lett. **11**, 1268–1275 (2020)
26. A. Taube, J. Judek, A. Łapińska, M. Zdrojek, ACS Appl. Mater. Interfaces **7**, 5061–5065 (2015)
27. J.D. Gupta, P. Jangra, B.P. Majee, A.K. Mishra, Nanoscale Adv. **5**, 2756–2766 (2023)
28. Y. Yue, J. Zhang, X. Wang, Small **7**, 3324–3333 (2011)
29. Z. Lin, W. Liu, S. Tian, K. Zhu, Y. Huang, Y. Yang, Sci. Rep. **11**, 1–9 (2021)
30. S. Najmaei, P.M. Ajayan, J. Lou, Nanoscale **5**, 9758–9763 (2013)
31. P.G. Klemens, Phys. Rev. **148**, 845 (1966)
32. Z. Lin, W. Liu, S. Tian, K. Zhu, Y. Huang, Y. Yang, Sci. Rep. **11**, 7037 (2021)
33. A. Singh, B.P. Majee, J.D. Gupta, A.K. Mishra, J. Phys. Chem. C **127**, 3787–3799 (2023)
34. Z. Li, Y. Wang, J. Jiang, Y. Liang, B. Zhong, H. Zhang, M. Zou, Nano Res. **13**, 591–595 (2020)

Chapter 5
Polarized Raman Spectroscopy

Sunil Kumar Srivastava and Pankaj Kumar Kushwaha

5.1 Introduction

Raman spectroscopy is a versatile and non-destructive analytical tool that has found applications in a wide range of scientific and industrial fields, contributing to our understanding of molecular structures and interactions [1–8]. In material science, the characterization techniques for determining orientation have proven critical in understanding the link between structure and performance. X-ray diffraction (XRD), Raman spectroscopy, transmission electron microscopy (TEM), scanning electron microscopy (SEM), and atomic force microscopy (AFM) are some of the methods that have been used to characterize microscopic structures [9–11]. Among these techniques, Raman spectroscopy is a quick, efficient, and nondestructive technique that has been used frequently to characterize materials. Raman spectroscopy has revolutionized our understanding of molecular structures and vibrational modes. However, in many cases, additional information is required to fully elucidate molecular symmetry and anisotropic properties. This is where Polarized Raman Spectroscopy comes into play, offering a deeper insight into molecular orientations and anisotropy.

The foundation of Raman spectroscopy lies in the inelastic scattering of light, which reveals molecular rotations and vibrations. When coupled with polarized light, a new dimension of information becomes accessible. By introducing specific polarization orientations for both incident and scattered light, we can glean insights into the orientation of molecular vibrations and the symmetry of molecules. Polarized Raman spectroscopy provides vital information about a samples inherent molecular structure as well as functioning as an effective probe for investigating orientation in highly ordered systems such as crystals, oriented polymers, and carbon nanotubes [12–14]. In this technique, the polarized light is used to excite vibrations in molecules

S. K. Srivastava (✉) · P. K. Kushwaha
Department of Physics, Mahatma Gandhi Central University, Motihari, Bihar 845401, India
e-mail: sksrivastava@mgcub.ac.in

D. K. Singh et al. (eds.), *Raman Spectroscopy*, Springer Series in Optical Sciences 248,
https://doi.org/10.1007/978-981-97-1703-3_5

and then detecting and analyzing the scattered light to obtain information about the sample. By varying the polarization of the incident and scattered light, it is possible to probe different molecular orientations and anisotropic properties of the sample. Polarized Raman spectroscopy finds applications across various scientific and industrial fields, offering unique insights into molecular symmetry, anisotropic materials, and crystallography [15]. It can be used to study the orientation and alignment of molecules in polymers, the conformational changes in proteins, and the crystallographic orientation of minerals. Recently, significant advances have been made in the instrumentation and methodology of polarized Raman spectroscopy. For example, using confocal microscopy and Raman imaging has enabled high-resolution mapping of molecular orientation and anisotropic properties in complex materials [16–19]. Additionally, the development of polarization-resolved Raman spectroscopy techniques [20], such as polarization-dependent resonance Raman spectroscopy, has enabled the study of electronic excitations and charge transfer processes in materials. In this chapter, we focus the principles, techniques, and potential applications of Polarized Raman Spectroscopy (Fig. 5.1).

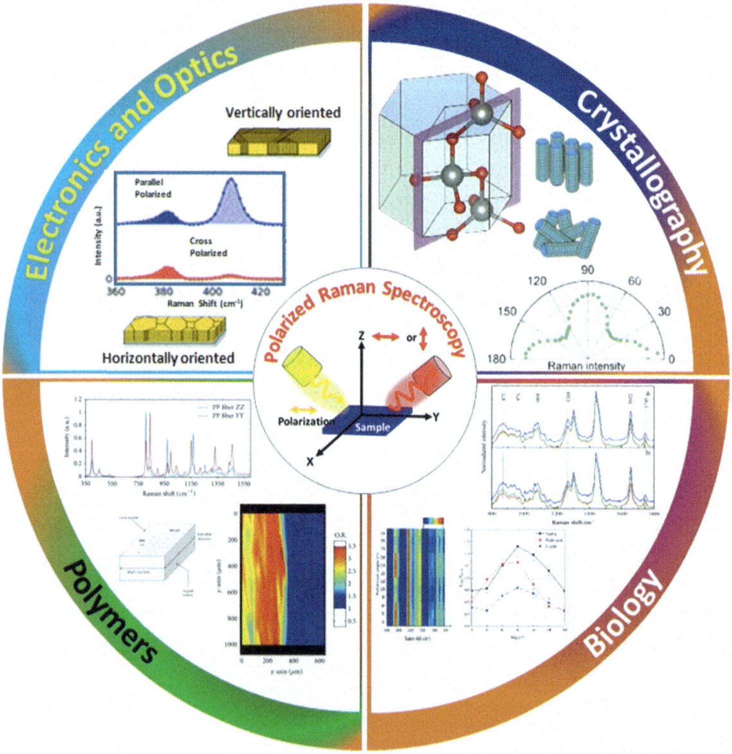

Fig. 5.1 Potential application of polarized Raman spectroscopy. Adapted from Vandalon et al. [21] (CC BY-NC 3.0). Adapted from Thyr et al. [22] (CC BY-NC 4.0). Adapted from Chen et al. [23] (CC BY 3.0), and adapted from Shang et al. [24] (CC BY 4.0)

5.2 Polarization of Light

Polarization of light refers to the orientation of the oscillations of the electric field vector as light propagates through space. Understanding polarization is essential in various fields, including optics, physics, and technology, where it plays a crucial role in manipulating light for various applications. Light waves are transverse waves, which means that the electric and magnetic field oscillations occur perpendicular to the wave propagation's direction. Natural or unpolarized light consists of electromagnetic waves with electric field oscillations occurring in random orientations perpendicular to the direction of propagation. In other words, the oscillations are in all possible planes perpendicular to the light's path. However, when light is polarized, the electric field oscillations occur in a specific plane or direction. There are different types of polarization, each with its own characteristics and applications. Here are some of the main types of polarization (Fig. 5.2).

Linear Polarization: In linearly polarized light, the electric field oscillates in a single plane. Light waves that pass through certain filters or materials can be selectively polarized in a specific direction.

Circular Polarization: Circularly polarized light consists of two electric field components that are perpendicular to each other, with equal amplitudes, but have a phase difference of $\pi/2$. It can be clockwise (right-hand) or counterclockwise (left-hand) circular polarization.

Elliptical Polarization: Elliptically polarized light consists of two electric field components that are perpendicular to each other, with unequal amplitudes, but have a phase difference of $\pi/2$.

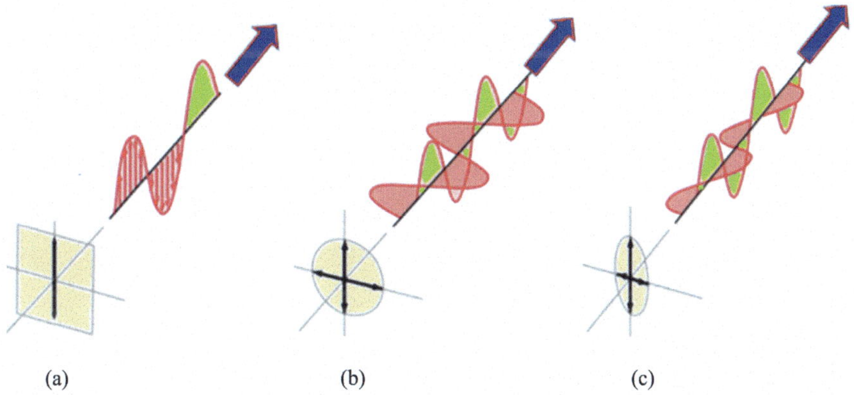

Fig. 5.2 Types of polarization of light. **a** Linear polarized, **b** circular polarized, and **c** elliptical polarized. Adapted from Nadeem et al. [25] (CC BY 4.0)

5.3 Isotropic and Anisotropic Samples

In many real-world scenarios, materials can have mixed behavior, showing some degree of isotropy in some properties and anisotropy in others. Engineering designs, simulations, and experiments often need to account for these properties to ensure the best performance and reliability of the materials being used. Therefore, understanding the isotropic or anisotropic nature of the material is crucial for accurate analysis and predictions as it directly influences how the material will perform and react under different conditions. Let's delve into what these terms mean.

Isotropic: An isotropic material is one that has the same physical properties in all directions. This means that if you were to measure a certain property (e.g., mechanical, thermal, electrical) of the material along any axis, you would obtain the same result. In other words, isotropic samples are those that do not have a defined orientation and the properties of an isotropic material are independent of the direction in which they are measured. Examples of isotropic materials include most gases and liquids, as well as some crystalline materials with a cubic crystal structure.

Anisotropic: An anisotropic material is one that has different physical properties in different directions. This means that the material behaves differently depending on the axis along which the measurement is taken. Anisotropy can arise due to the arrangement of molecules, atoms, or crystals within the material. Crystalline materials with non-cubic crystal structures (e.g., hexagonal, orthorhombic) often exhibit anisotropic properties. Wood, some minerals, and many engineering materials like composites, fibers, and layered materials are typically anisotropic.

In anisotropic samples, the molecules are fixed in specific positions and therefore their bonds can align with the polarization of the excitation light. When the bonds align with the polarization of the light, they interact more strongly with the light and produce a stronger Raman signal. This alignment of the bonds has a significant impact on the Raman peaks and their intensity. This property can be used to identify the different orientations of the molecules in the sample. The distinct spectra that are produced by anisotropic samples can offer valuable information about the sample. The spectra can be used to determine the crystal structure of the sample, the orientation of the molecules in the sample, and the presence of defects in the sample.

5.4 Polarized Raman Spectroscopy

Polarized Raman spectroscopy is an extension of traditional Raman spectroscopy that takes advantage of the polarization properties of light to gain additional information about molecular vibrations and the symmetry of a sample. In this technique, polarized light is used to probe the vibrational modes of molecules, and the resulting Raman scattering is analyzed based on its polarization characteristics. When light interacts with a molecule and undergoes Raman scattering, the scattered light can maintain the polarization of the incident light or change its polarization state. This polarization information can provide insights into the orientation and symmetry of molecular bonds and vibrations within the sample. By studying how the intensity and polarization of the scattered light change with different incident and detection polarizations, the information about the molecular structure, crystallographic orientation, and anisotropic properties of materials can be deduced. There are two main components to consider in polarized Raman spectroscopy.

Incident Polarization: The polarization state of the incident light can be adjusted using polarizers. By changing the angle of the incident polarization relative to the sample, certain vibrational modes can be selectively excited that are sensitive to specific polarization directions.

Detection Polarization: The scattered light can also be analyzed for its polarization properties. By using a polarizer in the detection pathway, the intensity of the Raman scattering in different polarization directions can be measured. This information helps in understanding the orientation of molecular vibrations.

Polarized Raman spectroscopy is particularly useful in the study of anisotropic materials, such as crystals and oriented molecular films, where the molecular vibrations are influenced by the crystallographic axes or orientation of the molecules [26–28]. It can provide insights into:

Molecular Symmetry: By analyzing the Raman scattering intensity in different polarization configurations, the symmetry properties of the vibrations can be determined. The relative intensities and polarization characteristics of the Raman bands provide valuable information about the molecular or crystal symmetry. which is necessary for understanding its structure and properties.

Anisotropic Properties: Polarized Raman spectroscopy is a versatile tool for studying anisotropic properties in a wide range of materials, including crystalline solids, polymers, liquid crystals, biological samples, and composite materials. Understanding anisotropic behavior is crucial for tailoring the properties of these materials for specific applications.

Sample Orientation: The orientation of the sample with respect to the incident and scattered light is crucial in polarized Raman spectroscopy. Depending on the angle and orientation of the sample relative to the polarized light, different Raman spectra can be obtained. This allows researchers to determine the orientation of specific molecular or crystal vibrations.

5.5 Experimental Set-Up: Polarized Raman Spectroscopy

The experimental set-up for polarized Raman spectroscopy (Fig. 5.3) involves controlling the polarization of the incident and scattered light to study the polarization properties of Raman-scattered light that results from the interaction between incident laser light and the sample. Here's a basic outline of an experimental setup for Polarized Raman spectroscopy.

Laser Source: A monochromatic laser source is used to illuminate the sample. Common lasers include those emitting light at visible or near-infrared wavelengths. The choice of laser depends on the sample and the desired Raman shift range.

Polarization Control: Polarizers and analyzers are used to control and measure the polarization state of light. Polarizer allows light of a specific polarization direction to pass through to interact with the sample. An analyzer, which is another polarizer placed in the path of scattered light to determine its polarization state. The analyzer's orientation can be adjusted to selectively pass light of a certain polarization direction. By rotating the analyzer, the intensity of scattered light at different polarization angles can be measured, which is essential for understanding the sample's polarization properties.

Sample Holder: The sample is securely placed on a sample stage. The sample can be in various forms, such as a solid, liquid, or gas. The sample holder may also allow for temperature control to study the temperature-dependent behavior of the material.

Optical Setup: The incident laser light is focused onto the sample using a microscope objective. The scattered light from the sample is collected by the same objective and directed to the monochromator.

Monochromator: A monochromator is a critical component of a Raman spectroscopy setup that is responsible for separating and selecting specific wavelengths (colors) of light from the scattered Raman spectrum. This separation allows the Raman-shifted frequencies, which correspond to vibrational modes of molecules in the sample, to be analyzed individually.

Detector: A sensitive detector, such as a charge-coupled device (CCD) or a photomultiplier tube (PMT), captures the dispersed Raman signals.

Data Analysis: The collected Raman spectra are analyzed using specialized software. By comparing the incident and scattered frequencies, the vibrational modes and molecular characteristics of the sample can be determined.

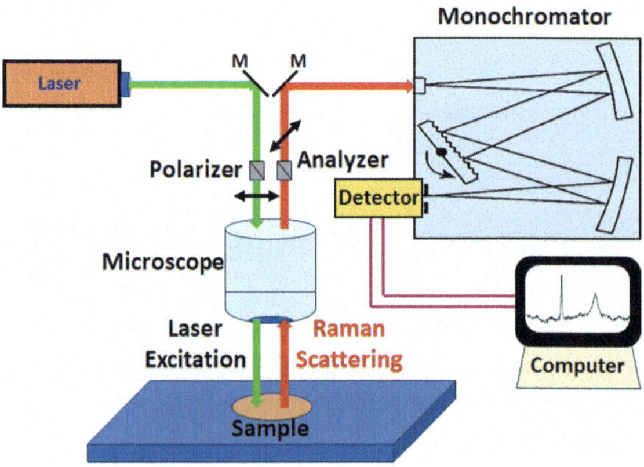

Fig. 5.3 Experimental set-up for polarized Raman measurement

5.6 Polarization Configurations

To ensure a consistent interpretation of the results when presenting polarization measurements, it is essential to keep track of the polarization circumstances. A method is developed to document the orientation of a sample in relation to the polarization of the incident light and the orientation of the outgoing light after the sample using the notation called Porto notation [29]. The notation consists of combination of four terms, that is as follows [30]:

$$\mathbf{X\,(A\,B)\,Y}$$

where, X and Y represent the direction of propagation of incident and outgoing light, respectively. Whereas, A and B represent the polarization direction of incident and outgoing light, respectively.

Figure 5.4 shows an example of an experimental setup and the corresponding Porto Notation in back-scattering geometry (180° configuration) using microscope. For isotropic sample, Raman spectrum is independent of the orientation of the sample with respect to the incident light. Examining the experimental setup in Fig. 5.3, there are four distinct combinations of excitation and scattering polarizations that are possible.

1. $\overline{Z}(XX)Z$
2. $\overline{Z}(XY)Z$
3. $\overline{Z}(YY)Z$
4. $\overline{Z}(YX)Z$.

In a polarized Raman spectroscopy setup, choice of polarization configuration depends on the material's properties and the specific information being sought from the Raman measurements. Two polarized measurements can be made, with the directions of the excitation and scattering polarizations being parallel and perpendicular, respectively.

Parallel Polarization (XX or YY): Incident and scattered light polarizations are parallel to each other. This configuration is used to probe vibrations polarized along the plane of the sample surface.

Perpendicular Polarization (XY or YX): Incident and scattered light polarizations are perpendicular to each other. This configuration is used to probe vibrations polarized out of the plane of the sample surface.

When dealing with anisotropic samples, it becomes crucial to document the samples orientation in relation to the stage or microscope. This can be accomplished by utilizing the coordinate system of the stage or sample (such as crystal structure), or by capturing a white light image of the sample if its orientation is unknown.

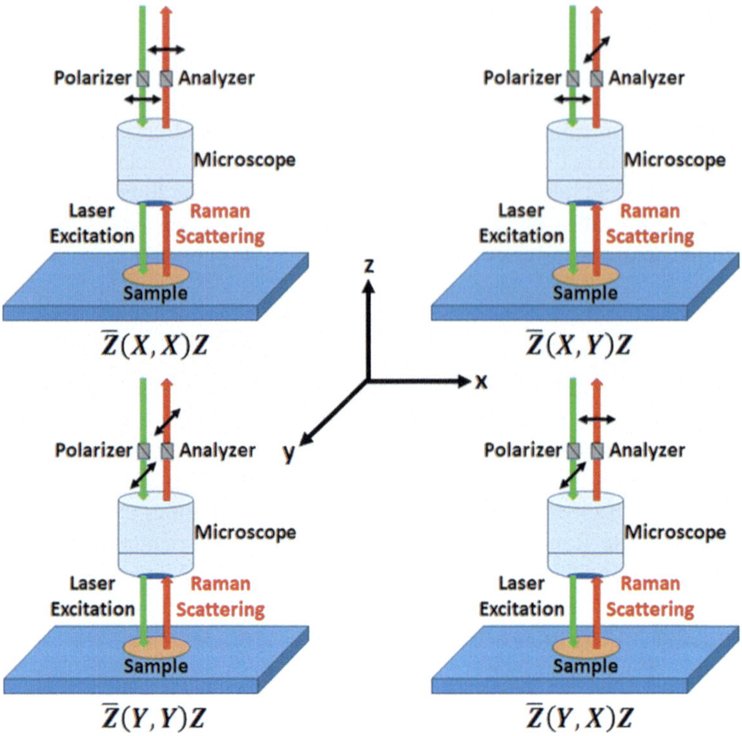

Fig. 5.4 Experimental set-up for polarization measurement using Porto notation

5.7 Depolarization Ratio

The depolarization ratio is an important parameter used to describe the polarization state of light as it interacts with various materials or substances. When light interacts with a medium, its polarization state can change. The depolarization ratio is a measure of the degree to which the initial polarization of light is changed after interacting with a material. The depolarization ratio in Polarized Raman Measurement is defined as the ratio of the intensity of Raman scattered light with a specific polarization state to the intensity of the incident polarized light [31, 32]. Depolarization ratio (ρ) is calculated using the following equation:

$$\rho = \frac{I_{\perp}}{I_{||}}$$

$I_{\perp}$ is the intensity of Raman scattered light with its polarization perpendicular to the plane of polarization of the incident light.

$I_{\parallel}$ is the intensity of Raman scattered light with its polarization parallel to the plane of polarization of the incident light. The depolarization ratio can take values between 0 and 1, where:

$\rho = 0$ indicates that the scattered light remains polarized and has the same polarization state as the incident light. This means that the Raman-active vibrational mode causing the scattering is symmetric along the direction of polarization.

$\rho = 1$ indicates that the scattered light is completely depolarized and has lost its initial polarization state. This suggests that the Raman-active mode is oriented in a direction perpendicular to the polarization of the incident light.

The depolarization ratio provides information about the symmetry properties of molecular vibrations. It helps to distinguish between different types of Raman-active vibrational modes and provides insights into the molecular structure and orientation of the sample [16, 33]. It is known that for a totally symmetric vibrational mode, ρ is less than 0.75. This is called a polarized band. If ρ is greater than or equal to 0.75, then the band is called a de-polarized band (i.e., not totally symmetric) [34, 35]. Totally symmetric vibrations are those in which the symmetry of the molecule does not change during the vibrational motion. For such vibrations, $\boldsymbol{\rho}$ values near zero are observed for highly symmetric molecules, such as CO_2, SF_6, and CCl_4. By measuring the depolarization ratio for various Raman bands, one can gain insights into the molecular orientation, symmetry, and anisotropy of the sample's vibrational modes. This information can be particularly useful in studying crystalline materials, polymers, and other complex molecular systems where molecular orientation and symmetry play a crucial role in their properties and behavior.

5.8 Applications

Polarized Raman spectroscopy finds applications across various scientific and industrial fields, offering unique insights into molecular symmetry, anisotropic materials, and crystallography [20, 28]. Here are some notable applications of Polarized Raman Spectroscopy:

5.8.1 Materials Science and Crystallography

Crystal Orientation Analysis: Polarized Raman spectroscopy is extensively used to determine the crystallographic orientation of materials [20]. It helps identify crystal axes and orientations in crystalline samples, aiding in material characterization and quality control.

Strain Analysis: By studying changes in vibrational modes under different polarization orientations, one can quantify strain in materials [36]. This is valuable in understanding mechanical properties and stress distribution in materials.

Phase Transitions: Polarized Raman spectroscopy is employed to investigate phase transitions in materials, providing insights into structural changes and transitions between different crystal phases [37].

5.8.2 Molecular Symmetry and Structural Analysis

Molecular Symmetry Determination: Polarized Raman spectroscopy is a powerful tool for identifying molecular symmetry groups and symmetry elements [16]. The analysis of polarization-dependent vibrational modes aids in characterizing the symmetry properties of molecules.

Biomolecular Studies: In biological systems, polarized Raman spectroscopy helps elucidate the orientation of biomolecules such as proteins, nucleic acids, and lipids [38]. It provides insights into their structural changes, interactions, and functions.

5.8.3 Optoelectronic and Photonic Devices

Anisotropic Semiconductors: Polarized Raman spectroscopy aids in the analysis of anisotropic properties in semiconductors, offering information about electronic band structures and optical anisotropy [39].

Liquid Crystals: Liquid crystal materials exhibit strong anisotropic properties. Polarized Raman spectroscopy is used to study their molecular orientation and phase transitions, crucial for liquid crystal display (LCD) technologies [40].

5.8.4 Pharmaceutical and Drug Development

Crystal Polymorphism: In pharmaceutical research, Polarized Raman Spectroscopy aids in identifying different polymorphic forms of drug molecules, which can affect their stability, solubility, and bioavailability [41].

Drug-Excipient Compatibility: Polarized Raman spectroscopy helps determine the compatibility between drugs and excipients, shedding light on their interactions and potential stability issues [42].

5.8.5 *Advanced Materials and Nanotechnology*

Nanomaterials: Polarized Raman spectroscopy is employed to study the orientation and properties of nanomaterials, such as nanotubes and nanoparticles, which often exhibit anisotropic behavior.

Photonic Crystals: Anisotropic photonic crystals are investigated using polarized Raman spectroscopy to understand their optical properties and interaction with light.

5.9 Examples

5.9.1 *Symmetry of Vibrational Modes in Isotropic Samples*

5.9.1.1 Cyclohexane (C_6H_{12})

The polarized Raman spectra of cyclohexane is shown in Fig. 5.5. The figure contains the parallel polarized ($I_{||}$) and perpendicular polarized ($I_\perp$) Raman spectra. It is evident from the spectra that the vibrational modes at 383 and 801 cm^{-1} whose intensity ratio or depolarization ratio ρ ($I_\perp / I_{||}$) is less than 0.75 are polarized and they are assigned to totally symmetric modes. The most prominent peak in the spectrum, located at 801 cm^{-1}, arises from the totally symmetric ring mode. On the other hand, the depolarization ratio ρ of the vibrational modes observed at 425 cm^{-1} is ≥0.75 shows that these modes are depolarized.

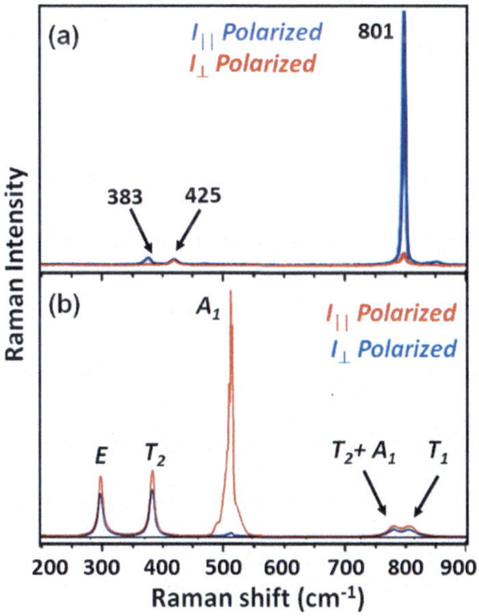

Fig. 5.5 Polarized Raman spectra of **a** Cyclohexane. Adapted from application note on polarized Raman spectroscopy of Edinburgh Instruments, retrieved from https://www.edinst.com/us/polarised-raman-spectroscopy/ [46] and **b** Polarized Raman spectra of CCl_4. Adapted from spectroscopyonline.com (2014), retrieved from https://www.spectroscopyonline.com/view/practical-group-theory-and-raman-spectroscopy-part-ii-application-polarization [31]

5.9.1.2 Carbon Tetrachloride (CCl_4)

Carbon tetrachloride (CCl_4) is a standard example for polarized Raman spectroscopy concept. Through group theory CCl_4 is described by the T_d point group [43]. There is only one totally symmetric stretching vibration for CCl_4, A_1, and the remaining vibrations for CCl_4 are asymmetric and transform as E and T symmetry. From the theory of the Raman effect [44]. It is well determined that the depolarization ratio, $\rho = I_{\perp}/I_{\parallel}$ where $I_{\perp}$ and $I_{\parallel}$ are the Raman intensities measured with a polarization perpendicular and parallel to the excitation polarization respectively, follows the relation of $0 \leq \rho < 0.75$ for totally symmetric vibrations and $\rho = 0.75$ for non-symmetric vibrations [45]. The CCl_4 Raman spectrum for parallel and perpendicular is shown in Fig. 5.4. The band peak at 459 cm^{-1} is the totally symmetric vibrational mode A_1 and is designated as totally polarized (i.e. $\rho = 0$), which means it produces Raman scattering in the same polarization orientation as the incident light [31].

The vibrational modes at 218, 314, 762 and 789 cm^{-1} are depolarized as we can see that the ratio of perpendicular to parallel Raman scattering intensity i.e., the depolarization ratio ρ is ≥ 0.75. The low vibrational energy modes at 218 and 314 cm^{-1} are attributed to bending modes of E and T_2 symmetry, respectively. The vibrational mode at 762 cm^{-1} is the combination of modes at 218 and 459 cm^{-1}, and is assigned to the $T_2 + A_1$ species. Similarly, the vibrational mode at 789 cm^{-1} is the combination of modes at 314 and 459 cm^{-1} is T_1 species. Combination and second order bands normally appear at energies slightly less than the values expected from addition or multiplication of the fundamental modes [47].

5.9.2 Characterization of Anisotropic Samples

5.9.2.1 Lithium Niobate ($LiNbO_3$)

Lithium niobate ($LiNbO_3$) is uniaxial crystalline material. It exhibits exceptional electro-optic, piezoelectric, and nonlinear optical properties. These properties make lithium niobate a crucial component in various technological applications, particularly in the field of photonics and electronics. Here again the polarized Raman spectroscopy is an important tool to characterized anisotropic material such as $LiNbO_3$. However, compared to the isotropic samples, the chemical bonds in crystals are all oriented in specific directions. As a result, while looking at a specific crystal face, particularly at an X, Y, or Z-axis face, one should not expect to see all of the Raman bands [48].

Figure 5.6 represent the polarized Raman spectra where the incoming laser beam is incident on the *X*-face of the crystal with the laser beam polarized along the Z-axis. Now keeping the analyzer parallel to the incident polarization selects the mode with A1 symmetry. Further, by rotating the analyzer by 90° with respect to the previous configuration selects the mode with E symmetry. The parallel and perpendicular polarized Raman spectra obtained from single crystals exhibit a striking difference.

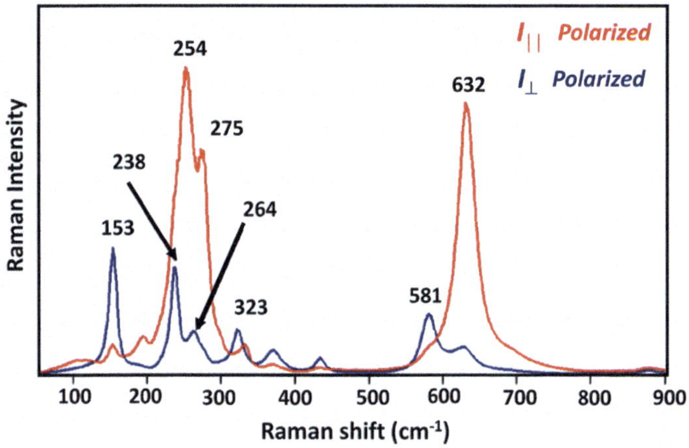

Fig. 5.6 Polarized Raman spectra of the *X*-face of single crystal $LiNbO_3$ with the incident laser polarized along the *Z*-axis. Adapted from spectroscopyonline.com (2014) [31], retrieved from https://www.spectroscopyonline.com/view/practical-group-theory-and-raman-spectroscopy-part-ii-application-polarization

Similarly, the symmetry properties of the anisotropic $LiNbO_3$ crystal can also be obtained by changing the direction of polarization of incident laser beam along Y-axis and the Raman spectra is obtained by keeping the analyzer once parallel to the incident laser beam polarization and then perpendicular to the laser beam polarization.

5.9.3 *Study of Raman Non-coincidence Effect*

Raman non-coincidence effect (NCE) is a phenomenon where the vibrational frequency of the isotropic and anisotropic components of Raman spectra are not coincide with each other. It is observed that in polar vibrational modes such as (C=O, C=S, etc.) the position of vibrational frequency of the isotropic and anisotropic components come at different position in Raman spectroscopy [15]. This phenomenon is attributed to transition dipole-transition dipole (TD-TD) interactions. The isotropic and anisotropic components of the Raman spectra from the polarized Raman measurement can be evaluated using the following relationships [15, 49]:

$$I_{iso} = I_{||} - \frac{4}{3} I_{\perp}$$

and

$$I_{aniso} = I_{\perp}$$

Figure 5.7a shows the example of NCE in the polarized Raman spectra of C=O stretching mode of neat acetonylacetone [50]. From the spectra, it is observed that there is a non-coincidence of ~8 cm^{-1} i.e., the difference between the isotropic and anisotropic Raman spectra in C=O stretching vibration. The NCE arises due to the vibration coupling between adjacent polar bonds which splits the degenerate vibrations into two vibrational modes one in phase vibration and other out of phase vibration. Figure 5.7b demonstrate the NCE in series of solvents having different static dielectric constant. It is shown that the NCE almost disappears in the solvent having high dielectric constant. This is because the solvent with high dielectric constant prevents the coupling or dimerization of acetonylacetone molecules.

5.9.4 Study of Thin Film Morphology

The electrical and optical properties of materials used in electronic devices are significantly influenced by their textures or morphology. Raman spectroscopy is a well-established technique for characterizing single-crystalline thin films, providing data on crystallinity and layer thickness, among other techniques. More specifically, the polarized Raman spectroscopy provide fundamental insight of the layer, both the local structure (e.g., shape, size) and morphology of the thin film surface. Vandalon et al. [21] has demonstrated the capability of polarized Raman spectroscopy to elucidate the texture of synthesized nanocrystalline MoS_2. Figure 5.8 presents the response of parallel polarized and cross polarized configuration to the horizontally and vertically aligned layers of MoS_2 on the surface.

The vertically oriented and horizontally oriented MoS_2 show a variation of approximately 3 times in the Raman intensity of A_{1g}/E^1_{2g}. The polarized Raman spectroscopy is used to determine the angular grain distribution and the monitoring the crystallization and texture of in-plane to out-of-plane film growth.

5.9.5 Study of Polymer Morphology

Polarized Raman spectroscopy is a valuable tool for characterizing polymer composition, crystallinity, orientation, and more, making it a crucial technique in polymer science and materials research. Chen et al. [23] has demonstrated the orientational mapping of the polypropylene based composite material using Polarized micro-Raman spectroscopy. The orientation of polymers within an oriented composite material with uniaxial orientation is characterize using the terms of orientation ratio (O.R.) by mapping the orientation across the entire cross-section of an oriented polymer composite. The O.R. is defined as:

$$O.R. = \frac{I_{1,ZZ}/I_{2,ZZ}}{I_{1,YY}/I_{2,YY}}$$

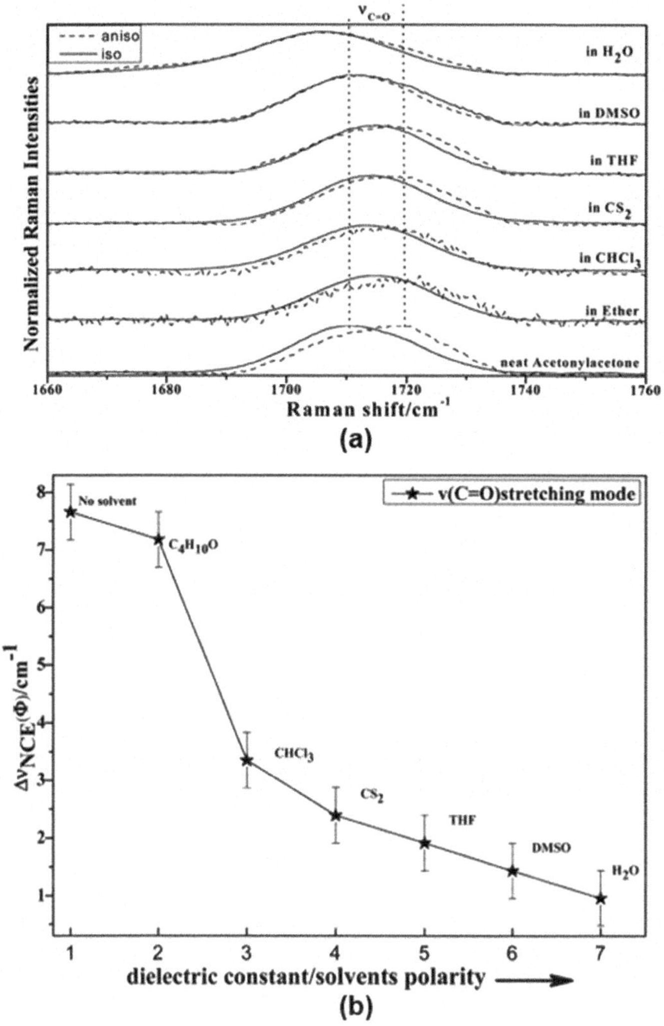

Fig. 5.7 **a** The isotropic and anisotropic parts of $\nu_{C=O}$ vibrational Raman spectra of acetonylacetone in the binary mixture with different solvents (equimolar); **b** variation in NCE of the C=O stretching mode of acetonylacetone as a function of the solvent dielectric constant. Adapted from Wang et al. [50] (CC BY-NC 3.0)

where I_1 corresponds to the intensity of a vibrational mode that is polarized while I_2 stands for the intensity of another vibrational mode that depolarized. The ZZ and YY in the subscript denote whether the spectrum is collected with the two polarizers oriented so that the electric field of the incoming laser beam and the outgoing signal beam are both parallel and perpendicular to the extrusion direction of the sample, respectively.

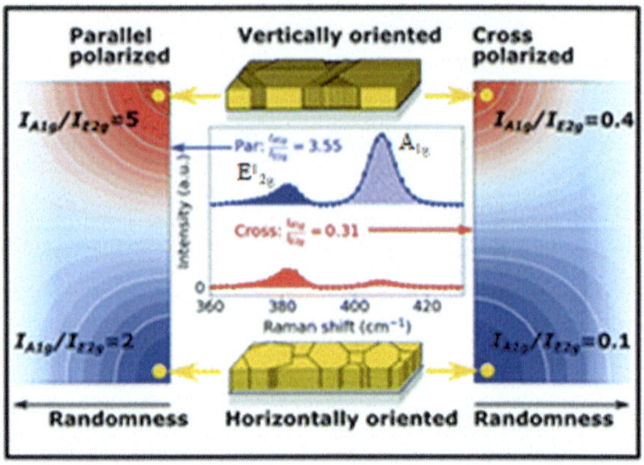

Fig. 5.8 The polarized Raman response of MoS_2 film measured in the parallel polarization (upper spectrum) and the cross polarized (lower spectrum) configuration. The mapping of the A_{1g}/E^1_{2g} ratio calculated for MoS_2 described by the Gaussian distribution for the parallel and cross polarized configuration. Adapted from Vandalon et al. [21] (CC BY-NC 3.0)

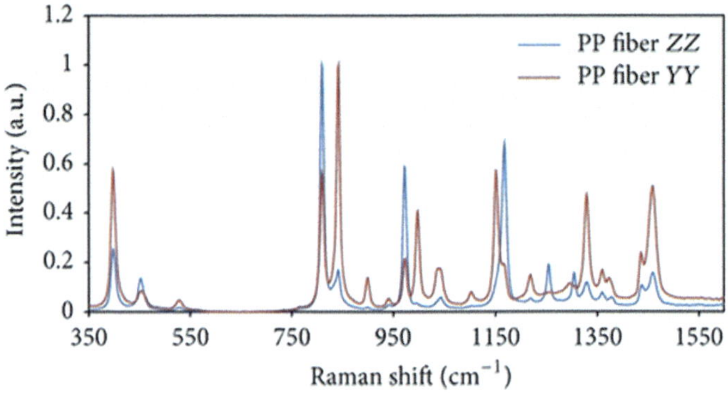

Fig. 5.9 Raman spectra polypropylene (PP) fiber Parallel (ZZ) and perpendicular (YY) to the Extrusion direction. Adapted from Chen et al. [23] (CC BY 3.0)

Figure 5.9 represent the polarized Raman spectra of polypropylene fiber. The line map methods were applied to acquire an area map of PP orientation. Using raster-scan method, spectra were recorded in the cross-section region of 600 × 1000 μm as shown in Fig. 5.10 right panel. The O.R. is mapped at 26 × 8 point of the selected region as shown in Fig. 5.10 left panel. The map clearly reveals that the orientation of skin layer is reasonably consistent, with a thickness of 200–250 μm.

5.9.6 Detection of Diseased Tissues

Raman spectroscopy is a powerful label-free technology with enormous biological potential for studying the chemical and structural features of biological samples ranging from individual molecules to cells and tissues. The non-destructive nature with an ability to reveal biochemical composition makes this technique valuable for

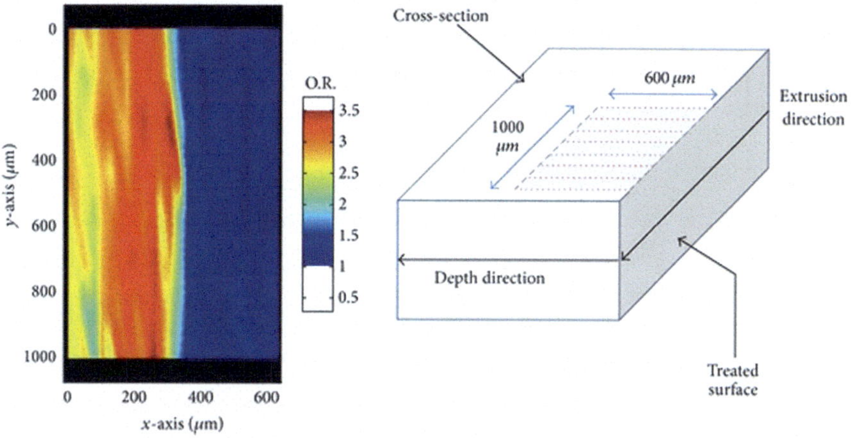

Fig. 5.10 Mapping of orientation for a 600 × 1000 μm area (left panel) as marked by the shaded area with red dots representing the spectral collection spots (right panel). Adapted from Chen et al. [23] (CC BY 3.0)

a wide range of biological applications. Collagen fiber distribution, which makes up the majority of connective tissue, is frequently optically anisotropic. In such case the Polarized Raman spectra can reveal important information about confirmation and orientation.

In medical science, the early detection of diseases is a critical aspect because it can lead to better treatment outcomes, increased survival rates, and reduced healthcare costs. Various methods and technologies are used for early disease detection across a wide range of medical conditions. Shang et al. [24] used polarized Raman spectroscopy in conjunction with deep learning technology to detect changes in the structure and components of tissues by improving the detection capability and automatic diagnosis of breast cancer with label-free probing and analysis.

The Raman spectra of protein and lipid present in breast tissue are shown in Fig. 5.11a, b, respectively. The structural changes of collagen fibers in breast cancer tissues and normal tissues are analyzed in terms of polarized Raman component. The amide III band with a unique bimodal structure of collagen band at ~1247 and ~1269 cm^{-1} show significant differences in the polarized spectra of normal and cancerous tissues. The bimodal ratio (I_{1247}/I_{1269}) of the polarized Raman spectra of normal and cancerous tissues with the polarization angle are shown in Fig. 5.12. It was observed that the bimodal ratio in cancerous cases is lower than normal and para-cancerous tissues irrespective of the polarization angle. This shows that the orientation of collagen fibers varies during the cancerization process. When malignant cells expand in breast tissue, the orientation of collagen fibers in the tissue gradually changes from horizontal to vertical, which is more favourable to the spread of cancerous cells.

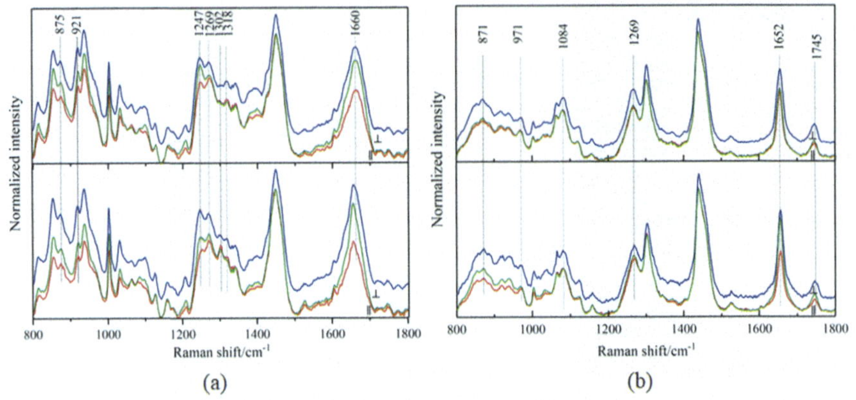

Fig. 5.11 Polarized (red lines for parallel and green lines for perpendicular) and conventional (blue lines) Raman spectra of **a** protein and **b** lipid collected from normal (top) and cancerous (bottom) tissues. Adapted from Shang et al. [24] (CC BY 4.0)

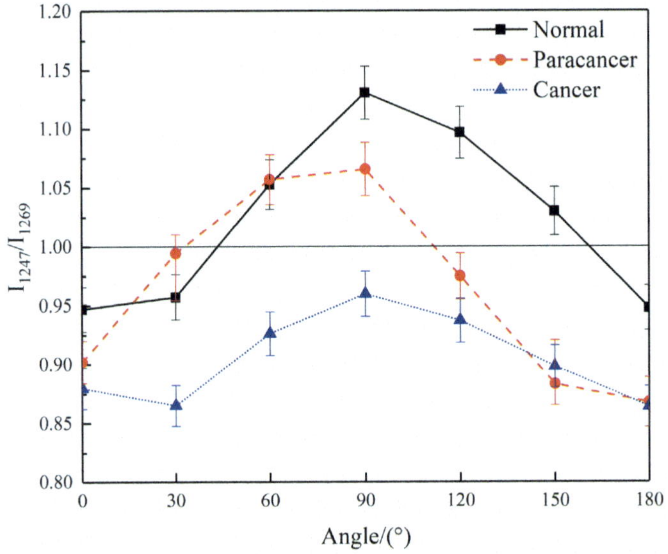

Fig. 5.12 The plot of the bimodal ratio (I_{1247}/I_{1269}) from the Raman spectra of cancerous, paracancerous and normal breast tissues with the polarization angle. Adapted from Shang et al. [24] (CC BY 4.0)

5.9.7 Study of Crystal Orientation

Polarized Raman spectroscopy is a powerful technique used in materials science and crystallography to determine the orientation of crystal planes in a sample. The orientation of crystal planes can be determined by comparing the measured polarization data with the expected polarization behavior for known crystal orientations. It is possible to compute the angles between the crystallographic axes and the incident polarization direction. Thyr et al. [22] used Polarised Raman spectroscopy as a method for determination of crystal growth and crystal plane orientation for nanomaterials. The orientation-dependent vibrational characteristics of ZnO single crystals with varied crystal orientations was investigated. Three crystals having distinct crystal facet (0001), $(1\bar{1}00)$ and $(11\bar{2}0)$ in wurtzite structured ZnO were used to exemplified to demonstrate the used methodology. The Raman spectra along these crystal facets show three distinct peaks at ~99, 332, and 438 cm^{-1} (Fig. 5.13) in all the crystals.

The polarization dependency of these vibrational modes was investigated by tracking their relative intensities at various azimuthal angles Φ and are presented in Fig. 5.14. From this study it was observed that the crystals having facet $(1\bar{1}00)$ and $(11\bar{2}0)$ show large polarization dependence whereas the crystal having facet (0001) show no polarization dependence at all for these major peaks.

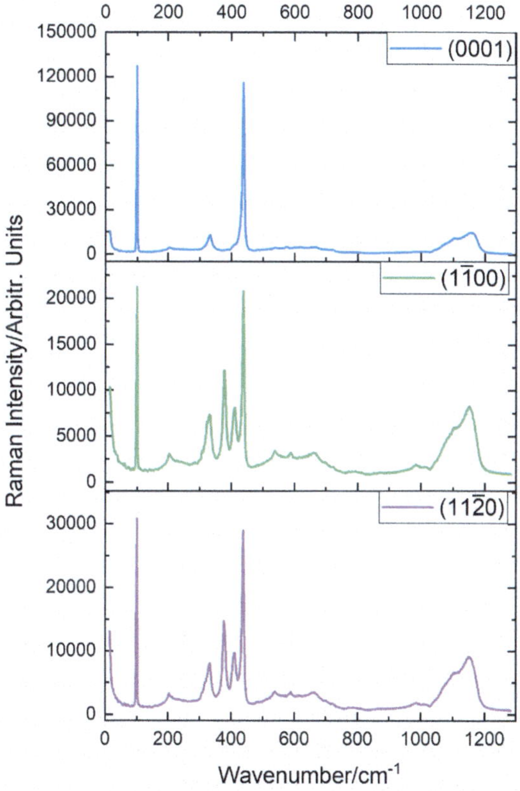

Fig. 5.13 Raman spectra of ZnO single crystals having crystal facets (0001), $(1\bar{1}00)$ and $(11\bar{2}0)$. Adapted from Thyr et al. [22] (CC BY-NC 4.0)

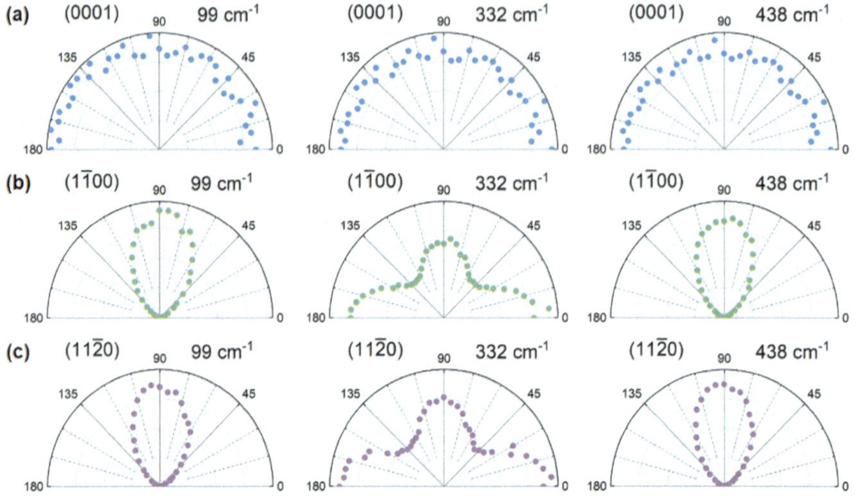

Fig. 5.14 Polarization dependence of the three distinct Raman peaks at 99, 332, and 438 cm^{-1} of the ZnO crystals having facet, **a** (0001), **b** (1$\bar{1}$00) and **c** (11$\bar{2}$0). Adapted from Thyr et al. [22] (CC BY-NC 4.0)

5.10 Conclusion and Future Perspectives

The chapter highlight just a few ways in which polarized Raman measurements are used to characterize the samples. Potential applications have been demonstrated in biology, polymer science, medicines, material science, and nanotechnology. It is demonstrated that polarized Raman measurements have a high potential for obtaining a wealth of information that would be difficult to obtain using other approaches. Advances in laser technology and detector sensitivity will continue to improve the capabilities of polarized Raman spectroscopy, which can be used to investigate a wide range of fields and materials. In materials science, polarized Raman spectroscopy will continue to be a valuable tool for characterizing novel materials, such as 2D materials, nanomaterials, and quantum dots to explore their unique electronic and structural properties. Integrating polarized Raman spectroscopy with other imaging techniques such as scanning electron microscopy (SEM), atomic force microscopy (AFM), or confocal microscopy can provide complementary information about sample properties, leading to more comprehensive material characterization. In the field of biomedicine, polarized Raman spectroscopy holds great promise for studying biological tissues and cells. Future developments may focus on non-invasive and label-free imaging techniques for disease diagnosis and drug development. Future developments may focus on achieving higher spatial and temporal resolution in Raman imaging and spectroscopy. This would enable researchers to study materials and biological systems at finer scales and capture dynamic processes with greater precision.

References

1. C.V. Raman, K.S. Krishnan, Nature **121**, 501–502 (1928)
2. A.C. Ferrari, D.M. Basko, Nat. Nanotechnol. **8**, 235–246 (2013)
3. M.S. Dresselhaus, G. Dresselhaus, R. Saito, A. Jorio, Phys. Rep. **409**, 47–99 (2005)
4. S. Zhang, N. Zhang, Y. Zhao, T. Cheng, X. Li, R. Feng, H. Xu, Z. Liu, J. Zhang, L. Tong, Chem. Soc. Rev. **47**, 3217–3240 (2018)
5. R.B. Godinho, M.C. Santos, R.J. Poppi, J. Raman Spectrosc. **47**, 579–584 (2016)
6. A.K. Ojha, S.K. Srivastava, N. Peica, S. Schlücker, W. Kiefer, B.P. Asthana, J. Mol. Struct. **735–736**, 349–357 (2005)
7. D.K. Singh, S. Mishra, A.K. Ojha, S.K. Srivastava, S. Schlücker, B.P. Asthana, J. Popp, R.K. Singh, J. Raman Spectrosc. **42**, 667–675 (2011)
8. R.K. Singh, S.K. Srivastava, A.K. Ojha, U. Arvind, B.P. Asthana, J. Raman Spectrosc. **37**, 76–84 (2006)
9. D. Wang et al., Phys. Rev. Lett. **116**, 126101(6) (2016)
10. H.G. Chae, T.V. Sreekumar, T. Uchida, S. Kumar, Polymer **46**, 10925–10935 (2005)
11. G. Xin, T. Yao, H. Sun, S.M. Scott, D. Shao, G. Wang, J. Lian, Science **349**, 1083–1087 (2015)
12. B. Vigolo, A. Penicaud, C. Coulon, C. Sauder, R. Pailler, C. Journet, P. Bernier, P. Poulin, Science **290**, 1331–1334 (2000)
13. J.N. Coleman, U. Khan, Y.K. Gun'ko, Adv. Mater. **18**, 689–706 (2006)
14. A.H. Gevorgyan, M.Z. Harutyunyan, Phys. Rev. E **76**(9), 031701 (2007)
15. S.K. Srivastava, A.K. Ojha, J. Koster, M.K. Shukla, J. Leszczynski, B.P. Asthana, W. Kiefer, J. Mol. Struct. **661–662**, 11–21 (2003)
16. M. Tanaka, R.J. Young, J. Mater. Sci. **41**, 963–991 (2006)
17. T. Liu, S. Kumar, Chem. Phys. Lett. **378**, 257–262 (2003)
18. S. Zhang, N. Mao, N. Zhang, J. Wu, L. Tong, J. Zhang, ACS Nano **11**, 10366–10372 (2017)
19. R. Loudon, Adv. Phys. **13**, 423 (1964)
20. B. Xu, N. Mao, Y. Zhao, L. Tong, J. Zhang, J. Phys. Chem. Lett. **12**, 7442–7452 (2021)
21. V. Vandalon, A. Sharma, A. Perrotta, B. Schrode, M.A. Verheijen, A.A. Bol, Nanoscale **11**, 22860–22870 (2019)
22. J. Thyr, L. Österlund, T. Edvinsson, J. Raman Spectrosc. **52**, 1395–1405 (2021)
23. X. Chen, M.A. Leugers, T. Kirch, J. Stanley, J. Spectrosc. **2015**(7), 518054 (2015)
24. L. Shang, J. Tang, J. Wu, H. Shang, X. Huang, Y. Bao, Z. Xu, H. Wang, J. Yin, Biosensors **13**(15), 65 (2023)
25. I. Nadeem et al., J. Phys. D. Appl. Phys. **55**(15), 033002 (2022)
26. D.A. Jarvis, I.J. Hutchinson, D.I. Bower, I.M. Ward, Polymer **21**, 41–54 (1980)
27. I.M. Ward, in *Advances in Polymer Science*. ed. by H.H. Kaush, H.G. Zachman (Springer, Berlin, Heidelberg, 1985), p.81
28. D.I. Bower, J. Polym. Sci. Part A-2 Polym. Phys. **10**, 2135–2153 (1972)
29. T.C. Damen, S.P.S. Porto, B. Tell, Phys. Rev. **142**, 570–574 (1966)
30. E. Kroumova, M.I. Aroyo, J.M. Perez-Mato, A. Kirov, C. Capillas, S. Ivantchev, H. Wondratschek, Phase Transit. **76**, 155–170 (2003)
31. D. Tuschel, Spectroscopy (2014). https://www.spectroscopyonline.com/view/practical-group-theory-and-raman-spectroscopy-part-ii-application-polarization
32. E.B. Wilson Jr., J.C. Decius, P.C. Cross, *Molecular Vibrations: the Theory of Infrared and Raman Vibrational Spectra* (McGraw-Hill, New York, 1955), p.285
33. J. Štokr, B. Schneider, D. Doskočilová, J. Lövy, P. Sedláček, Polymer **23**, 714–721 (1982)
34. D.C. Allemand, Appl. Spectrosc. **24**, 348–353 (1970)
35. J. Kiefer, M. Kaspereit, Anal. Methods **5**, 797–800 (2013)
36. C.-W. Huang, B.-J. Lin, H.-Y. Lin, C.-H. Huang, F.-Y. Shih, W.-H. Wang, C.-Y. Liu, H.-C. Chui, Nanoscale Res. Lett. **7**(6), 533 (2012)
37. F. Wu, B. Yang, E. Sun, R. Zhang, D. Xu, J. Zhou, W. Cao, J. Alloys Compd. **551**, 98–100 (2013)

38. H. Abramczyk, B. Brozek-Pluska, M. Kopec, J. Mol. Liq. **259**, 102–111 (2018)
39. S. Yang, F. Sui, Y. Liu, R. Qi, X. Feng, S. Dong, P. Yang, F. Yue, Nanoscale **15**, 13297–13303 (2023)
40. X. Yao, K. Nayani, J.O. Park, M. Srinivasarao, J. Phys. Chem. B **120**, 4508–4512 (2016)
41. B.A. Kolesov, M.A. Mikhailenko, E.V. Boldyreva, Phys. Chem. Chem. Phys. **13**, 14243–14253 (2011)
42. S.C. Pînzaru, I. Pavel, N. Leopold, W. Kiefer, J. Raman Spectrosc. **35**, 338–346 (2004)
43. D.L. Rousseau, R.P. Bauman, S.P.S. Porto, J. Raman Spectrosc. **10**, 253–290 (1981)
44. D.A. Long (ed.), *The Raman Effect: a Unified Treatment of the Theory of Raman Scattering by Molecules*, 1st edn. (Wiley, NJ, U.S.A, 2001), p. 640
45. B. Schrader (ed.), *Raman/Infrared Atlas of Organic Compounds* (VCH, Weinheim, 1989), pp. xiv+1118
46. T. Rm, R. Microscope, W. Raman, Technical Note of Edinburgh Instruments (2020). https://www.edinst.com/us/polarised-raman-spectroscopy/
47. I.R. Lewis, H. Edwards, *Handbook of Raman Spectroscopy: from the Research Laboratory to the Process Line*, 1st edn. (CRC Press, Boca Raton, 2001), p. 1072
48. D. Tuschel, Raman crystallography, in theory and in practice. Spectroscopy **27** (2012). https://www.spectroscopyonline.com/view/raman-crystallography
49. S. Singh, D.K. Singh, S.K. Srivastava, B.P. Asthana, Vib. Spectrosc. **56**, 26–33 (2011)
50. H. Wang, H. Xu, Q. Liu, X. Zheng, RSC Adv. **10**, 30982–30989 (2020)

Chapter 6
Optical Tweezers in Raman Spectroscopy

Bhumika Ray and Saurabh Raj

Abbreviations

OTRS	Optical tweezers and Raman spectroscopy
3D	Three-dimension
RBCs	Red Blood Cells
pN	Piconewton
fN	Femtonewton
NA	Numerical aperture
QPD	Quadrant photodiode
CCD	Charged coupled devices
CMOS	Complementary metal-oxide-Semiconductor
PCA	Principal component analysis
LDA	Linear discriminant algorithms
1NN	One nearest neighbor
SVM	Support vector machine
DMD	Digital micro-mirror device
AI	Artificial intelligence
ML	Machine learning
DL	Deep learning
DNN	Deep neural network
PLS-DA	Partial least squares discriminant analysis
CDA	Canonical discriminate function
ConvNet	Convolutional neural network

B. Ray · S. Raj (✉)
School of Biological Sciences, Indian Institute of Technology Delhi, Delhi 110016, India
e-mail: sraj@bioschool.iitd.ac.in

B. Ray
ASSIST Division, CSIR-Indian Institute of Toxicology Research, Vishvigyan Bhawan, 31, Mahatma Gandhi Marg, Lucknow 226001, Uttar Pradesh, India

D. K. Singh et al. (eds.), *Raman Spectroscopy*, Springer Series in Optical Sciences 248,
https://doi.org/10.1007/978-981-97-1703-3_6

ResNet	Residual network
PGGAN	Progressive growing of generative adversarial nets
NB	Nile blue
Rh6G	Rhodamine 6G

Mechanical forces play a critical role in many physiological processes, such as cell adhesion, migration, and differentiation. For example, mechanical forces can regulate gene expression and protein synthesis, and can also activate or inhibit signaling pathways that are important for cell proliferation and survival [1].

In addition to mechanical forces, biochemical signals also play a crucial role in regulating cellular behavior. Cells can receive and interpret signals from their environment, such as growth factors and cytokines, and respond by altering their gene expression, metabolism, and other cellular functions. Quite often, the mechanical forces induce chemical changes in the cells, thus coupling the mechanical and chemical properties of the cells [2]. The interplay between mechanical forces and biochemical signals is crucial to understand the cell's behavior at the molecular level [3].

Optical tweezers are a powerful tool for manipulating and studying individual biological molecules and cells, allowing us to probe their physical and chemical properties in real time. It allows us to manipulate the position and orientation of individual molecules, measure their mechanical properties, and even perform single-molecule biochemistry experiments. Raman spectroscopy, on the other hand, provides a highly sensitive and non-invasive method for analyzing the chemical composition of biological samples. With this technique, we can identify and quantify specific chemical species in our samples, providing valuable information about the molecular structure, composition, and interactions of our biological systems.

By combining these two techniques, researchers would gain a deeper understanding of the mechanical properties of cells and biomolecules and how they are influenced by molecular changes. For example, one could use optical tweezers to apply controlled mechanical stresses to cells, and then use Raman spectroscopy to monitor the molecular changes in response to these stresses. This could provide valuable insights into cellular mechanics, including the molecular basis of cellular deformation [2, 4], stiffness [5, 6], and viscoelasticity [7, 8]. There are many potential applications of this research, including the study of disease mechanisms and the development of new diagnostic and therapeutic strategies. Additionally, this approach could be used to study the mechanics of other biological systems, such as tissues, organoids, and extracellular matrix, providing a deeper understanding of their structure and function.

6.1 Introduction to Optical Tweezers

The ability of light to exert force on matter was known long before it was experimentally proven in the early 1900s [9]. Light carries both energy and momentum and whenever there is a change in momentum, force is invariably applied. Therefore, "radiation pressure" results as a consequence of the interaction of light with matter (or a particle) that causes a change in its momentum. This fundamental concept forms the basis of optical tweezers, also known as optical traps and/or laser tweezers.

History: The introduction of lasers in 1960 spurred experimental investigations into radiation forces, and a decade later in 1970, Arthur Ashkin (at Bell Laboratories USA) showed the confinement in motion of transparent polystyrene (latex) microspheres suspended in water drops using radiation forces from laser beam [9]. Further, in 1986, Ashkin et al. demonstrated a single-beam gradient optical trap (or optical tweezer) for stable confinement of micron-sized particles in three-dimension (3D) [10]. In the following years, Ashkin and his co-workers extended their optical trapping work to single biomolecule studies while trapping and manipulating viruses and bacteria [11], other unicellular organisms (like amoeba and protozoa), eukaryotic cells (red blood cells, RBCs) [12], and then to subcellular organelles, vesicles, and cytoplasm exhibiting pronounced viscoelastic behavior [13]. Subsequently, the research moved towards examining individual protein molecules, specifically, molecular motors such as kinesin and myosin [14, 15], RNA polymerases [15], and single/double-stranded DNA [16], presented and elucidated in Fig. 6.1 [17]. Since then, the optical trapping (optical tweezers) field has evolved exponentially, offering unprecedented micromanipulation possibilities. Today, the capabilities of optical traps have extended far beyond mere manipulation of individual cells/biomolecules to the measurement of molecular-scale lengths, motions, and forces in the piconewton (pN) and femtonewton (fN) ranges with unparalleled accuracy, precision, and velocity. Furthermore, the integration of spectroscopic techniques such as fluorescence and Raman, can greatly enhance the capabilities of optical tweezers. It is quite straightforward to incorporate spectroscopic techniques into optical tweezers since they are, in general, inverted microscopes [6, 18–20].

Nobel Prize in Physics: In 2018, it was awarded for *revolutionary innovations in the field of laser physics.* Arthur Ashkin received half of the award for his pioneering and extensive work on '*optical tweezers and their applications to biological systems*', and the other half was jointly awarded to Gérard Mourou and Donna Strickland '*for their method of generating high-intensity, ultra-short optical pulses*', i.e., for inventing chirped pulse amplification [21].

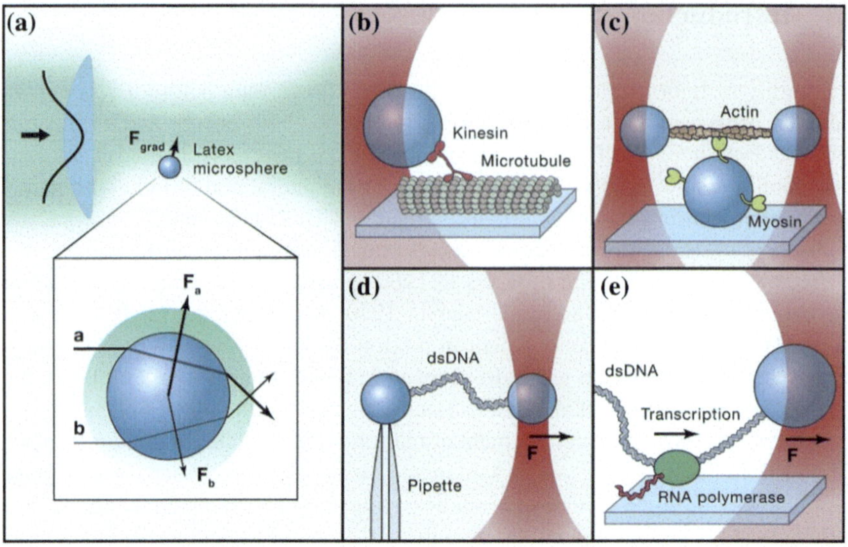

Fig. 6.1 A schematic illustrating basic principles and examples (in pictorial form) of early single-molecule experiments on optical tweezers. **a** During the initial experiment on optical forces, Ashkin observed that latex microspheres present near a weakly-focused laser beam moved towards the center of the beam having high intensity. This phenomenon was a consequence of the net gradient force (F_{grad}) resulting from the refraction of light through the transparent spheres in an asymmetric optical field. This was later used to create a 3D single-beam gradient optical trap, known as optical (or laser) tweezers, **b** Svoboda et al. [15] trailed the motion of a single kinesin motor molecule walking along a microtubule using an optically-trapped sphere. Results suggested that kinesin molecule takes 8 nm steps and can generate 5 pN of force [15], **c** Finer et al. [22] employed a dual optical trap to study the mechanical effects of ATP hydrolysis by a single myosin molecule, revealing an 11 nm power stroke and the generation of around 3–4 pN of force per hydrolysis event [22], **d** Smith et al. [16] studied elastic properties and behavior of DNA under high forces using optical tweezers [16], **e** Wang et al. [23] quantified the force produced by RNA polymerase during transcription and examined its behavior under different mechanical loads [23]. Adapted with permission from Killian et al. [17]. Copyright 2018, published by Elsevier

6.2 Principles of Optical Tweezers

Optical trapping (or single-beam gradient force trap, the original term given in 1970 by Arthur Ashkin) [9] is the crucial component of any optical tweezers setup. Optical tweezers operate based on the fundamental observation that photons possess momentum, given by $p = h/\lambda$, where h represents Planck's constant and λ is the wavelength of light. This momentum transfer of photons gives rise to an optical force, which becomes significant when interacting with microscopic or mesoscopic objects, such as a bacterium, a virus, a biomolecule, or a microsphere (microbead) upon exposure to a focused laser beam.

The optical tweezers primarily rely on two key optical forces: (a) the scattering force, and (b) the gradient force. When particles are hit by light, photons scatter in different directions, while some may be absorbed, causing momentum to transfer from the photons to the particles. If the scattering is isotropic, the forces cancel out in all directions except for the forward direction. As such, both absorption and scattering contribute towards the scattering component of the force on the particle. On the other hand, the gradient force arises from variations in light intensity (or spatial gradient) along its path and acts in the direction of the spatial light gradient. In the case of a focused laser beam in an aqueous suspension containing particles with a higher refractive index than the surrounding medium (like water), the gradient force directs these particles toward the focal point having the highest intensity, which is typically the center of the focus. In most cases, the scattering force is dominant, but in situations with a steep intensity gradient (such as near a laser's focus), the gradient force must also be considered. The scattering force works against the gradient force of the trap. For a three-dimensional stable trap, the force pulling the particle toward the laser focus must be stronger than the force pushing it away. To establish a stable trapping configuration, it is essential for the gradient force, influenced by the numerical aperture (NA) of the objective, to surpass the scattering force. An objective with a high NA creates a steeper gradient and, therefore, a better trap. The equilibrium position of the trapped particle is slightly beyond the focal point due to the balance between the gradient force and the scattering force [24].

The mechanism of optical tweezers in the simplest form can be explained with the help of ray optics, as shown in Fig. 6.2. Figure 6.2a demonstrates the scattering force and gradient forces within a focused beam. In the subsequent figures (Fig. 6.2b, c), the forces acting on a transparent dielectric sphere having a higher refractive index positioned near the beam focus are presented. When light (laser) encounters the sphere, it undergoes refraction according to Snell's law, leading to a change in its direction [25, 26] followed by a change in its momentum. As a result, optical forces are exerted on the sphere. Given the Gaussian intensity profile of the laser, we can see if the sphere (also referred to as a bead or particle) is displaced from the center of the incoming laser beam or not. If yes, then there will be a net force that is exerted on the sphere towards the center of the beam (Fig. 6.2c). To create a 3D trap, we need to hold the sphere in the direction of propagation of the beam, which is achieved by using a high numerical aperture (NA) objective that can generate a steep gradient in an axial direction. As seen in Fig. 6.2b, forces F_a and F_b pointing towards momentum change are entirely due to refraction, if reflection from the surface is considered zero. The restoring force is then directed back toward the focus when the sphere is displaced from the beam focus for small distances, describing a stable gradient force in the axial direction. In both instances, it becomes evident that the sphere undergoes a restorative force, drawing it back towards the center of the focus of the incoming light beam for small displacements.

While the above explanation using ray optics is illustrative, it is limited to objects that are considerably larger than the wavelength (λ) of the trapping laser (light) beam. This regime is often referred to as Mie or ray optics regime and the effects of diffraction can be neglected. Theoretical models have also been developed for objects that are much smaller than λ (Rayleigh regime), as well as for objects whose sizes are almost comparable to λ. Further, the calibration of an optical tweezer setup is

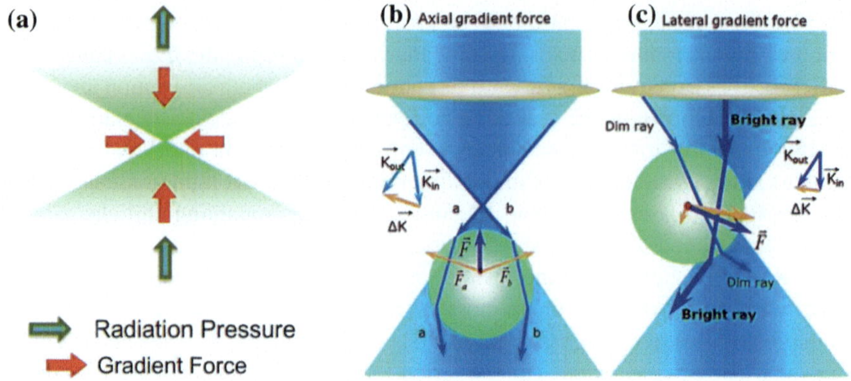

Fig. 6.2 Illustration of the trapping force mechanism. Here, **a** depicts the directions of gradient force and radiation pressure, **b** demonstrates the axial gradient force acting toward the center focus of the trapping beam, and **c** represents the lateral gradient force arising from a non-uniformly distributed laser beam

crucial for determining trapping forces exerted on particles. It involves determining the optical trap stiffness, which serves as a microscopic force transducer in the fN and pN range [27]. Various active methods and passive methods including video-based position detection, viscous drag method, and power spectral density method, are generally used for trap calibration [18, 28–30]. Recently, machine learning (ML) is gaining traction as a promising method for calibrating optical traps, especially for short or noisy trajectories [31]. While discussion on theories [32] and calibration methods is beyond the scope of this chapter; nevertheless, interested readers can find excellent and thorough explanations on the aforementioned topics in the references provided [33–35].

6.3 Introduction to Raman Spectroscopy

Raman spectroscopy is based on the Raman effect (or inelastic scattering) and named after Sir C. V. Raman. In 1928, Sir Raman and Dr. K. S. Krishnan found that ~1 in 10^7 of total scattered photons exhibit inelastic scattering when electromagnetic radiation (or beam) interacts with matter [36]. Usually, a small fraction of the incident radiation gets scattered from material in all directions besides absorption and transmission scattering. Most of the scattered radiation (in the form of photons hν. Here, 'h' represents Planck's constant, 6.62×10^{-34} m^2 kg/s, and 'ν' stands for the frequency of the radiation) has the same frequency as that of the incident radiation (photons) and is referred to as Rayleigh scattering (or elastic scattering). While in the Raman scattering process, the frequency of the scattered photons is different from the frequency of the incident photons, and the energy of scattered radiation by a molecule either decreases (Stokes line, longer wavelength) or increases (anti-Stokes line, shorter wavelength) depending on the initial state of the molecule. Both stokes lines and anti-stokes lines are equally wavelength shifted from Rayleigh lines and contains structural information of the molecules under investigation. Both Raman (due to inelastic scattering) and infrared (based on infrared absorption) are complementary vibrational spectroscopic methods yet have different selection rules [37]. In Raman spectroscopy, the Raman bands arise from a molecule whose bond vibrations can alter its polarizability. Here, polarizability refers to the ease with which the electron cloud around the molecule can be distorted by an external electric field. While, in infrared spectroscopy, molecules absorb infrared radiation, causing their chemical bonds to vibrate at specific frequencies. This generates a molecular vibrational spectrum with each band representing a specific frequency and amplitude. Unlike Raman spectroscopy, for a molecule to be infrared active, it must have a net change in the dipole moment. Therefore, molecules that are infrared active may not be Raman active, and vice versa [37–39]. Besides, the mid-infrared region contains characteristic molecular vibrational modes of biomolecules, hence, this region is mainly considered for vibrational spectroscopic studies.

The Raman scattering process is inherently weak, often necessitating highly concentrated analytes (or samples), particularly in biological systems [40]. However, this limitation has been overcome by the advent of surface-enhanced Raman spectroscopy (SERS). It capitalizes on local surface plasmon resonance phenomena in the presence of metallic nanostructures (conventionally employed as SERS substrates) [41, 42]. In SERS, the analyte of interest or under investigation is either adsorbed onto or placed in proximity to metallic nanostructures [43, 44], resulting in substantial improvements in scattering cross-section, often reaching up to 10^5–10^{11} orders of magnitude. This significant improvement in the Raman signal arises primarily from two mechanisms: electromagnetic enhancement (major contributor) and chemical enhancement [45–47]. So, the exceptional sensitivity and versatility of SERS makes it an indispensable technique in modern scientific research, encircling fields of biology, medicine, environmental monitoring, material science, and beyond [48–50]. In biology, SERS enables label-free detection and characterization of the structure and conformation of biomolecules along with associated alterations in them. It realizes complex inter- and intra-biomolecular interactions and biomolecule-ligand interactions [51] even at low analyte concentrations, and non-invasive probing of cellular processes such as investigating intracellular dynamics. In medicine, this technique has shown promise in early disease diagnosis and therapeutic monitoring, opening new avenues for personalized medicine [52, 53].

6.4 Raman Tweezers: OTRS Coupled Setup

The integration or coupling of optical tweezers with Raman spectrometer also referred to as Raman tweezers, was first realized by Lankers et al. in 1994 [54]. Subsequently, other studies were followed that utilized this combination with great success. But the first application of Raman Tweezers in the field of biology transpired in 2002 when a team from East Carolina University demonstrated its capability to distinguish living yeast cells from dead yeast cells [55]. It facilitates both chemical and physical analysis of optically trapped particles individually via molecular vibrational fingerprints. In a Raman tweezers setup, a focused beam is employed to select, capture and confine a single cell/biomolecule, while the same or different laser beam simultaneously excites and generates Raman signals, either in back-scattered or in forward-scattered geometries. Mostly, Raman tweezer setups are indigenously customized and developed. These are preferred over commercial ones because they offer precise structural information from biomolecules/cells in their physiologically native state (i.e., a controlled environment). Also, these can be easily upgraded and modified to meet the needs of the experiments. A typical experimental setup for a single-trap or dual-trap Raman tweezer includes an inverted microscope that couples an optical tweezer setup with a Raman spectrometer setup.

6.4.1 Instrumentation

Over the past decade, numerous versions of the Raman Tweezers system have been developed. Technological advancements have enabled us to successfully trap even metal nanostructures, which were previously challenging. Some of these variations have been discussed in later Sections. Traditionally, Raman Tweezers systems typically consist of several key components, as demonstrated in Figs. 6.3 and 6.4. At the center of the setup, is an inverted microscope with a high NA objective used for focusing the laser and collecting Raman signal. The Raman signal is then observed using a spectrograph/spectrometer or monochromator. Various filter designs are implemented to separate Raman scattered beam from the Rayleigh beam. A detector such as [CCD (charged coupled device), CMOS (Complementary Metal-Oxide-Semiconductor), or QPD (Quadrant Photodiode)] is used to record the position of a trapped particle and ascertain the force acting on it. Many Raman Tweezers systems use a trapping laser as the Raman excitation light source, as the trapped particle is at the focus of the beam and hence, can provide maximum scattering. But the use of separate lasers for trapping and Raman excitation gives better control in particle manipulation.

6.4.1.1 Dual-Trap Setup

The detailed instrumental layout and working of a simple dual-trap Raman tweezers setup are described below [2].

The optical tweezer setup utilizes a laser beam (e.g., 1064 nm diode laser, near-infrared region) for trapping. The trapping beam is expanded and collimated using lenses L1 and L2, then divided into two parts by polarizing beam splitter BS1. To ensure proper trapping, the beam is expanded to overfill the objective (inverted, oil immersion, high NA). Overfilling of the objective is to ensure maximum gradient

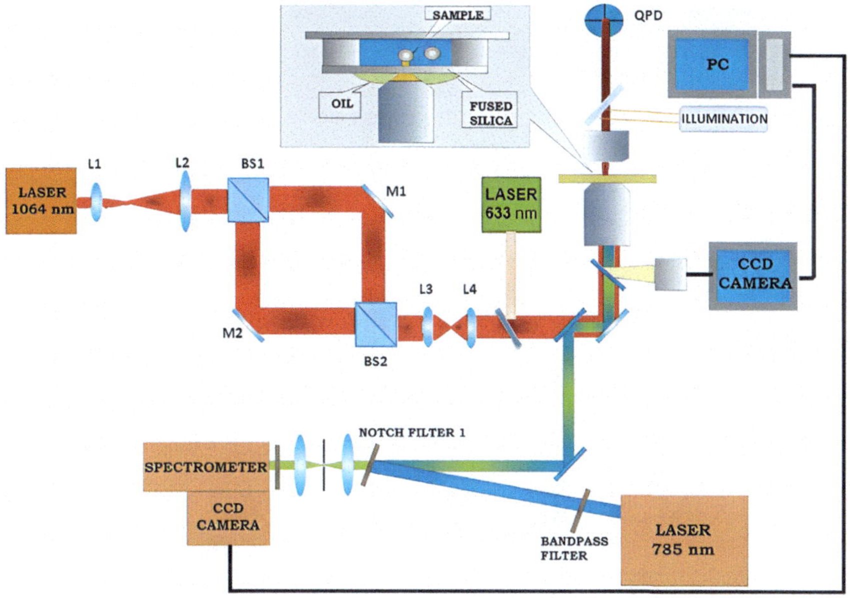

Fig. 6.3 Schematic of a typical dual-trap Raman tweezers setup. Here, a 785 nm (near-infrared) wavelength laser beam is used for Raman excitation, and a 633 nm (visible) wavelength laser beam is employed for microbead position detection and force measurement

for a stable trap. The two beams are recombined using beam splitter BS2 after being reflected by mirrors M1 and M2, which are positioned in conjugation with the objective's back focal plane. Another collimator formed by lenses L3 and L4, following the procedure suggested in [56], helps in achieving the conjugation. Adjusting the position of either lens (L3 or L4) controls the trap's movement along the beam propagation direction, while the movement of the mirrors (M1 or M2) results in the x–y movement of the trap without altering its intensity and shape, thereby maintaining a consistent trapping potential. One of the mirrors is connected to a motorized optical mount to render precise control of the trap movement. This trap movement manifests as the change in applied force on trapped cells and biomolecules, that can be controlled precisely. Both BS1 and BS2 polarize the beams passing through them, ensuring orthogonal polarization of the beams after exiting BS2. As a result, the two beams do not interfere with the microscope objective's focus, creating a stable dual trap. Another laser beam (e.g., 633 nm, visible) co-axially aligned with one of the trapping beams, is utilized for position detection and force measurement. The forward scattered beam is directed onto a QPD, providing precise information on the trapped object's displacement from the focus.

The next step involves setting up a confocal Raman microspectroscopy system and coupling it with the optical tweezers within the same inverted microscope of the optical tweezer setup. To achieve Raman excitation, a diode laser operating at a certain wavelength (such as 785 nm, i.e., a near-infrared wavelength that minimizes

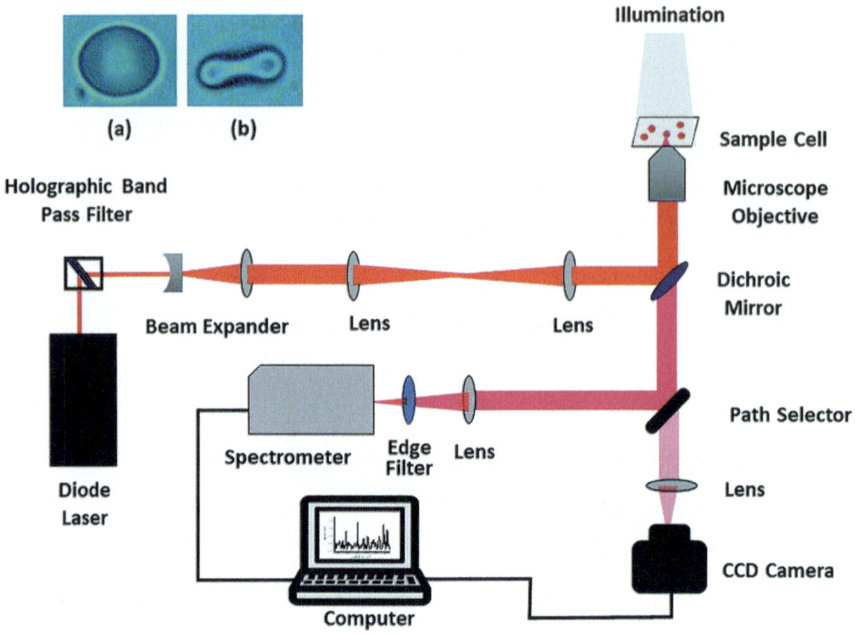

Fig. 6.4 Schematic of a typical single-trap Raman tweezers setup. A laser beam of a specific wavelength (e.g., near-infrared wavelength, 785 nm in this experimental study) is utilized for both trapping and Raman measurements. The inset embodies RBC **a** before and **b** after optical trapping in saline. Adapted with permission under a Creative Commons Attribution 3.0 Unported Licence (CC by 3.0) from Lukose et al. [57]. Copyright 2019, published by The Royal Society of Chemistry

autofluorescence from biological samples and photodamage from the laser intensity) is utilized. The laser power is adjusted based on the specific measurements being conducted, aiming for a high signal-to-noise ratio while ensuring minimal impact on the sample. To eliminate broad fluorescence emitted by the diode laser and narrow down its bandwidth, a bandpass filter (780 nm, if 785 nm laser is used) is employed to filter the beam. The filtered beam is then collimated before being directed at a holographic notch filter at a slight angle. The notch filter reflects the light, allowing it to enter an inverted microscope where the target object is trapped. The backscattered light collected by the same objective consists of both Rayleigh photons and Raman photons. The beam passes through the notch filter and a confocal pinhole with a dimension of 100 μm (as shown in the setup, Fig. 6.3). The notch filter blocks Rayleigh light but allows Raman light to pass. The confocal system effectively rejects signals originating from out-of-focus regions of the objective. Finally, the beam is focused onto the spectrometer's slit using a lens, and then to a liquid-nitrogen cooled (up to –80 °C) CCD detector that amplifies the signal and Raman spectra would be collected using software or indigenously written programs. Further, in the setup explained here, an additional CCD color camera has been connected to the inverted microscope to capture images before, during, and after the measurements and also to record movies during experiments.

Note: CCDs tend to amplify both noise and signals originating from the active area of the CCD. One common type of noise is thermal noise occurs due to thermally excited carriers within the detector. To mitigate these noise effects, it is necessary to cool the camera to a sufficiently low temperature. By doing so, the thermal noise can be effectively suppressed.

6.4.1.2 Single-Trap Setup

The comprehensive instrumental layout and working of a single-trap Raman tweezers setup are elucidated below [57].

For a single-trap customized Raman tweezer setup, a laser beam (e.g., 785 nm in the case of Fig. 6.4) is employed for both trapping and excitation of the object under investigation. An inverted microscope equipped with a high NA and microscope objective (oil immersion) enabled precise focusing of the laser beam. The scattered Raman signals are collected using the same objective and directed towards a spectrometer and finally to a liquid nitrogen-cooled CCD detector. To minimize the risk of damage to the object (e.g., cells), laser exposure time can be optimized and low power is recommended. The spectra were acquired from trapped objects using suitable software (e.g., MATLAB) and processed (baseline correction, normalization, and smoothing spectra using the vector normalization method) before final analysis. Figure 6.4 illustrates the schematic of a single-trap Raman tweezers setup and describes the to examine RBCs before and after trapping.

As evident, the dual-trap Raman tweezers setup offers better control over force and exposure, as compared to the single-trap setup.

6.5 Applications and Recent Advancements in Raman/SERS-Coupled Tweezers Setup

Raman tweezers have emerged as a versatile tool in cancer biology, erythrocyte studies, microbes identification and sorting, organelles, and sub-cellular structures [58], as elaborated in this section. Previously, the entrapment and manipulation of a single-erythrocyte (RBC) via Raman tweezers have enabled the differentiation of healthy cells from cells suffering from β-thalassemia [59] and expounded their mechanical behavior and characteristics. Moreover, when coupled with principal component analysis (PCA) and linear discriminant algorithms (LDA), this technique offers an accurate prediction of ABO blood cell typing [60] and unveils distinct Raman signatures indicative of RBC injury during viral (malaria, dengue), and bacterial (leptospirosis) infections [61]. The underlying mechanisms of eryptosis in erythrocytes i.e., programmed cell death induced due to different chemical and stress factors, have been examined and comparatively assessed [62]. Chronic consumption and/or exposure to alcohol (e.g. ethanol) adversely affect erythropoiesis and coagulation mechanics of erythrocytes and alter hematological components [63]. In vitro studies have been conducted using Raman tweezers to examine the alcohol-induced impact on optically trapped erythrocytes (RBCs). The Raman spectral features reveal depletion in the cytoplasmic membrane of RBCs suspended in saline [57] that maintains cell shape, in–out flow of intracellular molecules, and degradation of hemoglobin from oxygenated to deoxygenated state in blood plasma [63] that eventually leads to impaired oxygen transport in the body. These studies offer molecular insight into alterations in the (a) morphology, (b) structural integrity, (c) functional abilities, and (d) biochemical characteristics of an erythrocyte. Similar studies have been reported to explore the influence of crystalloid fluids [64] and buffers/saline [57] on blood components. Further, Singh et al. have explored the impact of short-term hyperglycemic stress on RBCs ex-vivo [65]. The RBCs exposed to high glucose levels, mimicking postprandial hyperglycemia (linked with cardiovascular complications in diabetes patients), exhibit limited photoreduction of intracellular hemoglobin. It indicates the formation of an irreversibly oxidized hemoglobin, elevated oxidative stress, and a subsequent loss of elasticity in the RBC membrane (determined via an optical stretching experiment on RBC trapped under a laser). Similarly, Raj et al. investigated how mechanical deformation or stress induces biochemical changes in erythrocytes. They analyzed the Raman spectra of erythrocytes trapped via polystyrene microbeads attached at diametrically opposite ends of RBC in both relaxed and stretched positions using PCA and 2D correlation spectroscopy. The study findings revealed that mechanical deformations in blood cells alter spectral characteristics primarily associated with hemoglobin, the spectrin network, and membrane proteins (mainly ankyrin) i.e., responsible for anchoring the cytoskeleton [2].

Raman tweezers, integrated with microfluidic systems, enable quick assessment of bacterial sensitivity to antibiotics and the differentiation between antibiotic-resistant and antibiotic-susceptible strains [66, 67]. A study by Bernatova et al. exposed methicillin-resistant and susceptible strains of the bacterial pathogen *Staphylococcus aureus* to the antibiotic oxacillin, observing a real-time gradual decrease in the Raman signal of nucleic acids specifically in the antibiotic-susceptible strains [68]. Additionally, in their pilot study, the authors employed Raman tweezers with multivariate statistical analysis and machine learning classification method (one nearest neighbor-1NN and support vector machine-SVM) to identify and differentiate among 254 strains of 20 microbes associated with urinary tract infections, achieving an accuracy rate of up to 98.39% [69]. Accurate identification of the causative pathogen, along with assessing its response to antibiotics and resistance profiles, is vital in devising an effective therapy plan for bloodstream infections. Therefore, in another pilot study, 305 pathogenic strains were sourced from 28 microbial species, responsible for causing bloodstream infections. These microbial strains were directly captured through optical tweezers at the single-cell level within human sera, and their Raman spectra were acquired and statistically analyzed. Through this, the strains were uniquely characterized, differentiated, and classified [70]. These findings underscore the significance of Raman tweezers in clinical diagnostics as well as in the management of microbes-borne diseases (infectious or not). In recent studies, Sinjab et al. made a notable contribution to the optical trapping field with their video-rate holographic instrument that incorporates reflective pinholes on a digital micro-mirror device (DMD) for detection. By utilizing liquid-crystal spatial light modulators, they successfully generated multiple trapping beams simultaneously, allowing for real-time interactive control. This instrument enabled the manipulation and simultaneous Raman spectral measurement of various dynamic entities, including polymer beads, *Pseudomonas aeruginosa* (bacteria), *Saccharomyces cerevisiae* (yeast), RBCs, and mammalian immune cells. They recorded time- and space-resolved Raman spectra for the formation of an immunological synapse between T cells and dendritic cells [71]. Subsequently, in another study, this research group monitored and characterized (in real-time) the transfer of L-Phenylalanine (isotope labeled, D8) from human cerebrovascular endothelial cell (host) to an intracellular pathogen *Toxoplasma gondii* (protozoan parasite) during its initial phase of causing brain infection. Hence, probing molecular trafficking and the cell-to-cell signaling process when the pathogen first interacts with the host cell [72].

As the key to effective interventions in combating cancer and improving patient outcomes, the search for cutting-edge breakthroughs in cancer diagnosis and precision medicine (therapeutics) is the pressing need of the hour [73, 74]. In a recent study, researchers introduced a fiber-coupled [75] Raman tweezer setup to examine cutaneous melanoma cells (skin tumor) and conclusively classified them into three categories-wild-type, common mutant, and drug-resistant, with accuracy rates of 94%, 92%, and 97.5%, respectively. The authors employed PCA and SVM as classification model systems for the rigorous analysis of Raman spectral datasets, acquired extensively. The study recommends this method as a promising frontier for early identification and diagnosis of tumor cells [76]. In a similar vein, Qiu et al. conducted a study where confocal Raman spectroscopy is coupled with laser tweezers to investigate the effects of three chemo-drugs (gemcitabine, cisplatin, and paclitaxel) on nasopharyngeal carcinoma cell lines (CNE2 and C666-1) at the single-cellular level [77]. The Raman profiles obtained revealed specific biomolecular changes, including alterations in protein structure, nucleic acids content, and base pairing, as well as lipid reduction, in response to drug treatment. Again, the combination of Raman tweezers with statistical algorithms (PCA, LDA, and CRT-classification and regression trees) allowed accurate classification of untreated and treated cells, with discriminant accuracies exceeding 90%.

Amidst developments in Raman tweezers, a transformative shift is underway as artificial intelligence (AI), machine learning (ML), and deep learning (DL) methods [78] excel traditional statistical and classification model systems. Notably, these learning algorithms can automatically process and translate vast Raman spectroscopic data into computational models unraveling intricate information and overall elevating the analytical prowess of the OTRS system [79]. For example, Shuyun et al. combined the deep neural network (DNN) classification method with the Raman tweezer setup (single-beam, 785 nm wavelength) to discriminatively analyze normal hepatocyte cell lines (HL-7702) and different cancer cell lines of the liver. For this study, low differentiated (SKHep1, SMMC-7721, and Hep3B) and highly differentiated (Huh7, and HepG2) liver cancer cell lines were selected. With random selection of around 300 Raman spectra from each cell line, the DNN algorithm achieved outstanding accuracy, sensitivity, and specificity (99.2%, 99.2%, and 99.8%, respectively) in identifying and classifying liver cancer cells and hepatocyte cells. This provided thorough information about cellular heterogeneity and the degree of cell differentiation at the single-cell level. Besides cancer diagnosis, the combination of Raman tweezer setup with machine learning algorithms allows the diagnosis of chronic neurodegenerative conditions like Alzheimer's disease (AD). Current clinical methods relying on medical imaging, such as CT scans, often yield false-positive results, making early AD diagnosis challenging. Thus, to address this, Lin et al. designed a study and investigated in vivo AD models. Triple transgenic rats (3xTg-AD) were selected and cultivated to early (3 months), middle (6 months), and advanced (12 months) staged AD models. Platelets from these AD models were extracted, optically trapped, and their Raman signatures were acquired. The Raman spectral data were rigorously mined using partial least squares discriminant analysis (PLS-DA), SVM, and PCA-CDA (canonical discriminate function) classification models, attaining an accuracy rate of 91%, 68%, and 97%, respectively, in distinguishing normal and diseased platelets at 3-, 6- and 12 months AD stages, respectively [80]. On similar lines, several other research groups are utilizing artificial intelligence and related algorithms to analyze different Raman spectra to extract intricate information from large datasets. These studies include categorizing 14 microbial species at a single-cell level using a framework of convolutional neural network (ConvNet) model (accuracy rate: 95.64%) [81]; identifying 15 pathogenic bacterial species using a residual network (ResNet) model (accuracy rate: 94.53%) [82]; and employing microfluidic chamber combined with fiber-coupled Raman tweezer and artificial neural network system for single microbe detection and classification of different microbes (15 species, accuracy rate: 94.93%) [83]. Various marine bacteria have been classified with exceptional accuracy (99.80%) using Progressive Growing of Generative Adversarial Nets (PGGAN) and ResNet analysis [84]. In another study, Wang et al. conducted molecular composition analysis, inter-species identification, and classification of RBCs from 10 different species (09 non-human and 01 human whole blood samples taken) at the single-cell level. They employed two algorithms, PCA-LDA and random forest for data analysis, achieving accuracy rates of 96.60% and 96.84%, respectively [85]. Subsequently, a thorough examination of Raman signatures acquired from RBCs, sourced from seven distinct animal species, was carried out. The Boruta algorithm in conjunction with seven machine-learning classification model systems was employed. A comparison was drawn among these models, focusing on critical factors like classification accuracy, precision, and recall indicators. The aim of the study was to recognize the most efficient single-cell analysis and classification model system [86].

While the conventional Raman tweezers allow the manipulation and characterization of micrometer-sized objects/particles, the integration of surface-enhanced Raman scattering with optical tweezers has extended their capabilities, allowing for precise manipulation and amplified Raman signal even from nano-scale particles with exquisite control and sensitivity. In SERS, there is coupling between the incident beam of light and surface plasmons (e.g., from metal nanoshapes) that results in strong local confinement of the electromagnetic fields which in turn enhances the optical response. Thus, plasmonic excitations arising from metal nanosurfaces are used to boost the otherwise weak Raman process by several orders of magnitude [87, 88]. Previous studies have shown to manipulate SERS-active metal nanosurfaces and investigate different species/molecules using single-trap or dual-trap (gradient force) Raman tweezers [89], thus, a complex interplay of SERS phenomena and radiation forces on the trapped objects was unveiled by Rao et al. They examined how SERS can induce radiation (or mechanical) forces or influence the average position (Brownian motion) of an optically trapped dielectric sphere (~2 μm silica bead) partially covered with metal silver colloids (between 50 and 100 nm). In the presence of the SERS active probe molecules (dyes), crystal violet, and emodin, the authors have simultaneously recorded and measured the SERS spectrum and Brownian motion of the bead to draw the correlation between them in a dual-trap experimental setup [90].

Recent advancements have truly elevated the surface-enhanced Raman tweezers systems. Traditionally, optical trapping and characterization of nano-scale structures have been challenging due to the relatively weaker gradient force in comparison to the scattering force acting on the nanoparticles. Recently, a standing-wave optical trap has been coupled with a confocal Raman spectroscopy setup. This scheme balances the scattering forces, creating a stable trap at low laser power, even for reflective materials. Strong SERS signals of analyte (organic- and bio-molecules) in a liquid medium can be acquired [91]. In a study, Dai et al. demonstrated a method that combines optical tweezers and Raman spectroscopy (in situ Raman tweezers setup) to observe and manipulate the hotspot formed between two silica beads coated with silver (Ag) nanoparticles. This technique allows to create adjustable and consistent enhancements in SERS with the ability to detect single molecules (Fig. 6.5). When combined with a microfluidic chamber, the authors have characterized three globular proteins namely hemoglobin, lysozyme, and bovine serum albumin in their native state. Also, they have extended the investigation to comprehend the structural features of transient species of alpha-synuclein (Fig. 6.6). This generates a platform providing intricate details of heterogenous and complex biosystems with single molecule sensitivity [90, 92]. Nowadays, researchers are also exploring dielectric dimer metamaterial arrays for generating Raman signal enhancement (SERS), akin to plasmonic nanostructures. These dimer metasurfaces can minimize heating effects due to optical absorption and reduce damage to the substrate and molecules (or analytes) being probed. Kenworthy et al. have shown that dielectric dimer arrays immersed in an aqueous solution provide high-intensity electric-field hotspots near the surface for the optical trapping and localization of probed molecules. This substrate-analyte configuration offered higher SERS enhancement factors even at low concentrations of the analyte (herein, enhancement factor of ~1.5×10^5 after 10 ms trapping), which is approximately 275% greater than the SERS enhancement factor computed for the same metasurface array-analyte complex in air, i.e., fixed and dried configuration [93,

94]. The authors proposed their method's utility in the identification and detection of nano-sized single-cell organisms such as viruses and proteins, and for inducing or initiating the crystal nucleation process.

A research team has introduced a thermoplasmonic optical tweezer that uses a single noble metal nanoparticle drop-casted on a glass substrate for SERS measurements. To demonstrate its capabilities, the researchers employed a two-analyte method with Nile Blue (NB) and rhodamine 6G (Rh6G) as Raman active molecules and were able to obtain single-molecule SERS signatures from reversible assemblies of colloidal plasmonic nanoparticles [95, 96]. Gold (Au) and Ag of various sizes were trapped under a single 633 nm wavelength and acted as both an optical trap and Raman optical beam. The study also examined the effect of varying surfactants and salt concentrations in the solution that contained Raman active molecules and nanoparticle assemblies. This created an environment conducive to trapping at a very low power density (approx. 0.1 mW/μm^2) that effectively heated plasmons, aiding the formation of reversible assemblies. This opto-thermal setup demonstrated the capacity to transport the nanoparticle assemblies within the microfluidic channel optically. The authors propose its potential applications as plasmonic meta-fluids and for targeted trapping of biomolecules in vivo [96]. Again, Ma et al. developed an in situ SERS-tweezers setup using a hybrid nanostructure for trapping and exciting (Raman modes) a single polystyrene sphere functionalized with Rh6G. The hybrid nanostructure utilized in the setup consists of a single bow-tie aperture having bull's eye rings. The single sphere particle of merely ~15 nm in size, was effectively trapped at the center of the nanohole. The researchers meticulously designed experiments and controls, and acquired Raman characteristics from the trapped single sphere. This study confirmed the significant Raman signal (SERS) enhancement facilitated by the bull's eye rings [97].

6.6 Limitations and Experimental Considerations in Raman/SERS Tweezers Measurements

Optical tweezers offer contactless 3D nanopositioning and manipulation of micron- or nano-sized biological objects/samples. Unlocking the full potential of optical tweezers requires the realization of the limitations that co-exist with their extraordinary versatility. It encourages the researchers to explore alternative perspectives and innovative strategies to overcome the constraints and propel the field forward. The interaction of highly-focused trapping laser with biological objects induces photodamage (also termed opticution), thermal damage, and mechanical stress in them [18, 98]. Thus, the wavelength and applied power of the trapping laser need to be carefully considered. Extensive research has been conducted since the beginning to observe and understand the causes and mechanisms of damage associated with trapped laser illumination power, time, and wavelength on various biological systems (cells and intracellular organelles) and systematically reported [99, 100].

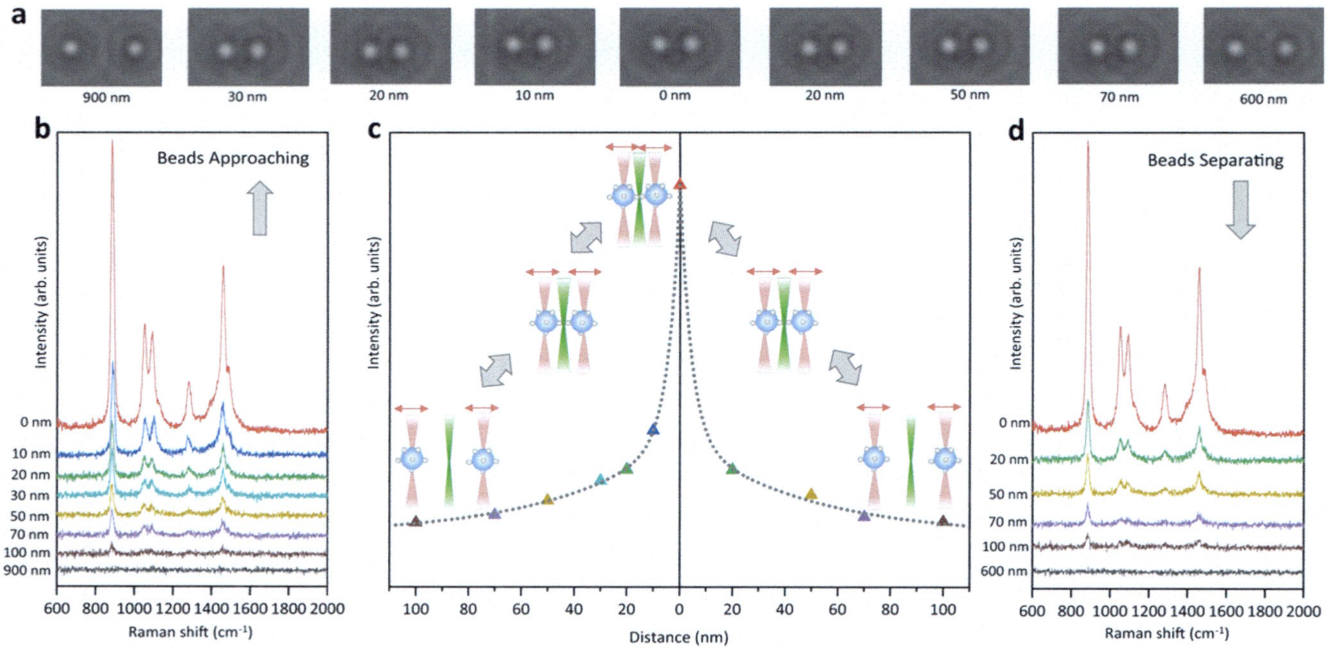

Fig. 6.5 Formation and control of dynamic hotspot with in situ SERS measurements of 1% ethanol aqueous solution. Here, **a** depicts real-time camera images of two silver nanoparticle-coated beads trapped at different distance, **b** presents the SERS spectral data of 1% ethanol aqueous solution recorded with 1 s acquisition time when two silver nanoparticle-coated beads are approaching, **c** describes the intensity of the characteristic peak of ethanol at 1458 cm^{-1} as a function of the distance between two silver nanoparticle-coated beads, i.e., from beads approaching each other to beads separating. The inset shows reversible bead positions, **d** shows the SERS spectral data of 1% ethanol aqueous solution recorded with 1 s acquisition time when two silver nanoparticle-coated beads are separating. Adapted with permission under a Creative Commons Attribution 4.0 International License (CC by 4.0) from Dai et al. [92]. Copyright 2021, published by Springer Nature

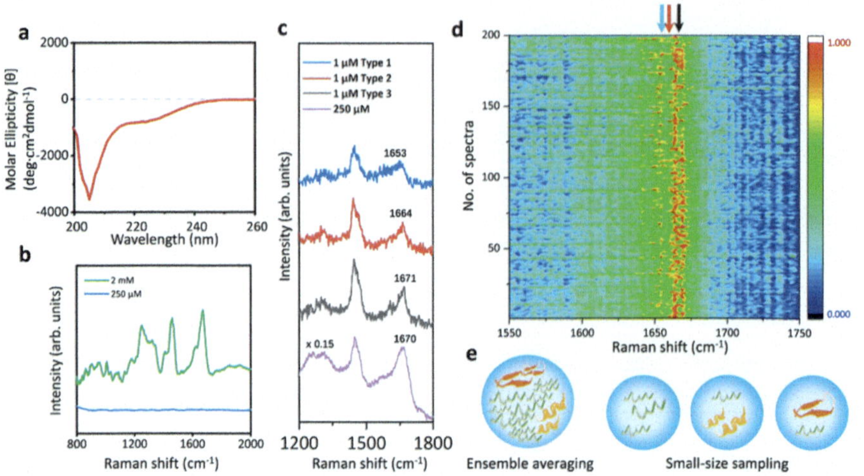

Fig. 6.6 Spectroscopic characterizations of intrinsically disordered alpha-synuclein protein. Here, **a** shows circular dichroism spectrum of alpha-synuclein (200 μM) in aqueous solution, **b** spontaneous Raman spectra of alpha-synuclein solution (2 mM, and 250 μM) recorded with 10 min acquisition time, **c** comparative display of three representative SERS spectra of alpha-synuclein (1 μM) with 1 s acquisition time, alongside SERS spectrum of alpha-synuclein (250 μM) with 5 min acquisition time, **d** mapping of 200 SERS spectra from alpha-synuclein (1 μM) using silver nanoparticle-coated beads (trapped when bead distance is 20 nm) with 1 s acquisition time. Color bar indicates normalized intensity from low (dark blue) to high (red), and **e** shows average measurements for alpha-synuclein ensembles: high concentration with long accumulation time and low concentration with short accumulation time. Adapted with permission under a Creative Commons Attribution 4.0 International License (CC by 4.0) from Dai et al. [92]. Copyright 2021, published by Springer Nature

Mainly these deleterious effects have been accredited to one-photon (linear excitation) absorption-associated damage that leads to fluorescence, heat generation and molecular fragmentation, and so on. Similar photodamage effects can also arise from two-photon (nonlinear excitation) absorption [98, 101]. Longer wavelengths in the near-infrared region have been opted to weaken one-photon absorption. Still, indirect sensitization and production of reactive oxygen and other reactive oxygen species, mainly superoxide $\left(O_2^-\right)$ have been observed across this wavelength range. The utilization of near-infrared wavelengths in optical tweezers has indeed mitigated the adverse impacts of local heating in trapped objects. However, it is important to note that the absorption of infrared photons, particularly by water molecules, can still lead to an increase in local temperature, and remains a critical concern [100]. To reduce the photodamage effect on cells, Yu et al. replaced a linear Gaussian beam with an azimuthally polarized beam (a structured donut-shaped vector beam to have better axial trapping capabilities) to trap and manipulate RBCs in vitro as well as in vivo with zebrafish as model organism [102]. Next, the stable trapping of nano-sized particles below 100 nm is challenging due to weak gradient force. To confine a particle, the gradient force must be stronger than, (a) the forward scattering force and (b) must overcome the particle's Brownian motion in liquid. The gradient force reduces significantly as the particle size decreases, requiring high optical power to counteract the destabilizing effects of Brownian motion and effectively confine nano-objects. The escape of nano-objects from the trap or unstable trapping of nano-objects hampers Raman and SERS spectral analysis and makes it difficult to obtain reliable data. Consequently, to improve the handling and manipulation capabilities of nano-objects, the researchers have introduced a standing-wave Raman tweezers setup, comprehensively elaborated in Sect. 2.2 [91].

6.7 Conclusion and Future Outlook

The usage of Raman Tweezers has become increasingly popular across various research fields, thanks to advances in technology and statistical analysis methods. Raman tweezers combine optical tweezers with Raman spectroscopy, enabling the non-destructive analysis of individual particles and/or biological cells trapped (confined in the 3D space) by optical tweezers. This approach allows the direct measurement of Raman spectral features from individual cells and/or biomolecules and facilitates a comprehensive insight into their mechanical and chemical properties. Thus, enhances our understanding of complex biological systems, chemical interactions, and, materials' characteristics. For instance, the viscoelastic properties and the chemical signatures of a cell can be obtained simultaneously. By examining individual particles rather than ensembles, inherent heterogeneity within a population can be precisely realized.

As Raman tweezers present an exciting frontier in various scientific disciplines, they offer a myriad of unexplored possibilities. In the biomedical field, this technique can facilitate label-free single-cell analysis, cell-type identification, monitoring of cellular pathways, and investigation into the underlying causes of diseases at the cellular level. Moreover, one can monitor the responses of cells to various treatments in real-time, e.g., the dynamics of interaction between drug molecules and individual cells could be analyzed at the molecular level, thus streamlining the drug development process. The continuous progress in Raman tweezers' accuracy, reliability, and ease of use would expedite their adoption in clinical settings in vivo, point-of-care disease diagnostics, personalized medicine, and monitoring of therapeutic outcomes and efficacies. Raman tweezers can be used in nanotechnology [103] and materials sciences to characterize individual nanostructures, eventually advancing the materials' design and fabrication. Further, combining this with artificial intelligence, machine learning, deep learning, and neural network models, leads to efficient handling and automated analysis of complex and vast datasets followed by rapid and accurate identification of patterns, connections, and relationships. Raman tweezers can be integrated with other advanced techniques, such as fluorescence imaging, mass spectrometry, super-resolution microscopy, and optogenetics to expand their overall analytical capabilities. The integration of Raman tweezers with fiber optics or their miniaturization would make them portable, thereby improving their practicality and accessibility in medicine and environmental monitoring. Lastly, the development of space-compatible Raman tweezers would uniquely contribute to space exploration, aiding the study of extraterrestrial materials, and understanding life beyond Earth [104, 105].

Despite its numerous advantages, Raman tweezers also face certain limitations. One key drawback is the need for high-intensity laser power, which can potentially cause sample damage or photobleaching in sensitive biological specimens. Additionally, Raman scattering signals are inherently weak, making it challenging to detect and analyze low-concentration samples accurately. Besides, Raman tweezers' spatial resolution is limited by the diffraction-limited spot size, restricting its ability to probe nanoscale features effectively. Finally, the technique is sensitive to the movement of particles/objects and vibrations, necessitating precise stabilization methods during measurements. Addressing these limitations has been crucial for improvements in the capabilities and applications of Raman tweezers.

As the technology continues to evolve, we can anticipate Raman tweezers to become more affordable, and capable of examining a broader range of particles (or objects). The setup would likely expand its horizons, revolutionizing diverse fields from fundamental sciences to quantum physics and beyond, ultimately leading to profound implications for human health, materials discovery, and our understanding of the universe.

References

1. A. Mammoto, T. Mammoto, D.E. Ingber, J. Cell Sci. **125**, 3061 (2012)
2. S. Raj, M. Marro, M. Wojdyla, D. Petrov, Biomed. Opt. Express **3**, 753 (2012)
3. C. Arbore, L. Perego, M. Sergides, M. Capitanio, Biophys. Rev. **11**, 765 (2019)
4. J.W. Chan, J. Biophotonics **6**, 36 (2013)
5. S. Raj, M. Wojdyla, D. Petrov, Cell Biochem. Biophys. **65**, 347 (2013)
6. S. Hu, J.-Y. Ye, Y. Zhao, C.-L. Zhu, Euro. Phys. J. Plus **137**, 1024 (2022)
7. Y. Xie, X. Liu, J. Biophotonics **15**, e202100315 (2022)
8. M. Tassieri, Curr. Opin. Colloid Interface Sci. **43**, 39 (2019)
9. A. Ashkin, Phys. Rev. Lett. **24**, 156 (1970)
10. A. Ashkin, J.M. Dziedzic, J.E. Bjorkholm, S. Chu, Opt. Lett. **11**, 288 (1986)
11. A. Ashkin, J.M. Dziedzic, Science **235**, 1517 (1987)
12. A. Ashkin, J.M. Dziedzic, T. Yamane, Nature **330**, 769 (1987)
13. A. Ashkin, K. Schütze, J. Dziedzic, U. Euteneuer, M. Schliwa, Nature **348**, 346 (1990)
14. S.C. Kuo, M.P. Sheetz, Science **260**, 232 (1993)
15. K. Svoboda, C.F. Schmidt, B.J. Schnapp, S.M. Block, Nature **365**, 721 (1993)
16. S.B. Smith, Y. Cui, C. Bustamante, Science **271**, 795 (1996)
17. J.L. Killian, F. Ye, M.D. Wang, Cell **175**, 1445 (2018)
18. C.J. Bustamante, Y.R. Chemla, S. Liu, M.D. Wang, Nat. Rev. Methods Prim. **1**, 25 (2021)
19. L. Kong, P. Zhang, G. Wang, J. Yu, P. Setlow, Y.-Q. Li, Nat. Protoc. **6**, 625 (2011)
20. H. Wang, Y. Guo, R. Zou, H. Hu, Y. Wang, F. Wang, L.A. Ju, Front. Phys. **10**, 771111 (2022)
21. D. Goswami, Resonance **23**, 1333 (2018)
22. J.T. Finer, R.M. Simmons, J.A. Spudich, Nature **368**, 113 (1994)
23. M.D. Wang, M.J. Schnitzer, H. Yin, R. Landick, J. Gelles, S.M. Block, Science **282**, 902 (1998)

24. K.C. Neuman, S.M. Block, Rev. Sci. Instrum. **75**, 9 (2004)
25. A. Kwan, J. Dudley, E. Lantz, Phys. World **15**, 64 (2002)
26. J.W. Shirley, Am. J. Phys. **19**, 507 (1951)
27. J. Gieseler et al., Adv. Opt. Photonics **13**, 74 (2021)
28. W.M. Lee, P.J. Reece, R.F. Marchington, N.K. Metzger, K. Dholakia, Nat. Protoc. **2**, 3226 (2007)
29. K. Visscher, S.P. Gross, S.M. Block, IEEE J. Sel. Top. Quantum Electron. **2**, 1066 (1996)
30. S.F. Tolić-Nørrelykke, E. Schäffer, J. Howard, F.S. Pavone, F. Jülicher, H. Flyvbjerg, Rev. Sci. Instrum. **77**, 103101 (2006)
31. A. Argun, T. Thalheim, S. Bo, F. Cichos, G. Volpe, Appl. Phys. Rev. **7**, 041404 (2020)
32. G. Pesce, P.H. Jones, O.M. Maragò, G. Volpe, Euro. Phys. J. Plus **135**, 1 (2020)
33. A. Ashkin, Biophys. J. **61**, 569 (1992)
34. K. Svoboda, S.M. Block, Annu. Rev. Biophys. Biomol. Struct. **23**, 247 (1994)
35. A. Mazolli, P. Maia Neto, H. Nussenzveig, Proc. R. Soc. London Ser. A: Math. Phys. Eng. Sci. **459**, 3021 (2003)
36. C.V. Raman, Indian J. Phys. **2**, 387 (1928)
37. E. Smith, G. Dent, *Modern Raman Spectroscopy: a Practical Approach* (Wiley, 2019)
38. N. Colthup, *Introduction to Infrared and Raman Spectroscopy* (Elsevier, 2012)
39. P. Larkin, *Infrared and Raman Spectroscopy: principles and Spectral Interpretation*, 2nd edn. (Elsevier, 2018)
40. H.J. Butler et al., Nat. Protoc. **11**, 664 (2016)
41. X.X. Han, R.S. Rodriguez, C.L. Haynes, Y. Ozaki, B. Zhao, Nat. Rev. Methods Prim. **1**, 87 (2021)
42. S. Agarwal, B. Ray, R. Mehrotra, Int. J. Biol. Macromol. **81**, 891 (2015)
43. P.L. Stiles, J.A. Dieringer, N.C. Shah, R.P. Van Duyne, Annu. Rev. Anal. Chem. **1**, 601 (2008)
44. S. Schlucker, Angew. Chem. Int. Ed. **53**, 4756 (2014)
45. B. Sharma, R.R. Frontiera, A.-I. Henry, E. Ringe, R.P. Van Duyne, Mater. Today **15**, 16 (2012)
46. E.C. Le Ru, P.G. Etchegoin, *Principles of Surface-Enhanced Raman Spectroscopy* (Elsevier, 2009)
47. R. Chang, *Surface enhanced Raman scattering* (Springer Science & Business Media, 2013)
48. J. Langer et al., ACS Nano **14**, 28 (2019)
49. D. Saviello, M. Trabace, A. Alyami, A. Mirabile, P. Baglioni, R. Giorgi, D. Iacopino, Front. Chem. **7**, 727 (2019)
50. A.I. Pérez-Jiménez, D. Lyu, Z. Lu, G. Liu, B. Ren, Chem. Sci. **11**, 4563 (2020)
51. B. Ray, R. Mehrotra, J. Photochem. Photobiol. B **211**, 111990 (2020)
52. S. Uskoković-Marković, V. Kuntić, D. Bajuk-Bogdanović, I. Holclajtner Antunović, in *Encyclopedia of Spectroscopy and Spectrometry* (2017)
53. K. Kho, U. Dinish, M. Olivo, in *Biophotonics for Medical Applications* (Elsevier, 2015), p. 101
54. M. Lankers, J. Popp, W. Kiefer, Appl. Spectrosc. **48**, 1166 (1994)
55. C. Xie, M.A. Dinno, Y.-Q. Li, Opt. Lett. **27**, 249 (2002)
56. E. Fällman, O. Axner, Appl. Opt. **36**, 2107 (1997)
57. J. Lukose, N. Mithun, G. Mohan, S. Shastry, S. Chidangil, RSC Adv. **9**, 7878 (2019)
58. R.D. Snook, T.J. Harvey, E. Correia Faria, P. Gardner, Integr. Biol. **1**, 43 (2009)
59. W. Jia, P. Chen, W. Chen, Y. Li, Medicine **97** (2018)
60. D. Lin et al., Opt. Express **24**, 24750 (2016)
61. S.S. Jacob et al., Front. Med. **9**, 858776 (2022)
62. S. Barkur, D. Mathur, S. Chidangil, J. Raman Spectrosc. **49**, 1155 (2018)
63. J. Lukose, M.N.G. Mohan, S. Shastry, S. Chidangil, J. Raman Spectrosc. **50**, 1367 (2019)
64. N. Mithun, J. Lukose, G. Mohan, S. Shastry, S. Chidangil, Spectrochim. Acta Part A Mol. Biomol. Spectrosc. **257**, 119726 (2021)
65. Y. Singh, A. Chowdhury, R. Dasgupta, S.K. Majumder, Eur. Biophys. J. **50**, 867 (2021)
66. Z. Pilát, S. Bernatová, J. Ježek, J. Kirchhoff, A. Tannert, U. Neugebauer, O. Samek, P. Zemánek, Sensors **18**, 1623 (2018)

67. L. Cui, H.-Z. Li, K. Yang, L.-J. Zhu, F. Xu, Y.-G. Zhu, TrAC, Trends Anal. Chem. **143**, 116415 (2021)
68. S. Bernatová et al., Euro. Phys. J. Plus **136**, 233 (2021)
69. K. Rebrošová, S. Bernatová, M. Šiler, M. Uhlirova, O. Samek, J. Ježek, V. Holá, F. Růžička, P. Zemanek, Anal. Chim. Acta **1191**, 339292 (2022)
70. K. Rebrosova et al., Microbiol. Spectr. **11**, e00028 (2023)
71. F. Sinjab, D. Awuah, G. Gibson, M. Padgett, A.M. Ghaemmaghami, I. Notingher, Opt. Express **26**, 25211 (2018)
72. F. Sinjab, H.M. Elsheikha, M. Dooley, I. Notingher, J. Biophotonics **13**, e201960065 (2020)
73. M. Liu et al., J. Biophotonics **12**, e201800332 (2019)
74. C. Li, S. Liu, Q. Zhang, D. Wan, R. Shen, Z. Wang, Y. Li, B. Hu, Spectrochim. Acta Part A Mol. Biomol. Spectrosc. **287**, 122049 (2023)
75. X. Zhao, N. Zhao, Y. Shi, H. Xin, B. Li, Micromachines **11**, 114 (2020)
76. X. Qiu, T. He, X. Wu, P. Wang, X. Wang, Q. Fu, X. Fang, S. Li, Y. Li, J. Biophotonics **15**, e202200158 (2022)
77. S. Qiu et al., Biomed. Opt. Express **11**, 1819 (2020)
78. H. Nguyen, A. Allaire, P. Diamandis, M. Bisaillon, M. Scott, M. Richer, BMC Med. **18**, 1 (2020)
79. W. Shuyun et al., Talanta, 124753 (2023)
80. M. Lin, H. Ou, P. Zhang, Y. Meng, S. Wang, J. Chang, A. Shen, J. Hu, Spectrochim. Acta Part A Mol. Biomol. Spectrosc. **280**, 121542 (2022)
81. W. Lu, X. Chen, L. Wang, H. Li, Y.V. Fu, Anal. Chem. **92**, 6288 (2020)
82. B. Zhou, L. Sun, T. Fang, H. Li, R. Zhang, A. Ye, J. Biophotonics **15**, e202100312 (2022)
83. C. Lin et al., BMEMat **1**, e12007 (2023)
84. B. Liu, K. Liu, N. Wang, K. Ta, P. Liang, H. Yin, B. Li, Talanta **244**, 123383 (2022)
85. Z. Wang, Y. Liu, W. Lu, Y.V. Fu, Z. Zhou, Biomed. Opt. Express **12**, 7568 (2021)
86. Y. Liu, Z. Wang, Z. Zhou, T. Xiong, Spectrochim. Acta Part A Mol. Biomol. Spectrosc. **277**, 121274 (2022)
87. K. Ajito, K. Torimitsu, Appl. Spectrosc. **56**, 541 (2002)
88. B. Ray, S. Agarwal, N. Lohani, M.R. Rajeswari, R. Mehrotra, J. Biomol. Struct. Dyn. **34**, 2518 (2016)
89. E. Messina et al., J. Phys. Chem. C **115**, 5115 (2011)
90. S. Rao, Š Bálint, P. Løvhaugen, M. Kreuzer, D. Petrov, Phys. Rev. Lett. **102**, 087401 (2009)
91. M.-Y. Wu, D.-X. Ling, L. Ling, W. Li, Y.-Q. Li, Sci. Rep. **7**, 42930 (2017)
92. X. Dai, W. Fu, H. Chi, V.S.D. Mesias, H. Zhu, C.W. Leung, W. Liu, J. Huang, Nat. Commun. **12**, 1292 (2021)
93. C.F. Kenworthy, L. Pjotr Stoevelaar, A.J. Alexander, G. Gerini, Sci. Rep. **11**, 6873 (2021)
94. J. Černigoj, F. Silvestri, L.P. Stoevelaar, J. Berzinš, G. Gerini, Sci. Rep. **8**, 15706 (2018)
95. E.C. Le Ru, M. Meyer, P.G. Etchegoin, J. Phys. Chem. B **110**, 1944 (2006)
96. S. Tiwari, U. Khandelwal, V. Sharma, G.P. Kumar, J. Phys. Chem. Lett. **12**, 11910 (2021)
97. J. Ma, X. Ning, Y. Lou, D. Wu, Q. Min, Y. Wang, Q. Zhang, Y. Pang, Opt. Lett. **48**, 1204 (2023)
98. A. Blázquez-Castro, Micromachines **10**, 507 (2019)
99. L. Tong, K. Ramser, M. Käll, Optical tweezers for Raman spectroscopy: datasheet from volume: "Raman Spectroscopy for Nanomaterials Characterization" in *SpringerMaterials* (Springer-Verlag Berlin Heidelberg). https://doi.org/10.1007/978-3-642-20620-7_18
100. K.C. Neuman, A. Nagy, Nat. Methods **5**, 491 (2008)
101. H. Zhang, K.-K. Liu, J. R. Soc. Interface **5**, 671 (2008)
102. P. Yu, Y. Liu, Q. Zhao, Z. Wang, Y.-M. Li, L. Gong, Appl. Phys. Express **13**, 032008 (2020)
103. Y. Minowa, X. Geng, K. Kokado, K. Sato, T. Kameyama, T. Torimoto, M. Ashida, Optica **9**, 139 (2022)
104. P. Polimeno et al., Euro. Phys. J. Plus **136**, 1 (2021)
105. J. Shane, R. Serati, H. Masterson, S. Serati, in *Optical Trapping and Optical Micromanipulation XIII* (SPIE, 2016), p. 220

Chapter 7
Chirality Revealed by Raman Optical Activity: Principles, Applications, Recent Developments and Future Prospects

Moumita Das, Andrii Kurochka, Petr Bouř, and Debraj Gangopadhyay

7.1 Introduction

Raman optical activity (ROA) spectroscopy is a powerful analytical technique that combines Raman scattering with optical activity and reflects molecular vibrations [1–9]. It belongs to the family of vibrational optical activity (VOA) techniques, of which ROA was the first phenomenon measured in 1973, followed by vibrational circular dichroism (VCD) [3, 10]. In some sense, ROA and VCD are complementary techniques, each of them having their weak and strong points [11–16]. In ROA, as in Raman scattering, incident light interacts with a molecule, and the scattered light has a different frequency. ROA measures the difference of scattered intensity between right and left circularly polarized light. On molecular level, this property is quantified through the Raman optical activity tensors, which include information about the spatial arrangement of atoms within the molecule and their chiral properties [4, 6, 17]. By analyzing ROA spectra, ideally it is possible to determine the absolute configuration, conformational changes, and intermolecular interactions in chiral systems [11, 16, 18, 19].

M. Das · A. Kurochka · P. Bouř · D. Gangopadhyay (✉)
Institute of Organic Chemistry and Biochemistry, Czech Academy of Sciences, Flemingovo náměstí 2, 16610 Prague, Czech Republic
e-mail: debraj.gangopadhyay@uochb.cas.cz

M. Das · P. Bouř
Department of Analytical Chemistry, University of Chemistry and Technology, Technická 5, 16628 Prague, Czech Republic

D. K. Singh et al. (eds.), *Raman Spectroscopy*, Springer Series in Optical Sciences 248,
https://doi.org/10.1007/978-981-97-1703-3_7

7.1.1 Chirality and Its Relevance in Various Scientific Disciplines

Chirality, or handedness, is a fundamental concept in nature, present in molecules, crystals, and biological systems. It refers to the property of an object that cannot be superimposed onto its mirror image [11, 20–28]. Chirality plays a crucial role in many scientific disciplines, including chemistry, biology, medicine, materials science, and pharmacology. Chiral molecules exhibit unique properties and behaviors, and their chirality can influence their reactivity, biological activity, and pharmacological properties [29–33]. Chiral materials exhibit unique optical, mechanical, and electrical properties that can be exploited in various applications. Chirality is inherent to several biomolecules such as proteins, nucleic acids, and carbohydrates. The chiral symmetry of biomolecules influences their structure, function, and interactions with other molecules [34–43]. By understanding chiral characteristics, researchers can develop new materials with improved performance for photonics, electronics, and catalysis [44–53]. Understanding the chiral nature of molecules is essential for drug design, asymmetric synthesis, and development of new materials with tailored properties, unraveling biological processes, insight into drug-receptor interactions, and design of biomimetic systems [54–68].

7.1.2 Motivation for the Development of ROA Technique

The development of ROA spectroscopy was motivated by the need for a technique that, for solutions, could provide structural information together with chiral sensitivity [10, 11, 69–73]. Older spectroscopic methods, such as circular dichroism (CD) and vibrational spectroscopy, also provide valuable insights, but with some limitations in terms of sensitivity, spatial resolution, and applicability [2, 7, 74–77]. Typical obstacle for CD, for example, is an absence of chromophore giving useful signal.

ROA spectroscopy overcomes some of these limitations. It combines the variability enabled by the Raman spectroscopy with the chiral sensitivity. All molecules provide vibrational transitions. Compared to CD, ROA bands are thus more numerous and better resolved. Wider range of molecules can be studied, and water solutions, problematic for IR and VCD, are not difficult for ROA, because of the weak Raman scattering of water [3, 71, 73, 78–80]. Development of more stable ROA spectrometers with advanced optics, detectors, artifact reduction and data analysis techniques has further propelled the field, enabling researchers to explore complex systems, their structural changes, intermolecular interactions, and dynamic behavior [3, 6, 11, 75, 81–88].

7.2 History and Fundamentals of ROA

Let us mention some key milestones and contributions that led to the establishment of ROA as a distinct spectroscopic technique [2–6, 9, 17].

To start with the Raman phenomenon, the foundations of Raman spectroscopy were laid by Sir C. V. Raman, who discovered the phenomenon of Raman scattering in 1928 [89–96]. Independently, this property of light and matter was reported by Landsberg and Mandelstam [97]. All of them observed that when light interacts with matter, a small fraction of the scattered light exhibits a frequency shift. Later it turned out that this provides unique information, such as that about the vibrational and rotational states of molecules. This groundbreaking discovery earned Raman the Nobel Prize in Physics in 1930.

History of the optical activity is even older. Early in the 19th century, Jean-Baptiste Biot made significant observations. In 1815, he discovered that certain substances, such as quartz crystals, rotated the plane of polarized light passing through them. He termed this phenomenon "optical rotation" and related it to chiral molecules [98–100]. The concept of chirality was further developed by Louis Pasteur. In the mid-19th century, Pasteur conducted meticulous studies on tartaric acid crystals and found that they existed in two distinct mirror-image forms [101–104]. This enabled understanding of stereochemistry of molecules and the concept of enantiomers. Other renowned 19th century scientists, Augustin Fresnel and Aimé Cotton, were successful in discovery of circular birefringence and circular dichroism [105–107]. Development of circular dichroism (CD) spectroscopy followed in the mid-20th century, as a powerful tool to probe electronic transitions of chiral molecules [108]. Circular dichroism occurs when a molecule exhibits differential absorption of left- and right-circularly polarized light. The first commercial instrument for measuring circular dichroism was designed in the 1950s, and allowed scientists to study optically active compounds in the ultraviolet and visible regions of the electromagnetic spectrum [105, 108].

The convergence of Raman spectroscopy and optical activity began in the late 1960s and early 1970s [1, 3]. In 1969, Atkins and Barron made theoretical predictions of the vibrational optical activity of chiral molecules, which were soon confirmed experimentally [69]. They described the chiral light scattering and introduced polarizability and optical activity tensors of molecules. They explored various ROA experimental setups, such as incident circular polarization (ICP) and scattered circular polarization (SCP) [3, 5, 12, 14, 73, 109–111]. In 1973, Barron et al. also conducted the first experimental measurements of this phenomenon on α-phenylethylamine and α-phenylethanol molecules [10]. In 1975, Hug independently verified these findings [112].

An important quantity in ROA spectroscopy is the (dimensionless) circular intensity difference ($\Delta = \mathrm{CID}$) defined as

$$\Delta = \frac{I_{\mathrm{ROA}}}{I_{\mathrm{RAM}}} = \frac{(I_{\mathrm{R}} - I_{\mathrm{L}})}{(I_{\mathrm{R}} + I_{\mathrm{L}})} \tag{7.1}$$

where I_R and I_L are the scattered Raman intensities for right- and left-circularly polarized scattered light, respectively [113, 114]. Molecular basis of ROA was elaborated by Barron and Atkins [69], Buckingham [114], and others. For Raman spectroscopy, we can imagine a molecule as an electric dipole interacting with the incoming and outgoing light. For ROA, we must additionally consider contributions from both the magnetic dipole and electric quadrupole. A molecule then behaves as an oscillating electric dipole, magnetic dipole, and electric quadrupole interacting with the radiation field. The electric dipole, μ_α, magnetic dipole, m_α, and electric quadrupole, $\Theta_{\alpha\beta}$, operators are:

$$\mu_\alpha = \sum_i e_i r_{i\alpha} \tag{7.2}$$

$$m_\alpha = \sum_i \frac{e_i}{2m_i} \varepsilon_{\alpha\beta\gamma} r_{i\beta} p_{i\gamma} \tag{7.3}$$

$$\Theta_{\alpha\beta} = \frac{1}{2} \sum_i e_i \left(3 r_{i\alpha} r_{i\beta} - r_i^2 \delta_{\alpha\beta}\right) \tag{7.4}$$

where particle i has a position vector r_i, charge e_i, mass m_i and linear momentum p_i. $\varepsilon_{\alpha\beta\gamma}$ is the Levi-Civita antisymmetric tensor and $\delta_{\alpha\beta}$ is the Kronecker delta, the Einstein summation convention is used.

Using the multipole expansion of the molecule-light interaction potential and the time-dependent perturbation theory, we arrive at the following quantum mechanical expressions for the dynamic molecular property tensors.

The electric dipole-electric dipole tensor (polarizability),

$$\alpha_{\alpha\beta} = \frac{2}{\hbar} \sum_{j \neq n} \frac{\omega_{jn}}{\omega_{jn}^2 - \omega^2} \mathrm{Re}\left(\langle n|\mu_\alpha|j\rangle\langle j|\mu_\beta|n\rangle\right) \tag{7.5}$$

the electric dipole-magnetic dipole tensor,

$$G'_{\alpha\beta} = -\frac{2}{\hbar} \sum_{j \neq n} \frac{\omega}{\omega_{jn}^2 - \omega^2} \mathrm{Im}\left(\langle n|\mu_\alpha|j\rangle\langle j|\mathrm{m}_\beta|n\rangle\right) \tag{7.6}$$

and the electric dipole-electric quadrupole tensor,

$$A_{\alpha\beta\gamma} = \frac{2}{\hbar} \sum_{j \neq n} \frac{\omega_{jn}}{\omega_{jn}^2 - \omega^2} \mathrm{Re}\ \left(\langle n|\mu_\alpha|j\rangle\right)\langle j|\Theta_{\beta\gamma}|n\rangle\right) \tag{7.7}$$

where $\omega_{jn} = \omega_j - \omega_n$ is the difference of angular frequencies of the excited and ground states, j and n respectively, and $\hbar$ is the reduced Planck's constant. Apart from ROA, these polarizabilities are important also for other spectroscopies. For example, the optical rotation tensor is tightly related to G'. Strictly speaking, these tensors are relevant to Rayleigh scattering only; for ROA, however, we need transition polarizabilities that are analogous to (7.5)–(7.7).

For isotropic samples (solutions, liquids, gases) the polarizabilities can be reduced to so called isotropic invariants [115]:

$$\alpha = \frac{1}{3}\alpha_{\alpha\alpha} = \frac{1}{3}(\alpha_{xx} + \alpha_{yy} + \alpha_{zz}) \tag{7.8}$$

$$G' = \frac{1}{3}G'_{\alpha\alpha} = \frac{1}{3}(G'_{xx} + G'_{yy} + G'_{zz}) \tag{7.9}$$

For anisotropic samples, the anisotropic invariants are:

$$\beta(\alpha)^2 = \frac{1}{2}(3\alpha_{\alpha\beta}\alpha_{\alpha\beta} - \alpha_{\alpha\alpha}\alpha_{\beta\beta}) \tag{7.10}$$

$$\beta(G')^2 = \frac{1}{2}(3\alpha_{\alpha\beta}G'_{\alpha\beta} - \alpha_{\alpha\alpha}G'_{\beta\beta}) \tag{7.11}$$

$$\beta(A)^2 = \frac{1}{2}\omega\alpha_{\alpha\beta}\varepsilon_{\alpha\gamma\delta}A_{\gamma\delta\beta} \tag{7.12}$$

Using them, we arrive to the expressions for the measured Raman and ROA intensities,

$$I_{\mathrm{RAM}} = I_{\mathrm{R}} + I_{\mathrm{L}} = 4K\left[D_1\alpha^2 + D_2\beta(\alpha)^2\right] \tag{7.13}$$

$$I_{\mathrm{ROA}} = I_{\mathrm{R}} - I_{\mathrm{L}} = \frac{8K}{c}\left[D_3\alpha G' + D_4\beta(G')^2 + D_5\beta(A)^2\right] \tag{7.14}$$

where the coefficients $D_{1\text{–}5}$ are dependent on scattering geometry as well as the modulation scheme employed in the ROA experiment [109, 116].

According to the angle between the excitation and analyzed radiation we recognize various experimental setups, such as the forward (0°), backward (180°), or right-angle (90°) scattering. For these, we obtain various expressions for CID [11, 113]:

$$\Delta(0^\circ) = \frac{4\left[45\alpha G' + \beta(G')^2 - \beta(A)^2\right]}{c\left[45\alpha^2 + 7\beta(\alpha)^2\right]} \tag{7.15}$$

$$\Delta(180^\circ) = \frac{24\left[\beta(G')^2 + \frac{1}{3}\beta(A)^2\right]}{c\left[45\alpha^2 + 7\beta(\alpha)^2\right]} \tag{7.16}$$

$$\Delta_x(90^\circ) = \frac{2\left[45\alpha G' + 7\beta(G')^2 + \beta(A)^2\right]}{c\left[45\alpha^2 + 7\beta(\alpha)^2\right]} \tag{7.17}$$

$$\Delta_z(90^\circ) = \frac{2\left[\beta(G')^2 - \frac{1}{3}\beta(A)^2\right]}{c\beta(\alpha)^2} \tag{7.18}$$

Within the harmonic approximation, derivatives of the tensors with respect to the nuclear coordinates are needed. In resonance, the theoretical expressions for ROA become even more complex, for example six tensors and 13 invariants are needed to describe the phenomenon [117, 118].

Computations of ROA intensities were implemented into common quantum-chemical software, such as the Gaussian program [119]. An important computational development was ROA implementation with the London (Gauge-invariant) atomic orbitals [120], ensuring independence of the calculated intensities of the origin of coordinates. An analytical implementation within the density functional theory made simulations of the spectra faster and more accurate [121]. Cartesian-coordinate transfer techniques [122, 123] enable to extend direct ab initio simulations of ROA to large molecules containing thousands of atoms [124]. It can be shown that unlike conventional Raman intensities, which remain the same in both forward and backward scattering, the ROA signal becomes amplified in backscattering geometry. As a result, ROA studies by employing the backscattering configuration are often preferred.

There are four most used modulation schemes of ROA measurements. For incident circular polarization (ICP), polarization of the incident laser beam alternates between the right circular polarization (RCP) and left circular polarization (LCP), while the detection is done for unpolarized scattered radiation. For the scattered circular polarization (SCP), unpolarized incident light enters the sample, and the difference between the RCP and LCP components of the scattered light is measured. The SCP instrument is considered to be more stable. The remaining two schemes involve the modulation of polarization in both the incident and scattered radiation and are known as dual circular polarization (DCP) ROA. They can be further separated into in-phase DCP_I and out-of-phase DCP_{II} measurements (Fig. 7.1) [82, 125]. The intensities are of course also dependent on the scattering angle, ξ. Using upper index for the incoming and lower one for the outgoing light polarization (u-unpolarized),

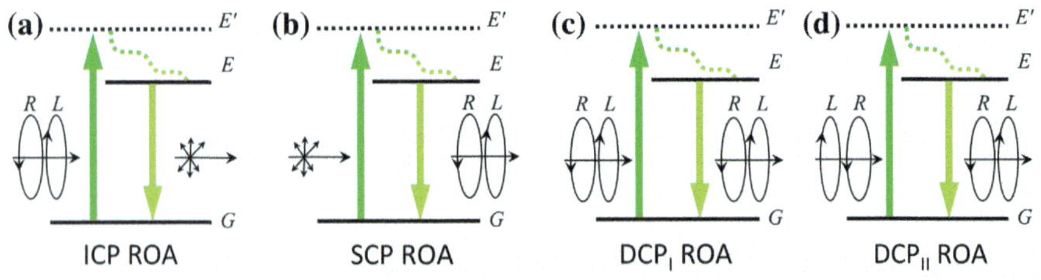

Fig. 7.1 Energy level diagram illustrating the four basic schemes of ROA. R and L stand for right-and left-handed circularly polarized light, respectively; unpolarized light is denoted by multi-directional arrows; the upward arrows represent the incident photon and the downward arrows the scattered photon. Adapted from Krupová et al. [126]

we can also write the ROA intensities as:

$$\text{ICP}: \Delta I_u(\xi) = I_u^{\text{R}}(\xi) - I_u^{\text{L}}(\xi) \tag{7.19}$$

$$\text{SCP}: \Delta I^u(\xi) = I_{\text{R}}^u(\xi) - I_{\text{L}}^u(\xi) \tag{7.20}$$

$$\text{DCPI}: \Delta I_{\text{I}}(\xi) = I_{\text{R}}^{\text{R}}(\xi) - I_{\text{L}}^{\text{L}}(\xi) \tag{7.21}$$

$$\text{DCPII}: \Delta I_{\text{II}}(\xi) = I_{\text{L}}^{\text{R}}(\xi) - I_{\text{R}}^{\text{L}}(\xi) \tag{7.22}$$

Until 2004, a backscattering ICP measurement strategy was widely used, such as the Glasgow backscattering biomolecular ICP ROA instrument (Fig. 7.2). Later, an instrument designed by Werner Hug was based on the SCP strategy [5]. This design significantly improved the reliability of ROA measurements by eliminating multiple artifacts including the flicker noise arising from dust particles, density fluctuations, laser power fluctuations, and other factors. A commercial instrument based on Hug's design, known as ChiralRAMAN from BioTools, Inc., was developed (Fig. 7.3a) [5]. Because of the suppression of polarization and other artifacts, it was capable of acquiring ROA spectra several times faster than the ICP ROA instruments. It also extended the ROA data acquisition to the low-wave number region, down to ~200 cm^{-1}.

Recently, Hug's design was further developed by Josef Kapitán at the Department of Optics at Palacký University Olomouc in collaboration with Czech companies ZEBR and Meopta-optika (Fig. 7.3b). The use of two CCDs and advanced optics

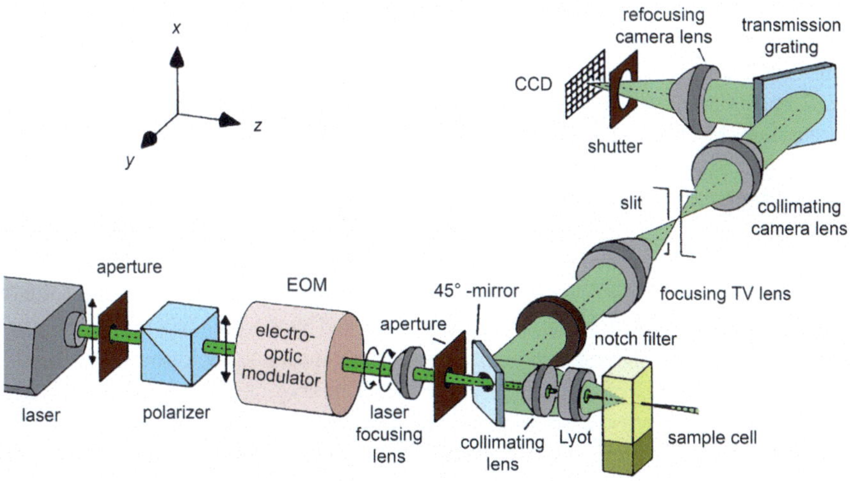

Fig. 7.2 The design of the Glasgow backscattering ICP ROA instrument. Adapted with permission from [127]. Copyright (2006) Elsevier

Fig. 7.3 **a** The ChiralRAMAN apparatus by BioTools, Inc. **b** ROA spectrometer developed at the Department of Optics at Palacký University Olomouc, Czech Republic in collaboration with ZEBR and Meopta-optika

enabled this spectrometer to record ROA spectra with improved artifact elimination in an extended spectral range 40–4500 cm^{-1} [128–133].

The possibility of DCP ROA was initially reported by Nafie and Freedman [134]. Subsequently, Nafie et al. conducted such measurements for several small molecules [135–139]. Although successful, these studies did not lead to immediate wider utilization of DCP ROA due to the longer time required to perform these measurements [140]. Later, advancements in signal throughput enabled the simultaneous measurement of ICP, SCP, DCP_I, and DCP_{II} ROA spectra, e.g. in the spectrometer shown in Fig. 7.3b, showing the potential of DCP ROA as a complementary approach to the more commonly used SCP and ICP configurations [115].

The choice of the excitation frequency also plays a crucial role in Raman and ROA spectra. Raman and Raman optical activity intensities are proportional to the fourth and fifth powers of the excitation frequency, respectively. The additional frequency power for ROA is seen, for example, in the expression of the optical activity tensor G' (as shown in (7.7) and (7.14)). Sometimes fluorescence emitted by impurities is stronger than Raman scattering, which can be avoided by working with different excitation wavelengths. The noise generated by the fluorescence background can easily obscure the desired ROA signal. Lasers that emit at 514 or 532 nm are commonly used and provide a reasonable balance between the intensity of Raman scattering and the signal from fluorescent impurities. Most ROA spectrometers employ these wavelengths, but other possibilities have been explored, too. For instance, Nafie et al. [141] pioneered the development of an instrument that utilized a 785 nm near-infrared light source. Also Unno et al. [142] constructed a near-infrared ROA spectrometer, incorporating the virtual enantiomer strategy of Hug [143], to reduce the artifacts.

In general, it is also good to complement the ROA spectroscopy with other analytical techniques, such as circular dichroism (CD) and other vibrational spectroscopies. The ongoing improvement of ROA spectrometers is aimed to improved sensitivity, broader spectral range, and faster data acquisition. Researchers continue to explore novel optical designs, detector technologies, and data analysis algorithms to enhance the performance and applicability of ROA spectroscopy.

7.3 Basic Components of ROA Spectrometers

This section summarizes some components necessary for functioning of ROA spectrometers and their functions. Figure 7.4 provides summary of the key components in the backscattering SCP ROA spectrometer, as designed by W. Hug and commercialized as "ChiralRAMAN" apparatus by BioTools, Inc. ICP systems are similar, the circular polarization analyzer in the scattered light would be replaced by a circular polarization modulator in the incident light, and DCP instruments can be created by combining the SCP and ICP building blocks. While a dual-arm system is not obligatory for ICP and DCP setups, employing two arms to switch the scattered light synchronized with rapid modulation of the exciting light can help reduce flicker noise.

7.3.1 Laser Source

The laser source is a fundamental component of ROA spectrometers. The choice of laser depends on several factors, including the sample's absorption properties, desired spectral range, and sensitivity requirements. The laser source should provide sufficient power. Commonly used sources include solid-state diode or gas lasers. The solid-state ones are cheaper and smaller, gas-ones might be more stable. Power and frequency-stability are very important for ROA measurements.

Historically, the 488 and 514.5 nm lines of the argon ion laser were commonly used, later they are more and more replaced by the 532 nm line of the frequency doubled YAG solid state laser. When using green light for excitation, laser power ranging from 100 to 400 mW at the sample is typically used for measuring organic liquids. However, in the case of aqueous solutions of proteins/peptides, a higher power of 1000 mW may be desired, which, however, might lead to degradation of the optics or of the sample. For absorbing/resonating compounds smaller powers are used to avoid sample overheating, etc.

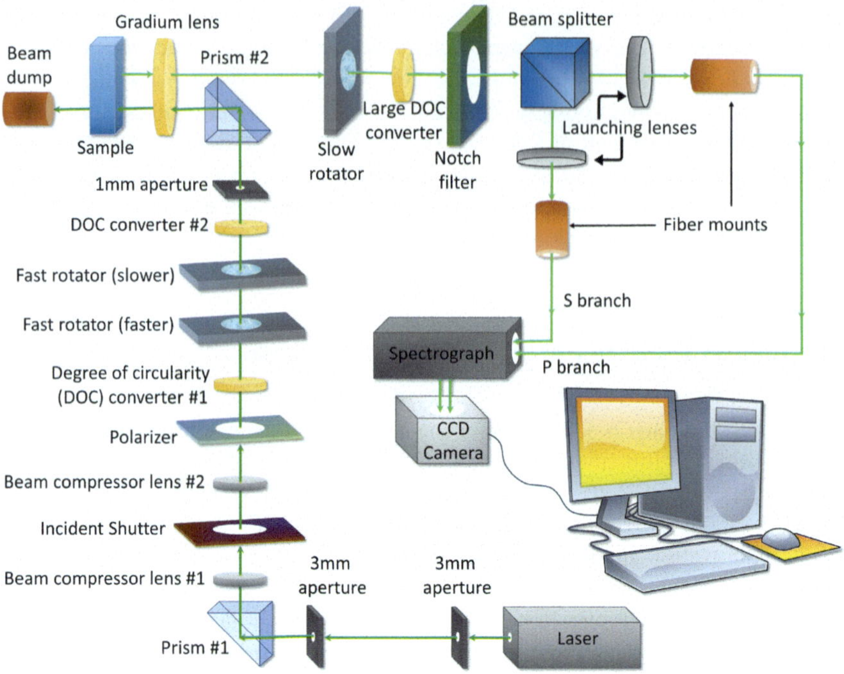

Fig. 7.4 Basic components of the ChiralRAMAN SCP ROA backscattering spectrometer from BioTools, Inc.

7.3.2 Sample Handling and Delivery Systems

Liquid samples are often measured in fused silica cells (cuvettes). It is desirable that no artifacts are introduced by the silica glass, e.g. due to internal stress. An anti-reflection coating increases efficiency of the measurement. Attempts were made using also micro-capillaries, where ROA could be determined with less than 1 μl of sample, but these are not much practical in most cases. Usually, rectangular cells that require 20–100 μl of sample are widely used (Fig. 7.5); cylindrical cells may produce data of similar quality. In Fig. 7.5, examples of such cylindrical cells are shown, including a quartz cell with a volume of 35 μl and a black Teflon cell with a pathlength of only 3 mm and a volume of 22 μl. These cells utilize small microscope cover glasses as windows, which have a thickness of only 0.15 mm to minimize the collection of Raman light from the glass. These cover glasses can be easily replaced if material gets burned onto their surface.

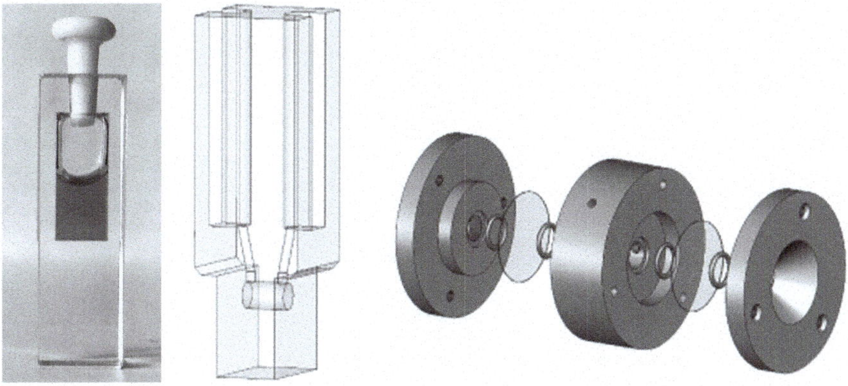

Fig. 7.5 (From left to right) A rectangular 100 μl quartz cuvette, (Right) cells made of quartz (with a volume of 35 μl) and black Teflon remountable cylindrical sample (with a volume of 22 μl). Adapted with permission from [5]. Copyright (2017) Elsevier

7.3.3 *Focusing and Optics*

Backscattering SCP ROA spectrometer typically consists of a lens for focusing and collecting backscattered light. More lenses may be employed to achieve a gentler focus of the laser on the sample and to eliminate stray light resulting from fluorescence and Raman scattering [5]. Figure 7.6 shows an advanced light collection optics that utilizes a holographic planar grating spectrograph.

7.3.4 *Polarization Control Elements*

Polarization elements in ROA spectrometers include half-wave plates, quarter-wave plates, and specialized optics. Waveplates, also known as retardation plates, are birefringent optical components that introduce a phase difference between the orthogonal polarization states of light. Half-wave plates and quarter-wave plates are used to generate circularly polarized light. Quarter-wave plates convert linearly polarized light into circularly polarized light.

An interesting way for eliminating ROA artifacts is the optical creation of a virtual enantiomer. This is achieved by scrambling the linearly polarized components in both the excited and scattered light. Figure 7.7 illustrates such optical configuration used in a modern SCP backscattering instrument. The half-wave plates can convert right to left circularly polarized light and vice versa and rotate the polarization plane of linear light. They are utilized in this context. Placing a molecule between aligned half-wave plates makes its chiroptical properties resemble those of its enantiomer. By using circularity converters, a half-wave cage is created around the sample. Two measurements are taken: one without and one with the half-wave plates. Treating

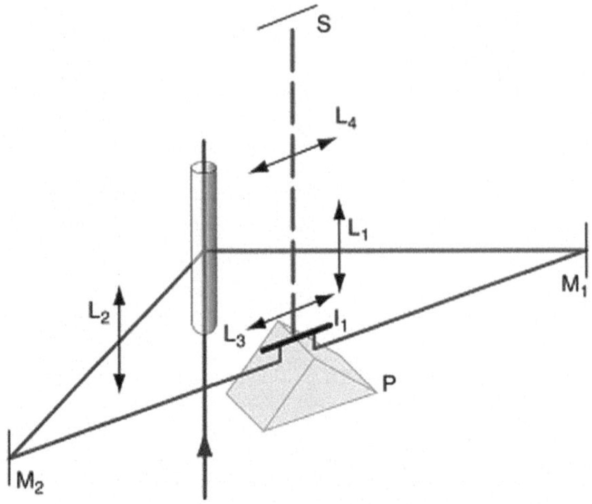

Fig. 7.6 A set of dual-lens light collection optics designed to compensate for offsets in right angle scattering, specifically for samples of less than 0.5 μl. L1 and L2 lenses work together to create an intermediate image, referred to as I1. This intermediate image is then projected onto the spectrograph entrance slit, denoted as S, using L4. Additionally, L3 functions as a field lens in this configuration. Adapted with permission from [5]. Copyright (2017) Elsevier

the second measurement as if it were for a real enantiomer and subtracting it from the first yields a doubled value. The actual process, based on this concept, involves a more complex sequence of four measurements. Offsets unrelated to the sample's chiroptical properties and optical imperfections between the circularity converters are eliminated. Linear rotators depolarize the exciting and scattered light, removing the effects of optical rotation and stray birefringence. This eliminates the need for aligning the circularity converters. Counter-rotating half-wave plates in the incident light increase the rotation speed of the polarization plane to over 50,000 revolutions per minute, which reduces the impact of imperfections.

7.3.5 Detection Systems

Detectors convert the scattered light into measurable electronic signals. The choice of detector depends on factors such as sensitivity, spectral range, and signal-to-noise ratio requirements. Photomultiplier tubes (PMTs) offer high sensitivity and fast response times. They are suitable for a broad spectral range and can be used in both single-channel and multichannel detection configurations. However, charge-coupled devices (CCDs) and CCD arrays are now more common types of detector employed in ROA spectrometers. CCD-based detectors offer excellent spectral resolution and the ability to capture multiple wavelengths simultaneously. They are coupled with

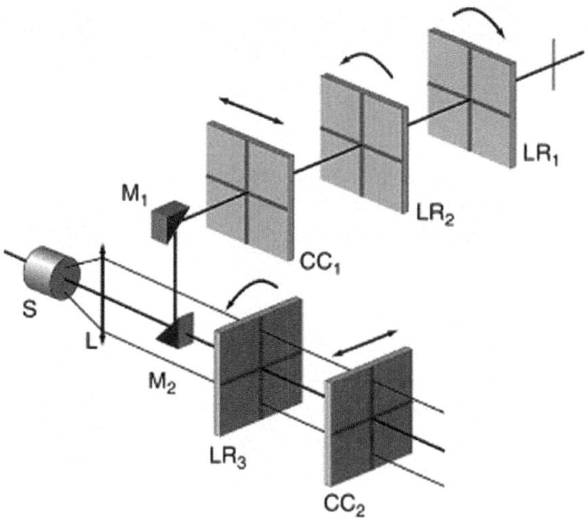

Fig. 7.7 Optics used to eliminate offsets in collinear scattering. The setup includes rotating half-wave plates (LR1, LR2, and LR3) and half-wave plates (CC1 and CC2) that are inserted into the optical path to create a virtual enantiomer. Mirrors (M1 and M2) are also present to deflect the beam while maintaining its polarization. Adapted with permission from [5]. Copyright (2017) Elsevier

spectrographs or monochromators. The spectrographs are equipped with planar holographic transmission gratings of nearly 100% efficiency for "s" polarization (perpendicular to the plane of incidence) and 60% for "p" polarization (parallel to the plane of incidence). However, due to right angle bends in perpendicular planes, the output light from the fiber optics becomes partially depolarized, resulting in similar efficiencies for light from both the s and p branches of the polarization analyzer.

Other detectors used in ROA spectrometers include photodiodes and intensified CCDs (ICCDs). Photodiodes offer fast response times and are suitable for detecting Raman signals in specific spectral ranges. ICCDs provide increased sensitivity by intensifying the detected signal through electron multiplication.

7.3.6 Signal Processing Tools

Signal processing tools and data analysis algorithms are essential for extracting and analyzing the ROA spectra obtained from the detection systems. In a circular intensity difference measurement, the fundamental acquisition cycle involves measuring and subtracting the Raman intensity for right circular polarization and left circular polarization. Synchronization is necessary for multiple aspects: illuminating the sample, reading the detector while the illumination is off, and in the case of ICP, switching the circular polarization of the exciting light. Each arm of the instrument measures the intensity of either the right or left circular component of the scattered light. By summing the measurements obtained from two switching states, the precision is significantly increased. Spectral deconvolution software is employed to separate overlapping Raman bands and extract the individual contributions. Data visualization programs allow for the interpretation and representation of the ROA spectra.

7.4 Recent Applications of ROA Spectroscopy

One of the primary applications of ROA is the discrimination of stereoisomers in solution. For example, complete absolute configuration of powerful natural anticancer agent, paclitaxel, was determined from four possible stereoisomers [144]. Additionally, the possibility to simultaneously measure Raman and ROA, provided by most of the instrumental architectures, allowed to quantify enantiomeric excess for α-pinene and alanine with incredibly low error of 0.22% [145]. Flexible organic molecules were always difficult case for spectroscopies due to numerous stable conformers in solution and ROA was shown to reliably determine the ratio of molecular populations as it was recently done for amphetamine derivatives [146] and cocaine [147]. In a similar way, predominant conformation of histidine complexes with zinc and nickel were determined by decomposition of experimental ROA spectra into calculated subspectra [148]. Another ROA study on monomeric nucleotides in solution revealed importance of a hydrogen bond between the sugar and the phosphate groups for the sugar puckering (Fig. 7.8) [149].

ROA spectroscopy is excessively utilized to access the chirality and three-dimensional structure of carbohydrates, providing valuable information that is not easily attainable through other techniques. By analyzing the ROA spectra of carbohydrates, researchers can gain knowledge about their ring conformations, glycosidic linkage configurations, and interactions with other molecules. For example, the conformational preferences of several monosaccharides [150] and oligosaccharides [151] in aqueous solution were determined by coupling experimental and computational ROA. ROA spectroscopy was also shown to be sensitive to the conformational

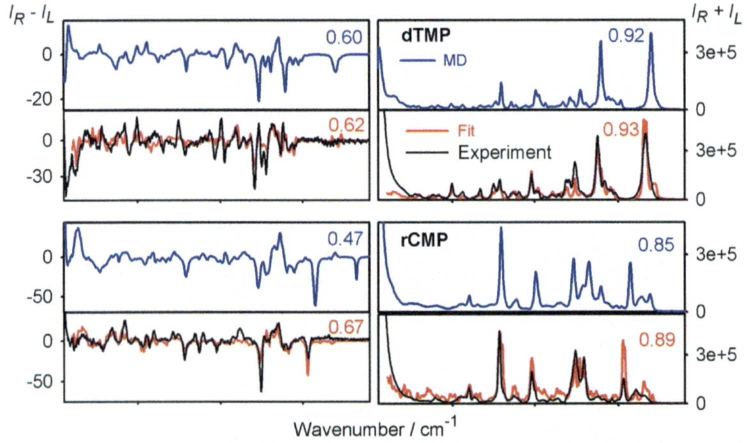

Fig. 7.8 Raman ($I_R + I_L$) and ROA ($I_R - I_L$) spectra of studied nucleotides, plain average from 700 molecular dynamics (MD) snapshots (blue), a fit [56], and experiment (black). The calculated spectra were normalized to the strongest Raman band, and similarity indexes to the experiment are indicated. Adapted from Schrenková et al. [149]

stability and rigidity of pyranose rings, the orientation of sugar hydroxyl groups, and the secondary structure of the CS's backbone of polysaccharide chondroitin sulfate [152]. Additionally, vibrational properties of glucose and sorbose were studied in extended wavenumber range (50–4000 cm^{-1}), which enabled to acquire details about CH stretching [132]. High frequency vibrations are rarely explored with ROA, but this is about to change—thanks to modern spectrometers [130] and computational algorithms that allow to simplify the anharmonic simulations [129].

Peptides and proteins can be conveniently studied with ROA too. Unlike Raman scattering, which is more sensitive to amino acid composition, ROA spectrum is simpler and, largely, represents only backbone conformation omitting the side chains [153]. Similarly to carbohydrates, experimental and computational ROA allow to determine ratio of peptide conformers in solution [131], which allowed to reconstruct whole Ramachandran plot for alanine dipeptide [154]. This technique was recently employed to investigate different conformations of α-synuclein protein associated with Parkinson's disease (Fig. 7.9) [133]. Structure of protein aggregates such as insulin amyloid fibrils was also studied with ROA [155]. It was also found, that α-helical peptides have intense structured ROA signal at low wavenumbers (around ~100 cm^{-1}) [156] that correspond to backbone deformations such as helix breathing, flattening and unwinding. Similar intense low frequency signals were observed at chiral liquids and they were assigned to molecular rotations and inter-molecular vibrations [157].

One of the big limitations of ROA is low sensitivity, which is typically addressed by the use of high concentration of the analyzed substance and long accumulation times. However, a complementary approach has been explored, known as surface-enhanced ROA (SEROA), that aims to combine the principle of surface-enhanced Raman scattering (SERS) with ROA. Similar to SERS, the ROA signal can increase to several orders of magnitude when the analyzed molecule is situated near the surface of a noble metal. Although the obtained SEROA data are often not reproducible, for carefully controlled experiments "mirror image" spectra of enantiomers could be obtained, which is considered a proof of data reliability. This approach operated on

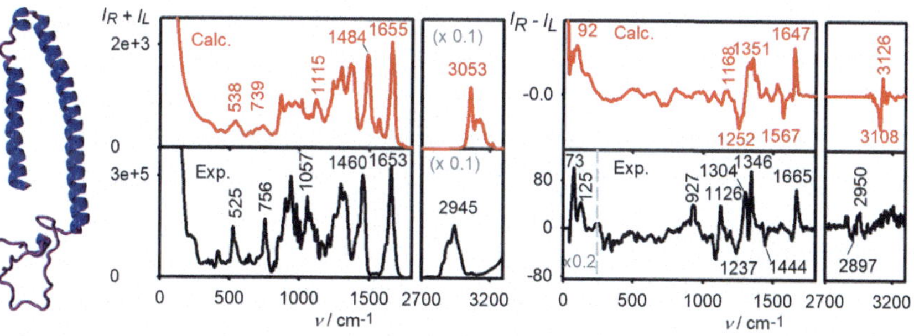

Fig. 7.9 The structure of partially α-helical αSyn, the (φ , ψ) angle distribution of free energy, calculated and experimental Raman ($I_R + I_L$) and ROA ($I_R - I_L$) spectra. Adapted from Kurochka et al. [133]

the principle of "sergeant and soldiers," where a chiral analyte induced chirality in non-chiral reporter molecules (Fig. 7.10) [158].

In presence of a chiral nickel complex, solvents were shown to exhibit extremely strong ROA signals due to "chirality transfer". The observations were related to the so-called quantum (molecular) plasmons that enable a strong chiral Rayleigh scattering of the resonating complex [159]. Later, this phenomenon was further explained by electronic circular dichroism (ECD) combined with Raman scattering of circularly polarized light from nickel and copper complexes (Fig. 7.11) [160]. It was found, that ECD strongly interferes with detection of "true" resonance Raman optical activity measurements. These findings opened the way for a new form of chiral Raman spectroscopy under resonance which is sensitive to natural chirality and structures of chiral molecules [161]. The study of vitamin B_{12} derivatives (cobalamins) confirmed that the ECD/ROA effect is proportional to the concentration and dependent on the optical pathlength of the light propagating through the sample [162]. The ECD-Raman signal can be subtracted from the recorded differential scattering in order to obtain "true" resonance ROA (RROA) [163]. This approach was utilized in the study on the model cobalt complex where RROA bands were shown to bear information on electronic excited states [164]. The ECD-Raman effect itself was also shown to allow chiral recognition of vanadium complexes at low concentration (10^{-4} mol/L) [165].

ROA spectrometers also allow for measurement of circular polarized luminescence (CPL), but there are some differences compared to conventional CPL spectrometers. For example, ROA instruments usually operate at fixed excitation wavelength, due to the use of lasers as the emission source, as well as fixed spectra accumulation range [8]. Nevertheless, ROA spectrometers proved suitable to measure induced CPL in some lanthanide metals, which can be used to probe the chemical environment [166]. The capabilities of the technique were demonstrated on recognition of

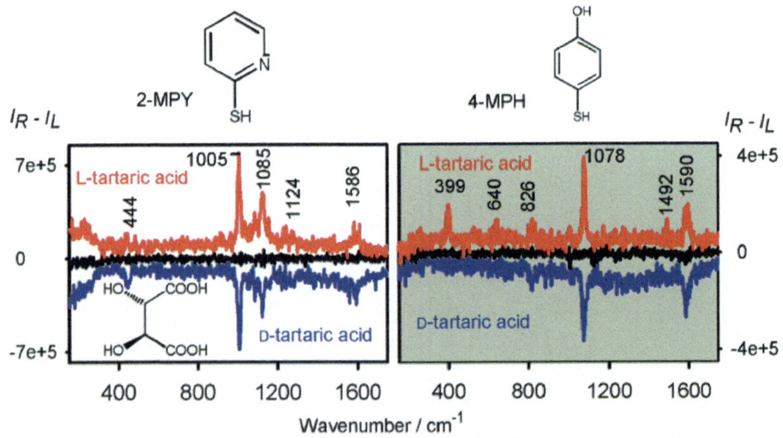

Fig. 7.10 SEROA spectra for 2-mercaptopyridine (left) and 4-mercaptophenol (right) linkers; noise spectra obtained without the chiral modifier (tartaric acid) are plotted by the black line. The concentration of the chiral acids was 6×10^{-6} M. Adapted from Das et al. [158]

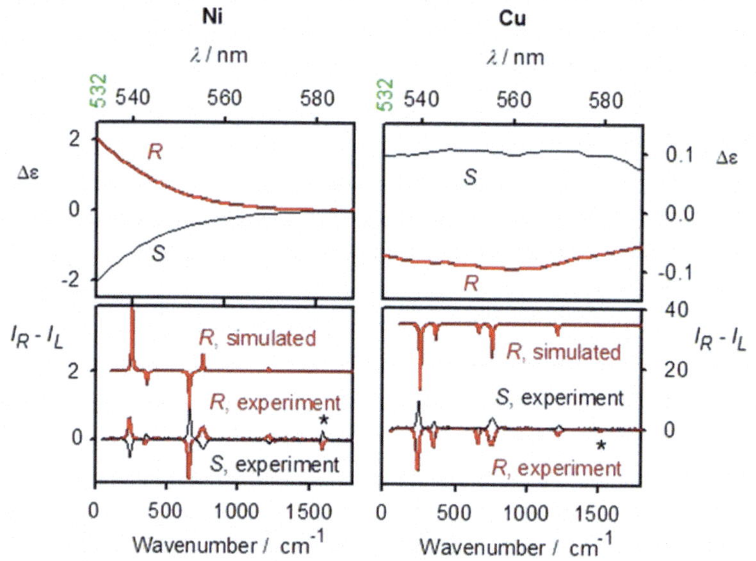

Fig. 7.11 ECD (Δε, L mol^{-1} cm^{-1} top) and ROA (ΔI, bottom, arbitrary units) spectra of Ni and Cu solutions in $CHCl_3$ within the region 0–1800 cm^{-1}, relative to the 532 nm laser, the asterisk marks the strongest band coming from the complex itself. Adapted from Wu et al. [160]

oligosaccharides with europium salts [108, 167]. In another study, CPL of three lanthanoids (Eu, Sm, and Er) were compared in presence of N-acetylneuraminic acid and Eu was found to exhibit the strongest CPL [168]. Amino acid composition of peptides was selectively probed with CPL of Eu [169]. The same metal provided specific CPL signals for single and double-stranded DNA [170]. Also, supramolecular chirality of fibrillar protein aggregates was found to be easily distinguishable using CPL of Eu complex with tris (dipicolinic acid) [126]. In case of rationally designed chiral ligand of Eu, the complex allowed to distinguish a number of phosphate groups of a nucleotide by the shape of CPL spectrum [171].

7.5 Conclusion and Future Perspective

Raman optical activity (ROA) spectroscopy has undergone a significant development in the last five decades, enabling its widespread applications in various scientific disciplines. The combination of Raman scattering and optical activity measurements provides unique insights into the chiral properties, structural characteristics, and dynamic behavior of molecules. Here, we have discussed the historical roots of ROA spectroscopy, highlighting its evolution from early theoretical concepts to the development of experimental techniques. We have described the basic components of ROA spectrometers, including light sources, optical systems, and detectors, which are crucial for obtaining high-quality ROA spectra. Selected applications of the ROA technique were presented, hopefully documenting possibilities of this type of spectroscopy for molecular studies.

Despite the progress, several challenges and opportunities lie ahead. Improving sensitivity, advancing data analysis and interpretation, broadening the range of applicable molecules, integrating with imaging techniques, exploring time-resolved and ultrafast chiral dynamics, combining with other techniques, embracing emerging technologies, and delving into chiral materials and nanotechnology are areas that require further research and development. These and other challenges will help the ROA spectroscopy to become even more useful tool in analytical chemistry, biological research, and material science.

References

1. S. Ostovar pour, L.D. Barron, S.T. Mutter, E.W. Blanch, in *Chiral Analysis*, 2nd edn., ed. by P.L. Polavarapu (Elsevier, 2018), pp. 249–291
2. V. Parchaňský, J. Kapitán, P. Bouř, RSC Adv. **4**, 57125–57136 (2014)
3. L.D. Barron, Biomed. Spectrosc. Imaging **4**, 223–253 (2015)
4. L.A. Nafie, in *Encyclopedia of Spectroscopy and Spectrometry*, 3rd edn., eds. by J.C. Lindon, G.E. Tranter, D.W. Koppenaal (Academic Press, Oxford, 2017), pp. 891–899
5. W. Hug, in *Encyclopedia of Spectroscopy and Spectrometry*, 3rd edn., eds. by J.C. Lindon, G.E. Tranter, D.W. Koppenaal (Academic Press, Oxford, 2017), pp. 881–890
6. J. Hudecová, P. Bouř, in *Vibrational Spectroscopy in Protein Research* (Elsevier, 2020), pp. 219–248
7. M. Krupová, J. Kessler, P. Bouř, ChemPlusChem **85**, 561–575 (2020)
8. T. Wu, Phys. Chem. Chem. Phys. **24**, 15672–15686 (2022)
9. G.G. Hoffmann, in *Encyclopedia of Spectroscopy and Spectrometry*, 3rd edn., eds. by J.C. Lindon, G.E. Tranter, D.W. Koppenaal (Academic Press, Oxford, 2017), pp. 853–862
10. L. Barron, M. Bogaard, A. Buckingham, J. Am. Chem. Soc. **95**, 603–605 (1973)
11. L.D. Barron, *Molecular Light Scattering and Optical Activity* (Cambridge University Press, 2009)
12. L.D. Barron, A.D. Buckingham, Chem. Phys. Lett. **492**, 199–213 (2010)
13. L.A. Nafie, Chirality **32**, 667–692 (2020)
14. P.L. Polavarapu, *Vibrational Spectra: principles and Applications with Emphasis on Optical Activity* (Elsevier, 1998)
15. P.L. Polavarapu, Compr. Chiroptical Spectrosc. **2**, 387–420 (2012)
16. P.L. Polavarapu, *Chiroptical Spectroscopy: fundamentals and Applications* (CRC Press, 2016)
17. L.D. Barron, S.T. Mutter, E.W. Blanch, in *Chiral Analysis* (Elsevier, 2018), pp. 249–291
18. C.D. Syme, E.W. Blanch, C. Holt, R. Jakes, M. Goedert, L. Hecht, L.D. Barron, Eur. J. Biochem. **269**, 148–156 (2002)
19. F. Zhu, J. Kapitan, G.E. Tranter, P.D. Pudney, N.W. Isaacs, L. Hecht, L.D. Barron, Proteins: Struct. Funct. Bioinform. **70**, 823–833 (2008)
20. M. Quack, in *Quantum Systems in Chemistry and Physics: progress in Methods and Applications* (Springer, 2012), pp. 47–76
21. S. Mason, Trends Pharmacol. Sci. **7**, 20–23 (1986)
22. G.H. Wagnière, *On Chirality and the Universal Asymmetry: reflections on Image and Mirror Image* (Wiley, 2007)
23. R.S. Cahn, C. Ingold, V. Prelog, Angew. Chem., Int. Ed. Engl. **5**, 385–415 (1966)
24. W.A. Bonner, Orig. Life Evol. Biosph. **25**, 175–190 (1995)

25. J.P. Riehl, *Mirror-Image Asymmetry: an Introduction to the Origin and Consequences of Chirality* (Wiley, 2010)
26. R. Popa, J. Mol. Evol. **44**, 121–127 (1997)
27. W.A. Bonner, in *AIP Conference Proceedings* (American Institute of Physics, 1996), pp. 17–49
28. A. Salam, J. Mol. Evol. **33**, 105–113 (1991)
29. J.V. de Julián-Ortiz, C. de Gregorio Alapont, I. Ríos-Santamarina, R. García-Doménech, J. Gálvez, J. Mol. Graph. Modell. **16**, 14–18 (1998)
30. B. Richard, C.N. Weiskopf, R.S. Gary, Anesthesiology **97**, 497–502 (2002)
31. B. Kasprzyk-Hordern, Chem. Soc. Rev. **39**, 4466–4503 (2010)
32. A.J. Hutt, S.C. Tan, Drugs **52**(Suppl 5), 1–12 (1996)
33. G.A. Hembury, V.V. Borovkov, Y. Inoue, Chem. Rev. **108**, 1–73 (2008)
34. L.D. Barron, Chirality **24**, 879–893 (2012)
35. C. Liang, K. Mislow, J. Am. Chem. Soc. **116**, 3588–3592 (1994)
36. S.D. Banik, N. Nandi, *Biochirality: origins, Evolution and Molecular Recognition* (2013), pp. 255–305
37. J.I. Kwiecińska, M. Cieplak, J. Phys.: Condens. Matter **17**, S1565 (2005)
38. J. Chela-Flores, Chirality **6**, 165–168 (1994)
39. R. Corradini, S. Sforza, T. Tedeschi, R. Marchelli, Chirality: the pharmacological, biological, and chemical consequences of molecular asymmetry **19**, 269–294 (2007)
40. D.B. Laurence, in *Strategies of Life Detection*, eds. by O. Botta et al. (Springer US, Boston, MA, 2008), pp. 187–201
41. A. Long, O. Perraud, M. Albalat, V. Robert, J.-P. Dutasta, A. Martinez, J. Org. Chem. **83**, 6301–6306 (2018)
42. H.-Y. Wang, S.A. Blaszczyk, G. Xiao, W. Tang, Chem. Soc. Rev. **47**, 681–701 (2018)
43. S. Gim, G. Fittolani, Y. Nishiyama, P. H. Seeberger, Angew. Chem. Int. Edit. **59**, 22577–22583 (2020)
44. M. O'Neill, S.M. Kelly, Adv. Mater. (Deerfield Beach, Fla.) **23**, 566–584 (2011)
45. C. Duan, Z. Cheng, B. Wang, Small **17**, 2007306 (2021)
46. S. Ma, J. Ahn, J. Moon, Adv. Mater. **33**, 2005760 (2021)
47. A. Tran, C.E. Boott, M.J. MacLachlan, Adv. Mater. **32**, 1905876 (2020)
48. M. Manoccio, M. Esposito, A. Passaseo, M. Cuscunà, V. Tasco, Micromachines **12**, 6 (2020)
49. P. Lv, X. Lu, L. Wang, W. Feng, Adv. Func. Mater. **31**, 2104991 (2021)
50. G.E. Fenoy, M. Rafti, W.A. Marmisollé, O. Azzaroni, Mater. Adv. **2**, 7731–7740 (2021)
51. Y. Du, D. Cao, B. Li, H. Lü, Y. Shen, Electrochim. Acta **395**, 139201 (2021)
52. S. Raza, X. Li, F. Soyekwo, D. Liao, Y. Xiang, C. Liu, Eur. Polym. J. **160**, 110773 (2021)
53. S. Nagahara, Y. Okada, Y. Kitano, K. Chiba, Chem. Sci. **12**, 12911–12917 (2021)
54. A.M. Garcia et al., Chem **4**, 1862–1876 (2018)
55. W.H. Brooks, W.C. Guida, K.G. Daniel, Curr. Top. Med. Chem. **11**, 760–770 (2011)
56. I.K. Reddy, R. Mehvar, *Chirality in Drug Design and Development* (CRC Press, 2004)
57. S.R. LaPlante, L.D. Fader, K.R. Fandrick, D.R. Fandrick, O. Hucke, R. Kemper, S.P. Miller, P.J. Edwards, J. Med. Chem. **54**, 7005–7022 (2011)
58. O. Mcconnell et al., Chirality: the pharmacological, biological, and chemical consequences of molecular asymmetry **19**, 658–682 (2007)
59. S.R. LaPlante, P.J. Edwards, L.D. Fader, A. Jakalian, O. Hucke, ChemMedChem **6**, 505–513 (2011)
60. C. Brown, *Chirality in Drug Design and Synthesis* (Academic Press, 2013)
61. I. Agranat, H. Caner, Drug Discov. Today **4**, 313–321 (1999)
62. J.R. Brandt, F. Salerno, M.J. Fuchter, Nat. Rev. Chem. **1**, 0045 (2017)
63. E. Karjalainen, D.F. Izquierdo, V. Marti-Centelles, S.V. Luis, H. Tenhu, E. Garcia-Verdugo, Polym. Chem. **5**, 1437–1446 (2014)
64. C.S. Wilcox, L.M. Greer, V. Lynch, J. Am. Chem. Soc. **109**, 1865–1867 (1987)
65. M. Khorloo, X. Yu, Y. Cheng, H. Zhang, S. Yu, J.W. Lam, M. Zhu, B.Z. Tang, ACS Nano **15**, 1397–1406 (2020)

66. J.A. Schmidt, J.A. Weatherby, I.J. Sugden, A. Santana-Bonilla, F. Salerno, M.J. Fuchter, E.R. Johnson, J. Nelson, K.E. Jelfs, Cryst. Growth Des. **21**, 5036–5049 (2021)
67. R. Eelkema, B.L. Feringa, Org. Biomol. Chem. **4**, 3729–3745 (2006)
68. K.E. Shopsowitz, H. Qi, W.Y. Hamad, M.J. MacLachlan, Nature **468**, 422–425 (2010)
69. P. Atkins, L. Barron, Mol. Phys. **16**, 453–466 (1969)
70. L.D. Barron, F. Zhu, L. Hecht, G.E. Tranter, N.W. Isaacs, J. Mol. Struct. **834**, 7–16 (2007)
71. L. Barron, L. Hecht, E. Blanch, A. Bell, Prog. Biophys. Mol. Biol. **73**, 1–49 (2000)
72. T.A. Keiderling, Chem. Rev. **120**, 3381–3419 (2020)
73. L.D. Barron, L. Hecht, I.H. McColl, E.W. Blanch, Mol. Phys. **102**, 731–744 (2004)
74. J.R. Escribano, L.D. Barron, Mol. Phys. **65**, 327–344 (1988)
75. F. Zhu, N.W. Isaacs, L. Hecht, L.D. Barron, Structure **13**, 1409–1419 (2005)
76. A.A. Bunaciu, H.Y. Aboul-Enein, Ş. Fleschin, Appl. Spectrosc. Rev. **50**, 176–191 (2015)
77. J.B. Aitken et al., Radiat. Phys. Chem. **79**, 176–184 (2010)
78. P.L. Polavarapu, J. Phys. Chem. **94**, 8106–8112 (1990)
79. L. Barron, A. Buckingham, Annu. Rev. Phys. Chem. **26**, 381–396 (1975)
80. W. Hug, in *Handbook of Vibrational Spectroscopy*, eds. by J.M. Chalmers, P.R. Griffiths (Wiley, 2006), pp. 745–758
81. L.A. Nafie, *Vibrational Optical Activity: principles and Applications* (Wiley, 2011)
82. L.A. Nafie, Annu. Rev. Phys. Chem. **48**, 357–386 (1997)
83. A. Melcrová, J. Kessler, P. Bouř, J. Kaminský, Phys. Chem. Chem. Phys. **18**, 2130–2142 (2016)
84. C. Mensch, L.D. Barron, C. Johannessen, Phys. Chem. Chem. Phys. **18**, 31757–31768 (2016)
85. N.R. Yaffe, A. Almond, E.W. Blanch, J. Am. Chem. Soc. **132**, 10654–10655 (2010)
86. A. Bell, L. Hecht, L. Barron, J. Am. Chem. Soc. **120**, 5820–5821 (1998)
87. A.F. Bell, L. Hecht, L.D. Barron, J. Am. Chem. Soc. **119**, 6006–6013 (1997)
88. F. Zhu, N.W. Isaacs, L. Hecht, L.D. Barron, J. Am. Chem. Soc. **127**, 6142–6143 (2005)
89. C.V. Raman, Proc. Math. Sci. **37**, 342–349 (1953)
90. C.V. Raman, Indian J. Phys. **2**, 387–398 (1928)
91. C. Raman, Trans. Faraday Soc. **25**, 781–792 (1929)
92. F.A. Miller, G.B. Kauffman, J. Chem. Educ. **66**, 795 (1989)
93. R. Singh, F. Riess, Curr. Sci. **75**, 965–971 (1998)
94. L. Nasdala, D.C. Smith, R. Kaindl, M.A. Ziemann, A. Beran, E. Libowitzky, Spectrosc. Methods Mineralo. **6**, 281–343 (2004)
95. D.A. Long, Int. Rev. Phys. Chem. **7**, 317–349 (1988)
96. R. Singh, Phys. Perspect. **4**, 399–420 (2002)
97. G.S. Landsberg, L.I. Mandelstam, J. Russian Phys.-Chem. Soc. Phys. Sect. **60**, 335 (1928)
98. P.L. Polavarapu, Chirality **14**, 768–781 (2002)
99. T.M. Lowry, *Optical Rotatory Power* (Dover publications, 1964)
100. H. Landolt, *The Optical Rotating Power of Organic Substances and Its Practical Applications* (Chemical Publishing Company, 1902)
101. J. Gal, Chirality: the pharmacological, biological, and chemical consequences of molecular asymmetry **20**, 5–19 (2008)
102. J. Gal, Nat. Chem. **9**, 604–605 (2017)
103. H. Flack, Acta Crystallogr. A **65**, 371–389 (2009)
104. J. Gal, Chirality **23**, 1–16 (2011)
105. G.D. Fasman, *Circular Dichroism and the Conformational Analysis of Biomolecules* (Springer Science & Business Media, 2013)
106. B. Kahr, Chirality **30**, 351–368 (2018)
107. S. Mitchell, Nature **166**, 434–435 (1950)
108. D.M. Rogers, S.B. Jasim, N.T. Dyer, F. Auvray, M. Réfrégiers, J.D. Hirst, Chem **5**, 2751–2774 (2019)
109. L.D. Barron, *Molecular Light Scattering and Optical Activity* (Cambridge University Press, 2004)

110. N. Berova, P.L. Polavarapu, K. Nakanishi, R.W. Woody, *Comprehensive Chiroptical Spectroscopy, Volume 2: applications in Stereochemical Analysis of Synthetic Compounds, Natural Products, and Biomolecules*, vol. 2 (Wiley, 2012)
111. N. Berova, P.L. Polavarapu, K. Nakanishi, R.W. Woody, *Comprehensive Chiroptical Spectroscopy, Volume 1: Instrumentation, Methodologies, and Theoretical Simulations*, vol. 1 (Wiley, 2011)
112. W. Hug, S. Kint, G.F. Bailey, J.R. Scherer, J. Am. Chem. Soc. **97**, 5589–5590 (1975)
113. L. Hecht, L.D. Barron, W. Hug, Chem. Phys. Lett. **158**, 341–344 (1989)
114. L.D. Barron, A.D. Buckingham, Mol. Phys. **20**, 1111–1119 (1971)
115. H. Li, L.A. Nafie, J. Raman Spectrosc. **43**, 89–94 (2012)
116. L.D. Barron, A.D. Buckingham, J. Am. Chem. Soc. **96**, 4769–4773 (1974)
117. L. Hecht, L.A. Nafie, Mol. Phys. **72**, 441–469 (1991)
118. A. Baiardi, J. Bloino, V. Barone, J. Chem. Theory Comput. **14**, 6370–6390 (2018)
119. M.J. Frisch et al., Gaussian 16 Rev. A.03 (2016)
120. T. Helgaker, K. Ruud, K.L. Bak, P. Jørgensen, J. Olsen, Faraday Discuss. **99**, 165–180 (1994)
121. K. Ruud, T. Helgaker, P. Bouř, J. Phys. Chem. A **106**, 7448–7455 (2002)
122. P. Bouř, J. Sopková, L. Bednárová, P. Maloň, T.A. Keiderling, J. Comput. Chem. **18**, 646–659 (1997)
123. S. Yamamoto, X. Li, K. Ruud, P. Bour, J. Chem. Theory Comput. **8**, 977–985 (2012)
124. J. Kessler, J. Kapitán, P. Bouř, J. Phys. Chem. Lett. **6**, 3314–3319 (2015)
125. L.A. Nafie, Appl. Spectrosc. **50**, 14A-26A (1996)
126. M. Krupová, J. Kapitán, P. Bouř, ACS Omega **4**, 1265–1271 (2019)
127. E.W. Blanch, L. Hecht, L.D. Barron, in *Chiral Analysis*, eds. by K.W. Busch, M.A. Busch (Elsevier, Amsterdam, 2006), pp. 545–594
128. M. Das, D. Gangopadhyay, J. Hudecová, J. Kessler, J. Kapitán, P. Bouř, Chempluschem, e202300219 (2023)
129. Q. Yang, J. Kapitán, P. Bouř, J. Bloino, J. Phys. Chem. Lett. **13**, 8888–8892 (2022)
130. P. Michal, R. Čelechovský, M. Dudka, J. Kapitán, M. Vůjtek, M. Berešová, J. Šebestík, K. Thangavel, P. Bouř, J. Phys. Chem. B **123**, 2147–2156 (2019)
131. M. Hope, J. Šebestík, J. Kapitán, P. Bouř, J. Phys. Chem. A **124**, 674–683 (2020)
132. V. Palivec, P. Michal, J. Kapitán, H. Martinez-Seara, P. Bouř, ChemPhysChem **21**, 1272–1279 (2020)
133. A. Kurochka, J. Průša, J. Kessler, J. Kapitán, P. Bouř, Phys. Chem. Chem. Phys. **23**, 16635–16645 (2021)
134. L.A. Nafie, T.B. Freedman, Chem. Phys. Lett. **154**, 260–266 (1989)
135. D. Che, L. Hecht, L.A. Nafie, Chem. Phys. Lett. **180**, 182–190 (1991)
136. G.-S. Yu, D. Che, T.B. Freedman, L.A. Nafie, Tetrahedron: Asymmetry **4**, 511–516 (1993)
137. G.-S. Yu, L.A. Nafie, Chem. Phys. Lett. **222**, 403–410 (1994)
138. G.S. Yu, D. Che, T.B. Freedman, L.A. Nafie, Biospectroscopy **1**, 113–123 (1995)
139. G.S. Yu, T.B. Freedman, L.A. Nafie, J. Raman Spectrosc. **26**, 733–743 (1995)
140. M. Vargek, T.B. Freedman, L.A. Nafie, J. Raman Spectrosc. **28**, 627–633 (1997)
141. L.A. Nafie, B.E. Brinson, X. Cao, D.A. Rice, O.M. Rahim, R.K. Dukor, N.J. Halas, Appl. Spectrosc. **61**, 1103–1106 (2007)
142. M. Unno, T. Kikukawa, M. Kumauchi, N. Kamo, J. Phys. Chem. B **117**, 1321–1325 (2013)
143. W. Hug, Appl. Spectrosc. **57**, 1–13 (2003)
144. V. Profant, A. Jegorov, P. Bouř, V. Baumruk, J. Phys. Chem. B **121**, 1544–1551 (2017)
145. P. Michal, J. Hudecová, R. Čelechovský, M. Vůjtek, M. Dudka, J. Kapitán, Symmetry-Basel **14**, 990 (2022)
146. K. Dobšíková, P. Michal, D. Spalovská, M. Kuchař, N. Paškanová, R. Jurok, J. Kapitán, V. Setnička, Analyst **148**, 1337–1348 (2023)
147. P. Fagan, L. Kocourková, M. Tatarkovič, F. Kralík, M. Kuchar, V. Setnička, P. Bouř, ChemPhysChem **18**, 2258–2265 (2017)
148. J. Hudecová, J. Kapitán, M. Dračínský, P. Michal, V. Profant, P. Bouř, Chem.-Eur. J. **28**, e202202045 (2022)

149. V. Schrenková, M.S.P. Kkadan, J. Kessler, J. Kapitán, P. Bouř, Phys. Chem. Chem. Phys. **25**, 8198–8208 (2023)
150. V. Palivec, V. Kopecký, P. Jungwirth, P. Bouř, J. Kaminský, H. Martinez-Seara, Phys. Chem. Chem. Phys. **22**, 1983–1993 (2020)
151. V. Palivec, C. Johannessen, J. Kaminský, H. Martinez-Seara, PLOS Comput. Biol. **18**, e1009678 (2022)
152. V. Profant, C. Johannessen, E.W. Blanch, P. Bouř, V. Baumruk, Phys. Chem. Chem. Phys. **21**, 7367–7377 (2019)
153. V. Andrushchenko, A. Kurochka, J. Kubelka, J. Kaminský, in *Reference Module in Chemistry, Molecular Sciences and Chemical Engineering* (Elsevier, 2022)
154. J. Jungwirth, J. Šebestík, M. Šafařík, J. Kapitán, P. Bouř, J. Phys. Chem. B **121**, 8956–8964 (2017)
155. J. Kessler, S. Yamamoto, P. Bouř, Phys. Chem. Chem. Phys. **19**, 13614–13621 (2017)
156. S. Yamamoto, S. Ishiro, J. Kessler, P. Bouř, Phys. Chem. Chem. Phys. **23**, 26501–26509 (2021)
157. P. Michal, J. Kapitán, J. Kessler, P. Bouř, Phys. Chem. Chem. Phys. **24**, 19722–19733 (2022)
158. M. Das, D. Gangopadhyay, J. Šebestík, L. Habartová, P. Michal, J. Kapitán, P. Bouř, Chem. Commun. **57**, 6388–6391 (2021)
159. G.J. Li, J. Kessler, J. Cheramy, T. Wu, M.R. Poopari, P. Bouř, Y.J. Xu, Angew. Chem. Int. Edit. **58**, 16495–16498 (2019)
160. T. Wu, G.J. Li, J. Kapitán, J. Kessler, Y.J. Xu, P. Bouř, Angew. Chem. Int. Ed. **59**, 21895–21898 (2020)
161. G.J. Li, M. Alshalalfeh, Y.Q. Yang, J.R. Cheeseman, P. Bouř, Y.J. Xu, Angew. Chem. Int. Ed. **60**, 22004–22009 (2021)
162. E. Machalska, G. Zajac, A.J. Wierzba, J. Kapitán, T. Andruniow, M. Spiegel, D. Gryko, P. Bouř, M. Baranska, Angew. Chem. Int. Ed. **60**, 21205–21210 (2021)
163. G. Zając, P. Bouř, J. Phys. Chem. B **126**, 355–367 (2022)
164. T. Wu, J. Kapitán, P. Bouř, J. Phys. Chem. Lett. **13**, 3873–3877 (2022)
165. E. Machalska, N. Hachlica, G. Zajac, D. Carraro, M. Baranska, G. Licini, P. Bouř, C. Zonta, A. Kaczor, Phys. Chem. Chem. Phys. **23**, 23336–23340 (2021)
166. T. Wu, J. Kapitán, V. Andrushchenko, P. Bouř, Anal. Chem. **89**, 5043–5049 (2017)
167. T. Wu, J. Kessler, J. Kaminský, P. Bouř, Chem.-Asian J. **13**, 3865–3870 (2018)
168. T. Wu, P. Bouř, Chem. Commun. **54**, 1790–1792 (2018)
169. E. Brichtová, J. Hudecová, N. Vršková, J. Šebestík, P. Bouř, T. Wu, Chem.-Eur. J. **24**, 8664–8669 (2018)
170. T. Wu, P. Bouř, V. Andrushchenko, Sci. Rep.-UK **9**, 1068 (2019)
171. J. Sahoo, T. Wu, B. Klepetářová, J. Valenta, P. Bouř, ChemPlusChem **85**, 694–700 (2020)

Chapter 8
Surface-Enhanced Raman Spectroscopy (SERS)

Alireza Mazaheri Tehrani, Faezeh Mohaghegh, and Arnulf Materny

8.1 Introduction

A considerable problem when measuring Raman signals is their rather small intensity, which makes it difficult to apply Raman spectroscopy to samples with low concentrations of relevant molecules. The Surface Enhanced Raman Scattering (SERS) effect results in significantly amplified Raman signals, which helps to overcome this problem in many cases. The amplification is mainly due to an enhancement of electromagnetic fields in close proximity to nanostructured coin metal surfaces. SERS was first observed in 1974 by Fleischman et al. [1], who studied the Raman signal of pyridine molecules adsorbed onto a silver electrode from an aqueous solution. They reported that the electrode surface area and its roughness played a crucial role in the signal enhancement.

Since SERS had been discovered accidentally, theories were developed to explain the mechanisms behind this enhancement when molecules are placed close to the metal surface. Two major contributions have been widely recognized: electromagnetic enhancement (EME) and chemical (electronic) enhancement (CE). CE results from the interaction between the electronic systems of molecules and the collective oscillations of the electrons on the metal surface (surface plasmons), while EME happens due to the enhancement of the electomagnetic field in close vicinity of the metal surface caused by those collective oscillations.

While simple SERS substrates including metal clusters in aqueous solutions or roughened metal films produce a moderate enhancement, more complex ones such as carefully designed nano-structures (and particles) have been reported to result in high enhancement factors of 10^{10}–10^{12} [2–4]. However, all these substrates suffer from poor reproducibility [5], which is crucial for studying the elementary processes contributing to SERS or for investigations of elementary properties of molecules

A. M. Tehrani · F. Mohaghegh · A. Materny (✉)
School of Science, Constructor University, Campus Ring 1, 28759 Bremen, Germany
e-mail: amaterny@constructor.university

D. K. Singh et al. (eds.), *Raman Spectroscopy*, Springer Series in Optical Sciences 248,
https://doi.org/10.1007/978-981-97-1703-3_8

by Raman spectroscopy. Flat and smooth surfaces would offer highly reproducible conditions, however, a direct coupling of electromagnetic waves to the plasmons is challenging in this case due to a phase mismatch between the light field and the surface plasmons. To overcome this, special setups can be used, utilizing total internal reflection conditions, originally proposed by Otto [6] and Kretschmann [7].

The aim of this chapter is to firstly present the basics of SERS. Then, the focus is on the so-called "Kretschmann configuration", which allows for reproducible SERS investigations. Since this is a rather special technique, the theoretical background is explained and basic studies are presented helping to understand different aspects not only relevant for this SERS technique. As an example, the application of the KC to the investigation of an antibiotics as contaminant of water is finally presented.

8.2 Major Mechanisms Contributing to Surface-Enhanced Raman Scattering

Currently, two mechanisms of surface enhancement are considered to be the most important basic contributions to the SERS effect. The first mechanism is known as electromagnetic enhancement (EME), where the analyte is in proximity to the (typically nano-structured) coin metal surface; this effect also plays a role when molecules are physisorbed and not in direct contact to the metal. In this case, the interaction between the light field and the plasmons of the metal surface leads to a field enhancement, which affects both the excitation light and the Raman signal. The second mechanism, known as chemical or electronic enhancement (CE), depends on the analyte and metal surface forming a hemical bond (chemisorption). Due to the contact with metal electrons, this bond enables the transfer of charges (electrons or holes) between the metal surface and the analyte, significantly increasing the molecule's polarizability. Additionally, as will be covered later, the emergence of new electronic resonances as a result of charge-transfer state formation can improve the Raman scattering. It is generally acknowledged that the combination of these mechanisms is what causes the SERS phenomena [8].

8.2.1 Electromagnetic Enhancement

Using a small metallic sphere in a dielectric surrounding as a model system provides a simple and straightforward approach to better understand the theory behind this mechanism. However, it is important to note that this simple sphere is merely an initial approximation, as more intricate systems, such as those involving well-designed nano-structures, can produce the SERS signal with extreme enhancements.

When a small metal sphere is exposed to an electric field, the resulting field at the surface can be described as follows [9]:

$$E_r = E_0\cos\theta + g\left(\frac{a^3}{r^3}\right)E_0\cos\theta \tag{8.1}$$

Here, a is the sphere's radius, E_r is the total electric field at a distance of r from its surface, θ is the angle with respect to the electric field's direction, and g is given by

$$g = \frac{\varepsilon_1(\nu_L) - \varepsilon_0}{\varepsilon_1(\nu_L) + 2\varepsilon_0} \tag{8.2}$$

where ν_L is the frequency of the incident radiation and ε_0 and ε_1 are the surrounding medium's and metal sphere's dielectric constants, respectively. At the minimum of the denominator, the value of g will be maximum. This will be the case when $\varepsilon_1(\nu_L)$ equals -2, since ε_0 is typically close to 1. The local field seen by the adjacent molecules will be stronger at the corresponding frequency ν_L, also known as the plasmon resonance frequency.

It is important to notice that the strength of the surface enhancement decreases rapidly with distance from the surface, following an inverse cubic relationship ($1/r^3$). Interestingly, the most significant enhancement is not observed around individual metal particles, but rather in regions where interactions between particles occur, known as "hot spots."

It is worth noting that the size, shape, and arrangement of particles within clusters greatly affect the enhancement of SERS. Therefore, providing an accurate description of the electromagnetic enhancement becomes a complex problem in many practical situations and requires computer simulations. Nonetheless, the basic functioning of the electromagnetic mechanism behind SERS is well understood.

8.2.2 *Charge Transfer*

It is crucial to differentiate between chemical effects, such as molecular adsorption or orientation, and the actual SERS enhancement factors. These chemical effects can influence the intensity of SERS but should not be considered as the primary source of enhancement. For instance, a negatively charged molecule would experience electrostatic repulsion and thus exhibit a significantly reduced or even absent SERS signal when attempting to adsorb it onto a negatively charged silver colloid. Conversely, in the same silver colloid solution, an intense SERS signal might be produced by a positively charged molecule. This "chemical effect" arises due to the properties of the sample and is unrelated to the SERS process itself.

Some studies propose the existence of a second enhancement mechanism known as chemical or electronic enhancement (CE). This phenomenon has been elucidated through a resonance Raman mechanism, which can be explained in two ways. The first assumption suggests that the interaction between the adsorbate and the metal

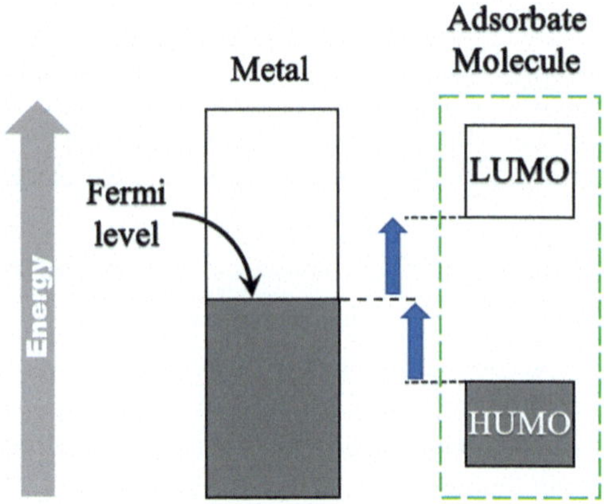

Fig. 8.1 Typical energy level diagram for a molecule adsorbed on a metal surface

surface causes shifts and broadening of the electronic states, resulting in new transition energies that align with the resonance conditions for Raman scattering at specific excitation wavelengths.

Another theory, illustrated in Fig. 8.1, is that the metal and the adsorbate create new electronic states that act as resonant intermediate states in Raman scattering. This theory states that charge transfer occurs from the adsorbate's highest occupied molecular orbital (HOMO) to the Fermi level, followed by a transfer to its lowest unoccupied molecular orbital (LUMO). This transfer is anticipated to occur at around half the energy of the adsorbate's typical excitations [10].

Distinguishing between these two effects is challenging. Based on the described processes, chemical or electronic enhancement (CE) will occur only for molecules directly in contact with the surface (chemisorbed molecules), meaning it is limited to monolayer coverage. Electromagnetic enhancement (EME) decreases when moving the molecules away from the surface, but it also contributes when molecules are not attached to the surface (physisorbed molecules). Consequently, EME in general plays a more significant role compared to CE. Nevertheless, CE must not be neglected, since it also has an influence on the vibrational energies and can therefore change the Raman spectrum itself.

8.3 Surface Plasmon Polaritons at Plane Interfaces

8.3.1 Theoretical Background

While for nano-structured metal surfaces (e.g. roughened surfaces, nano particles, etc.) an excitation of a plasmon resonance is possible when light is interacting with them, plane smooth metal surfaces do not allow for a phase-matched excitation without special arrangements. Let's examine a scenario involving a flat interface

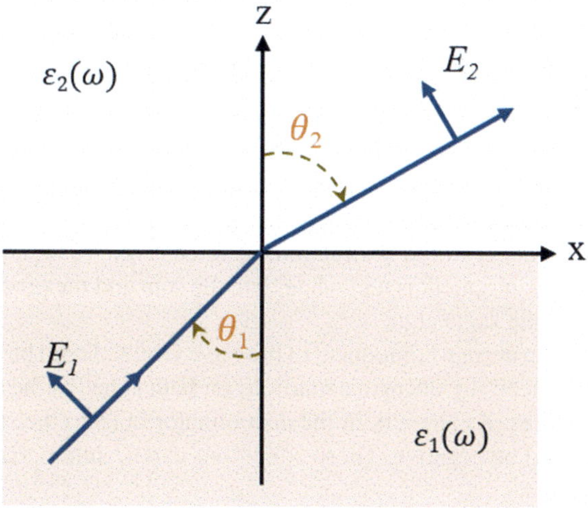

Fig. 8.2 Interface between two media with different dielectric functions

between two different media. The first medium is characterized by a complex dielectric function $\varepsilon_1(\omega)$, while the second medium is described by a real dielectric function $\varepsilon_2(\omega)$, as depicted in Fig. 8.2.

By employing a Cartesian coordinate system and setting the interface at the $z = 0$ plane, our objective is to find homogeneous solutions of Maxwell's equations that satisfy the wave equation at the interface [11]:

$$\nabla \times \nabla \times E(r, \omega) - \frac{\omega^2}{c^2}\varepsilon(r, \omega)E(r, \omega) = 0 \tag{8.3}$$

As there are no viable solutions for s-polarization, we can focus solely on p-polarized waves for both sides of the interface. The electric field vector must therefore be parallel to the plane of incidence determined by the wave vectors of the incident and reflected waves and the light must be polarized accordingly.

$$E_i = \begin{pmatrix} E_{j,x} \\ 0 \\ E_{j,z} \end{pmatrix} e^{ik_x x - iwt} e^{ik_{j,z}z}, \ j = 1, 2 \tag{8.4}$$

Taking into consideration the conservation of the wave vector parallel to the interface (k_x), the fact that there is no source in neither of the medium, i.e. ($\nabla D = 0$), and the corresponding boundary conditions at the interface leads to a dispersion relation and an expression for the normal component of the wave vector [11]:

$$k_x^2 = \frac{\varepsilon_1 \varepsilon_2}{\varepsilon_1 + \varepsilon_2} k^2 = \frac{\varepsilon_1 \varepsilon_2}{\varepsilon_1 + \varepsilon_2} \frac{\omega^2}{c^2} \tag{8.5}$$

$$k_{j,z}^2 = \frac{\varepsilon_j^2}{\varepsilon_1 + \varepsilon_2} k^2, j = 1, 2 \tag{8.6}$$

Here, the discussion about the existence of the surface modes will be of interest. To simplify the analysis, we assume that the imaginary component of the complex dielectric function is considerably smaller than its real component. In order for a propagating wave along the surface to occur, it is necessary for the wave vector component in the x-direction, denoted as k_x, to be a real value. Referring to (8.5), this requirement is met when both the sum and the product of the dielectric functions have the same sign.

The normal component of the wave vector should be purely imaginary to guarantee exponentially decaying solutions on both sides, as the definition of the surface wave requires. As a result, in the denominator in (8.6) the sum of the dielectric functions should be negative. These conditions can be summarized as follows [11]:

$$\varepsilon_1(\omega)\varepsilon_2(\omega) < 0 \tag{8.7}$$

$$\varepsilon_1(\omega) + \varepsilon_2(\omega) < 0 \tag{8.8}$$

This implies that one of the dielectric functions must have a negative value, with its absolute value being greater than that of the other dielectric function. In practical terms, metals like gold and silver are well-suited for this purpose, as they possess a significantly negative real component and a small imaginary component in their dielectric constants. Consequently, at the interfaces between metals and dielectrics, localized modes can be supported. For a detailed mathematical description, which is out of the scope of this chapter, we refer the interested reader to Chap. 12 of [11].

8.3.2 Realization of Excitation of Surface Plasmon Polaritons

To understand the requirements for exciting surface plasmon polaritons, it is essential to consider energy and momentum conservation. Analyzing the dispersion relation of surface waves becomes crucial in meeting these conditions. To facilitate this analysis, (8.5) is graphically represented as a solid blue line in Fig. 8.3.

The dashed-dotted line in Fig. 8.3 represents the light line in air, described by the equation $\omega = ck_x$. The solid line depicts the dispersion relation of surface plasmon polaritons (SPPs).

At lower energies, the SPP dispersion relation and the light line approach each other closely. However, it is evident that for any given energy $\hbar\omega$ the wave vector of light in free space never crosses the dispersion line representing the wave vectors of the SPPs. Consequently, when incident from free space onto flat surfaces, light

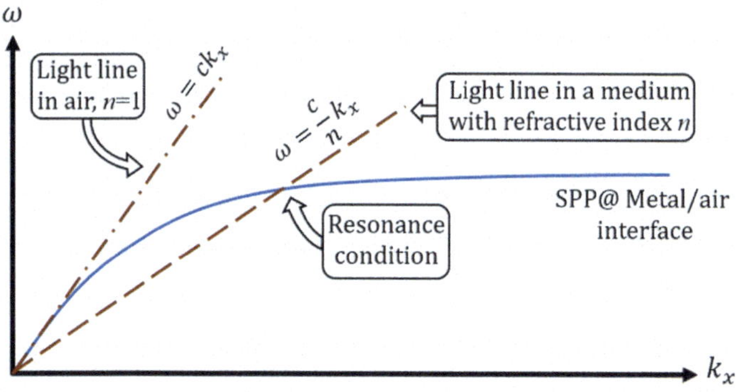

Fig. 8.3 The dispersion relation of SPPs together with the light lines of free-space and in glass. The latter has a smaller slope due to the higher index of refraction

cannot excite SPPs due to the phase-mismatch. To excite SPPs, the wave vector of the incident light needs to have a crossing point with the SPP curve [11].

There are various methods to achieve this. One approach involves tilting the light line by a factor of n, as illustrated by the dashed line in Fig. 8.3. This condition can be achieved by exciting SPPs with evanescent waves generated at the interface of a medium with a refractive index greater than 1, such as in total internal reflection.

In 1968, Otto proposed an experimental setup to fulfill this condition, known as the Otto configuration [6], which is one of the most commonly used setups. As shown in the left panel of Fig. 8.4, the evanescent wave (with wave vector Kevan) generated at the glass/air interface reaches the metal/air interface and supports SPPs. Adjusting the angle of incidence allows for satisfying both the total internal reflection condition and the resonance condition for SPP excitation. The resonance is characterized by a minimum intensity of the reflected beam, which is used to experimentally determine the angle. However, the utilization of the Otto configuration presented challenges due to the complexity of controlling the air gap between the two interfaces.

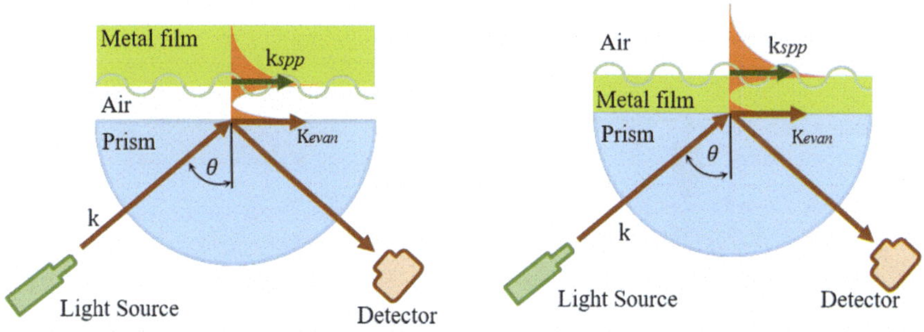

Fig. 8.4 Excitation of surface plasmons; left: Otto configuration, right: Kretschmann configuration (KC)

In 1971 Kretschmann [12] suggested another way to make use of the same effect using a simpler setup by depositing a thin metal film directly on the prism surface. The right panel of Fig. 8.4 illustrates the generation of evanescent waves at the glass/metal interface, which in turn excites surface plasmon polaritons (SPPs) at the metal/air interface. By plotting the reflectivity of the incident beam as a function of the angle of incidence, a resonance condition can be observed as a dip at a specific angle. This decrease in reflected light intensity indicates the coupling of light to the SPPs.

Gratings, or periodic nanostructures, also provide a means to couple light into SPPs [13, 14]. The shape, size, and spacing of metallic nanostructures play a crucial role in optimizing SERS signals. These parameters have been extensively studied and optimized and can be found in the literature [15–19].

8.3.3 Experimental Setup

As an example for the experimental setup of a KC experiment, we describe our realization in the following. In our laboratory, we employ a homemade microscope Raman setup illustrated in Fig. 8.5. A microscope objective focuses the incident laser onto the sample and collects the scattered light in the backward direction. The scattered light is then filtered by a notch filter and focused onto the entrance slit of the spectrometer. There, the signal light is dispersed and finally detected by a CCD camera.

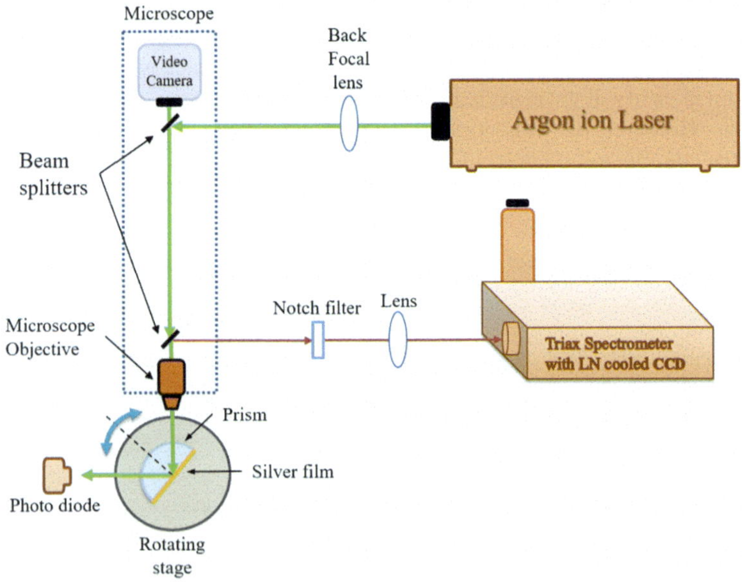

Fig. 8.5 Experimental setup

A rotating stage is necessary to change the angle of incidence when using the KC. Through the prism, the scattered light is collected in the backward direction. A photo diode connected to a voltmeter is used in the arrangement along with the rotating stage. It is used to measure the reflection intensity. Under optimized SERS conditions, the spectrometer can detect the Raman signal when the reflection reaches its minimum, indicating resonant excitation of plasmons (SPPs).

The excitation of the SPP is significantly influenced by the incidence angle. Only a small amount of the incident beam will be at the proper angle to contribute to enhanced Raman scattering if the beam is focused into the prism. Because it increases the fraction of the beam with the appropriate angle and thus maximizes the coupling efficiency, a collimated beam is more favorable. Therefore, a second lens with a long focal length (e.g., $f = 30$ cm) is added to the laser light path, but the signal light does not pass through it; this compensates the focusing effect of the microscope objective on the laser beam. The incident laser beam is precisely focused by this extra lens at the back focal point of the microscope objective. Consequently, it generates a nearly parallel beam on the sample surface, which fulfills the desired conditions [20].

In our experimental setup, we utilized the 514.5 nm emission line of an Ar-ion laser with a total power of 100 mW (4 mW at the sample). The prism coated with a 45 nm thick silver layer was attached to a motorized rotating stage, allowing for precise adjustments of the angle of incidence with 0.1° increments. Ideally, a hemispherical or hemi-cylindrical prism is used to avoid diffraction-related changes of the angle of incidence to the metal-coated prism side.

8.4 Basic Characterization

Numerous applications of SERS can be found in fields such as chemistry, biochemistry, and trace analysis. It serves as a valuable tool for the detection and quantification of various substances. In the medical field, SERS is utilized for the detection of molecules like glucose [21], proteins [22], DNA [23], a wide range of medicinal drugs [24–26], as well as substances relevant to forensic science [27].

Surface plasmons, generated using different configurations such as Kretschmann or Otto geometries, have played a pivotal role in bio and chemical sensing applications. For instance, surface plasmon resonance (SPR) sensors [28] and colorimetric detectors based on gold nanoparticles [29] have been developed for DNA analysis. Surface plasmon microscopy [30] and other related techniques have also been employed in this context.

Instead of giving few examples here, the book chapter will continue to present some basic concepts mainly related to the Kretschmann configuration. The findings presented in the following are of importance also for other SERS techniques and for the interpretation of the resulting SERS spectra, which must be used with some caution since they might differ from regular spontaneous Raman spectra.

8.4.1 *Identifying the Chemical-Enhancement Contribution to SERS Using a Kretschmann Arrangement*

While, as mentioned above, the two main mechanisms contributing to SERS are known [10, 31–35], there are still many details that remain unclear specifically for the CE part of the enhancement. The poor reproducibility of standard SERS substrates, such as roughened metal surfaces or metal nano-clusters, adds complexity to systematic studies. Therefore, the use of flat metal coatings as in the Kretschmann configuration offers significant advantages, which can also be used for accessing the contribution of the CE mechanism.

It is widely accepted that the largest enhancement in SERS stems from the EME mechanism, which is attributed to the generation of local surface plasmon polaritons (SPPs) [36]. In contrast, the CE mechanism arises from the modification of the electronic structure of molecules adsorbed on the metal surface, resulting in a stronger dependence on the particular molecule [37–39].

An inherent drawback of SERS, particularly for analytical applications, is the issue of substrate irreproducibility, which directly affects the Raman enhancement. Over the past decade, significant progress has been made in developing techniques to control nanostructures, leading to improved analytical applications of SERS [40–44]. However, the challenge of substrate irreproducibility persists, as higher enhancement levels typically correspond to greater unpredictability. While flat metallic surfaces offer simplicity, control, and reproducibility as plasmonic systems, their lack of roughness renders them ineffective for practical SERS applications. Roughness plays a vital role in SERS by facilitating coupling with surface plasmons. This limitation can be overcome by exciting propagating surface plasmon polaritons at metal-dielectric interfaces.

As introduced above, the Kretschmann (KC) and Otto configurations are the most commonly employed setups utilizing prisms and evanescent waves to excite SPPs [6, 7].

The KC, in particular, is widely used in commercial applications. The pioneering work of Pettinger, Tadjeddine, and Kolb [45] utilized the Kretschmann geometry in a Raman experiment, and since then, SERS in the KC has been extensively investigated both experimentally and theoretically [46–49]. Various aspects of surface plasmon-enhanced Raman scattering in attenuated total reflection experiments have been studied, including the effect of metal film thickness [50], as well as angular and polarization dependence [51].

In an experiment presented here, the excitation of surface plasmon polaritons (SPPs) can be tuned by varying the angle of incidence, and the observed results align well with theoretical predictions. Figure 8.6 illustrates the relationship between reflectivity and angle, with the resonance condition indicated by a dip in the reflectivity curve. The theoretical data points are represented by red dots, while the experimental data points are shown as blue dots.

Notably, there is good agreement between the two sets of data. When a monolayer of a sample is added to the prism surface, the resonance angle experiences a slight

Fig. 8.6 Normalized reflection intensity (theory and experiment) as function of the angle of incidence. Adapted with permission from [20]. Copyright © 2016 John Wiley & Sons Ltd.

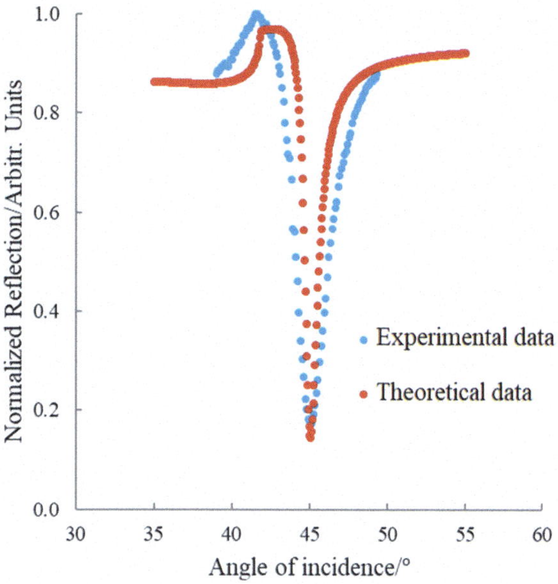

shift. This shift occurs due to the change in the refractive index caused by the presence of the sample molecules on the surface. This observation demonstrates the sensitivity of the resonance condition to changes in the local environment and highlights the potential for studying sample properties using this setup.

A few drops of a 100 mM solution of the analyte Nile Blue (NB) were applied to the prism's silver-coated surface to prepare the sample, which was then left for roughly 10 min. To assure the creation of a monolayer of NB on the surface, the prism was then washed with ultrapure water.

During the experimental measurements, each spectral window was recorded for a duration of 10 s. To improve the signal-to-noise ratio, five recordings were averaged, resulting in a total measurement time of 50 s.

A phenomenon known also from standard Raman spectroscopy is the decreasing of signal intensity with time. To study this bleaching effect for the given SERS setup, Raman spectra were collected at the resonance angle, which corresponds to the maximum interaction between the incident beam and the sample molecules. The goal was to observe any changes in Raman intensity over time. Figure 8.7 displays three spectra out of the nine that were recorded at different time points. The blue spectrum represents the 120 s mark, while the green and red spectra correspond to 680 and 1240 s, respectively. The time interval between each step was 560 s.

It is evident from the figure that the Raman intensity of the various peaks decreases over time. The most significant change in signal intensity occurs during the first 560 s, and the subsequent decrease becomes less pronounced in the following 560 s interval. Eventually, the decay process slows down, and an almost stable spectrum is observed. This indicates that the experimental conditions have reached a state of stability.

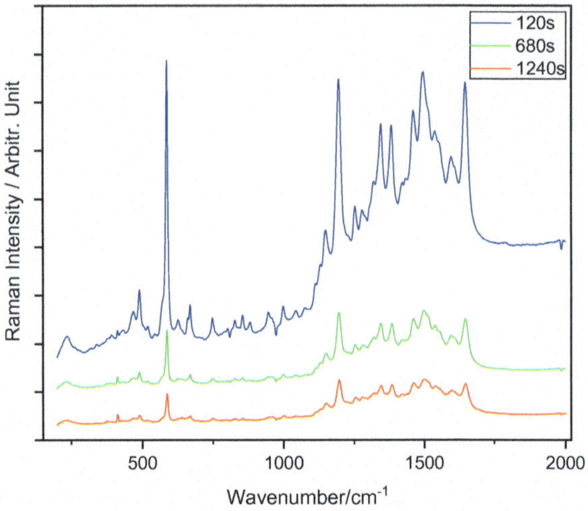

Fig. 8.7 Change of the SERS spectra of a monolayer of NB observed over approx. 21 min. Spectra are only shown for three out of nine steps. Adapted with permission from [20]. Copyright © 2016 John Wiley & Sons Ltd.

As mentioned above, bleaching effects have to be considered also in regular Raman spectroscopy, but due to the field enhancement, in SERS they play an even more important role. Therefore, the discussion of this phenomenon is of general importance when introducing SERS.

Based on the observation that the line positions in the Raman spectra remain unchanged, it can be inferred that the molecules did not undergo any damage or chemical changes. Despite the laser intensity remaining constant, the Raman signal intensity eventually reaches a plateau after some time. This behavior could be attributed to the reorientation of the molecules induced by the interaction with the electromagnetic (EM) field. Consequently, a decrease in signal intensity specific to each vibrational mode is expected.

The intensity of each peak was plotted as a function of time after Lorentzian peak fitting and background subtraction. Out of the 23 Raman bands examined, Fig. 8.8 shows the degradation of signal intensities for five distinct vibrational modes (585, 668, 1149, 1383, and 1645 cm^{-1}). To enhance visibility, an offset has been applied to each adjacent graph. The results reveal a significant mode dependence in the signal decay.

These findings support the hypothesis that the molecules undergo reorientation due to the interaction with the EM field. The mode-specific decrease in signal intensity suggests that certain vibrational modes are more susceptible to this reorientation effect than others. By studying the mode dependence of signal decay, valuable insights can be gained into the underlying mechanisms driving the interaction between the molecules and the EM field in the SERS process.

In order to quantify the decay behavior of the different Raman lines, the ratios of the peak intensities from the first (120 s) and ninth (1240 s) experiments in the time series were calculated. This analysis reveals interesting findings, indicating that the low-wavenumber Raman lines (below ~1100 cm^{-1}) are more significantly affected

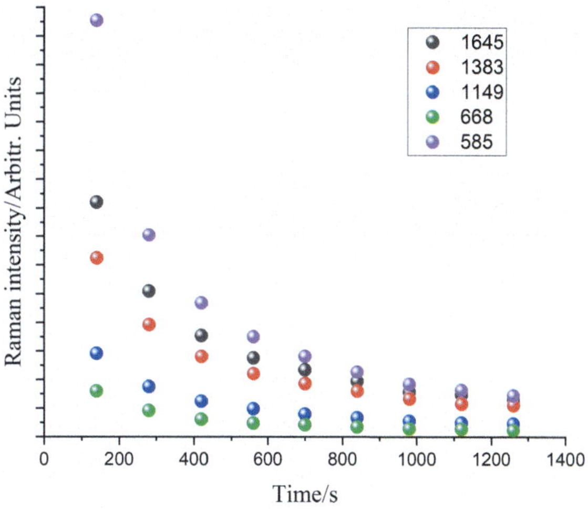

Fig. 8.8 The SERS intensities for various NB vibrational modes recorded at different times were obtained from the SERS spectra in Fig. 8.7

compared to the higher-wavenumber lines. Figure 8.9 presents the calculated ratios for all Raman lines, with the two distinct groups of data points highlighted. It is important to note that these experiments were repeated multiple times, allowing for the calculation of a statistical error of less than 10%. The observed trend, where the low-wavenumber Raman lines exhibit a higher rate of decay compared to the higher-wavenumber lines, provides valuable insights into the mode-specific behavior of the molecules in response to the interaction with the EM field. This mode dependence emphasizes the significance of considering the vibrational characteristics of the molecules when studying SERS phenomena. By quantifying the decay behavior and determining the mode-specific effects, researchers can gain a better understanding of the underlying processes involved in SERS and refine their experimental approaches accordingly.

After reaching a stable condition through longer irradiation, the molecules are expected to no longer experience orientation effects or degradation. To ensure that the subsequent experiments for the characterization of the SERS mechanisms are not influenced by further dynamical processes, a sufficient waiting time was allowed before conducting the angle-dependent measurements. This was confirmed by repeating a series of angle-dependent measurements and verifying the consistency of the results.

Without further bleaching effects occurring, it can be shown that by utilizing the KC arrangement and varying the angle of incidence, the strength of excitation of surface plasmons can be controlled. Figure 8.6 demonstrates the tunability of SPP excitation by angle variation. If the increase in SERS intensity were solely due to the EME mechanism, all modes would respond uniformly, resulting in a homogeneous increase in the overall spectrum. However, if the CE mechanism dominates, different modes would be affected with varying efficiency, leading to an inhomogeneous variation of SERS line intensities.

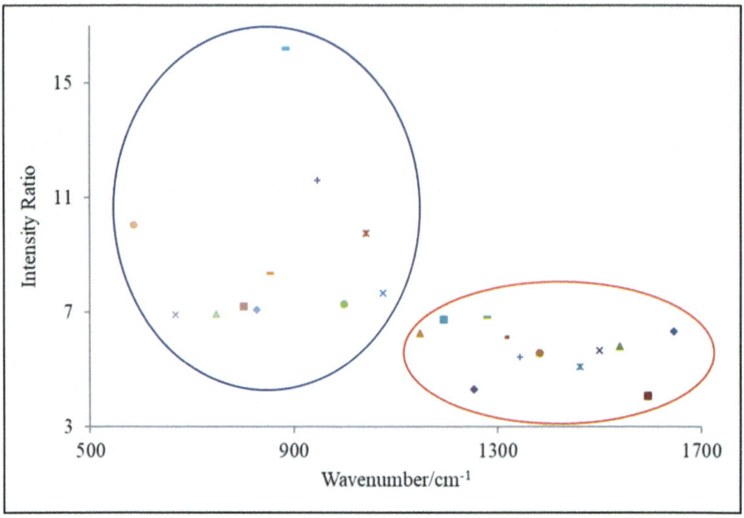

Fig. 8.9 The ratio of the Raman mode intensities measured at 120 and 1240 s for all Raman bands of NB that have been observed. Adapted with permission from [20]. Copyright © 2016 John Wiley & Sons Ltd.

Figure 8.10 presents the SERS signals of a monolayer of NB in air on the Ag-coated prism surface for different angles of incidence (θ). The spectra are compared under resonance and non-resonance conditions. In this figure, results for four angles (40°, 43°, 45°, and 47°) out of the nine measured angles are shown. Starting at 40°, which is relatively far from resonance, the Raman signal is weak. As the angle approaches 45°, the resonance angle, a strong enhancement is observed due to resonant excitation of SPPs. At 47°, where only partial resonance is expected (similar to 43°), the signal intensity decreases.

One can explore the influence of SERS intensity on resonance conditions and further analyze the contributions of the EME and CE mechanisms by systematically changing the angle of incidence and observing the resulting SERS signals. The intensity of Raman lines for various modes is shown in Fig. 8.11a plotted against the angle of incidence. It should be emphasized that weaker Raman lines might not be clearly recovered from spectra recorded at angles distant from resonance, even though identical behavior can be seen for all modes. Furthermore, it should be noted that the SERS spectra can display different intensity patterns for different modes compared to "normal Raman spectra."

The angle-dependent line intensities exhibit the expected behavior, with the resonance angle exhibiting the highest intensity. However, the improvement is considerably mode-dependent, showing that the CE mechanism—as mentioned above—contributed significantly to the SERS effect. An upper limit for the EME contribution can be established by determining the enhancement ratio for the bond with the least enhancement. This research highlights the significance of the CE mechanism,

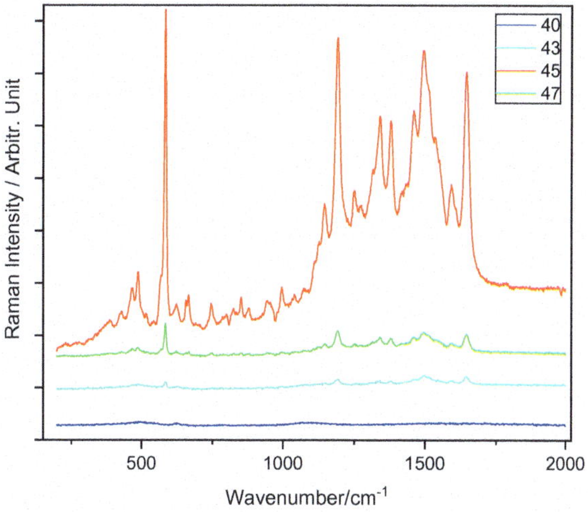

Fig. 8.10 KC SERS spectra collected from a monolayer of NB for various laser beam incidence angles. The values for each color indicate the angle of incidence for that color. Out of the nine angles that were selected for the experiment series, the spectra are only displayed for four of them. Adapted with permission from [20]. Copyright © 2016 John Wiley & Sons Ltd.

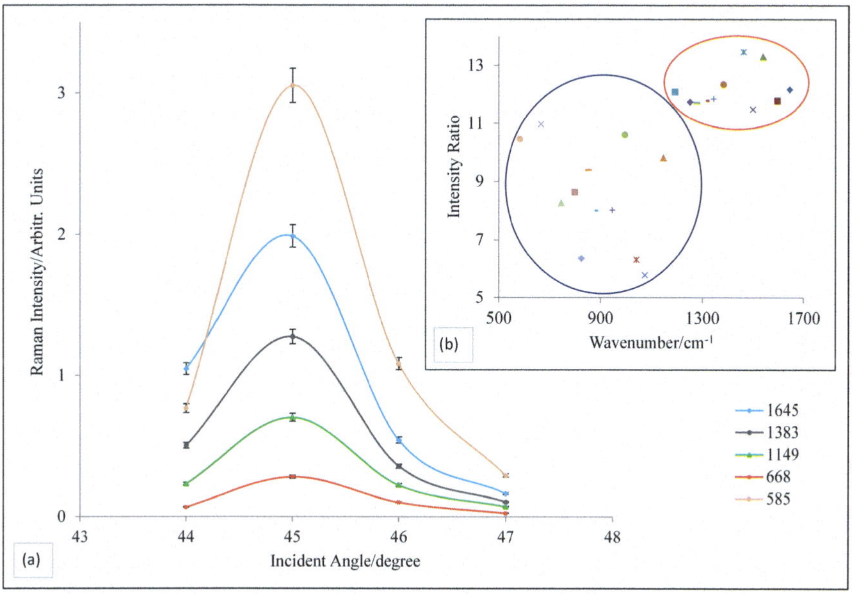

Fig. 8.11 **a** The Raman intensity of various NB vibrational modes for just five modes as examples. The wavenumber of the modes is provided in the legend. **b** Ratio of the Raman mode intensities at incoming beam incidence angles of 45° (resonant) and 47° (partially resonant) for various NB vibrational modes. Adapted with permission from [20]. Copyright © 2016 John Wiley & Sons Ltd.

particularly in the KC experiment when molecules are chemisorbed and hot spots and surface roughness are absent.

The ratio between the maximum intensity achieved at the resonance angle (about 45°) and the intensity recorded at partial resonance (47°) was computed for all observed lines and is presented in Fig. 8.11b to further demonstrate the mode-specific amplification. Lines at roughly 1200 cm^{-1} and above show a stronger enhancement than those at lower wavenumbers, demonstrating the variation of enhancement factors. Interestingly, as seen in Fig. 8.9, the ratio of enhancement is larger for vibrational modes that show less photo-bleaching. This finding suggests that the CE, which is based on the coupling between the molecule electronic system and the metal electrons, not only causes a larger enhancement of mode contributions in the SERS spectrum, but also plays a role in stabilizing these modes during the photo-bleaching process.

The experiments were conducted multiple times, including new sample preparations, to ensure the reliability of the data. The findings were successfully reproduced with a standard deviation of approximately 10%, indicating consistent results.

The complexity of NB as a molecule presents challenges for theoretical modeling. However, one finds that modes with higher enhancement ratios and slower deactivation involve in-plane ring vibrational motion, while modes with lower enhancement ratios and faster deactivation correspond to out-of-plane ring oscillation (based on the assignment of Raman lines in the literature). This observation provides insight into the vibrational motions contributing to the SERS effect in NB.

8.4.2 The Importance of the Chemical Enhancement Mechanism in SERS Using the Kretschmann Arrangement

In addition to demonstrate the mode-specific enhancement, the findings mentioned above also suggest that CE plays a substantial role in SERS experiments that use the Kretschmann configuration (KC) with flat metal layers. Here, further experiments were done to turn off the CE while guaranteeing that the EME continued to contribute adequately in order to confirm the critical role of CE. This was accomplished by preventing direct contact between the metal coating and the analyte molecules by adding a nanometric non-metallic intermediate layer between them. In this case, the molecules experience the EME and remain just a few nanometers from the surface, but their direct connection to the metal electrons necessary for the CE process is prevented. To avoid any significant changes to the electromagnetic field's strength and range, which are substantially influenced by its proximity to the surface, very thin intermediate layers were used.

A hemi-cylindrical BK7 prism with a radius of 1.25 cm and a refractive index of approx. 1.5 was used in the KC. The angle of incidence was changed in steps of 0.1° using a motorized rotating stage. The prism's cross-sectional height was 1 mm less than the cylinder radius. The substrate for all the coatings was a rectangular glass plate manufactured of BK7 with a thickness of 1 mm. The glass plate was coated with the various layers and then refractive-index-matching oil was used to place it on flat surface of the prism.

This configuration resulted in a hemi-cylinder setup, as illustrated in Fig. 8.12. This approach offered the replacement of surface layers without leaving any residues on the prism surface after cleaning.

Three distinct monolayers were used in the studies, and the outcomes with and without the intermediate layer were compared. Since self-assembled monolayers (SAMs) of al-kanethiols adsorbed onto various metal surfaces provide stable surfaces in a simple manner, SAMs were used to produce nanometer-thick intermediate layers [53, 54]. A sketch of the prism with the different layers coated on its surface is shown in Fig. 8.12.

The experimental setup was specifically designed to provide accurate conditions for both signal detection and excitation. The scenarios for excitation of the sample at the resonance angle (on the right) and pre-resonant excitation (on the left) are depicted in Fig. 8.13. A collimated beam can be produced by adding the "back-focal lens" between the laser and microscope setup as mentioned above, allowing a well-defined interaction with the coated surface of the prism (shown as a red line in Fig. 8.13). With an efficiency determined by the angle of incidence, which is constantly greater than the critical angle for total internal reflection, this light beam produces surface plasmon polaritons (SPP). The generated Raman signal is then coupled back by the surface plasmon polaritons, producing signal light inside the

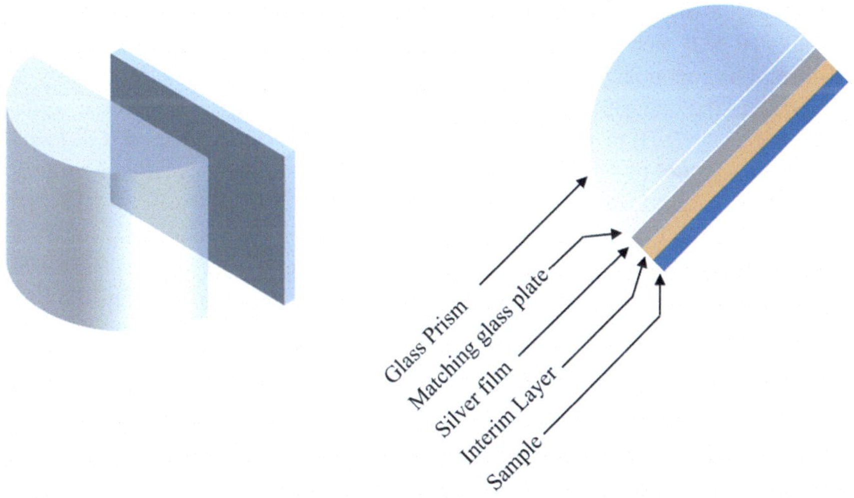

Fig. 8.12 (left) Prism-glass-plate arrangement of KC prism with (right) different layers on the flat glass-plate surface. Adapted with permission from [52]. Copyright © 2021 American Chemical Society

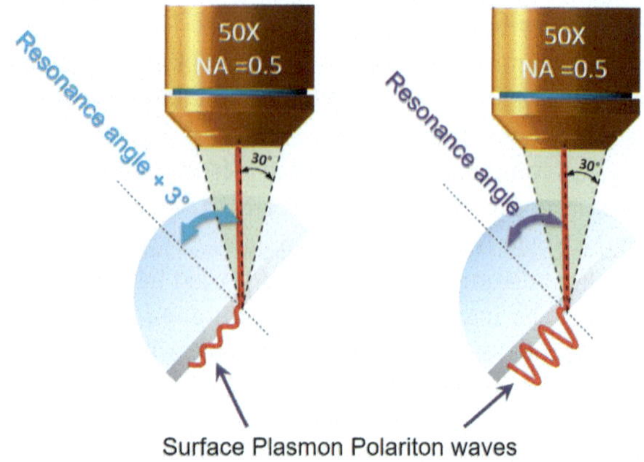

Fig. 8.13 Illustration of the excitation of SPPs using a collimated laser beam (red line) in resonance (left) and detuned by +3° from the resonance angle (right). The angular aperture of the microscope objective, which captures the Raman signal, is around 30°. Adapted with permission from [52]. Copyright © 2021 American Chemical Society

"Kretschmann cone" in the opposite direction of detection, surrounding the laser beam.

The signal path within the cone differs somewhat from that of the exciting laser light because of the red-shift of the dispersed light. For instance, a vibrational mode with a Stokes shift of 1650 cm^{-1} would just slightly alter the angle by about 0.7°. This setup's microscope objective (shown in Fig. 8.13) has a rather wide angular aperture of around 30°, thus any dependence on a particular vibrational mode for collecting effectiveness can be disregarded. Furthermore, angle detuning of the excitation will result in a less effective excitation of the surface plasmons in the silver layer, but won't have an impact on the effectiveness of Raman signal collection. An example of this is a shift of +3° from the resonance angle shown in the left side of Fig. 8.13.

Thiols (R-SH) were used as the interim layers because they can be made into self-assembling monolayers (SAMs) on metal surfaces very fast and efficiently. In this instance, the opposite end of the thiol molecule faces upwards, and the sulfur-containing group (SH) is drawn to the metal surface. 1-Octadecanethiol (ODT) and 4-Nitrobenzenethiol (4-NBT), two distinct thiols, were utilized exactly as received after being purchased from Sigma-Aldrich.

ODT and 4-NBT were pre-prepared in ethanol at a concentration of 5 mM, and the solutions were then sonicated for 10 min to create the thiol solutions. The metal-coated surfaces of the silver-coated glass plates were then immersed in the thiol solutions for one hour. A monolayer of thiol molecules developed on the surface throughout this period. After that, the metal surface was cleaned three times with ethanol and deionized water to get rid of any loose or weakly attached molecules.

Using ellipsometry, the ODT monolayer's thickness was determined to be (3.4 ± 0.1) nm. To ensure precision, this measurement was carried out three times for each preparation.

Glass plates were coated with 45 nm of silver and then 5 nm of molybdenum (Mo) by sputtering film deposition to test another kind of interim layer. A Turbo-Pumped Sputter Coater (Quorum Q150ts) was used to deposit the film at a rate of 25 nm/

min. A surface profiler (Dektak 3) was employed to measure and confirm the total thickness of the deposited layers.

An ultrathin layer of MoO_2 was created by exposing the Mo layer to the air for some time. Four-point resistance measurements were used to determine the film's surface resistance to verify that the layer is an insulator. Then, a few drops of the analyte, a 100 mM Nile blue (NB) solution, were applied to the silver-coated glass plate and allowed to dry.

The SPP resonance curve as a function of angle of incidence was showing the behavior shown in Fig. 8.6 also when the silver-coated glass plate was covered by an interlayer with a minor angular shift in the resonance angle compared to the case where no interlayer was added. This is not surprising since also the addition of sample layers slightly changes the resonance condition as mentioned above. The resonance angle for ODT has been found to be close to 45°. The angle was purposefully detuned by +3° from the precise resonance angle to achieve pre-resonance conditions.

As discussed in the introduction (Sect. 8.1), the CE mechanism causes a significant dependence of the enhancement on the vibrational modes. For the EME one would expect a uniform field enhancement of all modes. The interim layer should remove the CE since now no analyte molecules can chemisorb to the silver surface.

The SERS spectra of a dried droplet of NB in the presence of air on a 45 nm silver film are shown in Fig. 8.14. The experiments were performed at two different angles of incidence of the laser beam, one of which met the resonance requirements for SPPs and the other was purposefully detuned by +3°. The identical series of measurements was carried out under resonance and non-resonance conditions with NB isolated from the silver layer by an ODT monolayer applied on top of the silver film.

After subtracting the background and fitting the peaks, the intensities of various vibrational modes were examined for the two angles of incidence. Subsequently, the ratios between the intensity at the resonance angle (45°) and the intensity at the pre-resonance angle (48°) were calculated for each mode. These ratios are depicted in Fig. 8.15a for the SERS spectra obtained when NB is in contact with the metal layer (silver) and in (c) for the SERS signal obtained when an intermediate ODT layer was present between the analyte (NB) and the silver. It is evident that the mode-dependent variation of the ratios for the experiment without interim layer is much bigger than that observed for the case where ODT is acting as separation layer.

Experiments have been performed also with the other interim layers introduced above. As panels (b) and (d) of Fig. 8.15 show, also the interlayers MoO_2 and 4-NBT result in a suppression of the CE, respectively. All modes are approx. equally enhanced, which can be expected from the EME mechanism. Also performing these experiments with different analyte molecules results in the same observation. This was e.g. demonstrated for crystal violet on an ODT interim layer.

To provide a clearer understanding of how we derived the enhancement ratios for the various modes of NB and CV, we have outlined the procedure in Fig. 8.16 and its accompanying caption, using the data from CV without interlayer as an example.

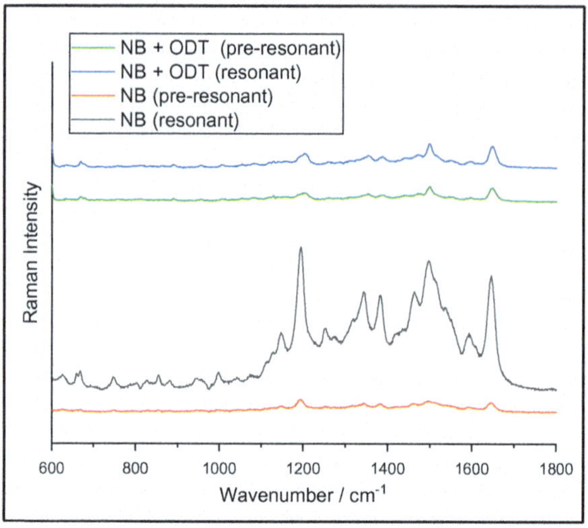

Fig. 8.14 SERS spectra of NB in air on 45 nm Ag film were recorded at two angles, one yielding resonance for the SPP excitation (black line) and the other resulting in non-resonance conditions (red line). Resonance (blue line) and pre-resonance (green line) comparable spectra of NB with ODT as interim layer are displayed. Adapted with permission from [52]. Copyright © 2021 American Chemical Society

It is important to note that these enhancement ratios do not represent the absolute enhancement factors associated with the KC SERS experiment conducted.

The experiments show that the CE effect can be removed by introducing very thin separation layers on top of the metal surface. This effect will not be restricted to SERS in the KC approach, but should be valid for SERS in general. Since however, for nano-structured metal substrates, the EME is the dominating enhancement mechanism, the enhancement itself should not be affected too much. There, the interface layer might be useful when the typical shifts of the Raman lines very often accompanying SERS experiments have to be avoided or reduced in order to obtain spectra better comparable to regular Raman spectroscopy results.

However, for the KC, it is obvious that even a very thin interlayer introducing only a very small separation of the sample molecules from the metal layer generally leads to a significant reduction in signal enhancement. This can be seen from a comparison of panels (b), (c), and (d) showing the situation with interlayer with panel (a) of Fig. 8.15, which gives the enhancement ratios for NB chemisorbed on silver.

EME cannot account for the entire decrease in enhancement. The required decrease of the evanescent field of SPPs (see Fig. 8.17) expected to cause the observed drop would occur for distances on the order of hundreds of nanometers. The reduction of relative enhancement in our case is substantially greater despite only having an interim layer of 3–5 nm. This suggests that contrary to the commonly held view

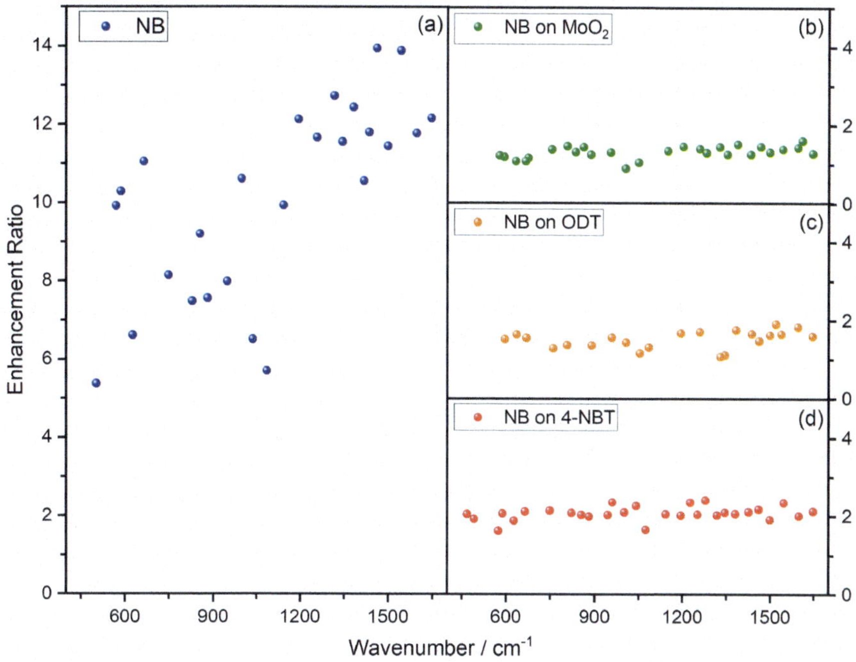

Fig. 8.15 Ratios between SERS intensities from different vibrational modes measured in and off resonance (compare Fig. 8.13). **a** NB on silver and NB separated from silver by **b** MoO_2, **c** ODT, and **d** 4-NBT layer. Adapted with permission from [52]. Copyright © 2021 American Chemical Society

that EME plays a dominating role in all SERS arrangements, CE is the primary contributor to surface enhancement in the KC utilized.

It should be noted that the introduction of surface roughness can alter these observations. Roughening the metal surface in the KC is known to increase the enhancement factor, but this may also be at least partially attributed to an increase in surface area and, consequently, an increase in CE. Additionally, it is important to acknowledge that the very high enhancement factors observed when nanostructured particles or surfaces are used as SERS substrates are likely due to the formation of hot spots, which are considered to play and important role in SERS using e.g. roughened metal surfaces. In these cases, molecules placed in resonator-like metal structures experience significantly higher enhanced fields, even enabling single molecule detection [55–57].

It is obvious that the scattering efficiency for different modes is affected to variable degrees by the excitation of metal electrons that couple to the analyte molecules, leading to significantly different Raman signals. The electromagnetic field, which in this case also changes with the angle, plays the key role here. Since nothing else changes besides the excitation of electrons and the SPP-triggered EME, we obviously are observing CE, which is a phenomenon brought on by the direct coupling of

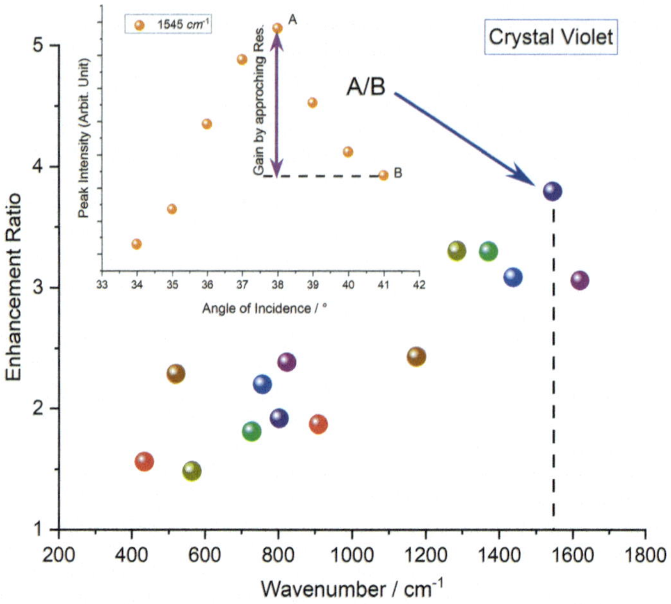

Fig. 8.16 The determination of the relative enhancement values shown for CV. The inset displays the 1545 cm^{-1} mode intensity as a function of the exciting laser's angle of incidence. For a detuning angle of +3°. A/B is the ratio derived from the intensities at resonance (**a**) and pre-resonance (**b**). This "enhancement ratio" is displayed in the plot for the different vibrational modes. Adapted with permission from [52]. Copyright © 2021 American Chemical Society

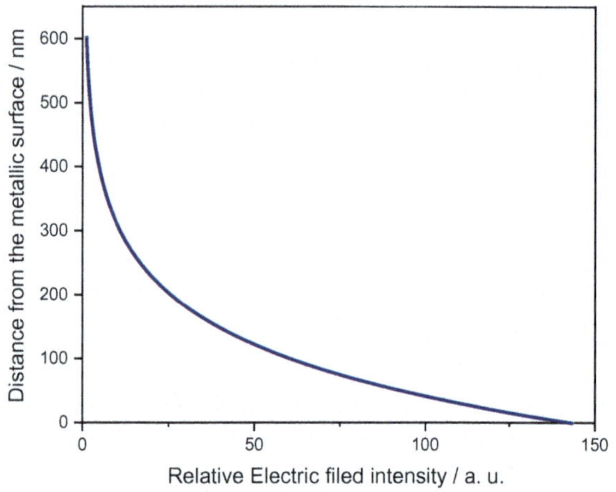

Fig. 8.17 As the distance from the metallic layer increases, the evanescent field decays. It becomes clear that no significant field change occurs for an interlayer thickness of 3–5 nm. Adapted with permission from [52]. Copyright © 2021 American Chemical Society

electronic systems. The resulting electromagnetic fields are only changed in intensity, not in direction or polarization, and the molecules keep their orientation. We believe that our work demonstrates the significant contribution of CE in the chosen KC (compared to EME) and, at the same time, also provides insight into the basic mechanism responsible for CE.

8.5 Example: Detection and Quantification of Moxifloxacin Using SERS in a Kretschmann Arrangement

The detection of substances in a biological medium is of great importance due to the potential environmental threats they pose to living organisms. Currently, there is significant research focused on quinolone and fluoroquinolone antibiotics, as they have been associated with health concerns such as heart problems, brain lesions [58], and various neuropathies [59].

Moxifloxacin (Moxi), also known as 1-cyclopropyl-6-fluoro-1,4-dihydro-8-methoxy-7-[(4aS,7aS)-octahydro-6H-pyrrolo[3,4-b]pyridin-6-yl]-4-oxo-3-quinoline carboxylic acid, belongs to the new fourth generation of 8-methoxy fluoroquinolone antibiotics. It exhibits broad activity against both Gram-negative and Gram-positive bacteria [60–62].

In clinical practice, Moxi is typically administered to patients in daily doses of 400 mg, and it is rapidly absorbed. After oral administration, the maximum concentration in plasma is reached between 1 and 4 h. The concentrations found in serum and in urine at the end are approx. 2–5 μg/ml and 30–60 μg/ml, respectively [63]. Consequently, it is expected that undesired antibiotic residues may be present in wastewater and ultimately contaminate groundwater [64–66]. Therefore, monitoring antibiotics capable of causing harm to human health, such as those associated with resistant bacteria, and the ecological environment is of utmost importance.

Various methods have been employed for the quantification of Moxi, including chromatography [62–67], capillary electrophoresis [68], voltammetry [69], spectrofluorimetry [70], square-wave adsorptive voltammetry [69, 71], and HPLC [62, 67, 72–75]. However, some of these methods have limitations, such as requiring expensive equipment, offering low sensitivity, and yielding lengthy waiting times for obtaining results.

Additionally, most of these techniques need extensive sample preparation and handling, which takes time and qualified personnel. As a result, these techniques are mostly used in sizable academic and medical institutions. Raman spectroscopy is emerging as a viable technology because it offers molecule-specific information with a reasonably basic setup and without the need for considerable sample preparation. To achieve reasonable signal intensities for low analyte concentrations, surface-enhanced Raman spectroscopy (SERS) is required. Our objective was to achieve satisfactory SERS signal enhancement while minimizing the complexity of the experimental process.

When quantification is desired, reproducibility becomes a critical factor. One of the main challenges with SERS is the difficulty in achieving reproducibility due to the varying quality and the inhomogeneity of nanoparticles or rough metal surfaces used as SERS substrates. The optimization of efficiency and enhancement requires careful consideration of factors such as the ratio of nanoparticles to analyte and the mixing time between them, since the activation time of "hot spots" has a considerable impact on SERS detection [76].

For instance, for the detection of residues of nitrofuran antibiotic, SERS substrates based on metal nanoparticles were utilized, requiring a highly intense laser source providing 300 mW at the sample to detect Raman spectra [77]. Alternatively, Popp and colleagues adopted a different approach by employing microfluidics to investigate antibiotics, including the recent detection of levofloxacin using a laser power of 35 mW at the sample [78].

In the subsequent experiments, we show the effectiveness of the KC utilizing flat silver layers as a rapid and straightforward SERS technique for the sensitive detection of Moxifloxacin (Moxi) even at low concentrations. The approach holds great appeal due to its ability to consistently generate highly reproducible results, as discussed earlier.

For our experiments, we employed a hemispherical sapphire prism coated with a 45 nm thick layer of silver. The sapphire prism had a refractive index of $n = 1.78$ and a radius of $r = 1.25$ cm. By utilizing a rotating stage, we were able to precisely control the angle of incidence with an accuracy of 0.02°. Each spectral window was recorded for a duration of 10 s, and to improve the signal-to-noise ratio, five accumulations of the signal were averaged, resulting in an overall measurement time of 50 s.

Moxifloxacin (Moxi) is a yellow powder with a relative molecular mass of Mr = 437.9. The Moxi used in our experiments was obtained from Sigma-Aldrich with a declared purity of over 99%. A stock solution with 10 mM Moxi and ultrapure water was prepared. Working solutions were created by appropriately diluting the stock solution with ultrapure water throughout the course of the experiment. These solutions were stable over time and showed no detectable alterations.

To facilitate the measurements, a few droplets of the Moxi solution were placed on the surface of the silver-coated prism. To prevent evaporation during the measurements, the droplets were covered with a coverslip, which also served to create a small measurement chamber in conjunction with the prism.

In Fig. 8.18, the chemical structure of Moxi is depicted. Its increased antibacterial action is caused by a quinolone ring with a methoxy group and a diazabicyclonyl ring [79].

In Fig. 8.19, we display the Raman spectra taken from a 10 mM aqueous solution of Moxi and Moxi powder. It is important to note that in normal Raman spectroscopy, this concentration marks the limit of detection. Given that most Raman bands involve the vibrations of several atoms, the vibrational assignments for Moxi have been studied in several papers. Without exact mathematical calculations, it can be difficult to identify certain vibrational modes given the complexity of the molecule. For instance, the C=C stretching vibration of the aromatic rings is considered to be

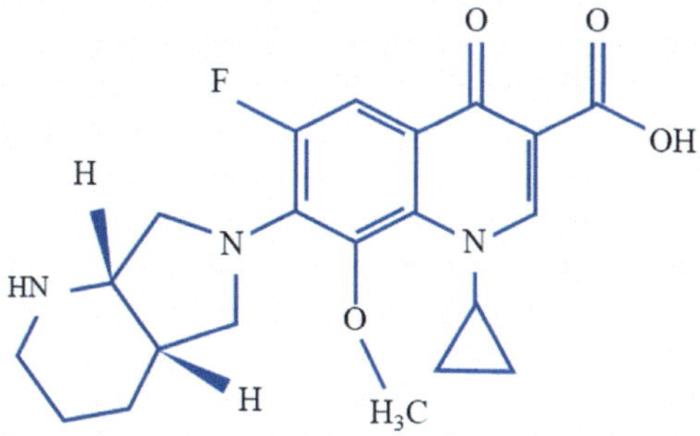

Fig. 8.18 Chemical structure of Moxifloxacin

responsible for the peak at 1620 cm^{-1} [80]. In this discussion, we will focus on a few key vibrational modes that are of particular importance.

It is obvious that the analyte signal in an aqueous environment is slightly different from the signal obtained from the powder form. This mismatch can be explained by the fact that whereas molecules do not form crystals in aqueous solutions, they do so in the case of powders. The Raman modes in the aqueous solution with the highest intensities are centered around 1620, 1373, and 1340 cm^{-1}; these modes will be taken into account in our following SERS analysis.

The detection of low concentrations, particularly in the context of water contamination, poses a challenge for conventional Raman spectroscopy due to the inherently weak signal intensities. In such cases, Surface-Enhanced Raman Spectroscopy (SERS) is a viable alternative as it can provide signal enhancements of several orders of magnitude compared to regular Raman spectroscopy. However, the detection of

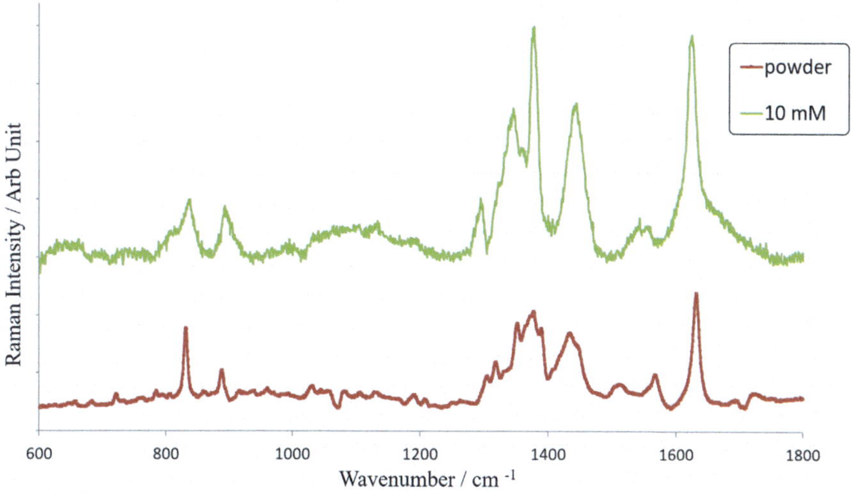

Fig. 8.19 Raman spectra taken of a 10 mM aqueous solution of Moxi and Moxi powder

Moxifloxacin (Moxi) at low concentrations ranging from 100 nM to 5 mM is of great interest, as these concentration ranges are relevant in clinical and environmental applications.

In the study presented as example for the application of SERS using the KC, multiple measurements at low Moxi concentrations were performed, and the associated errors were found to be less than 10%. The KC ensured reliable and reproducible results. The SERS experiment commenced with the application of a few droplets of an aqueous solution containing Moxifloxacin at an extremely low concentration of 10^{-9} M (1 nM) onto the prism surface. Subsequently, the concentrations were incrementally adjusted to 5, 100 nM, 1, 10, 100 μM, and 1 mM for successive measurements. Prior to capturing the SERS spectra at each concentration, the resonance angle was meticulously determined. Each Raman spectrum was acquired over a duration of 10 s and repeated five times to obtain an average. To ensure the absence of any residual sample on the prism surface, thorough rinsing with ultrapure water was performed prior to each new measurement. Furthermore, a test Raman measurement was conducted to confirm the absence of detectable sample remnants. At higher concentrations, washing with acetone was necessary to remove traces of the analyte from the prism surface.

Figure 8.20 illustrates the normalized SERS spectra obtained for various Moxifloxacin concentrations.

Moxifloxacin's Raman modes at 1340, 1373, and 1620 cm^{-1} are reliably detected at concentrations as low as 100 nM. The intensity of these bands shows saturation as the concentration rises, especially after reaching a concentration of 1 mM. This saturation phenomenon suggests that the illuminated active surface of the prism has reached equilibrium with the surrounding aqueous environment and is completely covered with Moxifloxacin molecules. Because of this, further molecules are unable to contribute to the SERS signal.

Figure 8.21 presents the spectrum obtained from the 100 nM concentration of Moxifloxacin after background subtraction and peak fitting. A dashed line is used for the overall fit to the spectrum, while the green, blue, and red lines correspond to the contributions at 1340, 1373, and 1620 cm^{-1}, respectively.

The intensities of the peaks were plotted against the logarithm of the concentration ranging from 100 nM to 1 mM to study the concentration dependency of the SERS signal. The results are shown in Fig. 8.22. The peak intensities and the logarithm of the concentration can be seen to have a nearly linear relationship, albeit with somewhat varying slopes for the different Raman lines. This result is consistent with our earlier research and shows that chemical enhancement is the main factor for SERS in the KC arrangement with flat metal layers. Furthermore, this mechanism is responsible for the observed mode-specific enhancement.

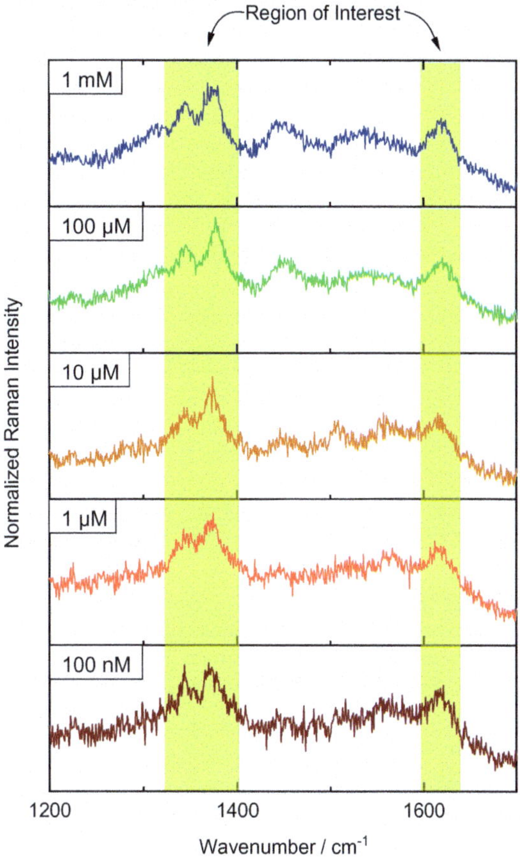

Fig. 8.20 SERS spectra obtained from different concentrations of Moxi using KC. Adapted with permission from [81]. Copyright © 2021 John Wiley & Sons Ltd.

The enhancement factors for the SERS signals [82] can be calculated using:

$$\text{Enhancement factor} = \frac{I_{SERS} \times C_R}{I_R \times C_{SERS}} \tag{8.9}$$

where C_{SERS} and C_R represent the concentrations of molecules, and I_{SERS} and I_R are the peak intensities for the SERS and normal Raman experiments, respectively.

Figure 8.23 displays the SERS spectra of 100 nM Moxi (blue line) on the prism and the Raman spectra of 10 mM (green line) in a standard cuvette. Using (8.9), one determines that the SE factor for the band at 1620 cm^{-1} is on the order of 10^4.

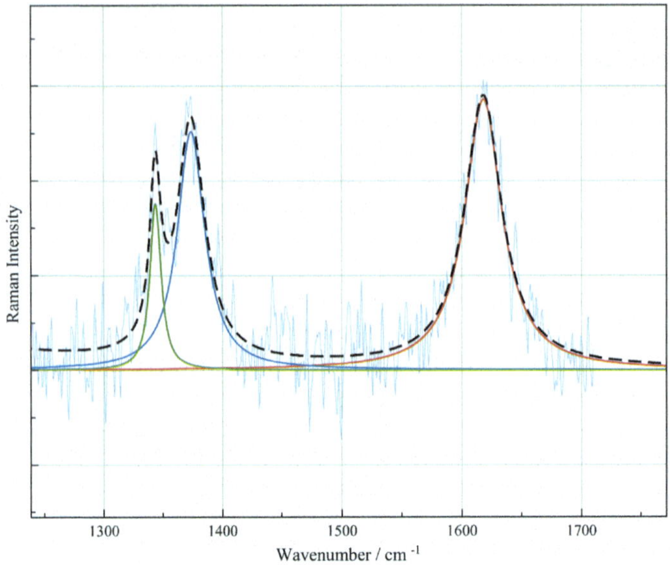

Fig. 8.21 Following background removal, the three Raman peaks utilized to detect Moxi were measured at the lowest concentration (100 nM); the dashed line represents the overall spectrum fit, while the full lines represent the Raman lines at 1620 cm^{-1} (red), 1373 cm^{-1} (blue), and 1340 cm^{-1} (green). Adapted with permission from [81]. Copyright © 2021 John Wiley & Sons Ltd.

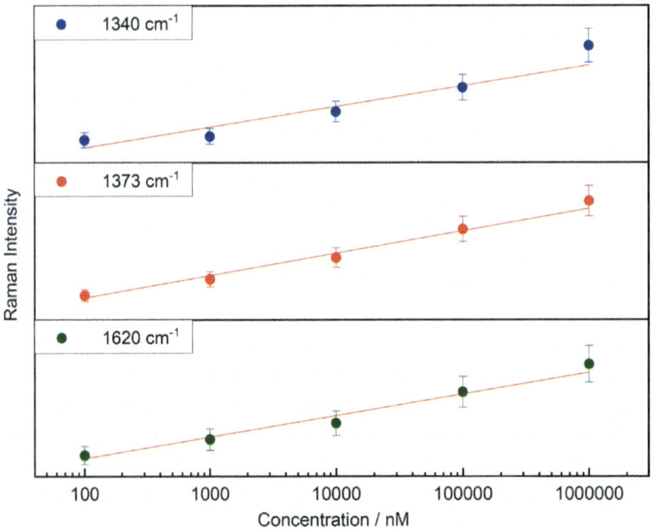

Fig. 8.22 Intensities of the SERS lines as function of Moxi concentration for three different modes. The data points were fitted by straight lines in the semi-logarithmic presentation. Adapted with permission from [81]. Copyright © 2021 John Wiley & Sons Ltd.

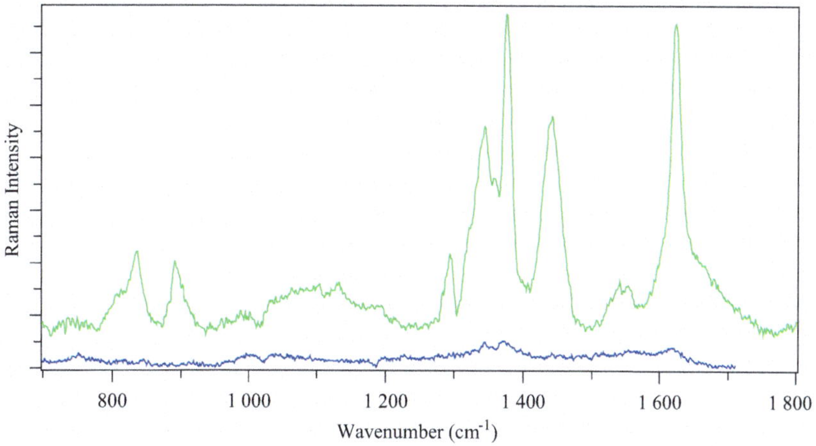

Fig. 8.23 Comparison of the normal Raman spectrum, obtained from the 10 mM solution of Moxi, with the SERS spectrum, obtained from the 100 nM solution of Moxi (blue line)

8.6 Conclusion

In this chapter, SERS was introduced with a focus on basic properties and a special technique utilizing the so-called Kretschmann configuration (KC). The two main mechanisms governing the enhancement effect were explained. In order to better understand basic properties of SERS spectroscopy, the discussion focused on a technique, in which a flat metal surface could be used in order to achieve highly reproducible SERS results. While an excitation of the surface plasmon polaritons (SPPs) is not possible when laser light is directly falling onto the flat surface due to phase-mismatch, in the KC this can be achieved. There, the evanescent field resulting from total internal reflection couples to the SPPs on a thin metal layer, on which sample molecules can be positioned. The experimental results presented in the chapter demonstrated important features, which have to be taken into consideration when performing SERS experiments. Due to the intense fields bleaching effects occur, which most probably are due to re-alignment processes of the adsorbed molecules. The chemisorbed layer contributes to a contribution of the so-called chemical enhancement (CE) mechanism, which considerably affects the resulting spectra. These spectra can be different from standard Raman spectra regarding line position and intensity ratios. Experiments with interlayers separating the sample molecules from the metal and preventing direct interaction demonstrated that for the KC the CE contribution is dominating while for nano-structured surfaces typically the electromagnetic enhancement (EME) mechanism plays the most important role. Depending on whether the strong enhancement or the reproducibility of SERS is more important, the best technique can be selected when taking into consideration the basic properties described in this chapter. As example for the usefulness of the KC-SERS approach the detection of very low concentrations of the antibiotics Moxifloxacin in water have been presented.

SERS applied in its different forms of realization is a useful technique to detect Raman signals also from samples at low concentrations. The KC-SERS offers high reproducibility. The basic findings presented in the chapter can help to better understand and interpret SERS spectra in general.

References

1. M. Fleischmann, P.J. Hendra, A.J. McQuillan, Chem. Phys. Lett. **26**, 163 (1974)
2. M.G. Albrecht, J.A. Creighton, J. Am. Chem. Soc. **99**, 5215 (1977)
3. L.A. Dick, A.D. McFarland, C.L. Haynes, R.P. Van Duyne, J. Phys. Chem. B **106**, 853 (2002)
4. D.L. Jeanmaire, R.P. Van Duyne, J. Electroanal. Chem. Interfacial Electrochem. **84**, 1 (1977)
5. J.P. Camden, J.A. Dieringer, J. Zhao, R.P. Van Duyne, Acc. Chem. Res. **41**, 1653 (2008)
6. A. Otto, Z. Phys. **216**, 398 (1968)
7. E. Kretschmann, H. Raether, Z. Naturforsch. A **23**, 2135 (1968)
8. T.E. Furtak, *Optical and Electronic Resonance: the Underlying Sources of Surface Enhanced Raman Scattering* (Wiley, United States, 1983)
9. H.G.M. Edwards, E. Smith, G. Dent, *Modern Raman Spectroscopy—A Practical Approach* (Wiley, 2005), p. 118
10. A. Campion, P. Kambhampati, Chem. Soc. Rev. **27**, 241 (1998)
11. B. Hecht, L. Novotny, *Principles of Nano-Optics*, eds. by B. Hecht, L. Novotny (Cambridge University Press, Cambridge, 2006), p. i
12. E. Kretschmann, Zeitschrift für Physik A Hadrons and nuclei **241**, 313 (1971)
13. W.L. Barnes, A. Dereux, T.W. Ebbesen, Nature **424**, 824 (2003)
14. W.L. Barnes, W.A. Murray, J. Dintinger, E. Devaux, T.W. Ebbesen, Phys. Rev. Lett. **92**, 107401 (2004)
15. H.T. Phan, A.J. Haes, J. Phys. Chem. C **122**, 14846 (2018)
16. C.J. Murphy, T.K. Sau, A. Gole, C.J. Orendorff, MRS Bull. **30**, 349 (2005)
17. M.E. Stewart, C.R. Anderton, L.B. Thompson, J. Maria, S.K. Gray, J.A. Rogers, R.G. Nuzzo, Chem. Rev. **108**, 494 (2008)
18. B. Wiley, Y. Sun, J. Chen, H. Cang, Z. Li, X. Li, Y. Xia, MRS Bull. **30**, 356 (2005)
19. J. Maria, T.T. Truong, J. Yao, T. Lee, R.G. Nuzzo, S. Leyffer, S.K. Gray, J.A. Rogers, J. Phys. Chem. C **113**, 10493 (2009)
20. F. Mohaghegh, A. Mazaheri Tehrani, A. Materny, J. Raman Spectrosc. **47**, 1029 (2016)
21. O. Lyandres, N.C. Shah, C.R. Yonzon, J.T. Walsh, M.R. Glucksberg, R.P. Van Duyne, Anal. Chem. **77**, 6134 (2005)
22. M. Larsson, J. Lindgren, J. Raman Spectrosc. **36**, 394 (2005)
23. A. Rasmussen, V. Deckert, J. Raman Spectrosc. **37**, 311 (2006)
24. S. Cinta-Pinzaru, N. Peica, B. Küstner, S. Schlücker, M. Schmitt, T. Frosch, J.H. Faber, G. Bringmann, J. Popp, J. Raman Spectrosc. **37**, 326 (2006)
25. J. Binoy, I.H. Joe, O.F. Nielsen, J. Aubard, Laser Phys. Lett. **2**, 544 (2005)
26. A.V. Szeghalmi, L. Leopold, S. Pînzaru, V. Chis, I. Silaghi-Dumitrescu, M. Schmitt, J. Popp, W. Kiefer, Biopolymers **78**, 298 (2005)
27. J.M. Sylvia, J.A. Janni, J.D. Klein, K.M. Spencer, Anal. Chem. **72**, 5834 (2000)
28. B. Liedberg, C. Nylander, I. Lunström, Sens. Actuat. **4**, 299 (1983)
29. C.A. Mirkin, R.L. Letsinger, R.C. Mucic, J.J. Storhoff, Nature **382**, 607 (1996)
30. B. Rothenhäusler, W. Knoll, Nature **332**, 615 (1988)
31. P.C. White, C.H. Munro, W.E. Smith, Analyst **121**, 835 (1996)
32. K. Kneipp, H. Kneipp, I. Itzkan, R.R. Dasari, M.S. Feld, Chem. Rev. **99**, 2957 (1999)
33. S. Link, M.A. El-Sayed, Annu. Rev. Phys. Chem. **54**, 331 (2003)
34. J.M. Chalmers, P.R. Griffiths, *Handbook of Vibrational Spectroscopy* (Wiley, New York, 2002), p.759

35. L. Jensen, C.M. Aikens, G.C. Schatz, Chem. Soc. Rev. **37**, 1061 (2008)
36. H. Xu, E.J. Bjerneld, M. Käll, L. Börjesson, Phys. Rev. Lett. **83**, 4357 (1999)
37. J.R. Lombardi, R.L. Birke, J. Phys. Chem. C **112**, 5605 (2008)
38. J.R. Lombardi, R.L. Birke, T. Lu, J. Xu, J. Chem. Phys. **84**, 4174 (1986)
39. L. Zhao, L. Jensen, G.C. Schatz, J. Am. Chem. Soc. **128**, 2911 (2006)
40. C. Noguez, J. Phys. Chem. C **111**, 3806 (2007)
41. D. Graham, G. McAnally, J.C. Jones, W. Ewen Smith, Chem. Commun. **11**, 1187 (1998)
42. A.F. McCabe, C. Eliasson, R. Arun Prasath, A. Hernandez-Santana, L. Stevenson, I. Apple, P.A.G. Cormack, D. Graham, W.E. Smith, P. Corish et al., Faraday Discuss. **132**, 303 (2006)
43. D. Lim, K. Jeon, H.M. Kim, J. Nam, Y.D. Suh, Nat. Mater. **9**, 60 (2010)
44. N.M.B. Perney, J.J. Baumberg, M.E. Zoorob, M.D.B. Charlton, S. Mahnkopf, C.M. Netti, Opt. Express **14**, 847 (2006)
45. B. Pettinger, A. Tadjeddine, D.M. Kolb, Chem. Phys. Lett. **66**, 544 (1979)
46. S. Ushioda, Y. Sasaki, Phys. Rev. B **27**, 1401 (1983)
47. R.M. Corn, M.R. Philpott, J. Chem. Phys. **80**, 5245 (1984)
48. K. Kurosawa, R.M. Pierce, S. Ushioda, J.C. Hemminger, Phys. Rev. B **33**, 789 (1986)
49. J. Giergiel, C.E. Reed, J.C. Hemminger, S. Ushioda, J. Phys. Chem. **92**, 5357 (1988)
50. N. Primeau, J.L. Coutaz, L. Abello, J. Appl. Phys. **73**, 5158 (1993)
51. J.P. Goudonnet, T. Inagaki, E.T. Arakawa, T.L. Ferrell, Phys. Rev. B **36**, 917 (1987)
52. F. Mohaghegh, A.M. Tehrani, A. Materny, J. Phys. Chem. C **125**, 5158 (2021)
53. D. Mandler, S. Kraus-Ophir, J. Solid State Electrochem. **15**, 1535 (2011)
54. C.K. Klutse, A. Mayer, J. Wittkamper, B.M. Cullum, J. Nanotechnol. **2012**, 319038 (2012)
55. K. Kneipp, H. Kneipp, J. Kneipp, Acc. Chem. Res. **39**, 443 (2006)
56. K. Kneipp, Y. Wang, H. Kneipp, L.T. Perelman, I. Itzkan, R.R. Dasari, M.S. Feld, Phys. Rev. Lett. **78**, 1667 (1997)
57. M. Rycenga, X. Xia, C.H. Moran, F. Zhou, D. Qin, Z. Li, Y. Xia, Angew. Chem. Int. Ed. **50**, 5473 (2011)
58. H. Stass, A. Dalhoff, D. Kubitza, U. Schuhly, Antimicrob. Agents Chemother. **42**, 2060 (1998)
59. B. Bozkurt, G. Karakaya, A.F. Kalyoncu, Allergol. Immunopathol. **33**, 38 (2005)
60. J.A.B. Balfour, L.R. Wiseman, Drugs **57**, 363 (1999)
61. M. Donati, M.R. Fermepin, A. Olmo, L.D' Apote, R. Cevenini, J. Antimicrob. Chemotherapy **43**, 825 (1999)
62. K. Vishwanathan, M.G. Bartlett, J.T. Stewart, J. Pharm. Biomed. Anal. **30**, 961 (2002)
63. R. Wise, J.M. Andrews, G. Marshall, G. Hartman, Antimicrob. Agents Chemother. **43**, 1508 (1999)
64. E.M. Golet, I. Xifra, H. Siegrist, A.C. Alder, W. Giger, Environ. Sci. Technol. **37**, 3243 (2003)
65. D. Fatta-Kassinos, S. Meric, A. Nikolaou, Anal. Bioanal. Chem. **399**, 251 (2011)
66. I. Senta, S. Terzic, M. Ahel, Water Res. **47**, 705 (2013)
67. T. Lemoine, D. Breilh, D. Ducint, J. Dubrez, J. Jougon, J.F. Velly, M.C. Saux, J. Chromatogr. B, Biomed. Sci. Appl. **742**, 247 (2000)
68. J.G. Möller, H. Stass, R. Heinig, G. Blaschke, J. Chromatogr. B, Biomed. Sci. Appl. **716**, 325 (1998)
69. M.A.G. Trindade, G.M. da Silva, V. Souza Ferreira, Microchem. J. **81**, 209 (2005)
70. J.A. Ocana, F.J. Barragan, M. Callejon, Analyst **125**, 2322 (2000)
71. N. Erk, Anal. Bioanal. Chem. **378**, 1351 (2004)
72. H.A. Nguyen, J. Grellet, B.B. Ba, C. Quentin, M. Saux, J. Chromatogr. B **810**, 77 (2004)
73. B.B. Ba, R. Etienne, D. Ducint, C. Quentin, M. Saux, J. Chromatogr. B Biomed. Sci. Appl. **754**, 107 (2001)
74. H. Liang, M.B. Kays, K.M. Sowinski, J. Chromatogr. B **772**, 53 (2002)
75. S. Tatar Ulu, J. Pharm. Biomed. Anal. **43**, 320 (2007)
76. W. Ji, W. Yao, Spectrochim. Acta Part A Mol. Biomol. Spectrosc. **144**, 125 (2015)
77. Y. Xie, X. Zhu, Y. Sun, H. Wang, H. Qian, W. Yao, Eur. Food Res. Technol. **235**, 555 (2012)
78. I.J. Hidi, M. Jahn, K. Weber, D. Cialla-May, J. Popp, Phys. Chem. Chem. Phys. **17**, 21236 (2015)

79. P. Neves, A. Leite, M. Rangel, B. de Castro, P. Gameiro, Anal. Bioanal. Chem. **387**, 1543 (2007)
80. U. Neugebauer, A. Szeghalmi, M. Schmitt, W. Kiefer, J. Popp, U. Holzgrabe, Spectrochim. Acta Part A: Mol. Biomol. Spectrosc. Honour Issue—Jim Durig **61**, 1505 (2005)
81. F. Mohaghegh, A. Mazaheri Tehrani, D. Rana, M. Winterhalter, A. Materny, J. Raman Spectrosc. **52**, 1617 (2021)
82. C. Jing, F. Shaolong, G. Fang, G. Edward, X. Jie, W. Shuo, H. Qian, L. Xiaonan, J. Food Sci. **80**, N834 (2015)

Chapter 9
Electrochemical Surface-Enhanced Raman Spectroscopy (EC-SERS): Techniques, Applications, and Future Perspectives

N. K. Murugasenapathi and Tamilarasan Palanisamy

9.1 Introduction

Electrochemistry is an interdisciplinary field in modern science that finds a wide range of applications, including energy storage, electroplating and sensors [1, 2]. Recent developments in electrochemical research have led to a growing interest in exploring these systems at molecular level, aiming to achieve a more comprehensive understanding of the complex processes occurring at the electrode-electrolyte interface [3–5]. To comprehend electrochemical interfaces at molecular level, it is imperative to employ sophisticated characterization techniques that can provide dynamic and real-time information about changes at the interface. Vibrational spectroscopy is one such powerful analytical technique that provide valuable chemical information of molecules of interest adsorbed at the electrode surfaces [6, 7]. Due to the dynamic nature of the electrochemical systems, in situ or *operando* measurements are often mandatory to capture the real-time chemical changes.

Integrating vibrational spectroscopy with electrochemical techniques has emerged as a powerful approach for comprehensively understanding the intermediate electrochemical processes occurring at the complex electrode-electrolyte interface. Vibrational spectroscopic techniques, such as infrared (IR) [8, 9], Raman [10, 11], and sum frequency generation (SFG) spectroscopies [12, 13], are particularly useful for identifying interfacial molecules and the nature of chemisorption bonds. Among these, Raman spectroscopy is the more convenient and powerful technique due to its ability to provide molecule fingerprint spectra using a wide range of excitations, its

N. K. Murugasenapathi · T. Palanisamy (✉)
Electrodics and Electrocatalysis Division (EEC), CSIR-Central Electrochemical Research Institute (CECRI), Tamil Nadu, Karaikudi 630 003, India
e-mail: tamilan@cecri.res.in

Academy of Scientific and Innovative Research (AcSIR), Ghaziabad 201 002, India

D. K. Singh et al. (eds.), *Raman Spectroscopy*, Springer Series in Optical Sciences 248,
https://doi.org/10.1007/978-981-97-1703-3_9

insensitivity to water, and its suitability for in situ measurements of electrochemical systems [14, 15]. However, a significant limitation of Raman spectroscopy is its low signal-to-noise ratio when detecting ultralow concentration of molecules due to low Raman scattering probability (~10^{-6}), which makes probing the intermediates highly challenging [14, 16].

The discovery of SERS in the 1970s revolutionized by enhancing the Raman signal as high as several billion times with the help of plasmonic nanostructures [16]. It also helped to overcome the limitations of conventional Raman spectroscopy and found applications not only in probing chemical changes, but also in sensors, pharmaceuticals, forensic etc. [17, 18]. The integration of this powerful technique with electrochemical methods offers numerous advantages, further expanding the potential of SERS for enhanced understanding and exploration in various fields.

In this chapter, a brief introduction along with historical developments, fundamentals of EC-SERS and its application towards sensors, electrocatalytic studies, fuel cells, batteries, corrosion, and electrodeposition are discussed. This chapter also highlights the advanced EC-SERS techniques, such as electrochemical shell-isolated nanoparticle-enhanced Raman spectroscopy (EC-SHINERS) and electrochemical tip-enhanced Raman spectroscopy (EC-TERS).

9.2 History of EC-SERS

The first observation of SERS was made exclusively through the application of electrochemical techniques, thus rendering it more suitable to designate this as the first discovery of EC-SERS [19–21]. Fleischmann, Hendra, and McQuillan from the University of Southampton developed a technique to enhance the adsorption of molecules on a roughened silver electrode by utilizing potential-controlled oxidation and reduction cycles [19]. To investigate the surface sites on catalysts, pyridine was chosen as an adsorbate due to its large Raman cross-section. By applying the aforementioned method in an aqueous electrolyte containing 0.1 M KCl + 0.05 M pyridine, they were able to increase the Raman signal strength of pyridine. The resulting Raman spectrum exhibited high-quality signals, where the signal strength varied with potential, indicating a potential-dependent variation in the adsorbed pyridine coverage (Fig. 9.1). These preliminary results were the first ever observation of EC-SERS, despite it not being named so at that time. The roughened silver electrode was the first nanostructure to exhibit SERS activity.

Following this observation, Van Duyne and Jeanmaire from Northwestern University conducted further experiments to test the hypothesis of signal enhancement by increased surface area through surface roughening [20]. Surprisingly, it was found that the Raman scattering intensity was increased with decreased surface roughness. The surface enhancement factor also has been calculated by comparing the signal strength of the molecule on the roughened surface with that in a solution. From the observations, Van Duyne et al., have suggested that the electrode surface or electrode/solution interface may enhance the effective Raman scattering cross-section, leading

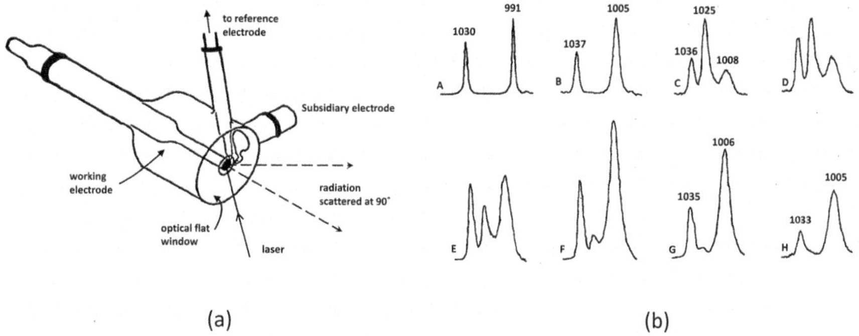

Fig. 9.1 **a** The spectroelectrochemical cell used for the electrochemical Raman scattering analysis. **b** The Raman spectra of **A** and **B** liquid pyridine and **C-H** pyridine adsorbed on the silver electrode with potentials ranging from 0.0 V to −1.0. Adapted with permission from [19]. Copyright, 1974, Elsevier

to unexpectedly high Raman scattering intensity from adsorbed pyridine molecules. In the same year, Albrecht and Creighton from the University of Kent independently reported a similar result from colloidal silver suspensions where the signal enhancement has been imputed to charge-transfer between the adsorbed molecule and silver [21].

Over the years, researchers have made remarkable progress in developing novel materials and methodologies for surface modification, such as functionalized metal nanoparticles, plasmonic nanostructures, and engineered substrates, enabling the precise control over SERS enhancement [22–24]. Combined with electrochemistry, this vibrational spectroscopy offers additional control over the adsorption and redox states of the analytes at the surface, leading to improved analytical performance, particularly in the case of charged analytes. EC-SERS has traditionally been used for investigating electrochemical reactions [25], corrosion processes [26], and their inhibition [27, 28], as well as spectroelectrochemical probing of nanostructured metallic surfaces and their interactions with molecular species [29]. Despite its potential, EC-SERS is still a developing technique compared to traditional SERS analysis.

9.3 Features of EC-SERS

EC-SERS combines the unique attributes of nanostructured electrodes and electrolyte solutions to enable simultaneous electrochemical and spectroscopic measurements. The typical EC-SERS instrumentation involves meticulous integration of electrochemical and Raman spectroscopy at the plasmonic SERS substrate, providing a powerful platform for the comprehensive analysis of analytes in complex electrochemical systems. The spectroelectrochemical cell provides a controlled environment for electrochemical reactions while enabling access to spectroscopic analyses

(Fig. 9.2a). The working electrode, typically fabricated from nanostructured plasmonic materials such as gold or platinum, serves a dual purpose as the site for electrochemical processes and as the substrate for SERS enhancement. In a typical EC-SERS measurement, the working electrode with plasmonic nanostructures is immersed in an electrolyte and biased with an appropriate potential to initiate electrochemical reactions. Concurrently, a laser beam is focused onto the SERS-active substrate, and the resulting Raman scattered light is meticulously collected and analyzed using a spectrometer. The instrumentation for EC-SERS encompasses a laser source, spectrometer, and a highly sensitive detector tailored for capturing and quantifying the Raman signal. Generally, a potentiostatic/galvanostatic method is used to precisely control the electrochemical conditions.

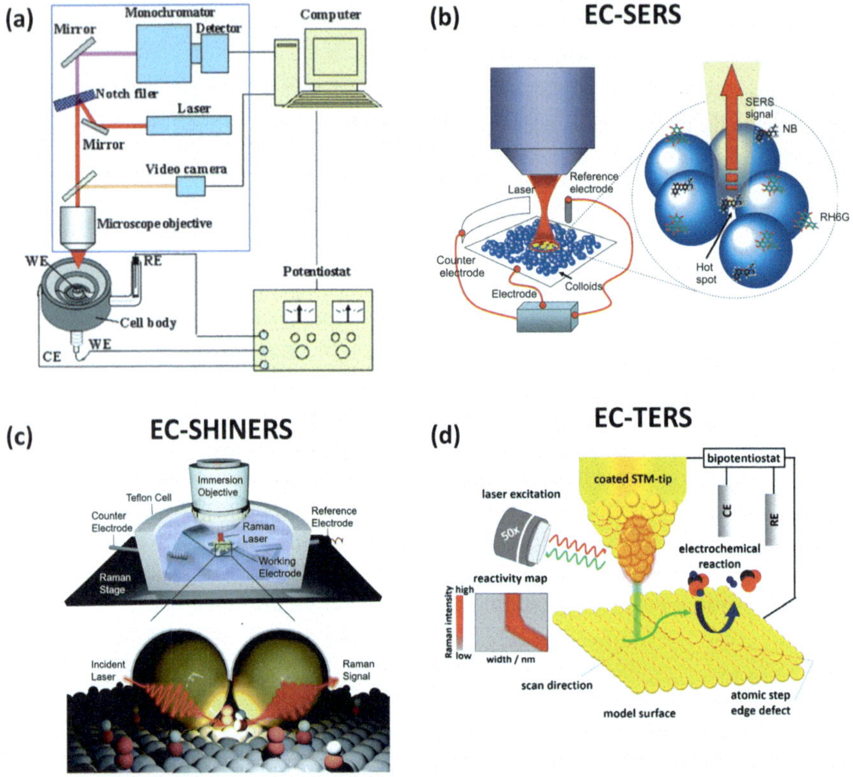

Fig. 9.2 **a** Illustration of the standard experimental setup for electrochemical Raman analysis, consisting of a Raman spectrometer, potentiostat, data acquisition system, and a spectroelectrochemical cell. The working electrode (WE), counter electrode (CE), and reference electrode (RE) are indicated. Schematic representation of the experimental setup for different techniques which includes EC-SERS, EC-SHINERS and EC-TERS are shown in (**b**), (**c**) and (**d**), respectively. Adapted with permission from [25, 31–33]. Copyright, 2008, 2020, The Royal Society of Chemistry. Copyright, 2010, American Chemical Society. Copyright, 2018, Elsevier

The SERS enhancement is widely attributed to two main mechanisms: electromagnetic enhancement (EM) and chemical enhancement (CE) [20, 21]. The EM is resulted from the electric field enhancement where the excitation of surface plasmon, their localization and coupling at the surface of noble metal nanostructures. In most SERS systems, electromagnetic field enhancement is the primary contributor (10^3–10^8 times) to the observed SERS signals. The CM is mainly due to the charge-transfer and complex formation which are determined by the Fermi level of the substrates and the molecules. Generally, CM contribution is relatively weaker (up to 10^3 times).

In electrochemical systems, the chemical enhancement mechanism also plays a significant role, particularly in the characterization of chemical species. The chemical enhancement mechanism is intricately linked to the chemical properties of surface species and the substrate, making it an essential factor in understanding the behavior of analytes in electrochemical SERS studies. By considering both the electromagnetic and chemical enhancements, a more comprehensive understanding of the SERS phenomenon in electrochemical systems can be achieved, facilitating the characterization and analysis of chemical species in various electrochemical environments.

In addition to its analytical capabilities, EC-SERS has garnered significant interest in elucidating valuable insights into adsorption configurations and reaction mechanisms in electrochemistry. An electrochemical double layer is formed at the interface between the working electrode and the electrolyte in an electrochemical system, making it a crucial and complex interfacial region in electrochemical SERS studies. The electrochemical double layer consists of two layers: the inner layer called the Stern layer, which is composed of tightly adsorbed ions, and the outer layer called the diffusion layer, which contains loosely adsorbed ions [30]. The behavior of analytes within this double layer significantly influences the SERS signal and its enhancement mechanisms. The interactions between the analyte molecules, the surface species, and the ions in the double layer are highly intricate and can give rise to changes in the chemical environment, charge distribution, and electric field at the electrode-electrolyte interface. Understanding the dynamics and complexities of the electrochemical double layer is therefore essential for elucidating the interplay between the chemical enhancement mechanism and the electrochemical processes in electrochemical SERS, facilitating the interpretation of SERS spectra and characterization of chemical species at the electrode-electrolyte interface.

Potential-dependent SERS intensity is another important and common feature of EC-SERS experiments. In the field of electrochemistry, the electrode potential refers to the applied electric field on an electrode immersed in a solution. By manipulating the electrode potential, it is possible to manipulate the behavior and arrangement of molecules that are adsorbed onto the electrode surface. In EC-SERS, these changes, in turn, have a direct impact on obtained SERS signal. The intensity of SERS signals in EC-SERS experiments can also be influenced by other factors, including the interaction between the molecule and the surface and the light-driven charge transfer. An alteration of the electrode potential can change the interaction between the molecule and the electrode surface which has direct impact on the geometric and electronic structure of the molecular arrangement, ultimately leading to changes in the spectral SERS signals. On the other hand, the charge transfer process can influence the chemical enhancement effects, which significantly impact the SERS signals.

Despite transition metals, such as Pt, Ru, Rh, Pd, Fe, Co, Ni and their alloys, hold greater practical significance in electrochemistry, only Ag, Au, and Cu are widely recognized for SERS signal enhancement [25, 34, 35]. The former set of transition metals exhibit markedly different electronic band structures. Particularly, the inter-band excitation across the entire visible light region significantly diminishes the surface plasmon resonance (SPR). Consequently, the field strength of SPR is suppressed in the visible excitation.

Moreover, the interaction of molecules with other transition metal surfaces is generally stronger than on coinage metals. This disparity in binding strength can be once again attributed to the distinct electronic structures and surface properties of these materials. Conclusively, the choice of electrode material and the excitation energy have a significant influence on the SERS enhancement.

The SERS intensity observed in EC-SERS systems is profoundly impacted by both the choice and concentration of the electrolyte solution which eventually decides the resulting compression (or expansion) of the electrochemical double layer. The choice of the electrolyte affects the occurrence of non-specific adsorption on the electrode surface while its concentration changes the diffusion resistance. Furthermore, the solvent directly impacts the chemical and physical properties of the adsorbed molecule of interest due to the respective solvation enthalpy and mechanism.

From the above discussion, it is evident that EC-SERS represent one of the most intricate and multifaceted analytical methods. However, it offers a higher degree of flexibility in exploring the interfacial structure and mechanisms of complex systems by manipulating various experimental degrees of freedom. These parameters include the choice of electrode material, solute, concentration, solvent, and electrode potential. By incorporating both SERS and electrochemistry, EC-SERS enables researchers to investigate the intricate interplay between molecular adsorption and electrochemical processes. This amalgamation of techniques provides valuable insights into the fundamental aspects of complex systems, facilitating a deeper understanding of their behavior and underlying mechanisms.

With time, significant progress has been made in the field of EC-SERS, leading to the development of several advanced strategies that have further expanded the scope of its applications. Notably, techniques such as borrowing SERS strategy, EC-SHINERS, and EC-TERS have emerged, effectively overcoming limitations imposed by electrode materials and enabling investigations of electrochemical interfaces with sub-atomic resolution. Furthermore, these advanced techniques have facilitated the exploration of non-metallic surfaces, opening new avenues for research, and expanding the potential of EC-SERS in diverse applications.

9.3.1 Borrowing SERS

Despite notable advancements in enhancing SERS signal through surface manipulation, the aforementioned EC-SERS approaches are constrained by the prerequisite that the targeted electrode must exhibit localized surface plasmon resonance. As mentioned earlier, transition metals hold immense significance in the realm of electrochemistry owing to their distinctive properties. Their interface with electrolyte has been extensively investigated, which has substantial implications in various electrochemical applications. This is accomplished through a proximity effect, whereby a SERS-active material is coated with a non-SERS-active electrode with a thickness on the order of a few nanometers [36, 37]. The latter can then "borrow" the indispensable electromagnetic enhancement, thereby expanding the applicability of materials with lower SERS activity. This methodology is referred to as borrowing SERS and has been successfully demonstrated as a viable alternative in numerous instances [38, 39].

9.3.2 EC-SHINERS

It is necessary to note that many of the aforementioned EC-SERS techniques primarily deploy nanostructured plasmonic electrodes. While enabling application on a broader spectrum of surfaces, the borrowing SERS approach encounters limitations in electrochemical interfaces due to potential interference arising from the plasmonic nanostructures. This interference adversely impacts the electrochemical response, as recently elucidated [40, 41]. In 2010, Tian and the team introduced a significant advancement to address this challenge. They developed a technique called SHINERS, where SERS-active nanoparticles were separated using an ultra-thin (~2 nm) dielectric layer (SiO_2 or Al_2O_3) [42]. This configuration allowed the metal cores to provide borrowed electromagnetic enhancement while being physically isolated from the deposited surface. SHINERS has revolutionized SERS applications by enabling measurements on any substrate with minimal interference from the "SHIN" resonators.

Furthermore, the versatility of SHINERS has greatly enhanced EC-SERS, leading to the exploration of a wide range of non-metallic electrode substrates. EC-SHINERS has facilitated successful investigations of diverse electrocatalytic reactions on a range of materials, including Ag, Au, Pt, Pd, Rh, Ni alloys, Cu, and glassy carbon (GC) [43, 44]. Additionally, EC-SHINERS allows detailed investigations of single crystal electrodes (Fig. 9.2c), which are of great interest as model systems in surface science [45, 46].

9.3.3 EC-TERS

While EC-SHINERS offers superior versatility compared to conventional SERS in terms of compatibility with electrochemical interfaces, it records spectra by averaging signals over a large, typical micron-sized, incident laser. This results in poor spatial resolution. While studying structural evolution and surface chemical changes of microstructures, comprehensive and precise chemical analysis with sub-nanometer scale resolution is mandatory. Fortunately, such a chemical sensitivity along with sub-nanometer spatial resolution has been realized by EC-TERS technique [33, 47]. EC-TERS facilitates detailed insights into the intricacies of electrochemical interfaces offering a nanoscale resolution [48]. This technique involves the utilization of a sharp metallic tip, shown in Fig. 9.2d, composed of a SERS-active material, which is positioned at the close proximity to the surface of interest, thereby creating localized hotspots that result in enhanced Raman signals.

In the following section, the notable applications of the EC-SERS have been discussed which manifest the versatility of EC-SERS in various research areas.

9.4 Applications of EC-SERS

The comprehensive analysis of EC-SERS spectral features requires integration of both experimental and theoretical perspectives. This combined approach facilitates the extraction of vital information regarding chemical bonding, molecular orientation, and even the formation of intermediates at the electrode surfaces. Despite extensive research on SERS in the past 4 decades, the focus on EC-SERS as an operando analytical tool remains limited. Particularly, previous studies have only emphasized the utilization of SERS in electrochemical studies, while the role and benefits of electrochemistry in SERS analyses are yet to be explored.

The following sections of this chapter will primarily focus on the utilization of EC-SERS in diverse applications, such as sensors, electrocatalytic investigations, corrosion analysis, batteries, fuel cells, and the detection of contaminants to highlight the emerging opportunities and potential applications of EC-SERS.

9.4.1 EC-SERS for Sensing

In the realm of sensors, both electrochemical and SERS methods have been extensively utilized for various chemical and biological analytes. Electrochemical transducers give potentiometric, amperometric, conductometric and impedimetric signals proportional to the analyte concentration [49]. On the other hand, SERS sensors identify and quantify based on the molecular fingerprints of the analyte [50]. While electrochemical sensor technologies have been successfully developed, SERS-based sensors have yet to be matured to adopt them as an alternative for the present complex procedures. The major challenge in implementation of SERS-based sensors is the reproducibility due the surface heterogeneity. Moreover, it leads to difficulties in scaling up laboratory-designed substrates.

These limitations can be overcome by hyphenating electrochemical aspects with SERS platforms. Generally, EC-SERS offers enhanced selectivity, facilitates controlled adsorption and desorption processes, improves signal reproducibility, and enables cross-validation of data during analyte quantification. These comprehensive advantages have resulted in increased attention towards EC-SERS-based sensors.

Zaleski et al. [52] demonstrated EC-SERS based rapid and sensitive therapeutic drug monitoring and drug compounding applications at clinical conditions. Here, gentamicin was successfully detected and quantified at clinically relevant range using standard and handheld Raman instrumentation. When normal Raman spectroscopy (NRS) alone was unable to detect, a label-free EC-SERS approach was developed to detect dobutamine, with 100 ng/mL (300 nM) limit of detection (LOD) along with good accuracy and selectivity using a low-cost SERS-active strip suitable for EC-SERS experiments. In addition, the same approach was deployed for the detection of secondary amines under acidic conditions.

Similar integrated EC-SERS strip was developed by Huang et al. [51] for the rapid detection of uric acid. Here, the working electrode was made with nanocone array polycarbonate (PC) substrate decorated with evenly distributed tightly packed gold nanospheres that showed enhancement factor up to 8.5×10^6 with rhodamine 6G (R6G). As a result of excellent uniformity, highly reproducible results were obtained as evidenced by remarkably low relative standard deviation of EC-SERS intensity (1.41%). The technology was further demonstrated with successful quantification of uric acid in synthetic urine (Fig. 9.3). Overall, this demonstration highlights the potential of EC-SERS in medical tests, food safety, and biotechnology.

For readers interested in exploring the potential of EC-SERS, a recent comprehensive review is recommended [53]. This review provides an extensive analysis of EC-SERS sensors, covering perspectives from both SERS-assisted electroanalysis and EC-assisted SERS. It offers valuable insights into the analytical applications of EC-SERS.

9.4.2 Electrocatalysis

Electrocatalytic reactions are of paramount importance in various industrial applications, such as energy conversion and storage, electro-synthesis and environmental remediation [54]. In this context, EC-SERS play a significant role in investigating intricate reaction pathways and comprehending the impact of reaction parameters, catalyst structure, and surface modifications on electrocatalytic efficiency.

EC-SERS techniques have gained widespread attention for investigating molecular adsorption and desorption kinetics at the electrode-electrolyte interface, particularly in the context of electrocatalytic reactions. This technique allows researchers

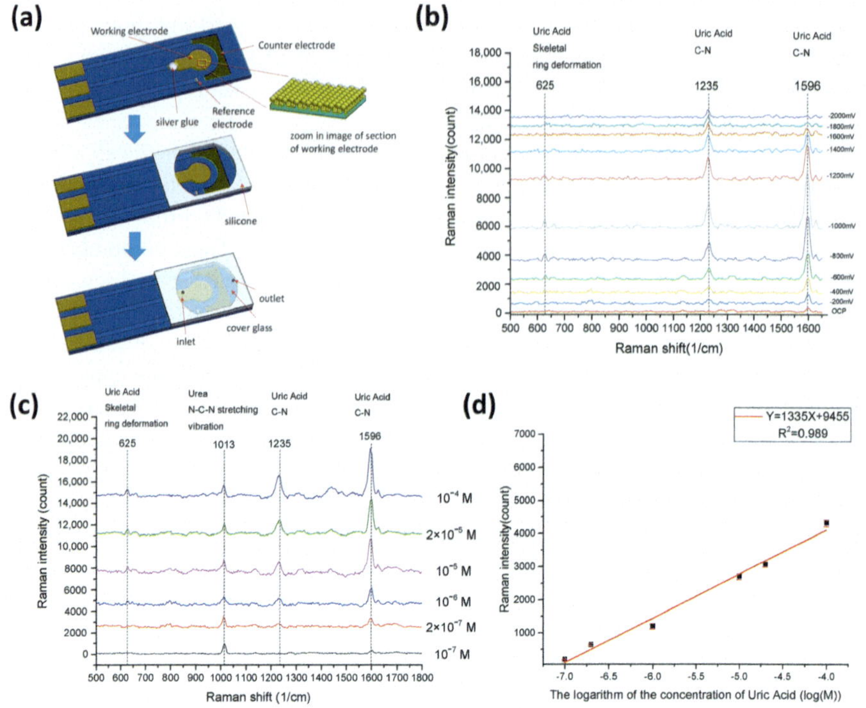

Fig. 9.3 **a** Integrated EC-SERS sensor strip. **b** EC-SERS dataset of uric acid at various applied potential. **c** Measured EC-SERS spectra of different uric acid concentrations in artificial urine. **d** Relationship between uric acid concentration and EC-SERS intensity at 1596 cm^{-1} Raman band, showing a proportional relationship. Adapted with permission from [51]. Copyright, 2020, MDPI

to obtain valuable insights into the behavior of molecules on the catalyst surface. In recent studies, researchers have pushed the boundaries of EC-SERS by employing it to quantitatively analyze interfacial pH [55]. By carefully monitoring the Raman signals of specific molecules, such as the bifunctional 4-mercapto pyridine (4MPy), researchers could accurately assess the thermoneutrality of proton transfer at the interfacial layer. This analysis provides crucial information regarding the regulation of interfacial pH and its consequential effects on the overall electrocatalytic activity. Similarly, Huang et al. [56] used EC-SHINERS to investigate the surface species formed during the electrochemical surface oxidation process of well-defined Pt(111) and Pt(100) surfaces (Fig. 9.4). The findings provide insights into the surface chemistry of Pt catalysts and offer potential implications for understanding catalyst degradation during the oxygen reduction reaction. In addition, the study demonstrates the effectiveness of in situ SHINERS for identifying intermediates on single-crystal surfaces, contributing to a deeper understanding of electrochemical surface processes at the molecular level.

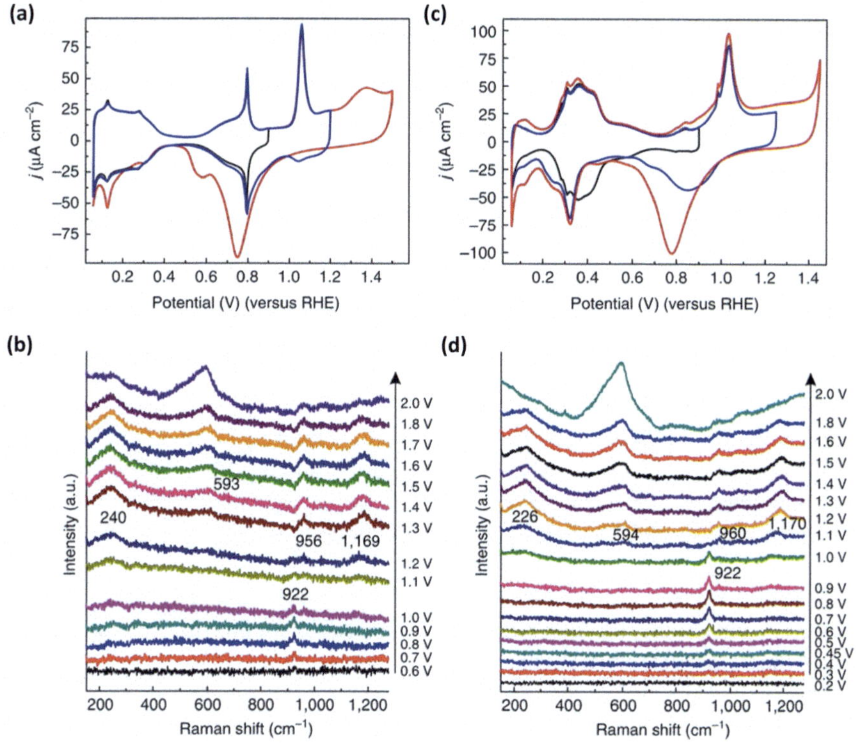

Fig. 9.4 Cyclic voltammogram of **a** Pt(111) and **c** Pt(100) electrodes in 0.1 M $HClO_4$ electrolyte at 50 mV s^{-1} scan rate. The corresponding SHINERS spectra of **b** Pt(111) and **d** Pt(100) at indicated potentials recorded with positive potential steps. Adapted with permission from [56]. Copyright, 2016, Nature publishing group

Ren and co-workers [47] integrated a commercial scanning tunneling microscope (STM) with a custom-built confocal Raman microscope in a horizontal illumination geometry to enable EC-TERS measurements. In their study, the behavior of 4-PBT((4-(pyridin-4-yl)biphenyl-4-yl)methanethiol) on Au(111) surface has been investigated.

Specifically, the molecule configuration changes as a function of applied potential have been examined and were attributed to the protonation and deprotonation processes of the adsorbed molecules. Further insights were gained with the help of density functional theory (DFT) calculations. The results revealed that the nitrogen end of the 4-PBT molecule underwent protonation at negative potentials or deprotonation at positive potentials. This observation provided valuable information about the electrochemical behavior of the system. Interestingly, the cyclic voltammetry analysis showed almost no discernible features within the explored potential limits which was attributed to the passivation of Au(111) surface by the 4-PBT monolayer. As the electrochemical data fails to give explicit indications, it was possible to analyze the local protonation and deprotonation only TERS signals. An excellent review by Pfisterer et al. [33] provides a comprehensive analysis of the versatility of EC-TERS in the field of electrocatalysis.

9.4.3 Electrodeposition

The application of EC-SERS in conjunction with electroplating processes can yield valuable information regarding the adsorption behaviors of various electroactive species in the bath. It can be extensively employed to investigate the adsorption behaviors of various electroactive complexes, ligands, leveling agents, and brighteners, commonly found in Au, Ag, and Cu electrodeposition baths, at the molecular level [57, 58].

Recently, EC-SHINERS has been successfully deployed for investigating intermediate electrochemical processes of cobalt electroplating [59]. Specifically, the chemical changes associated with transient potential oscillations during the electrodeposition of cobalt on a glassy carbon electrode in the presence of butynediol have been investigated. To facilitate the analysis, $Ag@SiO_2$ as an *operando* substrate was used for EC-SHINERS analysis. The carefully engineered SHINERS substrate showed a remarkable 10^7-fold signal enhancement according to finite-difference time-domain (FDTD) simulations.

This study gives direct spectroscopic evidence of significant processes occurring during the Co electrodeposition. A noteworthy finding was the observed adsorbed hydrogen scavenging by butynediol, affirming its efficacy as an inhibitor of the hydrogen evolution reaction. Additionally, the formation of $Co(OH)_2$ and its subsequent dissolution and reduction during the potential oscillation have also been observed [59]. These findings provided valuable insights into the intricate electrochemical reactions and species involved in the electrodeposition process. This branch of research is still in its infancy, offering a lot of opportunities for further development.

9.4.4 Corrosion Studies

The capability of EC-TERS to discern local nanoscale chemistry has been previously discussed. In a study by Pfisterer et al., the unique application of near-field optical nanoscopy, specifically EC-TERS, for investigating corrosion and mapping the reactivity of nanoscale defects under electrocatalytic reaction conditions has been illustrated [60]. This technique provides an exceptional chemical sensitivity at a spatial resolution of approximately 10 nm, facilitating studies on molecular level chemical changes.

The study revealed a direct correlation between the electro-oxidation reactivity of gold nano-defects and their surface topography. Specifically, the reactivity was found to be limited to an oxide depth of approximately 3 nm. Through local Raman

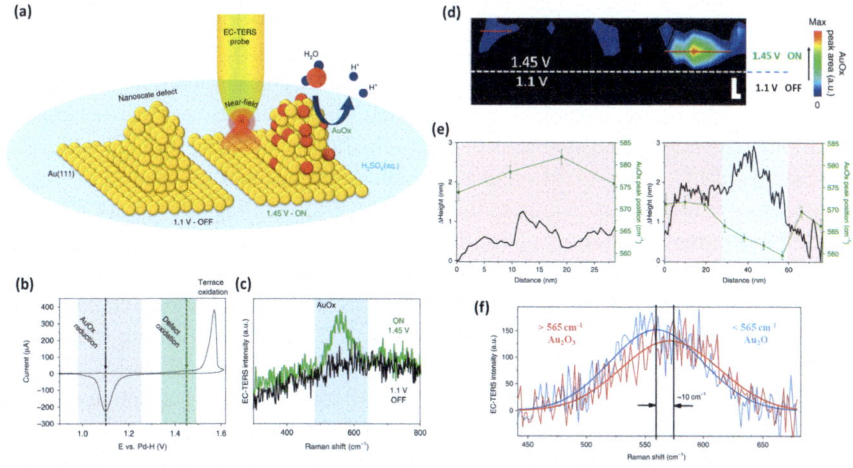

Fig. 9.5 **a–c** Schematic illustration of EC-TERS *operando* nanoscopy approach, highlighting the spatial and chemical resolution of defect oxidation states (OFF and ON), along with corresponding cyclic voltammogram and EC-TERS spectra. **d** EC-TERS map based on AuO_X band intensity and **e** corresponding height (Δ_{height}) profiles and peak positions correlation **f** EC-TER spectra from different AuO_X defect locations, displaying a peak position difference of ~10 cm^{-1}, with corresponding Gaussian peak fits. Adapted with permission from [60]. Copyright, 2019, Nature publishing group

analysis, the presence of at least two spatially separated AuO_X species was identified on the nano-defects. These species were characterized as Au_2O_3 and Au_2O (Fig. 9.5).

9.5 Challenges and Future Perspectives

EC-SERS faces several significant challenges that need to be addressed for its future development. One of the primary challenges is achieving stable and reproducible SERS signals in electrochemical conditions. The presence of electrolytes, ions, and potential-induced changes at the electrode surface can impact the stability and intensity of SERS signals. Ongoing efforts focus on developing strategies to optimize the stability and reproducibility of EC-SERS measurements, such as modifying the electrode surface and selecting suitable electrolytes.

Understanding and controlling the electric field distribution at the electrode surface during EC-SERS measurements presents another challenge. The electrochemical environment introduces additional complexity due to charge transfer processes and the presence of redox-active species, which can affect the enhancement of Raman signals. Advancing the understanding of the electric field distribution and its modulation in EC-SERS systems can enhance the sensitivity and reliability of the technique.

To further advance EC-SERS, there is a need for the development of new SERS-active substrates and materials specifically tailored for electrochemical applications. These substrates should exhibit excellent stability and reproducibility, high enhancement factors, ideally no interference and compatibility with various electrochemical conditions. The design and synthesis of such substrates are crucial for expanding the applications of EC-SERS to different electrochemical systems.

In terms of future perspectives, there are several exciting directions for EC-SERS. Firstly, the integration of EC-SERS with other analytical techniques can provide complementary information and more comprehensive understanding of electrochemical processes at the molecular level. This multimodal approach can offer unprecedented insights into complex electrochemical interfaces and their dynamics.

The interaction of photons, molecules, and metal surfaces is considered in the context of chemical enhancement. However, the influence of the surface photoelectric field on the probability of light-driven charge-transfer is not accounted in this framework. Therefore, developing a comprehensive theoretical model that incorporates both mechanisms and devising a quantitative calculation method poses significant challenges. These challenges constitute the central focus and direction for further research into the underlying mechanisms of SERS in a broader perspective.

The theoretical explanation of the chemical enhancement is particularly challenging as it requires accurate predictions of the vibrational and electronic eigenstates of molecules in specific geometric arrangements. It is worth noting that the SERS intensity is dependent on factors such as the size and shape of metal nanoparticles, as well as the distance between particles (nanogaps). Consequently, investigating the SERS mechanism is not only a long-standing topic in surface science and spectroscopy but also a challenge within the field of nanoscience.

Moreover, expanding the applications of EC-SERS beyond fundamental studies to practical applications, such as the analysis of real-samples or the monitoring of industrial electrochemical processes, holds greater promise. By addressing the challenges of stability, reproducibility, and robustness, EC-SERS can become a valuable tool in various fields, including energy storage and conversion, environmental monitoring, and electrochemical sensing.

9.6 Conclusions

This chapter provides a comprehensive overview of the early history, fundamentals, and significant applications of EC-SERS. The chapter highlights the key characteristics and capabilities of EC-SERS. It describes the advantages of electrochemistry part of EC-SERS that offers unique opportunities for in situ and *operando* studies, enabling the investigation of dynamic processes occurring at electrode-electrolyte interfaces. This chapter serves as a comprehensive resource to understand the foundations and advancements in EC-SERS. Furthermore, the chapter introduces recent advancements in the field, such as EC-SHINERS and EC-TERS. These emerging techniques expand the capabilities of electrochemical vibrational spectroscopy, allowing for the investigation of molecular structures and chemical processes with exceptional spatial resolution (sub-nanometer) and chemical sensitivity (sub-nanomolar). The potential applications and future prospects of these advanced techniques are briefly highlighted to describe their importance in the field of spectroelectrochemistry.

References

1. A.J. Bard, L.R. Faulkner, H.S. White, *Electrochemical Methods: fundamentals and Applications*, 3 edn. (Wiley, 2022)
2. J. Wang, in *Analytical Electrochemistry* (Wiley, 2006), pp. 1–28
3. M. Favaro, B. Jeong, P.N. Ross, J. Yano, Z. Hussain, Z. Liu, E.J. Crumlin, Nat. Commun. **7**, 12695 (2016)
4. F. Zaera, Chem. Rev. **112**, 2920–2986 (2012)
5. O.M. Magnussen, A. Gross, J. Am. Chem. Soc. **141**, 4777–4790 (2019)
6. A.J. Wain, M.A. O'Connell, Adv. Phys.: X **2**, 188–209 (2017)
7. J.J.A. Lozeman, P. Fuhrer, W. Olthuis, M. Odijk, Analyst **145**, 2482–2509 (2020)
8. M. Osawa, Bull. Chem. Soc. Jpn. **70**, 2861–2880 (1997)
9. G. Samjeske, M. Osawa, Angew. Chem. Int. Ed. Engl. **44**, 5694–5698 (2005)
10. Y.H. Wang, S. Li, R.Y. Zhou, S. Zheng, Y.J. Zhang, J.C. Dong, Z.L. Yang, F. Pan, Z.Q. Tian, J.F. Li, Nat. Protoc. **18**, 883–901 (2023)
11. Z.Q. Tian, B. Ren, Annu. Rev. Phys. Chem. **55**, 197–229 (2004)
12. A. Montenegro, C. Dutta, M. Mammetkuliev, H. Shi, B. Hou, D. Bhattacharyya, B. Zhao, S.B. Cronin, A.V. Benderskii, Nature **594**, 62–65 (2021)
13. Y. Tong, F. Lapointe, M. Thamer, M. Wolf, R.K. Campen, Angew. Chem. Int. Ed. Engl. **56**, 4211–4214 (2017)
14. R.R. Jones, D.C. Hooper, L. Zhang, D. Wolverson, V.K. Valev, Nanoscale Res. Lett. **14**, 231 (2019)
15. X.X. Li, K. Blinn, D.C. Chen, M.L. Liu, Electrochem. Energy Rev. **1**, 433–459 (2018)
16. X.X. Han, R.S. Rodriguez, C.L. Haynes, Y. Ozaki, B. Zhao, Nat. Rev. Methods Prim. **1**, 87 (2022)
17. L.P. Xie, H. Zeng, J.X. Zhu, Z.L. Zhang, H.B. Sun, W. Xia, Y.N. Du, Nano Res. **15**, 4374–4394 (2022)
18. C. Zong, M. Xu, L.J. Xu, T. Wei, X. Ma, X.S. Zheng, R. Hu, B. Ren, Chem. Rev. **118**, 4946–4980 (2018)
19. M. Fleischmann, P.J. Hendra, A.J. McQuillan, Chem. Phys. Lett. **26**, 163–166 (1974)
20. D.L. Jeanmaire, R.P. Van Duyne, J. Electroanal. Chem. Interfacial Electrochem. **84**, 1–20 (1977)
21. M.G. Albrecht, J.A. Creighton, J. Am. Chem. Soc. **99**, 5215–5217 (2002)
22. X.M. Lin, Y. Cui, Y.H. Xu, B. Ren, Z.Q. Tian, Anal. Bioanal. Chem. **394**, 1729–1745 (2009)
23. D. Cialla, A. Marz, R. Bohme, F. Theil, K. Weber, M. Schmitt, J. Popp, Anal. Bioanal. Chem. **403**, 27–54 (2012)
24. J. Langer et al., ACS Nano **14**, 28–117 (2020)
25. D.Y. Wu, J.F. Li, B. Ren, Z.Q. Tian, Chem. Soc. Rev. **37**, 1025–1041 (2008)
26. M. Fleischmann, I.R. Hill, G. Mengoli, M.M. Musiani, J. Akhavan, Electrochim. Acta **30**, 879–888 (1985)
27. M. Fleischmann, G. Mengoli, M.M. Musiani, C. Pagura, Electrochim. Acta **30**, 1591–1602 (1985)
28. A.V. Markin, N.E. Markina, J. Popp, D. Cialla-May, TrAC Trends Anal. Chem. **108**, 247–259 (2018)

29. S. Dick, M.P. Konrad, W.W. Lee, H. McCabe, J.N. McCracken, T.M. Rahman, A. Stewart, Y. Xu, S.E. Bell, Adv. Mater. **28**, 5705–5711 (2016)
30. O. Stern, Z. Elektrochem, Angew. Phys. Chem. **30**, 1014–1020 (1924)
31. E. Cortes, P.G. Etchegoin, E.C. Le Ru, A. Fainstein, M.E. Vela, R.C. Salvarezza, J. Am. Chem. Soc. **132**, 18034–18037 (2010)
32. K. Kuruvinashetti, Y.X. Zhang, J.N. Li, N. Kornienko, New J. Chem. **44**, 19953–19960 (2020)
33. J.H.K. Pfisterer, K.F. Domke, Curr. Opin. Electrochem. **8**, 96–102 (2018)
34. Z.-Q. Tian, B. Ren, D.-Y. Wu, J. Phys. Chem. B **106**, 9463–9483 (2002)
35. Z.-Q. Tian, Z.-L. Yang, B. Ren, D.-Y. Wu, in *Surface-Enhanced Raman Scattering: Physics and Applications*. ed. by K. Kneipp, M. Moskovits, H. Kneipp (Springer, Berlin Heidelberg, 2006), pp.125–146
36. L.W.H. Leung, M.J. Weaver, J. Am. Chem. Soc. **109**, 5113–5119 (1987)
37. S. Park, P. Yang, P. Corredor, M.J. Weaver, J. Am. Chem. Soc. **124**, 2428–2429 (2002)
38. Y.H. Wang, X.T. Wang, H. Ze, X.G. Zhang, P.M. Radjenovic, Y.J. Zhang, J.C. Dong, Z.Q. Tian, J.F. Li, Angew. Chem. Int. Ed. Engl. **60**, 5708–5711 (2021)
39. H. Ze, X. Chen, X.T. Wang, Y.H. Wang, Q.Q. Chen, J.S. Lin, Y.J. Zhang, X.G. Zhang, Z.Q. Tian, J.F. Li, J. Am. Chem. Soc. **143**, 1318–1322 (2021)
40. H. Zhang, S. Duan, P.M. Radjenovic, Z.Q. Tian, J.F. Li, Acc. Chem. Res. **53**, 729–739 (2020)
41. N.K. Murugasenapathi, K.A.E. Jebakumari, S.J. Mohamed, K. Giribabu, T. Palanisamy, J. Phys. Chem. Lett. **12**, 7046–7052 (2021)
42. J.F. Li, Y.F. Huang, Y. Ding, Z.L. Yang, S.B. Li, X.S. Zhou, F.R. Fan, W. Zhang, Z.Y. Zhou, D.Y. Wu, B. Ren, Z.L. Wang, Z.Q. Tian, Nature **464**, 392–395 (2010)
43. J.C.D. Ji-Yang, V.V. Kumar, J.F. Li, Z.Q. Tian, Curr. Opin. Electrochem. **1**, 16–21 (2017)
44. E. Cortes, R. Grzeschik, S.A. Maier, S. Schlucker, Nat. Rev. Chem. **6**, 259–274 (2022)
45. J.C. Dong, R. Panneerselvam, Y. Lin, X.D. Tian, J.F. Li, Adv. Opt. Mater. **4**, 1144–1158 (2016)
46. C.Y. Li, J.C. Dong, X. Jin, S. Chen, R. Panneerselvam, A.V. Rudnev, Z.L. Yang, J.F. Li, T. Wandlowski, Z.Q. Tian, J. Am. Chem. Soc. **137**, 7648–7651 (2015)
47. Z.C. Zeng, S.C. Huang, D.Y. Wu, L.Y. Meng, M.H. Li, T.X. Huang, J.H. Zhong, X. Wang, Z.L. Yang, B. Ren, J. Am. Chem. Soc. **137**, 11928–11931 (2015)
48. S.C. Huang, X. Wang, Q.Q. Zhao, J.F. Zhu, C.W. Li, Y.H. He, S. Hu, M.M. Sartin, S. Yan, B. Ren, Nat. Commun. **11**, 4211 (2020)
49. X.L. Luo, A. Morrin, A.J. Killard, M.R. Smyth, Electroanalysis **18**, 319–326 (2006)
50. P.H. Li, F. Long, W.L. Chen, J. Chen, P.K. Chu, H.Y. Wang, Curr. Opin. Biomed. Eng. **13**, 51–59 (2020)
51. C.Y. Huang, H.C. Hsiao, Sensors (Basel) **20**, 7066 (2020)
52. S. Zaleski, K.A. Clark, M.M. Smith, J.Y. Eilert, M. Doty, R.P. Van Duyne, Anal. Chem. **89**, 2497–2504 (2017)
53. R. Moldovan, E. Vereshchagina, K. Milenko, B.C. Iacob, A.E. Bodoki, A. Falamas, N. Tosa, C.M. Muntean, C. Farcau, E. Bodoki, Anal. Chim. Acta **1209**, 339250 (2022)
54. Y. Wang, X. Zheng, D. Wang, Nano Res. **15**, 1730–1752 (2021)
55. K.C. Deng, Z.X. Lu, J.J. Sun, J.Y. Ye, F. Dong, H.S. Su, K. Yang, M.M. Sartin, S. Yan, J. Cheng, Z.Y. Zhou, B. Ren, Chem. Sci. **13**, 10884–10890 (2022)
56. Y.F. Huang, P.J. Kooyman, M.T. Koper, Nat. Commun. **7**, 12440 (2016)
57. J.Q. Yang, L. Jin, Y.H. Xiao, H.H. Yu, F.Z. Yang, D.P. Zhan, D.Y. Wu, Z.Q. Tian, Langmuir **37**, 11251–11259 (2021)
58. L. Jin, Z.-H. Liang, J.-Q. Yang, A.-N. Zheng, Z.-Y. Wang, F.-Z. Yang, D.-Y. Wu, Z.-Q. Tian, D. Zhan, Electrochim. Acta **437**, 141494 (2023)
59. N.K. Murugasenapathi, M. Kiruba, K.A.E. Jebakumari, S.J. Mohamed, C. Jeyabharathi, T. Palanisamy, J. Phys. Chem. Lett. **14**, 3376–3383 (2023)
60. J.H.K. Pfisterer, M. Baghernejad, G. Giuzio, K.F. Domke, Nat. Commun. **10**, 5702 (2019)

Chapter 10
Surface-Enhanced Raman Excitation Spectroscopy: An Overview

Om Prakash

10.1 Introduction

Since the discovery of surface-enhanced Raman scattering (SERS) spectroscopy in 1974 and 1977 [1–3], scientists elucidate SERS in broad perspectives such as a fundamental interest in understating the concept and advancement of SERS and various applications [4–10]. In SERS, mainly two enhancement mechanisms namely; (1) electromagnetic (EM) enhancement and (2) Chemical enhancement/effect (CE) mechanisms, are attributed that causing the enhancement of the intensity of Raman bands. EM enhancement mechanism is based on the coupling of localized surface plasmon resonance (LSPR) in the vicinity of metallic nanostructures. The EM enhancement mechanism is a large-range effect. EM enhancement mechanism contributes dominantly to the amplification of Raman bands and leads to an efficient enhancement factor up to $\sim 10^8$ and even $>10^8$ in the case of highly confined EM fields near or at the sites of nanostructures and this vicinity of nanostructures is often referred to as "hot spots" [11, 12]. The chemical enhancement/effect (CE) mechanism is merely evolved from three distinct processes; (1) The charge-transfer (CT) process between the molecules and substrate, (2) molecular excitation resonances, and (3) nonresonant increase in the polarizability of the molecules after the adsorption onto the surface of metallic nanostructure. The chemical enhancement/effect (CE) mechanism contributes up to 10^3. Moreover, the CM is a short-range effect and is usually effective on the first layer of adsorbed molecules [13–15]. It is very difficult to disentangle the EM and CE contributions to the SERS signals due to many uncontrollable parameters arising in SERS sampling. At the condition of all three resonances with LSPR, charge-transfer (CT), and molecular resonance, enhancement of Raman bands is the largest with the highest enhancement factor and is usually referred to as surface enhancement resonance Raman scattering (SERRS) which provides an extremely

O. Prakash (✉)
Institute of Physical Chemistry Polish Academy of Sciences, 01-224 Warsaw, Poland
e-mail: omprakash5000@gmail.com

D. K. Singh et al. (eds.), *Raman Spectroscopy*, Springer Series in Optical Sciences 248,
https://doi.org/10.1007/978-981-97-1703-3_10

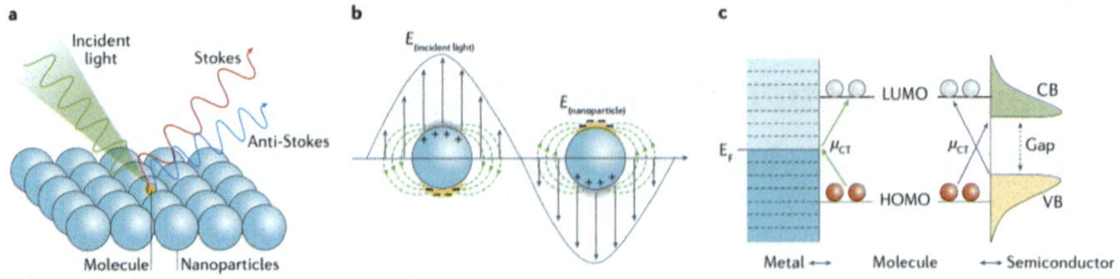

Fig. 10.1 Surface-enhanced Raman scattering (SERS) and enhancement mechanisms involved in SERS phenomena. **a** SERS of a probe molecule adsorbed on the surface of nanoparticles (SERS substrate) which results Raman scattered radiation of with a lower (red) or higher (blue) frequency compared to the frequency of incident light and this is defined as Stokes Raman scattering and anti-Stokes Raman scattering, respectively, **b** localized surface plasmon resonance (LSPR) role to SERS, and the amplified electrical field (E) on the metal nanoparticles surface results in the enhanced Raman signal of molecules, **c** contribution of charge transfer (CT) for enhancement in SERS at the interface of a metal nanoparticle–molecule or semiconductor nanoparticle–molecule; the CT transition (μ_{CT}) and arrows (green: metal–molecule; purple: semiconductor–molecule) represents the directions of CT. In the figure, red and white circles represent molecular orbitals. CB stands for conduction band; VB stands for valence band; EF stands for Fermi level; HOMO stands for highest occupied molecular orbital; LUMO stands for lowest unoccupied molecular orbital; Adapted with permission from [16]. Copyright (2022) (Nature Reviews Methods Primers)

high sensitivity which allows detecting an extremely low concentration of molecules even the single molecule. Figure 10.1a demonstrates a schematic diagram of the SERS; a molecule in the proximity of a hot-spot of nanoparticles showing Stokes and anti-Stokes lines. Figure 10.1b, c shows the schematic diagram of 'Electromagnetic enhancement' and 'charge-transfer enhancement mechanisms in SERS [16].

Observation and detection of the Raman signature of a single molecule are further mentioned as the single-molecule SERS (SM-SERS). The groups of Nie and Kneipp reported the single-molecule surface-enhanced Raman spectroscopy technique for the first time in 1997 [17, 18]. In single-molecule SERS, the two enhancement mechanisms; the electromagnetic and chemical effect work strongly and are in resonance with molecular resonance of adsorbed molecules, and that in combined form gives single-molecule level sensitivity. Ultimately, various studies reported suggest that an electromagnetic mechanism (EM) EF of 10^8 is an essential criterion for the observation of single molecular level surface-enhanced Raman spectroscopy. For example, for the case of SERS signal from the R6G adsorbed on Ag nanoparticles aggregate if excitation occurs at the molecular resonance [19, 20].

The intensity profile of a particular band in electronic molecular resonance Raman spectra with respect to the reference band (solvent or Si band) as the function of excitation wavelengths is referred as the Raman excitation profile. Similarly, surface-enhanced Raman excitation spectroscopy is basically to trace the profile of the intensity of a particular Raman band in SERS (relative to a reference band) excited with the different excitation wavelength obtained by a tunable laser source. This profile generally provides the extinction spectrum of complex structures (adsorbed molecules on the SERS substrate). And, thus the excitation profile of SERS very precise to reveal about the what extent of the particular enhancement mechanisms e.g. electromagnetic (LSPR), charger-transfer resonance, and molecular electronic resonance is involved and plays a role effectively in the enhancement of Raman bands in SERS. In other terms, Surface-enhanced Raman excitation spectroscopy or excitation profile involves monitoring the SERS enhancement for several different excitation wavelengths λ_{ex}. This technique is known as a valuable method for investigating effective SERS mechanisms. An obvious limitation of this method is to collect the number of data points that are obtained by tunable laser by tuning excitation wavelengths and detection system. The lack of such type of instrumental set-up and requirements have led to low data point density and/or limited spectral coverage. Overall, less number of publications on the study of SERS excitation spectroscopy or excitation profile are reported.

10.2 Brief Description of Instrumentation, and Sampling

A schematic diagram of the Raman spectrometer is shown in Fig. 10.2, using which SERS/SERES experiments have been performed [21]. A Nikon Eclipse TE300 inverted microscope (Fryer Co., Huntley, IL) with a 20 × objective (NA = 0.5) was used. Raman spectrometer has a piezoelectric stage (model P-517.3CD, Polytech PI, Auburn, MA), and this piezoelectric stage was used to place the substrates and also for positioning the samples. The piezoelectric stage was also used for the raster-scanning during spectral acquisition. A TriplePro three-stage spectrograph was used in the Raman spectrometer for the analysis of scattered light from the samples. This TriplePro three-stage spectrograph has a liquid nitrogen-cooled and deep-depletion Spec-10:400BR CCD detector (Roper Scientific, Trenton, NJ). For laser alignment and positioning of the samples, a color video camera was attached to the front port of the microscope of the Raman spectrometer. For the laser excitation to monitor the Raman spectra, the following systems were used: (1) a SpectraPhysics (Mountain View, CA) Milennia Xs (λ_{ex}) 532 nm), (2) a Spectra-Physics Tsunami with GWU harmonic generator (λ_{ex}) 350–500, 700–1000 nm), and (3) a Coherent (Santa Clara, CA) model 590 dye laser (λ_{ex}) 610–700 nm). PellinBroca prisms or a diffraction grating was used to filter the laser light from the tunable laser systems and make sure that sample was illuminated with monochromatic light.

10.2.1 Substrate and Hot-Spot

In SERS, excitation of the LSPR of a nanostructured surface or nanoparticle used as the SERS substrate is the most important factor. The control of various factors such as material, size, shape, interparticle spacing, and dielectric environment of substrates influences the LSPR to achieve the maximum signal strength with reproducibility. The laser light maximally excites the LSPR in the case of a "hot-spot"

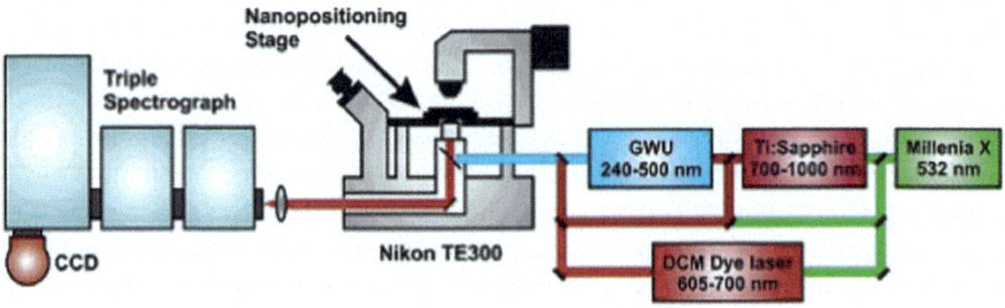

Fig. 10.2 Wavelength scanned surface-enhanced Raman excitation scattering (WS SERES) spectrometer. Adapted with permission from [21]. Copyright (2005) (The Journal of Physical Chemistry B)

of nanoparticles. The arrangement or geometry of nanoparticles in the form of gap structures or junction structures (e.g., dimeric) with the smallest possible gap distance such that one or a few molecules can adsorb in that gap structure. However, fabrication and design of such geometry is a difficult task, yet it is possible to produce using current synthesis and fabrication technology. Sub-monolayer of molecules on metallic nanoparticles provides an efficient SERS signal with better reproducibility.

10.2.2 Effect of LSPR, Charge Transfer Resonance, and Molecular Electronic Resonance

The largest EM enhancement occurs when metallic nanoparticles are excited with laser light that is in resonance with the metallic nanoparticles' LSPR frequency. The nanoparticle LSPR is generally a function of the particle size, shape, surrounding medium, and coupling effects between inter-particles. Since EM-enhancement dominantly contributes to the intensity amplification of Raman bands in SERS and SM-SERS. It was expected that the SERS signal as a function of excitation wavelength will be similar that of plasmon scattering (far-field e.g., transmission or dark-field) spectra. SERS is a near-field effect whereas plasmon scattering is a far-field therefore this explanation was not entirely correct. Van Duyne and co-workers reported a striking difference between the SERS excitation (near-field effect) and plasmon scattering (far-field effect) spectra of a single dimeric SERS structure as shown in Fig. 10.3, which contradicts the general concept on the excitation wavelength dependence in SERS and SMSERS [22]. Figure 10.3 shows the comparison of the SERS excitation spectrum and far-field scattering spectrum for a dimeric SERS hotspot.

The enhancement mechanism due to chemical effect (CHEM) comprises the characteristic features of the probe molecules and the new electronic properties produced by complex (adsorbate–metal nanostructure) formation. CHEM mechanism mainly results from the charge transfer (CT) between the metal nanoparticles and probe

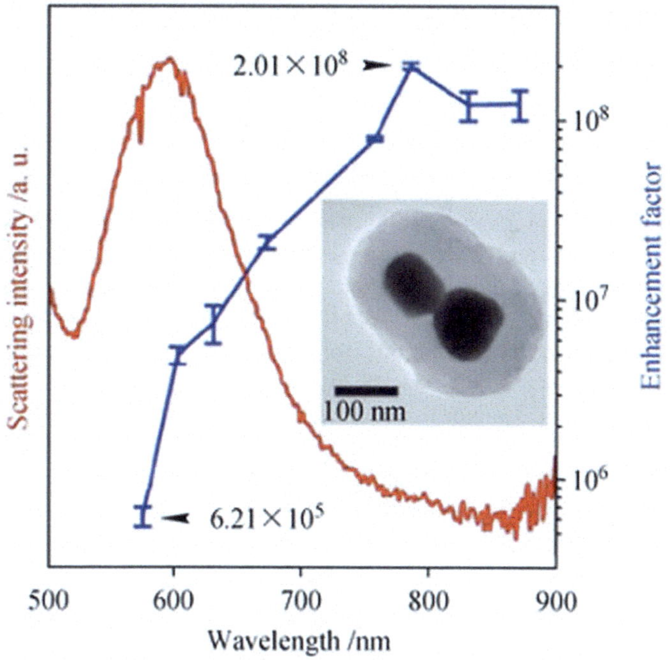

Fig. 10.3 The inset figure shows a dimeric hotspot. The red curve represents the far-field scattering spectrum of a dimeric SERS hotspot and the blue curve represents the SERS excitation spectrum for a dimeric SERS hotspot. Adapted with permission from [22]. Copyright (2013) (Journal of the American Chemical Society)

molecule or vice versa which provides about four orders of Raman bands enhancement in SERS. The charge transfer (CT) is a resonant effect. Other source for CHEM comes from a non–resonant effect (static charge transfer) and this non-resonant effect is produced by the adsorption of the adsorbate molecule in the electronic ground state of the molecule. And thus a new electronic state is not formed in this case due to the interaction of adsorbate–adsorbent/substrate; instead, only a renormalization of the molecular orbitals is formed which causes the modifications in the molecular polarizability of the molecules [23]. Since we know that in resonance Raman spectroscopy, the intensity of the Raman modes or bands that are amplified up to ~6 orders compared to the intensity of the band in normal Raman spectra. Therefore, the molecules adsorbed on SERS substrates showing the molecular electronic with excited wavelength for SERS provides enormous intensity enhancement in SERS.

10.3 Exemplification of Surface-Enhanced Raman Excitation Spectroscopy (or Excitation Profile)

In this section, we discussed some examples of SERS excitation profiles reported earlier. After that, we concluded that the contribution of the LSPR effect, charge-transfer resonance, and molecular resonance contribution in the SERS excitation profile. Most of the examples were reported on SERS of the ensemble of molecules and one example was on the excitation profile of a single molecule (R6G). Peter Hildebrandt et al. reported the SERRS excitation profile of R6G adsorbed on colloidal silver nanoparticles as shown in Fig. 10.4 [24].

Figure 10.4 shows the excitation profile showing normalized intensities of different SERRS bands versus SERRS excitation wavelengths. In this work, the spectral response of the detector was corrected to attain the SERRS spectra however frequency factor (υ^4) was not considered. The maxima of the SERRS excitation profiles almost coincide with the absorption band which was located at 530 nm. The excitation profile for the three stretching modes (1363, 1575, 1650 cm^{-1}) shows a similar pattern while the excitation profile for bending mode (773 cm^{-1}) shows less steeply decrement toward the red region compared to cases observed for the three stretching modes. This is because that the mode at 773 cm^{-1} attains its resonance

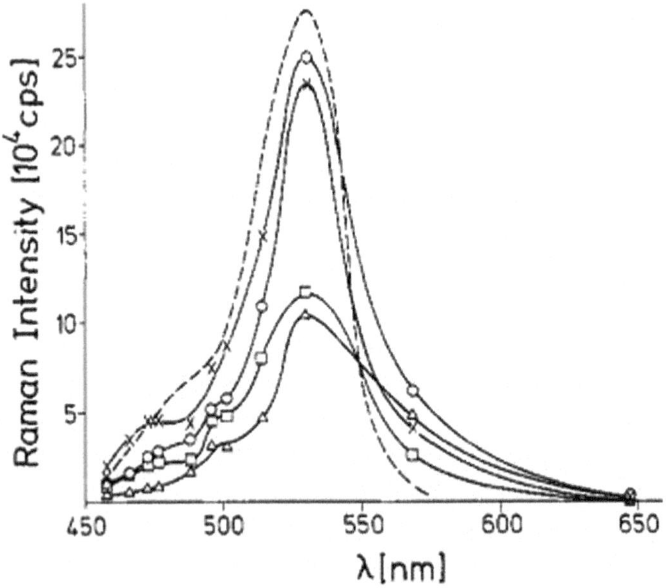

Fig. 10.4 SERS excitation profiles for different modes monitored from the R6G/Cl/Ag systems, Different signs represent different modes for example (X) 1650, (O) 1363, (Δ) 773, (□) 1575 cm^{-1}. The dashed line in the figure represents the absorption spectrum of the aqueous R6G solution. Adapted with permission from [24]. Copyright (1984) (Journal of Physical Chemistry)

Raman intensity via vibronic coupling. Finally, Peter Hildebrandt et al. concluded that SERRS excitation profiles in the presence of Cl^- anions showed a similar pattern as the molecular absorption of R6G molecules and that indicates that resonance behavior is dependent on the molecular resonance [24]. In the case of adsorbed R6G molecules on colloidal silver nanoparticles, surface enhancement for Raman intensity of the band is not significantly dependent on excitation wavelengths. The explanation of SERRS excitation profiles in this [24] report is passable for a particular level but for broad extents of explanation it is intricate to reproduce SERRS excitation profiles. In the same year, K. Kneipp et al. also reported SERRS excitation profile for molecule 3,3′diethyl-2.2′-benzthiatrimethinecyanine adsorbed on silver sol and explained qualitatively in terms of an "electromagnetic" model [25]. From the perspective of the electromagnetic model for the SERS excitation profile, it was demonstrated that the maxima of the SERS excitation profile are close to the LSPR band of the nanoparticle/substrate. In a similar explanation, in the forthcoming part we will see the example of plasmon scanned excitation profile i.e., relative Raman intensity in SERS spectra versus LSPR bands. M. Moskovits et al. showed the SERS excitation profile of Phthalazine adsorbed on single colloidal silver aggregates as a function of the cluster size of the single silver aggregate as shown in Fig. 10.5 [26]. Since the size of different silver aggregates shows different LSPR bands. In this paper, no systematic cluster-size-dependent differences were observed in the SERS excitation profiles.

Fig. 10.5 Excitation profiles obtained for different SERS bands; For a group of bands, **a**1382 cm^{-1}, **b** 946 cm^{-1}, **c**1230 cm^{-1}, **d** 1457 cm^{-1} excitation profiles represent a monotonic wavelength dependence. The terms CL1, CL2, and CL3 stand for the three clusters of nanoparticles. Adapted with permission from [26]. Copyright (1987) (Journal of Physical Chemistry B)

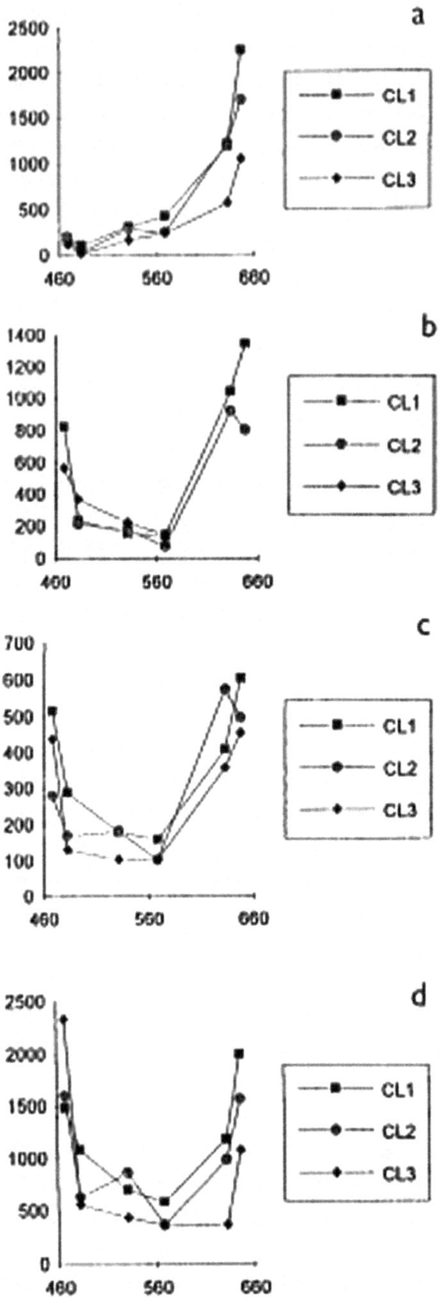

Like the profound contribution in the SERS field, Van Duyne research group has also contributed significantly in the direction of SERS excitation profile. Some examples; are wavelength-scanned (or dependent) SERS excitation profile, plasmon-scanned SERS excitation profile, wavelength-dependent SERRS excitation profile for a single molecule of R6G, etc. First in the work entitled "plasmon-sampled surface-enhanced Raman excitation (PS-SERE) spectroscopy", they explained the correlation between the maximum (λ_{max}) and SERS enhancement factor (EF) and showed the profile of enhancement factor of a particular band (located at 1575 cm^{-1}) of benzenethiol with respect to LSPR wavelengths excited with three different wavelengths 514.5, 532, and 632.8 nm as shown in Fig. 10.6 [27]. In the PS-SERE spectra, the maximum SERS EFs were observed when the LSPR λ_{max} lies between λ_{ex} and λ_{vib}. And it was observed that EF was not varied by more than a factor of 10 when the SERS signal is measurable above the background. The electromagnetic model was demonstrated in this work and for those electromagnetic effects without being hindered by issues of site-to-site adsorbate motion or nonensemble averaged spectral characteristics were investigated.

In the second work, Van Duyne group presented a detailed wavelength-dependent surface-enhanced Raman spectroscopic study of benzenethiol adsorbed on the arrays of Ag nanoparticle as shown in Figs. 10.7, 10.8, 10.9 and 10.10 [21]. They called this study as wavelength-scanned surface-enhanced Raman excitation spectroscopy (WS-SERES). The Ag nanoparticle arrays were fabricated by nanosphere lithography (NSL). In this study, the WS SERES spectra were spatially and spectrally correlated with the corresponding LSPR spectra of the arrays of Ag nanoparticles used. SERS spectra in this study were recorded in two excitation wavelength regions; (1) 425–505 nm, and (2) 610–800 nm. An excitation wavelength at 532 nm from a solid-state diode-pumped laser was also used for collecting the SERS spectra. The line shapes of the WS SERES spectra were similar to the line shapes of the LSPR spectra. It was observed that those excitation wavelengths that were blue-shifted with respect to the LSPR λ_{max} of adsorbate-covered Ag nanoparticle arrays showed the maximum enhancement of the bands in SERS spectra i.e. maximum SERS enhancement factor. Three vibrational modes 1575, 1081, and 1009 cm^{-1} of benzenethiol are selected as the probe band and investigated simultaneously for these bands on one (and the same) substrate, and it was observed that the smaller Raman shifted peak shows a maximum enhancement which was closer to the LSPR λ_{max} than that of a larger Raman shifted peak. This observation agreed with the observation calculated for the electromagnetic enhancement mechanism of SERS. In this study, the SERS enhancement factors of up to $\sim 10^8$ were attained.

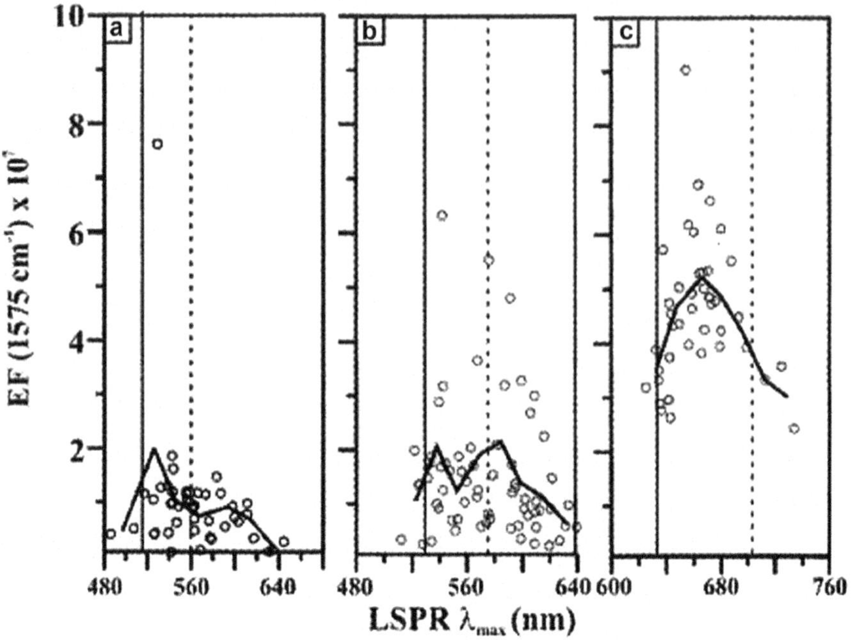

Fig. 10.6 Plasmon scanned surface-enhanced Raman excitation scattering (PS-SERES) profiles for a SERS band located at 1575 cm^{-1} (ν_{8a}) of the probe molecule benzenethiol for three different excitation wavelengths: **a** $\lambda_{ex} = 514.5$ nm, **b** $\lambda_{ex} = 532.0$ nm, and **c** $\lambda_{ex} = 632.8$ nm. In each figure, the solid line represents the wavelength location of the excitation and the dashed line represents the wavelength location of the scattering. The overlaid line shows the bin-averaged values of the LSPR λ_{max} and EF. Bin widths are different for all cases and are 24 nm, 16 nm, and 16 nm for this figure (**a**)–(**c**), respectively Adapted with permission from [27]. Copyright (2003) (Journal of Physical Chemistry B)

In another work, Van Duyne group has presented surface-enhanced Raman excitation spectroscopy of the ensemble-averaged Rhodamine 6G molecule adsorbed on Ag Island film (AgIF) and a single Rhodamine 6G molecule adsorbed on AgIF [28]. In this chapter, we discussed the surface-enhanced Raman excitation profiles of ensemble-averaged Rhodamine 6G molecule adsorbed on Ag Island film (AgIF). For monitoring the SERS excitation profiles of rhodamine 6G (R6G) adsorbed on Ag nanostructured surfaces, a tunable excitation source equipped with an adaptable detector system was used. This study is one of the well performed to understand the SERS excitation profile. The fine-tunable excitation source was such that it can be tuned near the electronic molecular resonance of R6G molecule (~500–575 nm). The absorption spectrum of R6G molecules adsorbed on Ag Ag Island film was recorded by a specular reflectance spectrometer in an integrating sphere. The absorption band of R6G in ethanol was observed at 18986 cm^{-1} (526.7 nm, fwhm ~974 cm^{-1}) corresponding to (0–0 transition) and this band was also accompanied by a vibronic

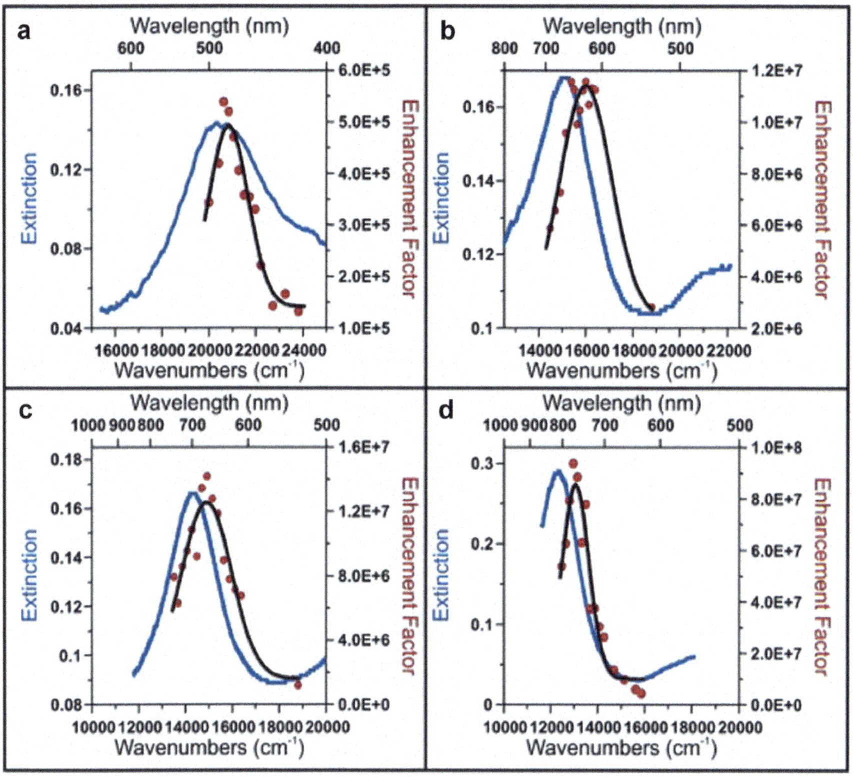

Fig. 10.7 SERS excitation profile of the band at 1575 cm^{-1} peak of benzenethiol with cyclohexane as intensity standard. An annealing temperature of 300 °C for 1 h was used for the preparation of the SERS substrate. Different LSPR ($\lambda_{max,LSPR}$) and profile fit maximum wavelengths ($\lambda_{ex,max}$) are used to draw the curves. **a** $\lambda_{max,LSPR} = 489$ nm and $\lambda_{ex,max} = 480$ nm and **b** $\lambda_{max,LSPR} = 663$ nm and $\lambda_{ex,max} = 625$ nm, **c** $\lambda_{max,LSPR} = 699$ nm and $\lambda_{ex,max} = 671$ nm, **d** $\lambda_{max,LSPR} = 810$ nm and $\lambda_{ex,max} = 765$ nm. Adapted with permission from [21]. Copyright (2005) (Journal of Physical Chemistry B)

shoulder at 19960 cm^{-1} (501.0 nm, fwhm ~1997 cm^{-1}). It was observed that none of the particular vibrational modes of R6G belongs to the energy difference between the 0–0 transition (absorbance maximum) and 0–1 transition (vibronic sideband). The electronic transition 0–1 showed broadening which indicates that vibronic coupling takes place in multiple vibrational modes. In Fig. 10.11a, a surface absorption band of R6G molecules adsorbed on Ag island film was observed, and this was deconvoluted into two Gaussian functions, one is with maxima and fwhm as follows $E_{max} = 18409$ cm^{-1} (543.2 nm) with fwhm = 1004 cm^{-1}, and other with maxima and fwhm as follows $E_{max} = 19246$ cm^{-1} (519.6 nm) with fwhm = 1822 cm^{-1}. These two deconvoluted bands correspond to the 0–0 and 0–1 transitions, respectively. Surface-enhanced Raman excitation profiles (SEREPs) of R6G molecules adsorbed on AgIF (ensemble averaged) were recorded and averaged the data of the ensemble of Raman spectra from molecules. Three prominent vibrational modes; 1647, 1504,

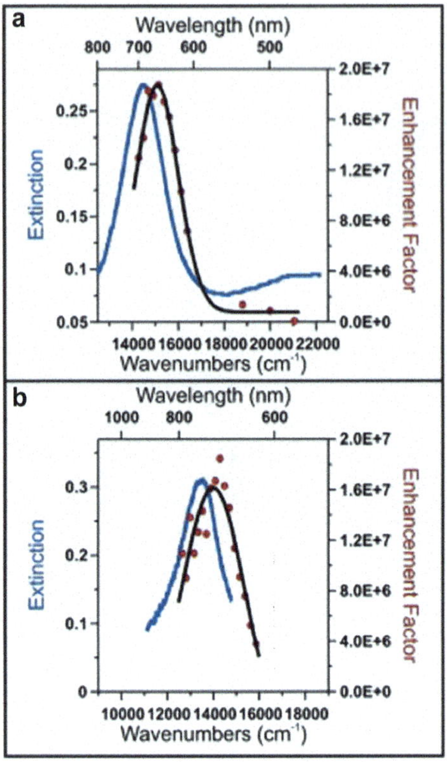

Fig. 10.8 SERS excitation spectra of the tha band located at 1081 cm^{-1} of benzenethiol and Si were used as intensity standard. **a** $\lambda_{max,\ LSPR} = 690$ nm, $\lambda_{ex,max} = 662$ nm. **b** $\lambda_{max,\ LSPR} = 744$ nm, $\lambda_{ex,max} = 715$ nm. Adapted with permission from [21]. Copyright (2005) (Journal of Physical Chemistry B)

and 610 cm^{-1} of R6G were selected for analysis of Surface-enhanced Raman excitation profiles. These three bands show a particular behavior in different spectral regimes demonstrating interesting excitation profiles. SEREPs of R6G adsorbed on AgIF are shown in Fig. 10.11b–d. The excitation profiles were deconvoluted into two Gaussian functions (bands) and as the sum of these bands/functions. The locations of fitted bands were at $E_{max} = 18409$ cm^{-1} (543.2 nm) with fwhm = 1004 cm^{-1}, and $E_{max} = 19246$ cm^{-1} (519.6 nm) with fwhm = 1822 cm^{-1}. In the deconvolution of the excitation profiles into two Gaussian functions, E_{max} and fwhm were set accordingly to fit the functions. As shown in Fig. 10.11b, the surface-enhanced Raman excitation profile for the band at 610 cm^{-1} of R6G molecules adsorbed on AgIF is dominated by a single peak at the fundamental transition. This profile shows a transition amplitude ratio of 4.9:1. An increase in surface-enhanced Raman intensity at the 0–1 transition which shows a transition amplitude ratio of 3.0:1 is observed for the band at 1504 cm^{-1} as shown in Fig. 10.11c. The surface-enhanced Raman excitation profile for the band at 1647 cm^{-1} (Fig. 10.11d) exhibits significant contributions from both

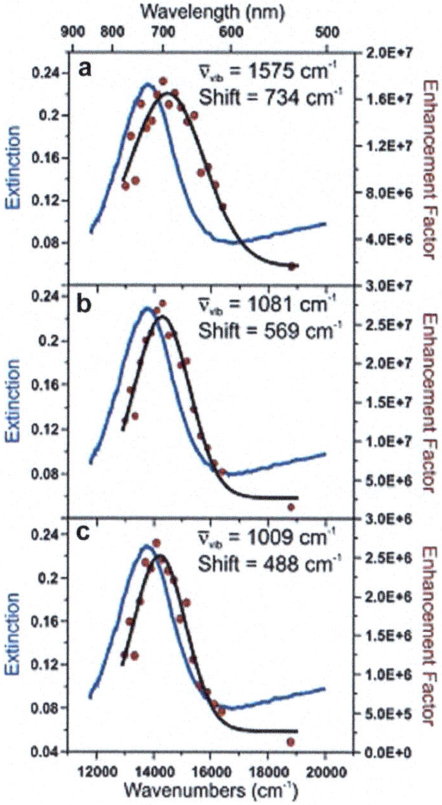

Fig. 10.9 In this figure, the distance between LSPR λ_{max} and excitation profile fit line $\lambda_{ex,max}$ are defined for each figure and have different values ($\lambda_{ex,max}$) and drawn for different enhancement factors (EF). **a** Profile of the band located at 1575 cm^{-1} of benzenethiol, $\lambda_{ex,max}$ = 734 cm^{-1}. EF = 1.8×10^7. **b** Profile of the band located at 1081 cm^{-1}, $\lambda_{ex,max}$ = 569 cm^{-1}, EF = 2.8×10^7. **c** Profile of the band located at 1009 cm^{-1}, $\lambda_{ex,max}$ = 488 cm^{-1}, EF = 2.7×10^6. Adapted with permission from [21]. Copyright (2005) (Journal of Physical Chemistry B)

transitions. The amplitude ratio for these two transitions was 1.1:1. In this work, it was concluded that the surface-enhanced Raman excitation profile of R6G adsorbed on AgIF comprises the contributions from the three effects; molecular resonance effect, electromagnetic enhancement effect, and other surface effects.

Pablo S. Londero et al. reported the SERS excitation profile of Cu phthalocyanine adsorbed on Ag nano-island film [29]. The electronic absorption of CuPc, plasmonic absorption of Ag nano-island film, and absorbance of complex Ag nano-island film + CuPc are shown in Fig. 10.12. The LSPR band of Ag nanoisland film was observed ~525 nm. The electronic absorption of CuPc ablated onto a quartz window exhibits a Q-band which generates the absorption after 550 nm. The absorption spectrum of CuPc adsorbed on Ag nanosiland film shows both features; plasmonic band of Ag

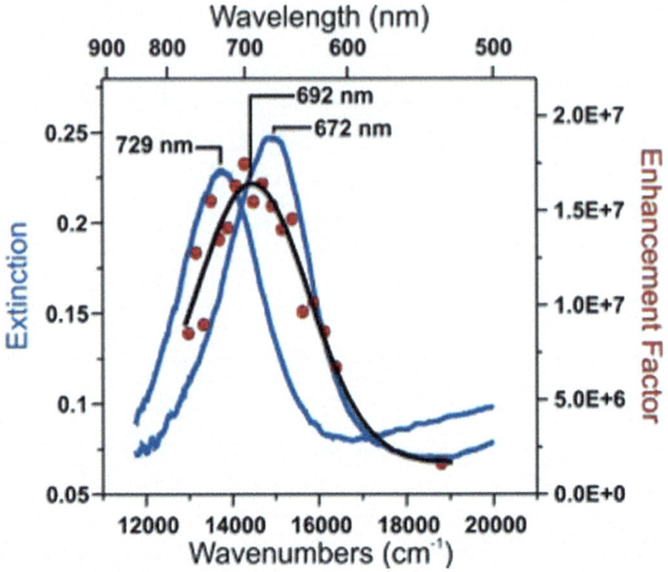

Fig. 10.10 For the band at 1575 cm^{-1} of benzenethiol, the LSPR shift and SERES profile are shown. The blue lines are shown for the LSPR extinction. The blue line designated with $\lambda_{max} =$ 672 nm is shown for the bare nanoparticle array and the blue line designated with $\lambda_{max} = 729$ nm is shown for the nanoparticle array with adsorbed benzenethiol monolayer. The black line designated with $\lambda_{ex,max} = 692$ nm is the best fit for the data points obtained for SERES. Adapted with permission from [21]. Copyright (2005) (Journal of Physical Chemistry B)

nanoisland film (with narrower shape) and Q-band of the molecule CuPc (with a bit red-shifted).

Excitation wavelengths ranging from 435 to 635 nm were used to monitor the SERS spectra of Cu phthalocyanine adsorbed on Ag nano-island film. SERS excitation profile has been obtained for totally symmetric (A_{1g}) bands and non-totally symmetric (B_{2g}) bands as shown in Fig. 10.13 and observed that SEP for A_{1g} mode attains maxima at 475 nm while SEP for B_{2g} mode attains maxima at 605 nm. The maxima for A_{1g} (1607 cm^{-1}) mode could be attributed either to a surface plasmon resonance alone or possibly the combination with the charge-transfer contribution while the maxima for B_{2g} (948 cm^{-1}) mode was obtained due to combined contributions from surface plasmon resonance and charge-transfer resonance. The maxima for A_{1g} (1607 cm^{-1}) mode also could be contributed from molecular electronic resonance in addition to surface plasmon resonance and charge-transfer contribution. The excitation plasmonic mode is clearly possible across the entire measurement region 435–635 nm. The scattering behavior of these modes; A_{1g} (1607 cm^{-1}) and B_{2g} (948 cm^{-1}) are different as a function of excitation wavelength. The maxima of scattering profile of A_{1g} (1607 cm^{-1}) were obtained at 475 nm while for B_{2g} (948 cm^{-1}) mode the maxima of scattering profile were at 605 nm.

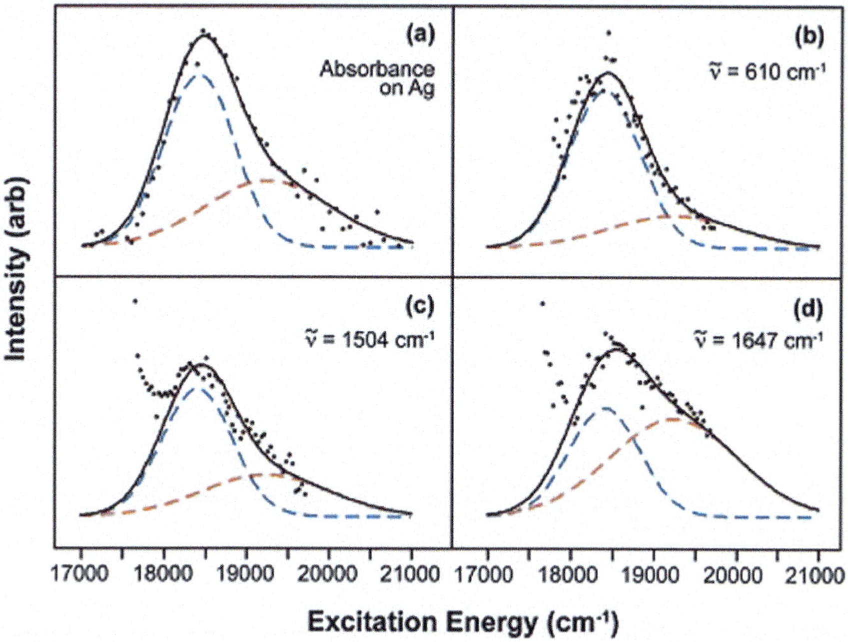

Fig. 10.11 **a** Absorption spectrum of R6G adsorbed on Ag island film. Ensemble-averaged SERS excitation profiles of different Raman modes of R6G on AgIF, **b** for wavenumber $\nu = 610\ \text{cm}^{-1}$, **c** for wavenumber $1504\ \text{cm}^{-1}$, and **d** for wavenumber $1647\ \text{cm}^{-1}$. The measurement was performed at $P_{\text{ex}} \approx 150\ \text{W cm}^{-2}$, and for $t_{\text{ac}} = 20$ s. A sum of two Gaussians is shown for the data points and those were fitted and shown by the black solid lines. However, deconvolution of the overall fit is shown by the dashed lines and demonstrates the relative contributions from the 0–0 and 0–1 vibronic transitions of R6G on AgIF. Adapted with permission from [28]. Copyright (2009) (Journal of the American Chemical Society)

Jing Long et al. reported the correlation between the LSPR of nanoantenna and SERS spectrum of the monolayer of MGITC molecules coated on nanoantenna as shown in the Fig. 10.14 [30]. A good correlation between the intensity ratios between the three major SERS bands and respective LSPR spectra was observed. In Figures (a–f), the localized surface plasmon resonance (LSPR) spectra and surface-enahnched Raman scattering spectra are shown and these spectra are monitored for the same nanoantenna. In this study, only those nanoantenna were chosen that showed a close-to-Lorentzian LSPR spectral profile.

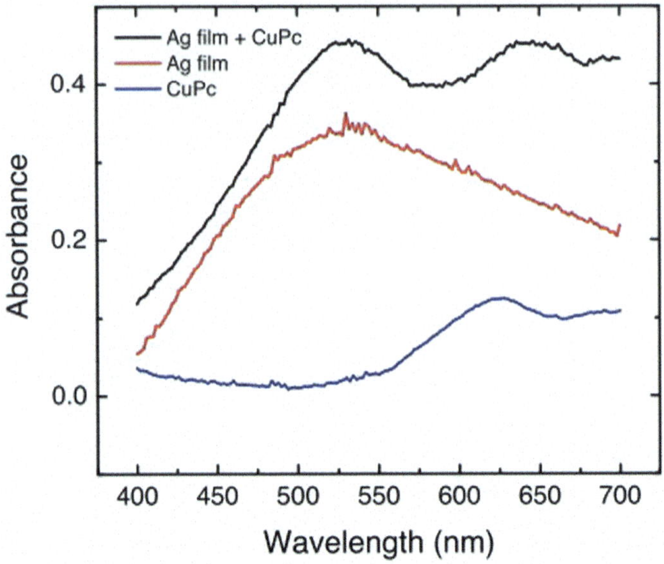

Fig. 10.12 UV-visible absorption spectra of CuPc (blue), Ag film (red) and CuPc on Ag film (black) with offsets scale. Absorption spectrum of the silver film represents the presence of localized surface plasmonic resonance. Absorption spectrum of CuPc ablated onto a quartz window (blue) represents the Q-band which is observed after 550 nm at around 625 nm. The absorption spectrum of the sample (CuPc + Ag film) shows both LSPR and Q-band. Adapted with permission from [29]. Copyright (2013) (Applied Physics Letters)

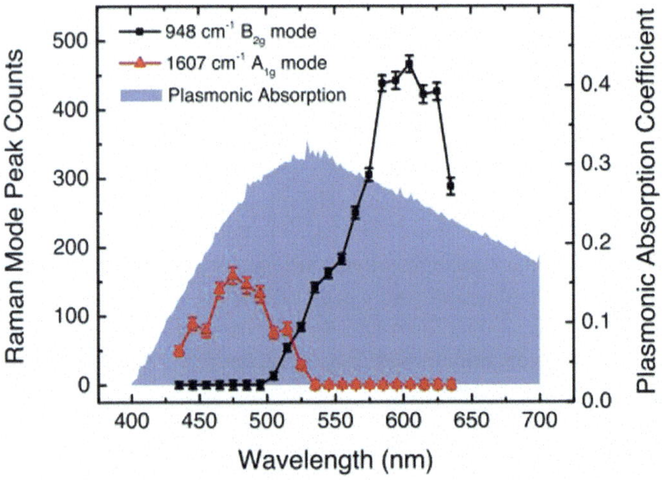

Fig. 10.13 Peak amplitudes for the modes at 1607 cm^{-1} A_{1g} mode and at 948 cm^{-1} B_{2g} mode are shown as the function of exciatation wavelength. The LSPR band of the nanostrucred silver film is shown in the background. Adapted with permission from [29]. Copyright (2013) (Applied Physics Letters)

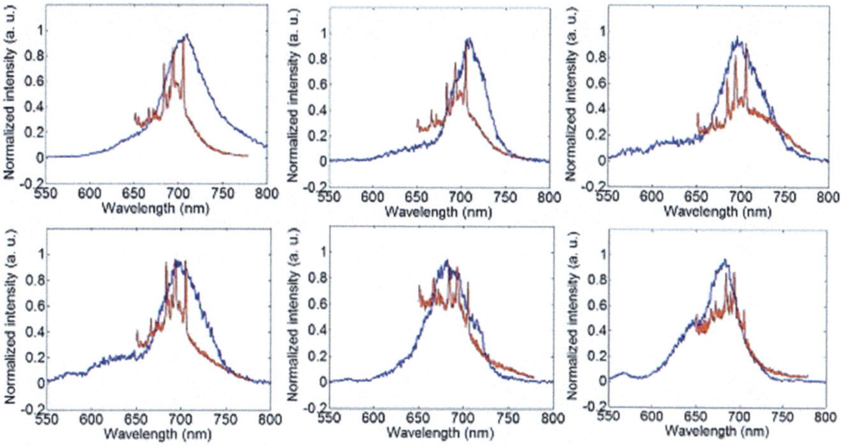

Fig. 10.14 Localized surface plasmon resonance (LSPR) band and surface-enhanced Raman scattering (SERS) spectra of six antennas which are coated with a monolayer of MGITC are shown. Blue shows LSPR. Red shows SERS. Adapted with permission from [30], Copyright (2016) (Scientific Reports)

10.4 Conclusion

In this chapter, we described the SERS and enhancement mechanisms very briefly and further instrumentation and sampling for experiments are described. Surface-enhanced Raman excitation spectroscopy is basically by definition to trace the profile of the intensity of a particular Raman band in SERS excited with the different excitation wavelengths obtained by a tunable laser source. Surface-enhanced Raman excitation profiles profile generally provides the extinction spectrum of complex structures i.e., molecules adsorbed on the SERS substrate. The excitation profile of SERS is very precise about the information that what extent of particular enhancement mechanisms (e.g., electromagnetic (LSPR), charger-transfer resonance, and molecular electronic resonance) contribute to the enhancement of Raman bands in SERS. Various examples of surface-enhanced Raman excitation profiles (SEREP); plasmon-scanned excitation profiles (PSEPs), and wavelength-scanned excitation profiles (WSEPs) are discussed along with corresponding conclusions. In PSEPs, it has been shown that the electromagnetic model explains the enhancement of Raman bands in SERS and also shown that the maximum contribution of enhancement is due to electromagnetic enchantment. In wavelength-scanned excitation profiles (WSEPs), it has been shown that all three resonances; LSPR, charger-transfer resonance, and molecular electronic resonance involved and suggested how these play the roles effectively in the enhancement of Raman bands in SERS.

References

1. M. Fleischmann, P.J. Hendra, A.J. McQuillan, Chem. Phys. Lett. **26**, 163 (1974)
2. D.L. Jeanmaire, R.P. Van Duyne, J. Electroanal. Chem. Interfacial Electrochem. **84**, 1 (1977)
3. M.G. Albrecht, J.A. Creighton, J. Am. Chem. Soc. **99**, 5215 (1977)
4. A. Campion, P. Kambhampati, Chem. Soc. Rev. **27**, 241 (1998)
5. R. Pilot, R. Signorini, C. Durante, L. Orian, M. Bhamidipati, L. Fabris **9**, 57 (2019)
6. P. Mandal, B.S. Tewari, Surf. Interfaces **28**, 101655 (2022)
7. A.I. Pérez-Jiménez, D. Lyu, Z. Lu, G. Liu, B. Ren, Chem. Sci. **11**, 4563 (2020)
8. J.R. Lombardi, R.L. Birke, Acc. Chem. Res. **42**, 734 (2009)
9. C. Zong, M. Xu, L.-J. Xu, T. Wei, X. Ma, X.-S. Zheng, R. Hu, B. Ren, Chem. Rev. **118**, 4946 (2018)
10. J. Langer et al., ACS Nano **14**, 28 (2020)
11. S.E.J. Bell, N.M.S. Sirimuthu, Chem. Soc. Rev. **37**, 1012 (2008)
12. S. Schlücker, **53**, 4756 (2014)
13. F.J. Adrian, J. Chem. Phys. **77**, 5302 (1982)
14. J.R. Lombardi, R.L. Birke, T. Lu, J. Xu, J. Chem. Phys. **84**, 4174 (1986)
15. M. Li, S.K. Cushing, G. Zhou, N. Wu, Nanoscale **12**, 22036 (2020)
16. X.X. Han, R.S. Rodriguez, C.L. Haynes, Y. Ozaki, B. Zhao, Nat. Rev. Methods Prim. **1**, 87 (2022)
17. S. Nie, S.R. Emory, **275**, 1102 (1997)
18. K. Kneipp, Y. Wang, H. Kneipp, L.T. Perelman, I. Itzkan, R.R. Dasari, M.S. Feld, Phys. Rev. Lett. **78**, 1667 (1997)
19. Y. Yang, X. Ting-Hui, W. Yunzhao, L. Wanjun, Z. Qing-Guang, L. Li, L. Zhi-Yuan, Adv. Photonics **2**, 014002 (2020)
20. J. Kneipp, H. Kneipp, K. Kneipp, Chem. Soc. Rev. **37**, 1052 (2008)
21. A.D. McFarland, M.A. Young, J.A. Dieringer, R.P. Van Duyne, J. Phys. Chem. B **109**, 11279 (2005)
22. S.L. Kleinman, B. Sharma, M.G. Blaber, A.-I. Henry, N. Valley, R.G. Freeman, M.J. Natan, G.C. Schatz, R.P. Van Duyne, J. Am. Chem. Soc. **135**, 301 (2013)
23. W.-H. Park, Z.H. Kim, Nano Lett. **10**, 4040 (2010)
24. P. Hildebrandt, M. Stockburger, J. Phys. Chem. **88**, 5935 (1984)
25. K. Kneipp, D. Fassler, Chem. Phys. Lett. **106**, 498 (1984)
26. B. Vlčková, X.J. Gu, M. Moskovits, J. Phys. Chem. B **101**, 1588 (1997)
27. C.L. Haynes, R.P. Van Duyne, J. Phys. Chem. B **107**, 7426 (2003)
28. J.A. Dieringer, K.L. Wustholz, D.J. Masiello, J.P. Camden, S.L. Kleinman, G.C. Schatz, R.P. Van Duyne, J. Am. Chem. Soc. **131**, 849 (2009)
29. P.S. Londero, M. Leona, J.R. Lombardi, Appl. Phys. Lett. **102** (2013)
30. J. Long, H. Yi, H. Li, Z. Lei, T. Yang, Sci. Rep. **6**, 33218 (2016)

Chapter 11
Tip-Enhanced Raman Spectroscopy

Wei Wang and Volker Deckert

11.1 Introduction

11.1.1 Raman Spectroscopy: A Brief Overview

When photons interact with matter, most of them are elastically scattered, which is known as Rayleigh scattering. Only a small portion of photons have energy exchange with the matter, resulting in a shift in frequency. Such frequency shift, known as Raman scattering or the Raman effect, carries information about the vibration and rotational states of the matter. Figure 11.1 depicts the schematic of Rayleigh and Raman scattering. The Raman effect is the physical foundation of Raman spectroscopy, which is a powerful analytical tool that provides information on the material composition, structure and interactions through optical signal. This technique is named after the Indian physicist Sir C. V Raman, who discovered the phenomenon in 1928 [1] and was awarded the Nobel Prize in Physics due to such groundbreaking work. Since its discovery, Raman spectroscopy has evolved into a powerful tool with diverse application in fields such as chemistry, biology, materials sciences, pharmaceuticals, forensics, and environmental analysis.

As a typical nonlinear optical process, Raman scattering can be classified into two types: first-order (m = 1) Raman scattering and second-order (m = 2) Raman scattering. In the first-order Raman process (shown in Fig. 11.1a), the scattered light undergoes a single energy exchange with the molecule, resulting in a shift in frequency, which will give rise to possible outcomes: Stokes and Anti-Stokes

W. Wang · V. Deckert (✉)
Leibniz Institute of Photonic Technology (IPHT), 07745 Jena, Germany
e-mail: volker.deckert@uni-jena.de

Institute of Physical Chemistry, Friedrich Schiller University Jena, 07743 Jena, Germany

V. Deckert
Abbe Center of Photonics, Friedrich Schiller University Jena, 07745 Jena, Germany

D. K. Singh et al. (eds.), *Raman Spectroscopy*, Springer Series in Optical Sciences 248,
https://doi.org/10.1007/978-981-97-1703-3_11

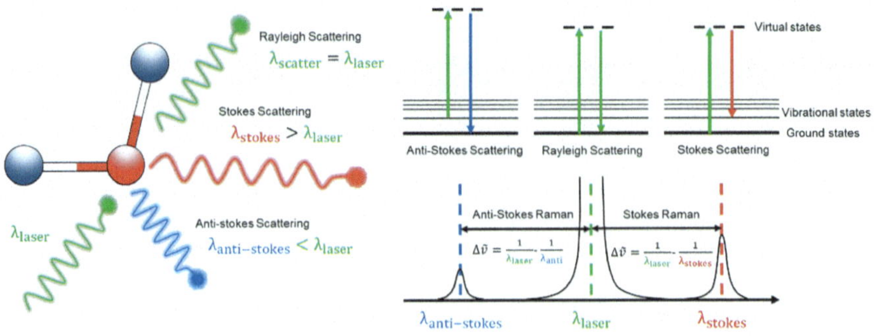

Fig. 11.1 **Raman principle**: **a** Raman scattering process. **b** Energy-level diagram of Raman process. Copy right (2023) (Wei Wang)

Raman scattering. In Stokes Raman scattering, the scattered light loses energy to the molecule, causing a shift to longer wavelengths (red shift/lower frequency). This process occurs when the molecule absorbs energy from the incident light and transitions to a higher vibrational energy state (shown in Fig. 11.1b). In anti-Stokes Raman scattering, the scattered light gains energy from the molecule, resulting in a shift to shorter wavelengths (blue shift/higher frequency). This process occurs when the molecule releases energy and transitions to a lower vibrational energy state ((shown in Fig. 11.1b). First-order Raman scattering is the most common type and typically dominates the Raman scattering spectrum. In the second-order Raman process [2], the scattered light interacts with two vibrational modes simultaneously, resulting in a more complex energy exchange. Second-order Raman scattering involves a double frequency shift and is typically weaker compared to first-order scattering. This process is less common than first-order Raman scattering as it only occurs under certain conditions. It's important to note that Raman scattering is a weak process compared to other forms of light-matter interaction, such as fluorescence. However, it provides valuable information about the vibrational properties of molecules and is widely used in various scientific and analytical techniques.

The modern Raman spectrometers consist of several components, including a laser as the excitation source, optical elements for the laser focusing and scattering light collection a dispersive optical element (such as prisms and blaze gratings), and a detector (such as CCD or CMOS arrays). If the laser beam is focused on the sample, then it falls into the category of confocal Raman microscopy [3], which is the combination of the standard Raman spectrometer of optical microscopy (see Sect. 11.3 for more detail). The confocal Raman allows high spatial resolution Raman analysis of the sample. Such synergy between the spectral and spatial information enables the chemical analysis of individual particles, discrete sample features or layers down to less than 1 μm in size (the diffraction limit of conventional optical systems, more detailed discussion can be found in Sect. 11.2.2.1 and Fig. 11.8). This advantage makes this particular technology very popular in scientific and industrial fields.

In recent years, tremendous advancements have been made in Raman spectroscopy instrumentation and associated technologies. The invention of portable and handheld Raman spectrometers enables on-site analysis and rapid identification of samples. Additionally, associated technologies such as surface-enhanced Raman spectroscopy (SERS) (Chaps. 8–10 in this book) [4–6], coherent anti-Stokes Raman scattering (CARS) (Chap. 13 in this book) [7], and tip-enhanced Raman spectroscopy (TERS) [8–10] have further enhanced the sensitivity, capability and resolution of Raman spectroscopy.

11.1.2 Evolution of TERS: Historical Development

In the 1970s, it was found that employing a rough silver (Ag) surface as a substrate can lead to the enhancement of Raman scattering by 5–6 orders higher than expected. According to the associated publication of David etc., [11] the authors assume some property of the Ag substrate surface is acting to enhance the effective Raman scattering cross section for the sample molecules. It took until the 1990s that this Raman enhancement mechanism was sufficiently understood [5, 6]. The major part of the enhancement is caused by the localized electrical field enhanced "hot spots" as a result of surface plasmon modes of the novel metal nanoparticles. Such localized electrical field enhancement, also known as near-field enhancement mechanism, forms the theoretical base of SERS and TERS. A detailed discussion of this enhancement mechanism is given in Sect. 11.2.2.1. The origins of TERS can be traced back to 1985 when Wassel proposed the intriguing idea that a single metal nanostructure could intensify the electrical field, resulting in an amplified Raman signal [12]. As depicted in Fig. 11.2i(b), this concept entailed replacing the rough metal film with a sharp metal tip, serving as the exclusive "hot spot". This tip would then be systematically scanned across the sample surface using Scanning Probe Microscopy (SPM). The experimental substantiation of this concept gave rise to what we now know as TERS. Remarkably, a similar notion emerged much earlier when in 1928,[1] Synge introduced the concept of utilizing a small aperture to achieve sub-wavelength resolution imaging with visible light [13]. His vision entailed an opaque metal screen equipped with an aperture smaller than the wavelength of light (e.g., 100 nm, see Fig. 11.2i(a). This screen would be illuminated from the rear and meticulously scanned over the sample surface for imaging. The resolution of such an image should be limited by the size of the aperture and not by the wavelength of the illuminating light. The practical realization of this technique, known as aperture-type Scanning Near-Field Optical Microscopy (aperture SNOM, Fig. 11.2ii(a) [14] took shape in the early 1980s [15]. In aperture SNOM, light is coupled into a fiber, culminating in a sharp tip that is scanned in close proximity to the sample surface (as depicted in Fig. 11.2ii(a).

---

[1] It is interesting to mention in here that the Raman effect was also discovered in 1928 [1]. It was not, however, until 70 years later that Raman and sSNOM techniques were finally combined as TERS.

However, a significant limitation of aperture SNOM was the considerable reduction in recorded intensity as the aperture size decreased, although this contributed to enhanced spatial resolution. This challenge was compounded by the exceedingly low Raman cross-sections of many samples and the limited light transmission and damage thresholds of optical fiber probes, which impeded the combination of aperture SNOM and Raman. The pioneering contributions in aperture SNOM-Raman were made by Tsai [16], Smith [17], and Hallen [18]. Subsequently, advancements in fiber probe fabrication and explorations into combinations of SNOM-Raman with resonance Raman scattering and SERS enabled the application of these techniques in the realms of chemistry and biology. Nonetheless, the restricted throughput of fiber probes remained a stumbling block for the widespread adoption of aperture-based near-field Raman imaging across various fields. A solution to this predicament emerged in the form of scattering-type SNOM (sSNOM, Fig. 11.2ii(b) [14], which involved the use of a metal-coated tip or even a massive metal tip crafted from suitable materials. The first report on sSNOM emerged in 1994 [19]. Inspired by the invention of the Scanning Tunneling Microscope (STM) [20] and the discovery of SERS [5, 6], Wassel predicted that the challenges associated with controlling the distance between the particle and the sample could be overcome by employing a metal STM tip in direct contact with a substrate as a field enhancer. He astutely anticipated the complexities involved in tip preparation and the integration of intricate mechanics, optics, and electronics that would be necessary for the successful realization of such an instrument. As it depicts in Fig. 11.2i(b), Wassel's envision for TERS resembles the concept of sSNOM. It would not be until 1999 before three independent research groups, led by Zenobi [22], Anderson [23], and Kawata [24], achieved the first successful TERS experiments. The theory base of the beyond Abbe limit imaging ability of SNOM and TERS techniques is illustrated in Fig. 11.2iii, which can be explained by the convolution of the spatial frequencies of sample ($\hat{T}$) and tip field ($\hat{\boldsymbol{E}}_{tip}$). A more detailed explanation will be given in Sect. 11.2.

11.2 Theory of Tip-Enhanced Raman Spectroscopy

11.2.1 Concepts of Optical Near-Field and Far-Field

As an effort to illustrate the concept of optical near-field and the working principle of SNOM[2] and TERS, we will use the optical scattering of objects in homogeneous media as example, which is shown in Fig. 11.3i(a). The angular spectrum[3] representation is a mathematical technique suitable for the description of the such light scattering problems. We assume $\boldsymbol{E}(\boldsymbol{r})$ is the electrical field at any point $\boldsymbol{r}(x, y, z)$ in the free space. As the electrical field is the dominant driving fore of light-matter

[2] The following discussion applies for both aperture and scattering SNOM.

[3] Also known as plane wave spectrum.

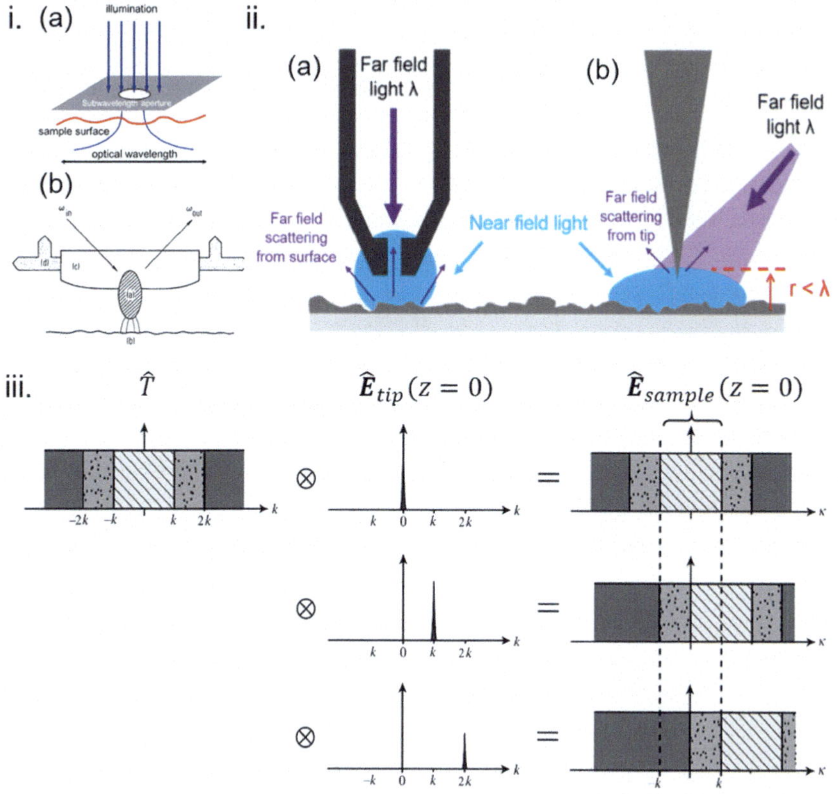

Fig. 11.2 i. The concepts of SNOM and TERS: **a** Synge's idea of scanning a backside-illuminated opaque metal screen with a small aperture in close distance over a sample for imaging with sub-wavelength resolution in 1928 [13]. **b** The concept of TERS proposed by Wassel in 1985 [12]. Adapted with permission from [12]. Copyright (1985) (Journal of the Optical Society of America). **ii. Types of SNOM** [14]. **a** Aperture SNOM where an evanescent light field created by a nanoscale opening at the end of a tip is scattered off of the surface. Light scattered from the surface is detected in the far-field. **b** Apertureless SNOM utilizes near-field light emitted from the surface due to external illumination that is scattered off of a sharp tip into the far-field to be detected. Adapted with permission from [14]. Copyright (2017) (Applied Sciences). **iii. Beyond diffraction limit** [21]. Convolution of the spatial spectra of sample transmission ($\hat{T}$) and tip field ($\hat{\boldsymbol{E}}_{tip}$). Three discrete spatial frequencies of $\hat{\boldsymbol{E}}_{tip}$ are shown. Convolution with $\delta(k_{||} - mk)$ shifts the spatial spectrum of T by mk. $m = 0$ corresponds to a plane wave at normal incidence, $m = 1$ to a plane wave at parallel incidence, and $m = 2$ is an evanescent wave. In the far-field, the resulting spectrum of $\hat{\boldsymbol{E}}_{sample}$ can only be detected in the range $k_{||} = [-k \ldots k]$. The figure illustrates that evanescent components in the source field shift the high spatial frequencies of the sample into the detection window. Adapted with permission from author of [21]

interaction, we will neglect the magnetic field $\boldsymbol{H}(\boldsymbol{r})$ in the discussion and following content. In the concept of an angular spectrum, $\boldsymbol{E}(\boldsymbol{r})$ can be described as a superposition of propogation waves and evanescent waves with variable amplitudes and propagation directions. As is depicted in Fig. 11.3i(a), $\boldsymbol{E}(\boldsymbol{r}) = \boldsymbol{E}_{inc} + \boldsymbol{E}_{scatt}$, where $\boldsymbol{E}_{inc}$ is the incident field (source) and $\boldsymbol{E}_{scatt}$ is the scattered field.[4] If we draw an axis-z and consider $\boldsymbol{E}(\boldsymbol{r})$ in the xy plane with $z =$ const. Thus $\boldsymbol{E}(\boldsymbol{r})$ can be represented with Fourier transform, which reads as:

$$\boldsymbol{E}(\boldsymbol{r}) = \iint_{-\infty}^{+\infty} \hat{E}(k_x, k_y; z) e^{i(k_x x + k_y y)} dk_x dk_y \tag{11.1}$$

where k_x, k_y are the spatial frequencies or reciprocal coordinates or projected wavevectors (see Fig. 11.3i(b)). For the field in homogeneous, isotropic, linear, and source free medium, a time-harmonic electrical field with angular frequency ω has to fulfill the Helmholtz equation:

$$(\nabla^2 + k^2)\boldsymbol{E}(\boldsymbol{r}) = 0 \tag{11.2}$$

where k is the spatial frequencies or reciprocal coordinates or wavevector defined by:

$$k = \left(\frac{\omega}{c}\right) n = 2\pi n / \lambda_0 \tag{11.3}$$

Where c is the speed (phase velocity) of light in vacuum, n is the real part of the refractive index, λ_0 is the wavelength of light in vacuum. According to (11.3), it is interesting to notice that the speed of light in the medium is determined by c/n, or the wavelength is reduced to λ_0/n.[5] Inserting the Fourier representation of $\boldsymbol{E}(\boldsymbol{r})$ (11.1) into the Helmholtz equation (11.2), we can get the expression of $\boldsymbol{E}(\boldsymbol{r})$ along the z-axis in the form of:

$$\hat{\boldsymbol{E}}(k_x, k_y; z) = \hat{\boldsymbol{E}}(k_x, k_y; 0) e^{\pm i k_z z} \tag{11.4}$$

With

$$k = \sqrt{k_x^2 + k_y^2 + k_z^2} \tag{11.5}$$

where k_z are the spatial frequencies or reciprocal coordinates or projected wavevector along the z-axis (see Fig. 11.3i(b)). The +sign in (11.4) refers the wave propagation

[4] $\boldsymbol{E}(\boldsymbol{r})$ Is also known as "total field". The concept of "total field" is very important in the numerical simulation of electromagnetic (EM) field.

[5] Thus, "wavelength" is not a good physical quantity to define light as it depends on the refractive index of medium. However, due to historical and customary reasons, "wavelength" is still the most common concept for the definition of light.

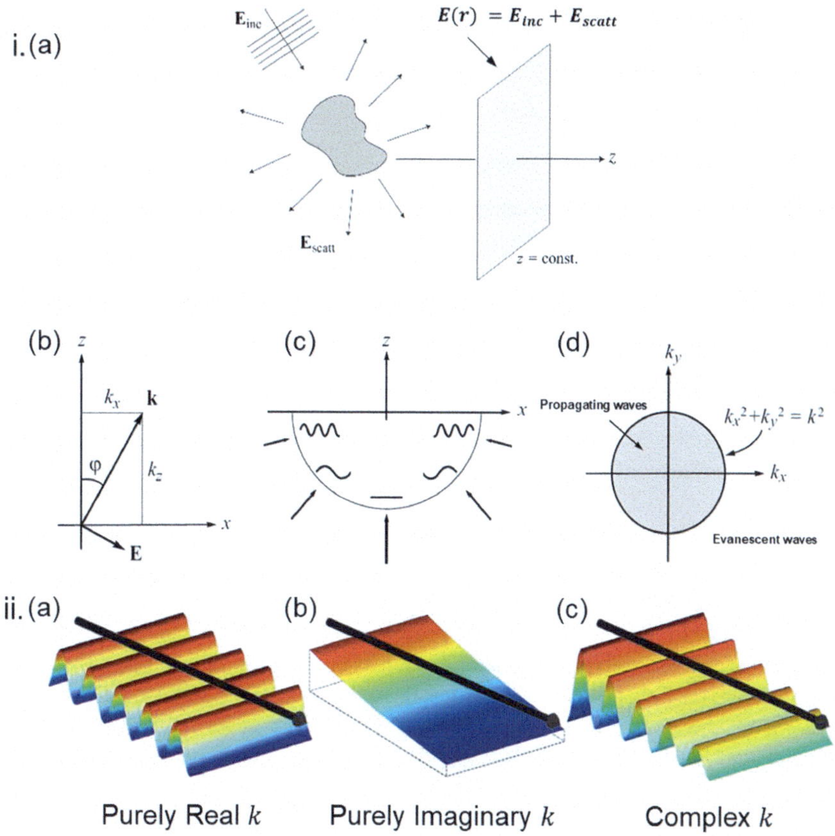

Fig. 11.3 **i. Object scattering in homogeneous media. a** In the angular spectrum representation, the fields are evaluated in planes at z = const. **b** Representation of a plane wave propagating at an angle φ to the z axis. **c** Illustration of the transverse spatial frequencies of plane waves incident from different angles. The transverse wavenumber $(k_x^2 + k_y^2)^{1/2}$ depends on the angle of incidence and is limited to the interval $[0...k]$. **d** The transverse wavenumbersk_x, k_y of plane waves are restricted to a circular area with radius$k$. Evanescent waves fill the space outside the circle. Adapted with permission from author of [21]. **ii. Wavevectors with a** Only real k component. Propagating waves (**far-field**) for example. **b** Only imaginary k component. Evanescent waves (**near-field**) for example. **c** Complex k. Surface plasmon polaritons for example. Adapted with permission from the author of [25]. Copyright (2018) (Raymond C. Rumpf, EMPossible)

into the space $z > 0$, while the −sign denotes the wave propagating into the space $z < 0$. We can learn from (11.4) that that the Fourier spectrum of $\boldsymbol{E}(\boldsymbol{r})$ in an arbitrary image plane located at z = const. can be calculated by multiplying the spectrum in the object plane at $z = 0$ by the factor $e^{\pm ik_z z}$, which is thus called the propagator[6] in reciprocal space. Equation (11.5) is known as the wavevector (momentum) conservation rule. Based on (11.4) and (11.5) we can obtain the angular

[6] Also known as optical transfer function (OTF).

spectrum representation as:

$$E(r) = \int\int_{-\infty}^{+\infty} \hat{E}(k_x, k_y; 0) e^{i(k_x x + k_y y \pm k_z z)} dk_x dk_y \tag{11.6}$$

In the case of a homogeneous dielectric medium with no loss,[7] the wavevector k_z is either real or imaginary, which will turn the propagator $e^{\pm ik_z z}$ into oscillatory or exponentially decaying function respectively. For a certain transverse wavevector $k_{||} = \sqrt{k_x^2 + k_y^2}$, we can obtain two different characteristic solutions:

$$\textbf{Propagating waves}: e^{i(k_x x + k_y y)} e^{\pm i|k_z|z}, \text{ with } k_x^2 + k_y^2 < k^2 \tag{11.7}$$

$$\textbf{Evanescent waves}: e^{i(k_x x + k_y y)} e^{-|k_z||z|}, \text{ with } k_x^2 + k_y^2 > k^2 \tag{11.8}$$

when the sum of k_x^2 and k_y^2 is smaller than k^2 (11.7), the k_z wavevector is purely real. In this case, the propagator term is just an oscillation function ($\cos(k_z z)$) along z direction, corresponding to propagating wave or a plane wave (see Fig. 11.3ii(a)). The term "**far field**" refers to such light propagates through space in an unconfined manner. For conventional optical systems (such as confocal microscopy), only propagating waves (far field) can be recorded as the detectors (photodiode, CCD camera, spectrometer, etc.) are located at sufficient long distance away from the objective plane. When the sum of k_x^2 and k_y^2 is larger than k^2 (11.8), the k_z wavevector has to be imaginary. In this case, the propagator term $e^{ik_z z}$ will decay exponentially along the z-axis, as is depicted in Fig. 11.3ii(b). Without special techniques (such as near-field probing or frustrated total internal reflection (FTIR) [26], the field of the propagator term $e^{ik_z z}$ cannot be collected by the detector. In theory, it's possible to attain an unlimited range of spatial frequencies, e.g., with the ideal point source function. Nevertheless, as the spatial frequencies $k_{||}$ of an evanescent wave increase, the rate of field decay along the z-axis also accelerates. The concept of "**near field**" refers to the evanescent wave, which is nonpropagating and only exists near the surface of an object at distances around one wavelength.

The concepts of propagation and evanescent fields play a central role in nano-optics, which are of major importance for the understanding of beyond diffraction imaging mechanism of SNOM and TERS techniques. According to the previous discussion, the propagator $e^{\pm ik_z z}$ is an oscillating function for $k_x^2 + k_y^2 < k^2$ (11.7) and an exponentially decreasing function for $k_x^2 + k_y^2 > k^2$ (11.8). Thus, if the image plane is sufficiently separated from the object plane (adv., in far field position), the contribution of the decaying parts (evanescent waves) is zero and the integration can be reduced to the circular area $k_x^2 + k_y^2 \leq k^2$. In other words, the image at z is a low pass filtered representation of the original field at z = 0 (see Fig. 11.3i(d)). The spatial

[7] Imaginary part of refractive index is zero.

frequencies ($k_x^2+k_y^2 > k^2$) of the original field are filtered out during propagation and the information on high spatial variations gets lost forever. Hence, there is always a permanent loss of information on propagating from near- to far-field.

Thus, the Fourier spectrum $\boldsymbol{E}(\boldsymbol{r})$ only in terms of far-field can be expressed as:

$$\boldsymbol{E}(\boldsymbol{r}) = \frac{ire^{-ikr}}{2\pi} \iint_{k_x^2+k_y^2 \le k^2} \hat{\boldsymbol{E}}_\infty(k_x, k_y) e^{i(k_x x + k_y y \pm k_z z)} \frac{1}{k_z} dk_x dk_y \tag{11.9}$$

Equation (11.9) neglects the contribution of near field light. In the paraxial approximation (assume $k_{||} \ll k$), $k_z \approx k$, $\boldsymbol{E}(\boldsymbol{r})$ and its far field $\hat{\boldsymbol{E}}_\infty$ form a perfect Fourier transform pair. The resolving power of an optical system depends on the bandwidth of spatial frequencies $\Delta k_{||} = \sqrt{\Delta k_x^2 + \Delta k_y^2}$ that are collected by the optical system. Simple Fourier mathematics leads to:

$$\Delta k_{||} \Delta r_{||} \ge 1 \tag{11.10}$$

where $\Delta r_{||}$ is the lateral spatial resolution (diffraction) limit. Equation (11.10) is similar to the Heisenberg uncertainty principle [27] in quantum mechanics. The product of $\Delta r_{||}$ and $\Delta k_{||}$ is minimized for a Gaussian distribution of spatial frequencies. The upper limit of $\Delta k_{||}$ is the wavevector value defined by (11.3). Combining (11.3) and (11.10) we can obtain the spatial resolution limit as:

$$\min(\Delta r_{||}) \approx \frac{1}{k} = \frac{\lambda_0}{2\pi n} \tag{11.11}$$

The resolution limit given by (11.11) is very optimistic. According to the famous Abbe's criterion [27], the resolution limit is expressed as:

$$\text{Abbelimit: } \Delta L = 0.6098 \frac{\lambda_0}{NA} \tag{11.12}$$

where NA is the numerical aperture of the optical system. The Abbe limit given in (11.12) is based on the paraxial approximation and applies to the special case of two parallel dipoles oriented perpendicular to the optical axis.

Now we know the resolution ability of the optical system is limited as a result of spatial frequency bandwidth limitation imposed by the propagating wave condition, i.e., we can only obtain optical information with $\Delta k_{||} \le k$. Then how do we break such a fundamental limit and achieve super solution imaging (with $\Delta r_{||} \sim \lambda/20$ for example)? One good approach is the collection of evanescent waves, which contain spatial frequencies with $\Delta k_{||} > k$. We can bring a small scattering objective[8] into the near field region of the sample. The scattering objective acts as a small probe, which allows the spatial frequency with $\Delta k_{||} > k$ to be transferred into the far field

[8] Or a sub wavelength aperture, which is actually Synge's idea [13]. See Fig. 11.2i(a).

positions. We can understand the scheme as the tip is illuminated by the near field of the sample (more precisely, total field) and re-scatters the light field to far field as a light source. Let's assume the sample is an infinitely small object, which can be described by a transmission function $T(x, y; 0)$, which should be in the form of the Dirac delta function (δ distribution). $T(x, y)$ is zero everywhere except at position (x, y). Under the illumination of the probe field $\boldsymbol{E}_{tip}(x, y; 0)$, the field contains the optical information of the sample $E_{sample}(x, y; 0)$, which can be expressed as:

$$\boldsymbol{E}_{sample}(x, y; 0) = T(x, y; 0)\boldsymbol{E}_{tip}(x, y; 0) \tag{11.13}$$

The expressions of T, E_{tip} are in the spatial domain (direct space), the multiplication of T and $\boldsymbol{E}_{tip}$ can be calculated by the convolution in Fourier space:

$$\hat{\boldsymbol{E}}_{sample}\left(\kappa_x, \kappa_y; 0\right) = \int\int_{-\infty}^{+\infty} \hat{T}\left(\kappa_x - k_x, \kappa_y - k_y; 0\right)\boldsymbol{E}_{tip}\left(k_x, k_y; 0\right)dk_x dk_y \tag{11.14}$$

Now we can propagate the sample field $\boldsymbol{E}_{sample}$ to the detector in far field position at $z = z_{\infty}$:

$$\boldsymbol{E}_{detector}(x, y; z_{\infty}) = \iint_{\kappa_{\|}^2 \le k^2} \hat{\boldsymbol{E}}_{sample}\left(\kappa_x, \kappa_y; 0\right)e^{i\left(\kappa_x x + \kappa_y y\right)}e^{i\kappa_z z_{\infty}}d\kappa_x d\kappa_y \tag{11.15}$$

where the transverse wavenumber $\kappa_{\|} = \sqrt{\kappa_x^2 + \kappa_y^2}$ still cannot exceed the propagating waves' bandwidth (11.7), which makes (11.15) appears just be a restatement of the resolution limit. How can we interoperate (11.15)? To answer this question, we should decompose the probe field $\boldsymbol{E}_{tip}$ as an infinite number of partial source fields with discrete spatial frequencies:

$$\boldsymbol{E}_{tip}\left(k_x, k_y; 0\right) = \int\int_{-\infty}^{+\infty} \boldsymbol{E}_{tip}\left(\tilde{k}_x, \tilde{k}_y; 0\right)\delta\left(\tilde{k}_x - k_x\right)\delta\left(\tilde{k}_y - k_y\right)d\tilde{k}_x d\tilde{k}_y \tag{11.16}$$

For each partial field $\delta\left(\tilde{k}_{x,y} - k_{x,y}\right)$ we can separately calculate the interaction with the sample and the resulting far-field at the detector (see Fig. 11.2iii). In the end, we may sum over all the individual responses. Recall that the convolution of $\hat{T}\left(k_x', k_x'; 0\right)$ (with $k_x' = \kappa_x - k_x$, $k_y' = \kappa_y - k_y$) and $E_{tip}\left(k_x, k_y; 0\right)$ was performed in (11.14). In Fourier mathematics, the convolution operation will only shift the transverse wavevectors of the sample $k_{\|}'$ as:

$$\kappa_{\|} = k_{\|} + k_{\|}' \tag{11.17}$$

i.e. it translates the spectrum $\hat{T}$ by $k_{||}$. Figure 11.2iii illustrates the shift in the sample spectrum $\hat{T}$ for three discrete transverse wavenumbers of the source field: $\delta(k_{||})$, $\delta(k_{||}-k)$, and $\delta(k_{||}-2k)$. A plane wave at normal incidence is represented by $\delta(k_{||})$, which does not alter the original spectrum. The plane wave with the largest transverse wavenumber is incident parallel to the surface, represented by $\delta(k_{||}-k)$. This wavenumber shifts the original spectrum by k, effectively bringing the spatial frequency range $k'_{||} = [k \ldots 2k]$ into the detection window $k_{||} < k$ where propagating plane waves exist. Finally, $\delta(k_{||}-2k)$ represents an evanescent wave, which shifts $\hat{T}$ by $2k$ and brings the frequency range $k'_{||} = [2k \ldots 3k]$ into the detection window.[9] Consequently, the merging of significant spatial frequencies within the sample and the probe field results in a different wavevector that aligns with an angular spectrum wave propagation directed towards the detector. As a result, employing a source field that is both confined and possesses a broad spectrum of spatial frequencies grants us access to the higher spatial frequencies within the sample in the far-field. It's worth noting that the more effectively we confine the source field, the finer the resolution we can achieve for the sample. For a confined source field with a characteristic lateral dimension D (aperture diameter of aperture SNOM, tip diameter of scattering SNOM and TERS, ...) the highest spatial frequencies can be expressed as:

$$k'_{||,max} \approx \left| \frac{\pi}{D} \mp \frac{2\pi NA}{\lambda} \right| \tag{11.18}$$

For $D \ll \lambda$ we can neglect the last term and find that the source confinement entirely defines the highest detectable spatial frequencies of the sample. We can obtain the new spatial resolution limit by inserting (11.18) into (11.11):

$$\min(\Delta r_{||}) \approx \frac{1}{k'_{||,\mathrm{max}}} = \frac{D}{\pi} \tag{11.19}$$

we can learn from (11.19) that the smaller the tip radius, the higher the resolution we can obtain, which formed the foundation of super spatial resolution SNOM and TERS.

In the end we would like to have a more detailed discussion on the evanescent waves. According to the above discussion, evanescent waves can be generated by light scattering from small objects in homogeneous and no loss media (Fig. 11.3i(a)). We can also generate evanescent waves in inhomogeneous and no loss media. The simplest case is at the interface between two media characterized by refractive index n_1 and n_2 (assume $n_1 > n_2$). Plane wave impinging on such an interface will lead to refracted waves and reflected waves with amplitudes and directions (see Fig. 11.4i(a)–(c)), which are expressed by Fresnel coefficients and Snell's law [21]. When the light source incidents from the optically dense medium (with $n = n_1$) and the incident angle θ of the plane wave is larger than the critical angle $\theta_c = \sin^{-1}(n_2/n_1)$, , an evanescent wave decays exponentially into to the optically less dense medium (with

[9] The Structured Illumination Microscopy (SIM) also relies on the same working principle.

$n = n_2$, with $n_2 < n_1$) is excited as a result of total internal reflection (TIR, as depicted in Fig. 11.4i(c)). For p-polarized[10] evanescent waves, the intensity of the evanescent wave can be larger than that of the input beam [21, 28]. The physical reason for this enhancement is a surface polarization that is induced by the incoming plane wave, which is also represented by the boundary condition. A similar enhancement effect, albeit a much stronger one, can be obtained when the glass/air interface is covered by a thin layer of a noble metal. Here, the so-called surface plasmon polaritons (SPP) can be excited. For s-polarized light[11] no field enhancement is observed. It is important to mention in here that the Kawata team, one of the three independent groups who performed the first TERS experiment, utilized the evanescent field based on TIR to excite the apex of the silver probe [24], which will be discussed in detail in Sect. 11.3. For p-polarized light, the decay constant γ is defined by:

$$\gamma = \frac{2\pi n_2}{\lambda_0}\sqrt{(\frac{n_2}{n_1}\sin\theta)^2 - 1} \tag{11.20}$$

Compared with (11.8), whose decay constant only depends on the value of $|k_z|$, γ is a much more complicated function that depends on refractive index and incident angle. For the glass/air ($n_1 = 1.518, n_2 = 1$ at 546 nm vacuum wavelength) interface and an angle of incidence of $\theta_c = 41.2°$ (with 546 nm vacuum wavelength plane wave), the decay constant is $\gamma = 2.22/\lambda = 0.0041\ \text{nm}^{-1}$. This means that already at a distance of $\approx\lambda/2 \approx 273$ nm from the interface, the time-averaged field is a factor of e smaller than at the interface. At a distance of $\approx 2\ \lambda = 1092$ nm the field becomes negligible. The larger the angle of incidence θ (adv., the larger the spatial frequencies $k_{||}$) the faster the decay will be, which behaves the same as the evanescent waves decay function defined by (11.8). Experimental verification for the existence of this rapidly decaying field in the optical regime relies on approaching another optically dense body (with $n > 1$) or plasmonic thin film or sharp metallic tip to within less than $\lambda/2$ of the interface that supports the evanescent field. These measurement techniques are frustrated total internal reflection (FTIR) [21, 26], photon scanning tunneling microscopy [26], surface plasmons excitation [29], and scattering SNOM [28, 30] (as shown in Fig. 11.4ii [28]).

F. H'Dhili experimentally demonstrated that local optical field enhancement can occur at the end of an apertureless SNOM tip [28]. As depicted in Fig. 11.4ii(a), a metallic SNOM tip is illuminated by evanescent light generated by TIR as a small light source to perform high-resolution photolithography on a photosensitive polymer that is placed in close proximity to the extremity of the SNOM tip. The tip diameter $D = 40$ nm. Figure 11.4ii(b) (AFM topography image) shows results of the experiment. A dot with ~70 nm width can be clearly seen in this figure. In this experiment, a 532 nm vacuum length laser was used as light source. Thus, according to the Abbe limit (11.11), the smallest possible spot size is ~216 nm (with NA = 1.5). The super resolution dot shown in Fig. 11.4ii(b) can only be explained by the confined field

[10] Also known as the Transverse Electric (TE) mode.

[11] Also known as the Transverse Magnetic (TM) mode.

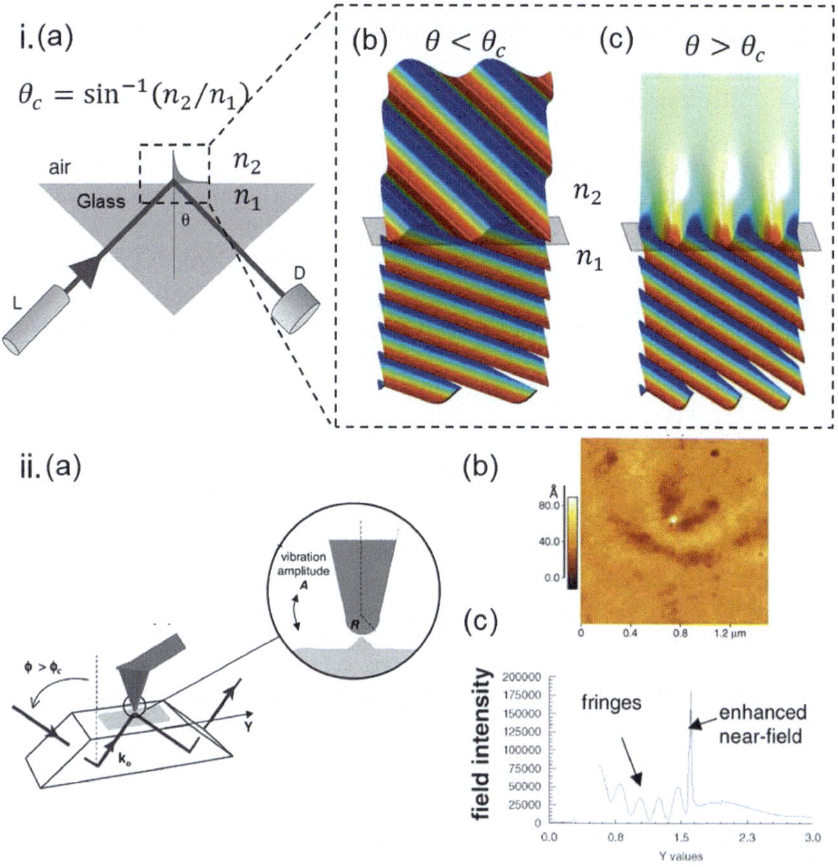

Fig. 11.4 **i. a** Experimental arrangement to realize total internal reflection. Wave front at glass/air interface with, **b** $\theta < \theta_c$, propagating waves, **c** $\theta > \theta_c$, evanescent waves by total internal reflection. **ii. Local field enhancement at the end of apertureless SNOM tips**. Adapted with permission from the authors of [21, 25]. Reference [25] Copyright (2018) (Raymond C. Rumpf, EMPossible). **a** Schematic diagram of the experimental configurations. Tip illuminated by evanescent light generated by total internal reflection (TIR). Inset: tip end. **b** TM-AFM images obtained after p-polarized illumination of a platinum tip. **c** Intensity distribution beneath the tip end from FDTD simulation. Adapted with permission from the author of [28]

at the tip apex, which is confirmed by the finite difference time domain (FDTD) simulation (Fig. 11.4ii(c)).

11.2.2 Enhancement Mechanisms of TERS

11.2.2.1 Electromagnetic Enhancement

It is widely accepted that TERS occurs as a result of electromagnetic (EM) enhancement and chemical enhancement mechanisms. When a nanometer sized metallic probe is illuminated with the EM field, the free electrons will collectively oscillate at the surface of the metal probe, giving rise to surface plasmon polaritons (SPPs), which will propagate along the surface of the probe. The dispersion curve for an SPP mode (shown in Fig. 11.5a) shows the momentum match must be fulfilled in order to couple free space light and SPP modes together, with the SPP mode always lying beyond the light line, that is, it has greater momentum (k_{SPP}) than a free space photon (k_o) of the same frequency ω. To satisfy momentum matching, Kretschmann configuration [31] (see Fig. 11.11ii(a)) and grating-assisted configuration [32, 33] (see Figs. 11.5 and 11.11ia) are two of the mostly adopted methods. Take the Kretschmann configuration for example, as shown in Fig. 11.5a, the light line is titled by a factor of n (the refractive index of the optically dense material), which makes the $k_{SPP} = k_o$ momentum matching possible.

SPPs at the interface between a metal and a dielectric material have a combined electromagnetic wave and surface charge character as shown in the insert of Fig. 11.5b. SPPs can also be understood as the hybridation of charge density oscillations in the metal and EM field [34]. SPPs are transverse magnetic (TM mode) in character (**H** is in the y direction, see insert of Fig. 11.5b), and the generation of surface charge requires an electric field normal to the surface.[12] This combined character also leads to the field component perpendicular to the surface being enhanced near the surface and decays exponentially with distance from the surface. The perpendicular field behaves just like the evanescent waves discussed in Sect. 11.2.1, which can be characterized by waves with a purely imaginary k component (see Fig. 11.3ii(b)). In the parallel direction (along metal surface), SPPs are also attenuate along the distance but quite differently, which can be described as waves with complex wavevector (see Figs. 11.3ii(c) and 11.5b). The attenuation is caused by the losses arising from absorption in the metal. However, the propagation distance of SPPs along metal surface can be in the range of 100–1000 μm [34]. To understand the field enhancement mechanism of sSNOM and TERS, the fundamental knowledge of the SPPs is indispensable. A more detailed discussion of the properties of SPPs can be found from the book of Novotny [21] and publication from Barnes [34].

As the SPPs travel in the deep-subwavelength regions (such as the apex area of a sharp metal sSNOM/TERS probe), only limited SPPs modes can survive and are spatially confined, known as localized surface plasmons (LSPs). Based on the dispersion relation with the free-space light line and SPPs (**given in** Fig. 11.5a), SPPs can only be excited with $k_0 = k_{SPP}$. According to the work of Esmann ect. [35], only lowest order ($n = 0$, 1

[12] Also known as p-polarized.

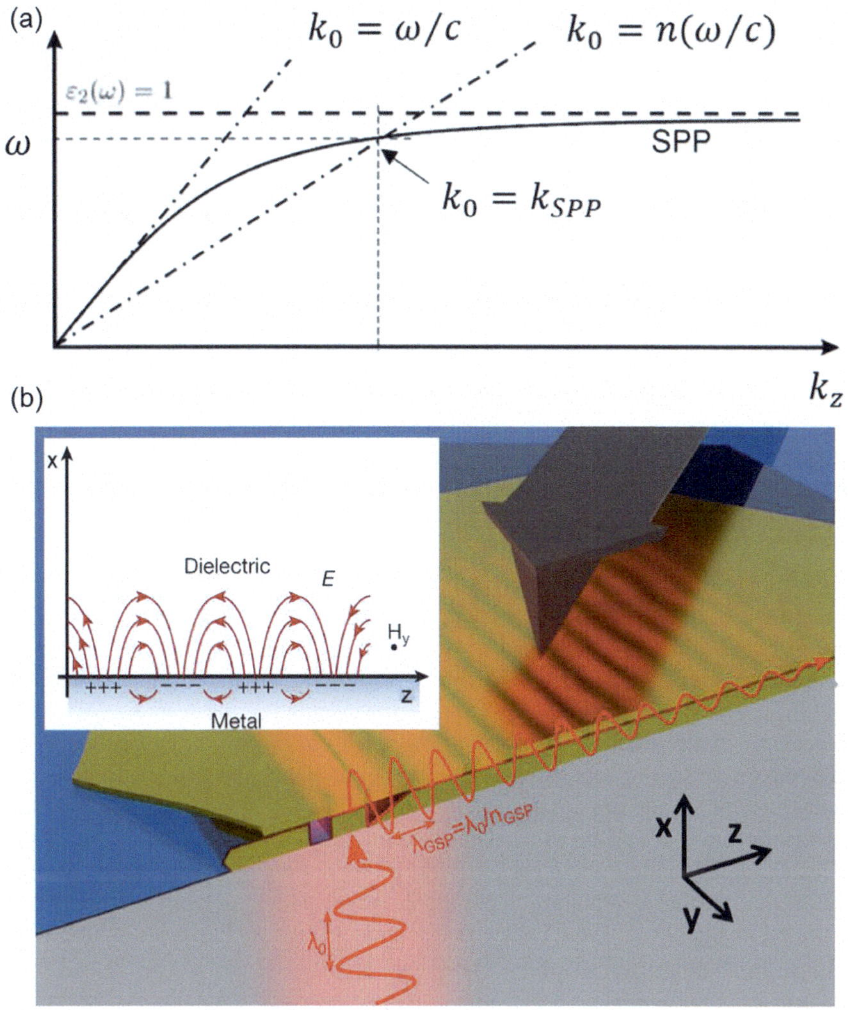

Fig. 11.5 **a** Dispersion relation with the free-space light line and SPP [21]. Adapted with permission from the author of [21]. **b** Schematic illustration of the SPP on Au–Al_2O_3–Au film [32]. The red curves schematically illustrate the electric-field profile of the excitation light (bottom left) and of the propagating SPP mode (in the sample) with the group-index n_{SPP} (corresponding to an SPP-wavelength λ_{SPP} at the free space wavelength λ_0). Insert: Sketch of the SPP at the interface between a metal and a dielectric medium [34]. Adapted from [32]. Insert is adapted from [34]

shown in (Fig. 11.6ii(a), (b), (d)) eigenmode LSPs can bound close to the end of the tip. Theoretical studies have shown that only the lowest, rotational symmetry (see Fig. 11.6ii(a)) eigenmode (n = 0 shown in Fig. 11.6ii(a), (d)) can exists at the very apex area of the tip and gets highly confined, which leads to the excitation of strong near-fields. The fundamental $n = 0$ mode is sustained as a bound mode even for vanishing wire radius, while higher order mode ($n = 1$ shown in Fig. 11.6ii(b), (d)) can only exist within the last micro from the tip apex. In contrast, for the $n = 1$ mode, the propagation constant approaches the light line, $k_0 = \omega/c$, for radii of about 70 nm, indicating that the mode loses its bound character. Such cut-off behavior originates from the momentum (phase) match between SPPs and EM field, shown in Fig. 11.5a. The SPP on a plane interface cannot be excited by light of any frequency that propagates in free space. Excitation of an SPP by light is only possible if a wavevector component of the exciting light can be increased over its free-space value with $k_z \geq k_0$, where k_z is the propagation constant of SPP. For $n = 1$ mode, the surface charge has an opposite sign (anti-symmetric) in respect to the tip axis, as shown in Fig. 11.6iii(b). As a consequence, the end of the tip remains uncharged. According to numerical simulation results, no field enhancement beneath the tip end is the case. On the other hand, for n = 0 mode (Fig. 11.6iii(a)), the surface charge is radially symmetric and has the highest amplitude at the end of tip. As a result, the field intensity beneath the tip end is strongly increased. The EM enhancement mechanism is a direct result of the field intensity enhancement of the $n = 0$ LSP mode.

To gain a more physically intuitive understanding for the optical response of the TERS tip and the sample, it is favorable to find an approximate model that can be treated analytically. The simplest model for the tip and sample interaction is to treat them as quasi-static spheres, as shown in Fig. 11.7i. This model is also known as "Keilmann model", which was proposed by Fritz Keilmann and Bernhard Knoll in 2000 [37]. The Keilmann model is suitable for calculating the response of single molecules or particles close to the probe. At small distances, with respect to probe and sample size, both probe and sample dipole moments appear enhanced by the mutual local fields and image charges. It is possible to rather similarly calculate the optical interaction of a probe with a planar sample surface based on this model. In the Keilmann model, both probe tip and sample are represented by polarizable spheres. Within small distances, tip and sample dipole moments are mutually enhanced. As illustrated in Fig. 11.7i, the TERS probe is modeled as a polarizable sphere with radius of a and polarizability of $\alpha = 4\pi r_0^3 \frac{\varepsilon_p - 1}{\varepsilon_p + 2}$,

where ε_p is the complex dielectric constant of the probe material. According to Mie theory [38], α is the well-known quasistatic polarizability of a small sphere sample. The sample with complex dielectric function of $\varepsilon(\omega)$ is positioned close to the probe with distance of d. As applying an incident electric field $\boldsymbol{E}_0(\omega)$ (plane wave) to the sample surface, the field in the vicinity of the tip can be accurately described by the irradiation of an effective dipole $\boldsymbol{\mu}(\omega)$ at the center of the tip apex. For the case $\boldsymbol{E}_0(\omega)$.

perpendicular (p-polarized illumination) to the sample surface (Fig. 11.7i)), the electric field of the induced probe dipole is in phase with the probe dipole ((n = 0 LSP mode shown in Fig. 11.11i(a), ii(a))), the corresponding effective polarizability is $\alpha_{zz}(\omega)$. As the incident electric field $\boldsymbol{E}_0(\omega)$ is parallel (s-polarized illumination) to the sample, the image dipole is out of phase to the tip dipole as shown in Fig. 11.7i(b) (n = 1 LSP mode shown in Fig. 11.11i(b) and ii(b)) and the corresponding effective

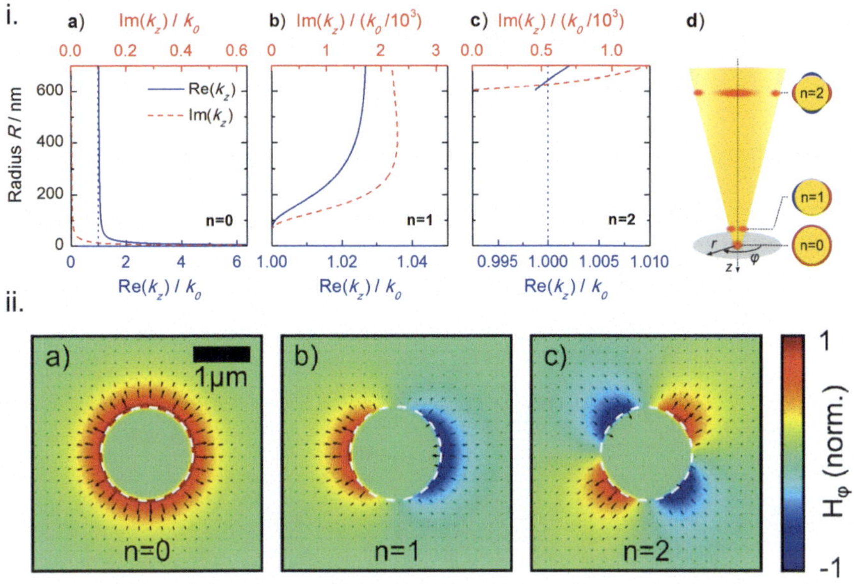

Fig. 11.6 **i. Dispersion relations of SPP wire eigenmodes** [35]. Propagation constants k_z of the three lowest eigenmodes of an infinitely long gold wire as a function of wire radius R. Displayed are the real (solid blue lines) and imaginary (dotted red lines) parts of k_z pointing along the taper axis for angular momenta $n = 0, 1, 2$ (panels a)–c)). The SPP frequency corresponds to a vacuum wavelength of 800 nm, as used in the experiment. For $n = 0$, both $Re(k_z)$ and $Im(k_z)$ diverge for small radii. For $n = 1$, k_z approaches the light line, $k_0 = \omega/c_0$, for small radii. A cutoff behavior at approximately 600 nm is observed for $n = 1$. . Panel d) summarizes these findings and schematically illustrates the mode profiles. Adapted with permission from author of [35]. Copyright (2013) (Bilstein journal of nanotechnology). **ii. Field distribution of the three lowest angular momentum SPP wire eigenmodes at the cross-section** [36]. Color-coded is the tangential in-plane component of the magnetic field H_φ, which is continuous across the gold-air interface. The position of the interface is indicated by the dashed white circles. The black arrows display the in-plane electric field $\boldsymbol{E}_{\parallel}$. Adapted with permission from author of [36]. Copyright (2017) (Martin Esmann)

polarizability is $\alpha_{xx}(\omega)$. Thus, the dipole induced by the exciting field $\boldsymbol{E}_0(\omega)$ and its magnitude can be expressed as [21, 37]:

$$\boldsymbol{\mu}(\omega) = \begin{bmatrix} \alpha_{xx} & 0 & 0 \\ 0 & \alpha_{yy} & 0 \\ 0 & 0 & \alpha_{zz} \end{bmatrix} \boldsymbol{E}_0(\omega) \tag{11.21}$$

Where the z-axis coincides with the tip axis. $\alpha_{xx,yy}$ and α_{zz} denotes the transverse (vertical to z-axis) and longitudinal (parallel to z-axis) effective polarizabilities defined as:

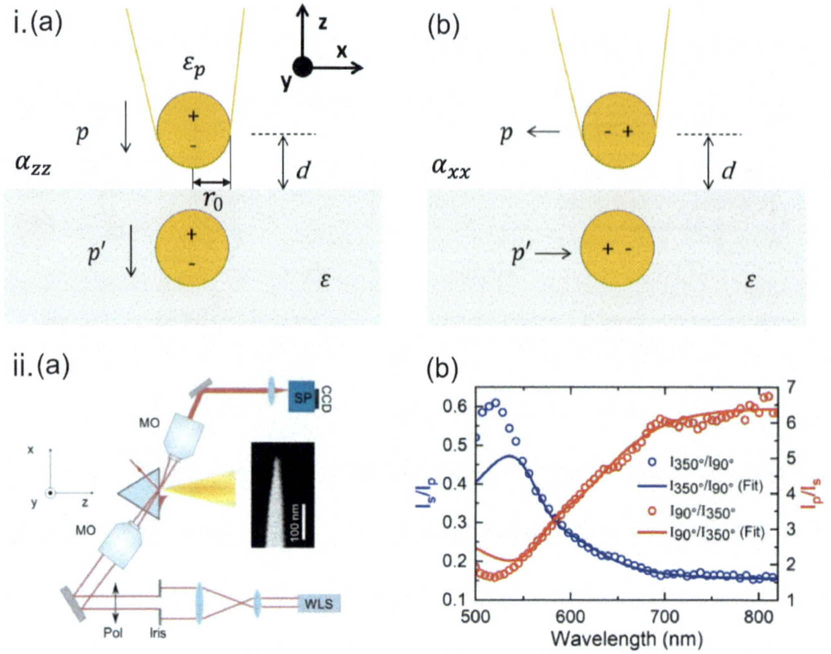

Fig. 11.7 i. **A typical probe-sample configuration of point dipole model in a uniform electric field** [37, 39], **a** perpendicular and **b** parallel to the sample surface. The probe tip dipole p and sample image dipole p' are separated with a distance of 2d. Adapted with permission from author of [39]. Copyright (2023) (Jinxin Zhan). ii. **Evanescent wave light scattering from conical gold tapers** [30]. **a** Schematic of the experimental setup and SEM image of a sharp gold taper with an apex radius of 9 nm used for the broadband scattering measurements. Light from a white light source (WLS) is focused by a microscope objective (MO) onto a prism in total internal reflection geometry under an incident angle of 50°, to excite evanescent fields at the prism-air interface. The polarization of the incident light is controlled by a polarizer (Pol) and half-wave plate. The scattered light from the tip apex is collected with an objective and focused on the entrance of a spectrometer. **b** Comparison of the experimental data (open circles) to the anisotropic tip model (solid lines), taking the ratio of the scattering spectra at 350° and 90° (blue line). This normalization cancels out the effects of the spectral variation of the excitation source and the light detection in the experiment. The inverse ratio of the spectrum measured at 90° to that at 350° (red line) emphasizes the enhanced broadband scattering by the longitudinal tip resonance for wavelengths beyond 600 nm. Adapted with permission from author of [30]

$$\alpha_{xx}(\omega) = \alpha_{yy}(\omega) = \frac{\alpha(1-\beta)}{1-\frac{\alpha\beta}{32\pi r^3}} \tag{11.22}$$

$$\alpha_{zz}(\omega) = \frac{\alpha(1+\beta)}{1-\frac{\alpha\beta}{16\pi r^3}} \tag{11.23}$$

where $\beta = \frac{\varepsilon-1}{\varepsilon+1}$, r_0 is the tip radius. Once the tip dipole is determined, the electric field E in the vicinity of the tip is calculated as:

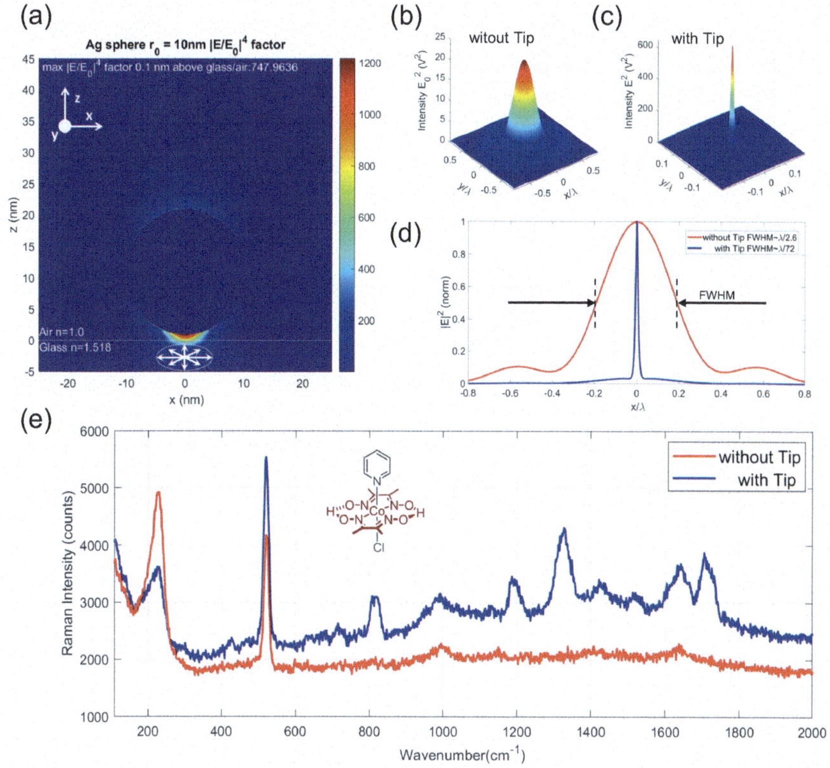

Fig. 11.8 FDTD simulation of a spherical TERS tip. a $|E/E_0|^4$ intensity distribution of the induced electric field $\boldsymbol{E}$ relative to the excitation field $\boldsymbol{E}_0$ near $r_0 = 10$ nm sized Ag sphere TERS tip placed 1 nm above the glass/air interface, calculated by 3D-FDTD method in a condition that a radial polarized EM field (with 532 nm vacuum wavelength) focused on the glass/air interface with NA = 1.4 lens. **b** The excitation field $\boldsymbol{E}_0$ (without tip) and **c** induced field $\boldsymbol{E}$ (with tip) distribution at 0.1 nm distance above the glass/air interface. (d) The comparison of the normalized intensity $(|\boldsymbol{E}|^2)$ at 0.1 nm distance above the glass/air interface along the x-axis without (red line) and with (blue line) tip. **e** Raman spectra of Co(dmgH)$_2$(py)Cl (insert) without (confocal Raman, red line) and with (TERS, blue line) tip. Copy right (2023) (Wei Wang)

$$\boldsymbol{E}(\boldsymbol{r}, \omega) = \boldsymbol{E}_0(\boldsymbol{r}, \omega) + \frac{1}{\varepsilon_0}\frac{\omega^2}{c^2} G(\boldsymbol{r}, r_0, \omega)\boldsymbol{\mu}(\omega) \quad (11.24)$$

where G is the dyadic Green's function. The detected scattered power is proportional to the square of the field $\boldsymbol{E}$ evaluated in the far-field. For parallel polarization, the probe and image dipoles are antiparallel so that they nearly cancel in the total scattering (Fig. 11.7i(b)). This cancellation is less perfect with dielectric substrates (such as silicon) than with plasmonic substrates (such as gold). Since perpendicular polarization yields much higher scattering due to in phase double dipoles interaction

(Fig. 11.7i(a)), we can restrict the following discussion to incident fields perpendicular to the sample surface. The electrical field arises from α_{xx}/α_{yy}, effective polarizabilities can be neglected α_{zz} in (11.25) follows the requirement that the field magnitude induced by $\boldsymbol{\mu}(\omega)$ at the surface of the tip is equal to $\sqrt{f_e(\omega)}\boldsymbol{E}_0$, $\sqrt{f_e(\omega)}$ is the complex enhancement factor. We can express the electrical field $\vec{E}$ (total field) at the tip apex ($x = 0$, $y = 0$, $z = r_0$) under p-polarized plane wave illumination as:

$$\boldsymbol{E}(x = 0,\, y = 0,\, z = r_0) = \sqrt{f_e(\omega)}\boldsymbol{E}_0 \tag{11.25}$$

The EM enhancement mechanism is clearly explained, now it is important to understand the role of the field enhancement factor in TERS. When the sample is directly illuminated, the sample surface will give rise to a far-field signal as background. The optical interaction between sample and light field can be considered as an *nth* order nonlinear process and only the surface is active. Then for first-order Raman scattering, the far-field background can be expressed as:

$$S_{ff} \sim AI_0 \tag{11.26}$$

where A is the illuminated area and I_0 is the illumination intensity. The near-field that is excited by the EM enhancement ($n = 0$ LSP mode) under the tip can be expressed as:

$$S_{nf} \sim a(f_e I_0) \tag{11.27}$$

where a is a reduced area defined by the tip size. If the near-field needs to be stronger than the background $\left(S_{nf} > S_{ff}\right)$, with $A = 40000\ \text{nm}^2$ and $a = 100\ \text{nm}^2$, the intensity enhancement factor $\left(f_e = \frac{|\boldsymbol{E}|^2}{|\boldsymbol{E}_0|^2}\right)$ should be larger than 400. According to the report of Martin, etc. [40], the enhancement factor with tip forms like thin cylinders, spheres, and ellipsoids, and with materials like silicon, tungsten, aluminum, and gold were calculated with FDTD method. Based on the simulation prediction, finite sized elongated shapes can provide very high enhancement factors. Under optimal plasmon resonance condition, an enhancement factor of $\left|\sqrt{f_e}\right| = 50$ can be achieved at the end of a small gold protrusion mounted on a wider silicon or glass, with $\lambda = 633$ nm.

Zhan etc. [30], adopted TIR evanescent wave light scattering method (Figs. 11.4 and 11.7ii), as an effort to determine the effective polarizability of a gold tip with 9 nm radius apex on glass substrate experimentally (as depicted in Fig. 11.6ii). In this work, both $\alpha_{xx}(\omega)$ and $\alpha_{zz}(\omega)$ components of the tip polarizability tensor were determined by a broadband (400–800 nm wavelength) polarization resolved evanescent wave excitation. Enhanced broadband scattering as a result of the longitudinal tip resonance was observed above 600 nm in this work. At 800 nm wavelength, the ratio of the vertical and parallel effective polarizability (α_{zz}/α_{xx}) is around 6, which is in good agreement with another independent report [41].

In addition to field enhancement, the LSP mode will also act as an antenna and thus re-radiate (as a dipole) the Raman scattering from molecules in near-field. According to the discussion in Sect. 11.2.1, the near field signal cannot be collected by the far field optical system, i.e., conventionally confocal Raman spectroscopy. The combination of field enhancement and Raman re-radiation induced by the metal tip is commonly referred as the $|\boldsymbol{E}/\boldsymbol{E}_0|^4$ approximation [42], where the Raman signal enhancement factor is defined as following:

$$\frac{|\boldsymbol{E}_{exc}|^2}{|\boldsymbol{E}_0|^2} \cdot \frac{|\boldsymbol{E}_{emm}|^2}{|\boldsymbol{E}_0|^2} \approx \left|\frac{\boldsymbol{E}}{\boldsymbol{E}_0}\right|^4 \tag{11.28}$$

In (11.28), The enhancement of the electric field that excites the sample ($\boldsymbol{E}_{exc}$) and Raman scattering ($\boldsymbol{E}_{emm}$) are normalized to the incident field ($\boldsymbol{E}_0$). The Raman signal enhancement factor can be approximated with the 4th power of the local field $\boldsymbol{E}$ and $\boldsymbol{E}_0$. The EM field enhancement factor is wavelength-dependent (see Fig. 11.10 in [43]), mainly following the plasmon resonance spectrum of the tip. Because the wavelength difference between the excitation (λ_{exc}) and emission (λ_{emm}) is generally small, it is often ignored. However, enhancement bandwidth can vary if the plasmon resonance is not broader than the difference in these wavelengths. As illustrated in Fig. 11.7i, the electromagnetic enhancement of a small spherical particle dependent upon illumination. For an isolated particle with excitation at plasmon resonance frequency, enhancement factors are a modest with $\left|\frac{E}{E_0}\right|^4 \sim 10^5$ [44]. According to the publication of Kim etc. [45], the maximum $|\boldsymbol{E}/\boldsymbol{E}_0|^4$ value can be as high as ~10^3 with a 40 nm radius silver sphere on top of a gold substrate under 568 nm illumination, which is enough for the enhancement of first-order Raman scattering (see (11.21) and (11.22)).

As an effort to demonstrate the field enhancement of TERS probes, we performed 3D-FDTD simulation[13] of $r_0 = 10$ nm sized Ag sphere with 1 nm distance above glass substrate, give in Fig. 11.8. In this simulation, $\lambda_0 = 532$ nm vacuum wavelength radial polarized light was focused on the glass/air interface (the focal plane) with NA = 1.4 virtual objective. For TERS application, the electric field without Ag sphere in the simulation model is incident field $\boldsymbol{E}_0$, while electric field with Ag sphere is excited field $\boldsymbol{E}$. The field distribution of tightly focused radial polarized light 0.1 nm above the glass/air interface (sample plane) is given in Fig. 11.8b, with full width at half maximum (FWHM) ~205 nm, which is around $\lambda/2.6$. The Abbe limit given in (11.12) with NA = 1.4 optical system is $\lambda/2.29 \sim 232$ nm, which is in good agreement with the simulation result. For radially polarized tightly focused light, a strong longitudinal field ($\boldsymbol{E}_z$) is present at the center of the focus and contributes 100% of the intensity $\left(I_{focus}(0,0) = |\boldsymbol{E}_z|^2\right)$ according to the simulation [46]. Thus, when the TERS tip is presented near the focus spot, strong field enhancement will

[13] If the readers are interested in the numerical simulation of EM field, the online course (https://empossible.net/) and latest publication "Electromagnetic and Photonic Simulation for the Beginner: Finite-Difference Frequency-Domain in MATLAB" by Raymond C. Rumpf are strongly recommended.

be excited due to the longitudinal field (see Fig. 11.7i(a)) and associated discussion in this section). The field distribution with TERS probe is given in Fig. 11.8c. Based on Fig. 11.8b, c, we can obtain the maximum enhancement factor at the sample plane $f_{sample} \approx 27$, leading to $\left|\frac{E}{E_0}\right|^4 \approx 747$ (see Fig. 11.8a). We have to stress in here that such a TERS enhancement factor is not very high, as the tip is not excited at resonance wavelength. According to the simulation of Taguchi [43], the resonance wavelength of a 10 nm radius Ag sphere should be in the UV range. Despite the low TERS enhancement factor, the FWHM of the field at the sample plane is reduced to 7.4 nm, which is around $\lambda/72$. Keilmann etc. [37], estimated the achievable FWHM based on probe and image dipoles interaction is $1.17\ r_0 = 11.7$ nm $\sim \lambda/45$ for a $r_0 = 10$ nm sized Ag sphere in contact with the substrate. Thus, the field at the sample is not only enhanced, but also strongly confined due to the presence of TERS probe, which explains the Raman signal enhancement mechanism and the sub diffraction limit imaging ability of TERS. Figure 11.8e shows the Raman spectra of $Co(dmgH)_2(py)Cl$ with (TERS) and without (confocal Raman) the probe approached.

11.2.2.2 Chemical Enhancement

Besides EM mechanism, the chemical mechanism can add additional enhancement and even alter the Raman spectra pattern considerably, such as peak intensities and positions. Like the EM enhancement mechanism, the chemical enhancement mechanism in TERS also has been extensively investigated both theoretically and experimentally [47–53]. Chemical enhancement is usually interpreted as a resonance Raman process between the ground electronic state of the molecule–metal complex and its new excited levels arising from charge transfer (CT) between the metallic surface and the adsorbed molecule (see Fig. 11.9ii). When a molecule is attached to a metal structure, new electronic states can be formed due to chemisorption. At the same time, if the Fermi level of the metal happens to be located between the lowest unoccupied molecular orbitals (LUMOs) and the highest occupied molecular orbitals (HOMOs), then CT excitations may occur at lower energy than intrinsic intramolecular excitations. Thus, the CT mechanism can be considered as a resonance Raman process happening between the ground electronic state (or LUMOs) of the of the molecule-metal complex and its excited electronic state (or HOMOs). Sun etc. have extensively investigated the chemical enhancement mechanism of SERS and TERS with molecule–Ag complex, in which the CT has been visualized with charge difference density [49–53].

Figure 11.9i illustrates the molecular-metal complexes formed by pyridine and Ag_{20}. According to the calculation of Sun et al. [53], the charge redistribution (CR) on the electronic intracluster collective oscillation excitation leads to a Föster excitation transfer from the Ag cluster to pyridine. For the tetrahedral Ag_{20} cluster, there are two binding sites (S-complex and V complex, see Fig. 11.9i(a), (b)). The S-complex is featured as on-top binding to the tetrahedral face, which is the (1 1 1) surface

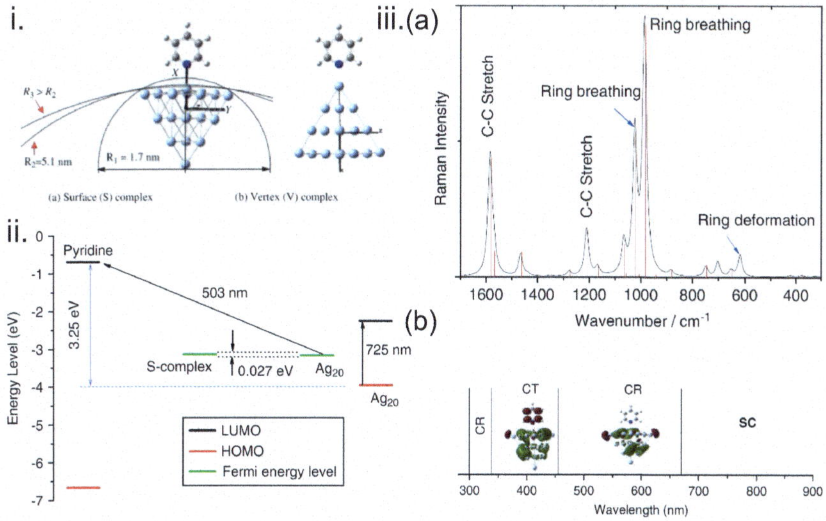

Fig. 11.9 **Chemical mechanism of TERS** [49]. **i. a** and **b** Configurations of the two pyridine–Ag_{20} complexes having C_S symmetry. **ii.** The Fermi energy level and energy levels of HOMO and LUMO for Ag_{20} cluster, the Fermi energy level of S-complex, and the energy levels of pyridine. **iii. a** The simulated SERS spectrum of pyridine adsorbed on large (S-complex) silver nanoparticles. **b** The chemical and electromagnetic field enhancement mechanisms at different wavelength regions. The two inserts are the charge difference densities for the intracluster excitation and CT excitation (charge transfer from Ag_{20} cluster to pyridine), respectively, where the green and the red stand for hole and electron, respectively. Adapted with permission from author of [49]

of face centered close packing (fcc) silver crystal. The tetrahedral Ag_{20} can well mimic a part of large nanoparticles, and that is probably the reason why the S-complex is normally used for SERS/TERS calculations for molecules absorbed on metal nanoparticles. The Fermi energy level and the energy levels of HOMO and LUMO for the Ag_{20} cluster can be seen from Fig. 11.9ii. The charge CR within the Ag_{20} cluster on the electronic intracluster collective oscillation excitation (EICOE) results in a Förster excitation transfer from the Ag_{20} cluster to pyridine. According to Fig. 11.9ii, the minimum energy of EICOE is 725 nm for the isolated Ag_{20} cluster. Because of the coupling interaction between pyridine and the Ag_{20} clusters (CT within the complex), the Fermi energy level of the S-complex is slightly increased by 0.027 eV. The minimum energy of the incident light is ~503 nm for the CT resonant transitions. The CT resonant transition with large oscillator strength is at ~380 nm. Due to the coupling interaction between pyridine and the Ag_{20} clusters, the CT of the complexes results in a large static electronic polarizability (see Table 11.1 in [49]). The Raman spectrum at zero frequency can be gathered from Fig. 11.9iii(a), which is contributed by the static chemical (SC) enhancement. There are three strongly enhanced modes, 984 cm^{-1} (ring breathing with N moving toward silver), 1064 cm^{-1} (ring breathing) and 1582 cm^{-1} (C–C stretching with the α carbon next to nitrogen vibrating toward the silver cluster).

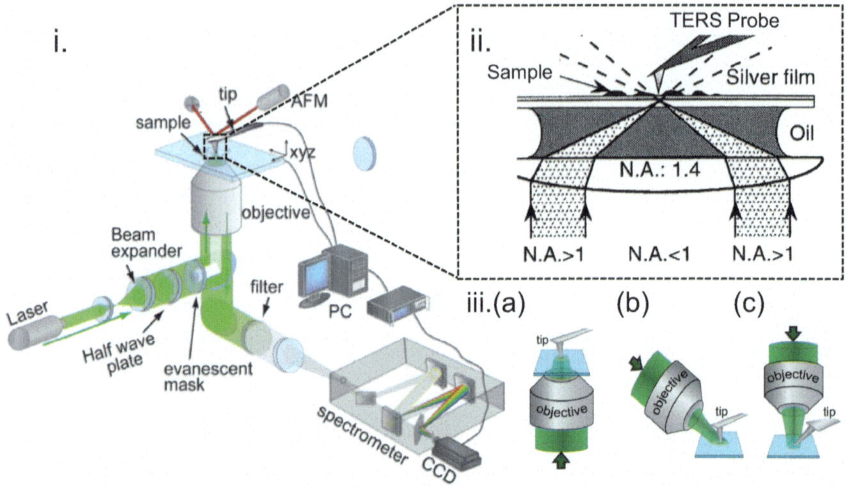

Fig. 11.10 **TERS Instrumentation** [8, 24]. **i.** Illustration of a typical AFM-based TERS measurement system that consists of a laser, a high-NA inverted objective, several optical and electronic components, a spectrometer, and a CCD detector. **ii.** **Evanescent field excitation geometry** [23]. A mask is used in the path of the expanded incident laser so that the NA < 1 component at the center of the beam is blocked and only high-NA component goes through the objective to create a total internal reflection at the sample substrate. This confirms that the incident light does not transmit through the substrate and the tip apex is illuminated only by the evanescent light, which drastically reduces the unwanted background signal from the propagating light. **iii.** **Illustration of the different possible illumination and detection geometries for TERS** [8]. **a** Standard bottom illumination with backscattering geometry. **b** Side illumination, often used for STM-based TERS, but can also be used for AFM-based TERS. **c** Top illumination, often used for AFM-based TERS with tilted tips. Adapted from [8, 23]

In summary, the spectrum enhancement mechanisms are illustrated in Fig. 11.9iii(b), where CT for chemical enhancement happens roughly in the region of 330–455 nm, and CR for EM enhancement roughly in the regions of 300–330 nm and 455–670 nm. The region of the SC enhancement is above 670 nm. It is interesting that the formal fragmented experimental studies on SERS/TERS of pyridine at different incident wavelengths can actually confirm our results in Fig. 11.9iii(b). Comparing the calculated Raman spectrum of pyridine in Fig. 11.9iii(a) with the experimental Raman spectrum at the incident light of 532 nm. Since SERS of pyridine at 532 nm is contributed by the EM mechanism in the enhancement spectrum of Fig. 11.9iii(b), and the EM enhancement should not change the Raman profile, the consistency between the experiment and the simulation can be well explained. Sun et al. also have developed a visualization method of CT for the chemical enhancement mechanism on SERS/TERS with various chemicals (Rhodamine 6G, pyrazine, ρ-aminothiophenol, etc.) and metal clusters (Ag_2, Al_3, Au_{20}, etc.). Readers can refer to the publications from Sun et al. for more information regarding the chemical enhancement mechanism of SERS/TERS [49–53].

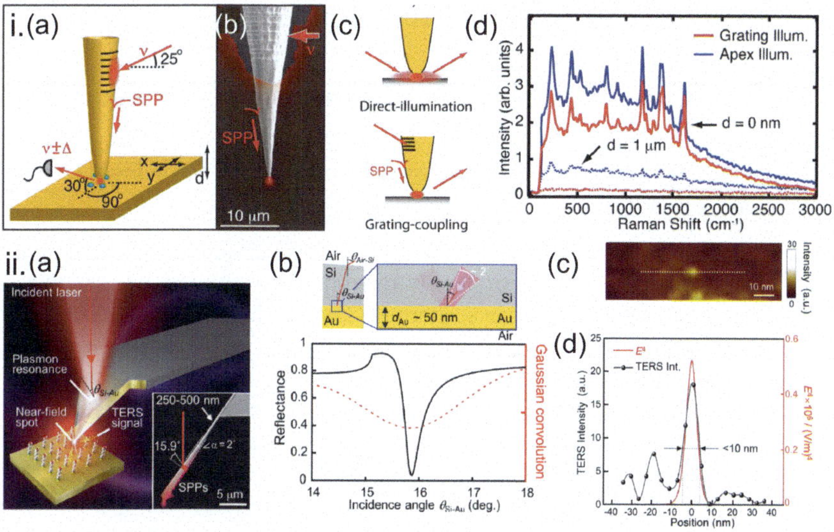

Fig. 11.11 TERS tips prepared with FIB and VPD. i. FIB [33]. **a** Schematic of the experiment. Incident light is focused onto the grating, and the Raman scattered light excited by the nanofocused SPP at the apex. **b** SEM micrograph of an electrochemically etched Au tip superimposed with the optical image of grating. **c** Illustration of grating-coupled and direct illumination TERS. **d** Corresponding TERS spectra taken with tip in contact (solid line) and tip retracted (dotted line) for grating-coupled TERS (red) compared with direct apex illumination (blue). Adapted with permission from author of [33]. Copyright (2010) (The Journal of Physical Chemistry Letters). **ii. VPD** [31]. **a** Working principle of the thin-film waveguide probe. The incident laser illuminates the waveguide at the plasmon resonance angle. Inset: SEM image showing the laser-excitation processes (red region), incident angle on the Si/Au interface, and thickness range of the thinnest part of the waveguide. $\alpha = 2°$ denotes the angle between the upper surface and lower surface of the Si layer after FIB cutting. **b** Illustration of an air/Si/Au/air four-layer system and its reflectance calculated with a 50 nm Au film as a function of the incident angle at a wavelength of 660 nm. **c** Part of a 200 × 200 nm^2 2D TERS map demonstrating spatial resolution. **d** Line profile (black dotted line) corresponding to the white line in **c** and $|\mathbf{E}|^4$ line profile (red line) for the gap between the tip apex and Au substrate. The FWHMs of both lines are <10 nm. For all measurements, the resolution was 128 × 128 pixels, and the line scan rate was 0.1 Hz. Adapted with permission from author of [31]. Copyright (2021) (Analytical Chemistry)

Table 11.1 TERS tips prepared by several strategic methods and enhancement factor

Preparation method	Materials	Enhancement factor ($\|E/E_0\|^4$)	Authors
FIB	Au	10^4	Berweger et al. [33]
Silver nanowire	Ag	1.1×10^6	Liu et al. [57]
Silver nansphere	Ag	2.7×10^3	Umakoshi et al. [61]
VPD	Au	$1.0 \times 10^6 \sim 10^7$	Zhang et al. [31]
VPD	Ag	5.5×10^3	Taguchi et al. [43]
ALD	Au/Ag	1.7×10^3	Huang et al. [59]
RF plasma	Ag	2.7×10^3	Capaccio et al. [60]

11.3 Experimental Setup and Techniques

11.3.1 TERS Instrumentation: Components and Procedure

The schematic illustration of the TERS experiment setup is shown in Fig. 11.10 with the following components:

a. **Light source**: Depending on the nature of the tips and sample, a laser source in the ultraviolet (UV), visible, or near-infrared range is selected, which serves as the excitation source for Raman scattering. Common lasers include diode lasers, solid-state lasers, or tunable lasers. If an Ag tip is used, a 532 nm laser is normally adopted. For Au tips, 633 nm lasers are often used as light source.
b. **Beam expander**: A beam expander is used to expand and collimate the laser beam. The entrance pupil (aperture) of the objective should be overfilled to satisfy the ideal illumination condition.
c. **Objective Lens**: An objective lens featuring a high numerical aperture (NA) plays a crucial role in focusing the laser excitation onto the sample and capturing the Raman scatter emitted by the sample. Theoretical lateral focusing resolution, adhering to the Abbe limit outlined in (11.12), is achievable and remains diffraction-limited. The utilization of a high NA is imperative for maximizing the efficiency of Raman signal collection. Different optical geometries, namely bottom, side, and top illumination (as depicted in Fig. 11.10iii), are adopted to address challenges posed by various sample types.

Figure 11.10iii(a) illustrates bottom illumination, also known as the transmission illumination mode. In this configuration, light is directed from beneath the sample through a transparent substrate onto the sample tip. The tip contacts the sample and interacts either directly with the laser focus or its evanescent field (as shown in Fig. 11.10ii). Often, high refractive index glass, such as #1.5H cover glass, is used as a sample support, though advanced support layers like thin gold films can provide additional field enhancement through surface plasmon polaritons (SPP). The use of a glass support necessitates the application of refractive index matching oil (such as Carl Zeiss 518F immersion oil with $n = 1.518$) between the substrate and a high numerical aperture objective with an NA of 1.3–1.45. As mentioned earlier, a higher NA enables tighter focusing, resulting in more concentrated laser energy and a smaller illuminated background area.[14] Due to the wide variety and versatility of AFM instruments and confocal microscopes, bottom-illumination setups are well-suited for investigating biological samples

Two alternative TERS focusing and collection configurations are side illumination (Fig. 11.10iii(b)) and top illumination (Fig. 11.10iii(c)), which are also widely adopted to overcome the limitations associated with transparent samples in the bottom illumination approach. In these setups, a laser beam is focused onto the tip end using

[14] It is worth to stress in here that the laser energy is further confined by the TERS probe (see Fig. 11.8 and associated discussion).

long working distance objectives. A long working distance is necessary to reach both the tip and the sample. Consequently, the NA of the objectives falls within the range of 0.28–0.55. Due to the angle between the focal plane and the sample surface, the laser focus assumes a non-symmetric shape, further expanding to approximately 2–3 mm. To compensate for the increased spot size, the laser power used inside illumination setups often ranges between 5 and 10 mW. In contrast, laser power can be as low as 10 μW for bottom illumination schemes [54]. The collection capabilities of these systems are constrained by the limited angle of collection characteristic of low NA lenses. However, the advantage lies in the absence of restrictions in selecting sample support materials and the ability to work with opaque samples. Despite the lower NA, these systems can harness significant enhancement within the gap between two dipoles formed by the metal or metallized TERS tip and a mirror dipole on the metal surface (as seen in Fig. 11.7i and the associated discussion).

For top illumination, in order to avoid the shadow cast by the TERS probe, a tilted tip is essential, as illustrated in Fig. 11.10iii(c). This configuration combines the benefits of both side and bottom illumination. The NA falls between that of the other two techniques and can reach as high as 0.9. Focusing is achieved symmetrically, leading to a more predictable and tighter focus compared to side illumination. Top illumination merges this approach with a reflective geometry, facilitating work on both opaque and transparent samples. In this orientation, one of the major challenges is the positioning of the tip between the objective and the laser focus. Depending on the tip's size and location, it may obstruct a portion of the incident light during excitation and the signal detection process.

d. **Evanescent mask**: According to Sect. 11.2.2.1, for the EM enhancement mechanism the maximum filed enhancement factor is achieved with the longitudinal polarization of the metallic tip. Such longitudinal polarization is induced as the result of the $\boldsymbol{E}_z$ component of the illumination field. When the illumination field is focused with high NA objective on glass/air interface (#1.5H glass with n = 1.518 at 532 nm vacuum wavelength), the $\boldsymbol{E}_z$ component is highest in the form of evanescent waves,[15] which is the result NA > 1 (plane waves incident above the total internal reflection angle θ_c, see Fig. 11.4i(c)) part of the illumination. The NA < 1 (plane waves incident below the total internal reflection angle θ_c) part of the illumination corresponds to the transmitted light and does not contribute much to the $\boldsymbol{E}_z$ component at focus (see Fig. 11.4i(b)). Thus, in practice, it is ideal to have only NA > 1 illumination (also known as annular illumination) for the TERS application. This reduces the unwanted background signal from the transmitted propagating light and also allows localized illumination of the tip apex. One method to achieve such illumination is the use of evanescent mask as shown in Fig. 11.10i. The evanescent mask is actually a ring pupil, which blocks the center part (with NA < 1) of the illumination. The Kawata team utilized the evanescent mask for the first successful TERS experiment [24]. However, it is important to point out in here that the evanescent mask is not a mandatory

[15] More specifically, $\boldsymbol{E}_z$ component is highest with p-polarized illumination with NA ~ 1. For more detail, please see the Sect. 11.2 *Evanescent fields* from *Principles of Nano-Optics* [21].

component of a TERS setup. The evanescent wave can also be achieved by using a larger filling factor or a higher NA objective [21]. For example, in applications based on solid immersion lenses [55] with numerical apertures of 1.8–2 over 90% of the beam power is reflected at the interface.

e. **Atomic Force Microscope (AFM) or Scanning Probe Microscope (SPM)**: TERS is often performed in conjunction with AFM or SPM. The AFM/SPM is used in a feedback loop to maintain a constant tip-sample distance, typically in the range of a few nanometers. This precise control is essential for maintaining the near-field enhancement necessary for TERS. The tip serves as both the Raman excitation source and the near-field probe for collecting scattered light (see Sect. 2.2.1 of this Chapter). The tip is typically a sharp metal or dielectric structure that is coated with noble metals (such as gold, silver, platinum, etc.) to enhance the Raman signal through surface plasmon polaritons. In practice, the excitation light is first focused on the sample through a high-NA objective. Then the tip is brought in the tightly focused laser spot, a strong electrical field enhanced and confined nano-scale light source is created in the close vicinity of the tip apex. A more detailed discussion of the TERS probe will be given in the following Sect. 3.2 of this Chapter.
f. **Filter**: The Raman scattered light is collected by the same objective. However, much stronger excitation wavelength signal as a result of reflection from the glass/air interface and Rayleigh scattering from the sample is also collected by the same objective. Thus, a filter should be placed along the reflection optical path as an effort to eliminate this signal. A long pass filter is most commonly used, which only allows optical signal with wavelength longer than the desired wavelength to pass through. The adoption of long pass filters would lead to the loss of anti-stokes Raman signal. If both stokes- and anti-stokes Raman scattering are desired, a notch filter [56] which is designed to transmit most wavelengths with little intensity loss while attenuating light within a specific wavelength range (around the excitation wavelength) to a very low level should be used.
g. **Raman Spectrometer**: The scattered light passing through the filter is directed to a Raman spectrometer for the final spectral analysis. The spectrometer disperses the light onto a detector, typically a charge-coupled device (CCD) or a photomultiplier tube (PMT).

TERS measurement is completed in two steps

The acquisition of Raman signals is initiated by bringing the TERS probe in close proximity to the sample through vertical control of the SPM. During this stage, an amplified Raman signal is generated from a minute volume of the sample positioned directly beneath the tip (see Fig. 11.8a). This amplification is brought about by the excitation of a nanoscale light source produced at the apex of the tip, referred to as the near-field signal.[16] Concurrently, Raman signals are also produced from the entire focal area due to the incident light, termed as the far-field signal (the concepts

[16] It should be "total field" (near-field plus far-field, see Fig. 11.3i(a)) more precisely. However, due to historical reason, it is commonly just referred to as "near-field".

of near-field and far-field were given in Sect. 2.1). Consequently, the signal collected during the initial step encompasses a combination of both near-field and far-field Raman signals.

In the subsequent step, the tip is retracted vertically from the sample, typically by several hundred nanometers, and Raman signals from the same region of the sample are captured. In this phase, since the tip is distanced from the laser's focal point, there is no nanoscale light source generated at the tip's apex. Even when TIR of the incident light is not perfect and some light leaks through, creating a nanoscale light source at the tip's apex, it remains at a reasonable distance from the sample to excite Raman signals. As a result, the Raman signals collected in the second step exclusively comprise far-field scattering originating from the focus point of the incident light. Actually, the system is just a standard confocal Raman microscopy [3] with tip retracted. The comparison of Raman spectra with and without probe can be seen in Figs. 11.8e, 11.11i(d), and 11.13ii in this Chapter.

By subtracting the signal measured in the second step from that obtained in the first step, it becomes possible to isolate the pure near-field Raman scattered signal. These two steps are iterated for each scanning point, ultimately culminating in the construction of a TERS microscopy image composed of purely near-field Raman signals.

11.3.2 Tip Fabrication and Characterization

11.3.2.1 TERS Probe Fabrication Methods

The tip is the key part of a TERS setup due to its ability of Raman signal enhancement from nanoscale samples. According to the Sect. 11.2.2, EM enhancement is considered the primary enhancement mechanism of TERS. Thus, the electric field enhancement factor $\left(|\boldsymbol{E}/\boldsymbol{E}_0|^4\right)$ is the most important criterion f the tip (see Sect. 11.2.2.1). A high $\boldsymbol{E}$ -field enhancement factor makes it possible for the investigation of samples with nanometer size and low Raman scattering cross section. Several nanoscale fabrication technologies have been adopted for the preparation of TERS tips, such as focused ion beam (FIB) milling [33], single silver nanowire attachment [57], thermal evaporation [58], electrodeposition [58], vacuum vapor deposition (VPD) [43], atomic layer deposition (ALD) [59], radio frequency (RF) plasma treatment [60], etc. Some representative tip preparation technologies and associated parameters are summarized in Fig. 11.11 and Table 11.1.

In TERS, the limited excitation efficiencies of plasmonic nanostructures often result in concentrating only a minute fraction of the incident power at the tip. Unfortunately, the uncoupled light can still initiate Raman scattering processes on the sample, albeit with low spatial resolution. This not only generates background noise, diminishing the signal-to-noise ratio, but also heightens the risk of sample photodegradation, especially considering the common use of strong lasers to pump the TERS signal. Addressing these challenges, plasmonic nano-focusing probes have emerged as a promising solution for achieving low background-noise TERS acquisition. In these innovative designs, incident light is transformed into SPPs through grating couplers [33] (see Fig. 11.11i) or thin-film waveguides [31] (see Fig. 11.11ii). The generated SPPs travel towards the probe apex along the tapered probe, undergoing further compression due to conical geometries, enabling high-spatial-resolution near-field imaging. Importantly, by separating the SPP excitation region from the Raman acquisition region, these designs significantly reduce Raman noise originating from the uncoupled incident light. This advancement marks a substantial improvement in the efficiency and precision of TERS, mitigating background interference and enhancing the overall quality of spectroscopic data.

Fig. 11.11i(a) presents the schematic representation of a TERS tip that has been fabricated using FIB. In this specific setup, a nanoemitter is created through the propagation of SPP along the shaft of a noble metal tip (also depicted in Fig. 11.6i) [35, 36]. The emergence of propagation-induced adiabatic SPP focusing within the tip's apex region is a result of the continuous modification of the surface mode's dimensions. This approach offers the advantage of spatially separated SPP coupling and subsequent excitation of the probe apex with a field confinement reaching tens of nanometers across a broad spectral range, while maintaining high focusing efficiency (for more detail see Sect. 11.2.2.1). To launch SPP modes onto the shaft of monolithic Au tips, a grating-coupler was manufactured using FIB. The conical tips, incorporating a two-stage optical mode matching, effectively enable far-field SPP coupling, and the mechanism responsible for the adiabatic SPP field concentration within the tip apex introduces a unique optical antenna concept for efficient far-field transduction into nanoscale excitation. In Fig. 11.11i(b) a comparison is made between direct apex illumination TERS (in blue) and nano-focusing Raman (in red) with the tip both in contact with and retracted from a sample of Malachite Green dye on an Au substrate under resonant excitation at $\lambda = 632.8$ nm. Although similar near-field signal levels are observed for both grating illumination and direct apex illumination TERS, the background signal, its origin, and the resultant contrast exhibit fundamental differences. In the case of direct apex illumination, the residual background in the TERS signal when the tip is retracted arises from the elliptical diffraction-limited far-field focus, as described in Sect. 11.2.1 and (11.12). Conversely, for grating illumination using the same tip, as depicted in Fig. 11.11i(c), no far-field Raman background is observed due to the intrinsic nanometer-scale field confinement achieved through SPP propagation-induced nano-focusing at the apex. The capability for nonlocal generation of a nanoscale excitation source with true nano-focusing efficiency and intrinsic background suppression[17] holds promise for a wide range of spectroscopic techniques. Potential applications of remote excitation near-field spectroscopy beyond TERS range from s-SNOM implementations without the need for signal demodulation [37] to background-free near-field luminescence spectroscopy, offering full spatial and temporal control over optical fields on the nanoscale for nonlinear and ultrafast nano-spectroscopy. Sputtering is a vapor

[17] Background suppression can also be achieved via evanescent field excitation, for more detail see Fig. 11.10.

deposition process used in the manufacturing of thin films and coatings on various surfaces. It is commonly employed in the semiconductor industry, as well as in the production of solar cells, optical coatings, and other thin film applications. This technique involves the removal of material from a solid target (the source material) through the bombardment of the target by energetic ions, which are typically generated in a vacuum chamber. In a notable study by Zhang et al. [31], sputtering took center stage for the creation of innovative indirect-illumination TERS tips, leveraging a thin-film plasmon resonance waveguide. This pioneering probe design facilitated the physical separation of far-field excitation and TERS signal-generation regions.

Exceptional reproducibility was achieved through an uncomplicated automatic FIB fabrication process, a stable coating material, and a distinctive 2D SPP excitation method. The design featured a flat Si/Au 2D interface configured in a Kretschmann arrangement positioned above the tip apex to form a triangular thin-film waveguide (see Fig. 11.11ii(a)). During TERS measurements, the far-field excitation laser traverses the air/Si interface and focuses on the Si/Au interface at an incident angle θ, inducing surface plasmon resonance (SPR) at the Si/Au interface. SPPs accumulate at the tip apex, generating a spatially and energetically confined near-field spot crucial for low-background, high-resolution TERS measurements. The SEM image in Fig. 11.11ii(a) reveals that the thinnest portion of the waveguide measures approximately 250–500 nm in thickness. Figure 11.11ii(b) illustrates the calculated reflectance of the three-layer Si/Au/vacuum system. The solid line signifies reflectance at different focusing angles through analytical calculation, with the optimized plasmon resonance angle identified at $\sim$15.9° for a wavelength of 660 nm. Consequently, the excitation-laser-induced SPPs are efficiently generated resonantly. This fabrication method, utilizing a thin-film waveguide probe, offers substantial technical advantages compared to other existing FIB-assisted or SEM-assisted tip-processing approaches. The thin-film waveguide probe achieves efficient SPP excitation and propagation, powerful near-field spot generation, and a low-background signal. To assess the spatial resolution of the thin film waveguide probe, a 2D mapping of the Raman signal at $\sim$1600 cm^{-1} was conducted on a 4-PBT thin-film sample. A local TERS map over a small area, depicted in Fig. 11.11ii(c), was instrumental in estimating the spatial resolution. Figure 11.11ii(d) presents the cross-sectional line profile, as indicated in Fig. 11.11ii(c), with the line profiles corresponding to $|\mathbf{E}|^4$ for the gap between the tip apex and Au substrate. Both the cross-sectional line and $|\mathbf{E}|^4$ profiles exhibit full width at FWHM of <10 nm, defining the spatial resolution of the thin film waveguide probe. This spatial resolution notably surpasses the radius of curvature for the tip apex (30–50 nm), emphasizing that the local near-field spot size on the tip apex is the primary determinant of the spatial resolution in the TERS setup.

In comparison to gold, silver proves more advantageous for TERS experiments owing to its stronger plasmonic enhancement and lower fluorescence background. Crystalline silver nanowires (AgNWs), synthesized chemically, exhibit unique characteristics that render them particularly well-suited for TERS applications. These nanowires possess nanoscale field confinement, mechanical robustness, and low plasmonic loss, all attributable to their polytwined crystalline nature. Notably, silver nanowire SPPs can be easily excited through various methods, including prism or grating couplers, near-field coupling, as well as simpler approaches such as tip, defect, and nanoantenna scattering. This versatility positions them as excellent candidates for remote-excitation TERS, offering a convenient alternative to the grating-assisted nanofocusing technique. A study by Ma et al. [62] demonstrated the feasibility of high-resolution remote-excitation TERS (RE-TERS) imaging using AFM-mounted sharp-tip AgNW. As illustrated in Fig. 11.12a, the RE-TERS probe was installed on a commercial AFM-TERS system with modifications to enable polarization adjustment of the incident laser. To maintain the remote excitation condition, the NC needs to be at least several micrometers away from the NW tip (Fig. 11.12b), so that the tip sits well outside the laser focus ($\sim$1 μm × 0.9 μm). These AgNCs display high size monodispersity and have an edge length of $\sim$200 nm (Fig. 11.12c). Given an average of ~3 μm distance between the AgNC and the AgNW tip, the propagation loss of the SPPs was around 1.2 dB. The conical tip is tapered to an ultrasharp apex (tip radius $\sim$15 nm, Fig. 11.12d). Apart from background suppression, the RE-TERS can also produce stronger signal compared to the conventional direct-excitation (DE-TERS) configuration under the same incident power. In the RE-TERS experiment, the incident laser is focused on the AgNC, whereas in DE-TERS, the laser is focused directly on the AgNW tip (Fig. 11.12e–i). To ensure substrate and sample homogeneity, a self-assemble monolayer (SAM) of 4-aminothiophenol (4-ATP) on an ultrasmooth Ag thin film as the standard sample was used. Figure 11.12e, f illustrate the RE- and DE-TERS measurement configurations. With both s- and p-polarized incident light, the background ($I_{retracted}$) is significantly reduced in RE-TERS, with a background reduction ratio ($I_{retracted,RE}/I_{retracted,DE}$) of 0.26 and 0.52, respectively. Similarly, for both polarizations, the enhancements of the TERS signal in RE opposed to DE excitation were observed, featuring a signal enhancement ratio ($I_{engaged,RE}/I_{engaged,DE}$) of 12.2 (s-pol) and 3.3 (p-pol).

In a study by Liu et al. [57], a realiable TERS probe fabrication technique was introduced. This method utilizes a micromanipulator to precisely attach a sharp-tip, surfactant-free AgNW onto a tungsten probe with micrometer-level accuracy. The contrast values for regular probes and sharp-tip probes are 2.7 and 7.7, respectively. All TERS spectra exhibit intense spectral features that correspond to well-defined graphene phonon modes, particularly the G-band at approximately 1580 cm^{-1} and the 2D peak at around 2680 cm^{-1}. The Raman enhancement factor, calculated using (11.27), is 1.1×10^6 for the sharp-tip AgNW probe (with Raman intensity contrast = 7.7), representing a substantial 66-fold increase compared to the regular AgNW probe ($\sim 1.6 \times 10^4$, Raman intensity contrast = 2.7). Since Raman intensity is proportional to the fourth power of the electric field $\left(|\boldsymbol{E}|^2\right)$, the Raman enhancement factor is indicative of electric field enhancement. This results in an electric field enhancement of approximately 32 for the sharp-tip AgNW and $\sim$11 for the regular AgNW. These promising findings suggest that the micromanipulation method for assembling sharp-tip AgNWs onto metallic tips, for STM-TERS applications, is a reliable fabrication technique with the potential to significantly simplify near-field Raman imaging research.

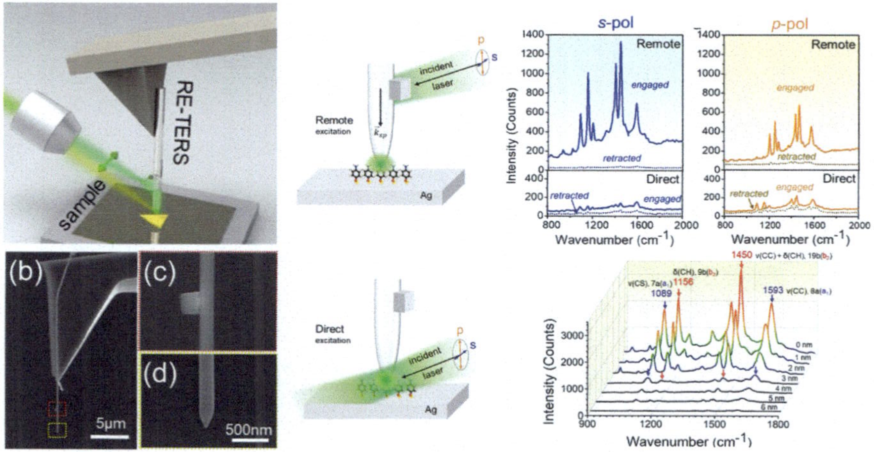

Fig. 11.12 Comparison between RE-TERS with DE-TERS and dependence of RE-TERS signals on tip-to-substrate distance [62]. **a** Schematic illustration of the RE-TERS setup. The green excitation laser beam (532 nm) is focused in on an AgNC to excite SPPs on the AgNW waveguide. The SPPs propagate toward the tapered tip to excite TERS signals, which are collected through the same objective lens. **b** A SEM image of a RE-TERS probe. **c**, **d** Close-up SEM images of the AgNC–AgNW junction **c** and sharp AgNW tip (**d**). **e**, **f** Schematics showing the remote **e** and direct, **f** excitation methods. In both cases, s- and p-polarizations were examined on a self-assembled monolayer of 4-ATP on an ultrasmooth Ag substrate. **g**, **h** 4-ATP Raman spectra measured with RE- and DE-TERS with incident lasers with s- **g** and p- **h** polarizations in both engaged (light color, solid lines) and retracted (dark color, dashed lines) modes. **i** As the tip gets closer to the 4-ATP SAM sample, the overall signal increases drastically (s-polarization). At large tip-substrate distances, the peaks marked by blue arrows (1089 and 1593 cm^{-1}) have the highest intensities, consistent with regular SERS spectra, whereas at smaller distances, the peaks marked by the red arrows (1156 and 1450 cm^{-1}) become dominating. Excitation: 532 nm, 0.2 mW at sample surface. Integration time = 2 s for **g**, **h** and 4 s for (**i**). Adapted from [62]. Copyright (2019) (Nano Letters)

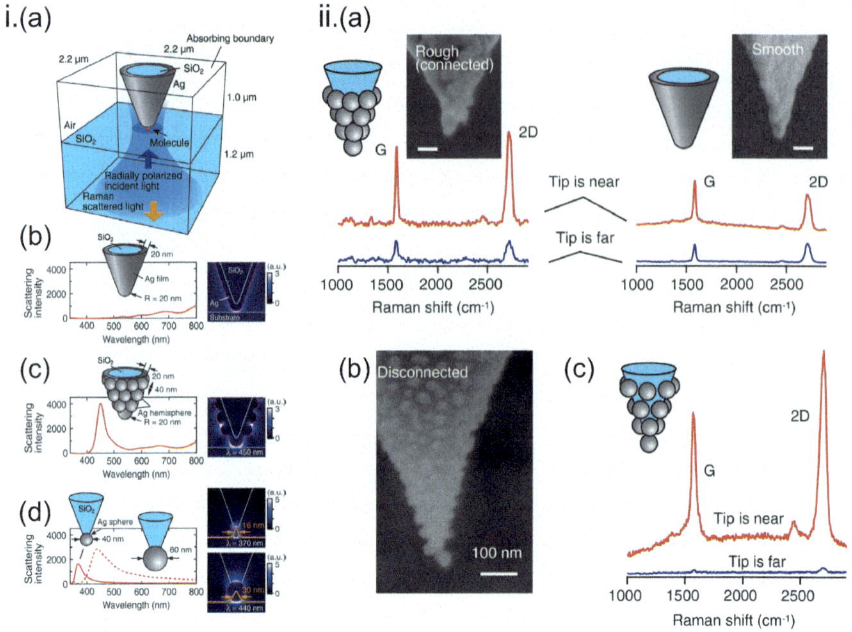

Fig. 11.13 **i. Simulated scattering spectra of Ag TERS probes with various shapes** [43]. **a** Model of TERS system used in simulations. **b** Smooth Ag film on SiO_2 cone. **c** Rough corrugated Ag film on SiO_2 cone. **d** Single Ag particle with a diameter of 40 nm (solid line) and 80 nm (dashed line) attached on SiO_2 cone. The physical parameters of the tips are shown in the insets. **ii. Experiment comparison of Ag TERS probes with various shapes. a** Raman spectra of graphene sheets scattered with Ag tip having rough (left panel) and smooth (right panel) surface. The spectra are vertically shifted for clarity. The insets show scanning electron micrograph (SEM) of the tip used. The scale bar is 100 nm. **b** SEM of disconnected Ag grains coated on Si cone probe [70]. The scale bar is 300 nm. **c** Raman spectra of a graphene sheet measured with the disconnected multiple grain tip. The spectra are vertically shifted for clarity. Adapted with the permission of author from [40]

Atomic Layer Deposition is a highly precise thin film deposition technique characterized by its self-limiting nature, making it versatile with respect to substrate material and morphology. This method allows for the meticulous control of deposition thickness down to the subnanometer or even atomic scale through deposition cycles. In their work, Huang and colleagues [59] harnessed the ALD process for the fabrication of TERS tips. They developed a universal ALD approach for crafting shell-isolated tips made of Au or Ag, which are designed to enhance the capabilities of tip-enhanced Raman and fluorescence spectroscopy. These tips are engineered with various types of shells, including silica, aluminum oxide, and titanium oxide, with precise control over shell thickness ranging from 1 to 20 nm. The result are tips that exhibit robust EM enhancement and offer significant advantages for TERS when applied in solution environments. To create these shell-isolated Au and Ag tips, the process involved an initial step of fabricating the tips through electrochemical etching. Subsequently, shells were deposited onto these tips using ALD. In TERS experiments, these tips were employed to investigate the detection of 4-PBT assembled on an Au (111) substrate in a solution containing pyridine. When using the bare Au tip (without the 2 nm silica shell), the TERS spectrum exhibited Raman signals from both 4′-(pyridin-4-yl)biphenyl-4-yl)methanethiol (4-PBT) and pyridine, with characteristic bands at 1185, 1286, and 1603 cm^{-1} and 1009 and 1034 cm^{-1}, respectively. This observation indicated that pyridine or other impurities in the solution could readily adsorb onto the bare tip, resulting in unwanted Raman signals from species besides the analyte on the substrate. In stark contrast, when utilizing the Au tip coated with a 2 nm silica shell, only the Raman signals of 4-PBT were evident. These findings underscore the efficacy of the ultrathin silica shell in preventing interference from impurities by inhibiting their interaction with the tip. Consequently, it facilitates the acquisition of pristine data pertaining to the analyte of interest. Therefore, these shell-isolated tips are well-suited for TERS investigations in complex solution environments, such as live cells, without being hindered by impurities adsorbing on the tip.

Although enhancement factor is the most important standard for the quality of the tip. The reproducibility of the preparation method is also very vital as the fluctuation of the enhanced signal makes it very hard to quantify the experiment result when several tips were used. The FIB [33], single nanowire attachment [57], and ALD [59] are suitable tip preparation technologies if constant tip performance is desired. What is more, in practice parts of the prepared tips are not TERS-active and cannot be used. Thus, the yield is also an important standard of the preparation method. VPD [43] and ALD [59] methods are ideal for high yield TERS tip fabrication. So far, there is no method combined with excellent enhancement factor, good reproducibility, and high yield at the same time.

11.3.2.2 TERS Probe Characterization

The characterization of tip performance is a crucial and intriguing subject. However, it is challenging to achieve this solely through experimental results due to the presence of numerous uncontrollable variables, such as sample properties, optical aberrations, and environmental fluctuations. Consequently, numerical simulations, with controlled variables, offer valuable insights into the prediction of TERS tip behavior. The Kawata research team identified an intriguing experimental phenomenon where the Raman signal exhibited a significant reduction when a smoother metallic layer was used. Building upon this observation, they conducted both simulations and experiments to investigate the relationship between surface morphology and the scattering efficiency of the TERS probe [43]. In contrast to previous numerical studies that focused on smooth conical tips [63, 64], the Kawata team delved into the role of metal grains in the plasmonic properties of a probe fabricated through VPD method. Surprisingly, the plasmon resonance properties of a tip covered with granularly structured metal had not been thoroughly explored in existing literature, despite the persistent hypothesis that metal grains within a metallized probe are significant factors [65]. Indeed, beyond factors like grain size and inter-granule spacing, the number of metal grains near the tip apex exerts a profound influence on the plasmon resonance properties of the tip. The arrangement of these metal grains with nanometer-scale separations is analogous to plasmonic nanoparticle chains previously studied by some researchers. For instance, Maier et al. [66, 67] explored coupled plasmon modes in nanoparticle chain waveguides and discussed how plasmon resonance is dependent on the number of particles and their distances from one another. In a prior study, Kawata et al. [68] examined the propagation of near-field light through stacked silver nanorods, demonstrating remarkable broadening in plasmon resonance, which has potential applications in multicolor image transfer at sub-wavelength spatial resolutions. Fan et al. [69] also showcased various plasmon modes, including magnetic and Fano-like resonances, using assembled clusters of plasmonic spheres. Consequently, one would anticipate a significant impact of grain arrangement near the tip apex on plasmon resonance, influencing parameters such as resonance wavelength, broadening, and enhancement. The key discovery made by the Kawata team is that Raman enhancement reaches its maximum when there's an optimal number of grains on a plasmonic tip [43]. Moreover, the nanoscale gaps between these grains play a pivotal role in tuning plasmon resonance and increasing the enhancement. This report is the first to comprehensively investigate the enhancement and spectral properties of plasmons at the tip's end concerning the number of grains and their arrangement. In an attempt to validate this hypothesis, the efficacy of the multi-grain structure in TERS was experimentally demonstrated through imaging carbon nanotubes and graphene.

The Kawata team adopts VPD method for the fabrication of Ag TERS probes, the role of metal grains on the plasmonic properties of the probe is thoroughly studied (see Fig. 11.13). Figure 11.13i(a) shows the model of tip-sample-substrate configuration used in the finite element simulation (COMSOL). The TERS probe is composed of SiO_2 conical core coated with Ag. The diameter and cone angle of the SiO_2 core are 10 nm and 40°, which are very close to the actual physical model. This substrate also composed of SiO_2 is located 4 nm beneath the tip. A radial polarized light is focused through the substrate by a NA 1.4 virtual objective.[18] The source covers the spectrum range between 300 and 800 nm. The tightly focused radially polarized light produces a field polarized parallel to the tip axis ($\boldsymbol{E}_{zz}$) at its focus [46], which facilitates the excitation of a longitudinal mode of plasmon along the tip axis.[19] As a result, strong Rayleigh scattering (Fig. 11.1) will be produced and detected by the monitor in the form of propagation wave (see Fig. 11.3). At the same time, the near-field component of the field (see Sect. 11.2.1, Figs. 11.3 and 11.4) will excite the low-order eigenmodes at the tip apex (see Fig. 11.6), which are mainly responsible for the excitation of the Raman scattering of the sample located between the tip and substrate. The Raman scattering is then rescattered by the tip apex and detected by the detector in the form of far-field projection (TERS).

---

[18] The beam profile ($|\boldsymbol{E}_0|^2$) of NA = 1.4 focused 532 nm wavelength light at glass/air interface is shown in Fig. 11.8.

[19] More precisely, the NA > 1 component of the focused light mainly contributes the excitation, see Sect. 11.3.1 and Fig. 11.10.

Figure 11.13i(b)–(d) shows the scattering spectra (300–800 nm) detected in the far-field with different metallic structures. The spectra are normalized to the situation without a tip $\left(|\boldsymbol{E}_0|^2\right)$, which can also be understood as the situation when the tip is far away from the substrate. When the tip is composed of a smooth film (Fig. 11.13i(b)), no pronounced resonance peaks can be seen in the visible spectrum, which indicates that a smooth infinite long tip does not act as a plasmonic antenna in the visible range. However, the field between the tip and substrate is still enhanced, as depicted in Fig. 11.13i(b). Such a phenomenon is attributed to the lighting rod effect, which describes the field enhancement following a strong potential gradient due to the curvature of a metallic interface [71]. Due to the non-resonant nature of the lighting rod effect, the enhancement factor is only weakly wavelength dependent, which can explain the good reproducibility enhancement in TERS using smooth Ag layer by ALD method [59]. Significant advancements have been observed in STM–TERS, primarily attributed to the crucial role of a metallic substrate in regulating tunneling current within STM. This mechanism generates what is referred to as a 'gap-mode' [72] in proximity to the tip's apex, effectively confining light within the gap between the tip and the substrate. However, the weak wavelength dependent nature of such tips makes it not possible to tune or improve plasmon resonance. When the tip is decorated with hemispherical Ag grains with 20 nm radius (see Fig. 11.13i(c)), a strong peak at 450 nm can be seen in the scattering spectrum. The enhancement factor is estimated around 4000 in intensity $\left(|\boldsymbol{E}/\boldsymbol{E}_0|^2\right)$. The enhancement mechanism in this case can be explained by the optical antenna theory [73]. The Ag grains function as resonant dipole antennas which receive/reradiate EM signal between far-field and near-field. The resonance wavelength of an optical antenna is primarily dictated by its size. Unlike radio wave or microwave antennas, which have dimensions that are integral multiples of half the signal wavelength, the length of an optical antenna is significantly shorter than the wavelength of light. This is because the wavelength of plasmon polaritons excited in the visible spectrum is much shorter than that of light propagating through free space.

When the tip is featured with a single Ag nanoparticle (see Fig. 11.13i(d)), the resonance can still be observed in the scattering spectrum, but much weaker compared to the one with many Ag grains (Fig. 11.13i(c)). This result indicates that a single metal particle can still be used for field enhancement, which is supported by the study of Kalkbrenner et al. [74, 75], who used a single gold nanoparticle mounted on the apex of an AFM probe. Figure 11.13i(d) also shows that the resonance peak can be tuned by the size of the Ag particle, which can be explained by the Mie scattering theory [38] or optical antenna theory [73].

In this study, it is indeed possible to use a single plasmonic particle for field enhancement. Surprisingly, it also reveals that a solitary particle doesn't yield the optimal enhancement. The resonance peak for the 40 nm particle, matching the size of the particles in Fig. 11.13i(c), is observed at 370 nm, exhibiting a significant blue shift compared to the probe with a rough corrugated film (Fig. 11.13i(c)). Since the probe exhibits virtually no enhancement in the visible range, it is clear that a single 40 nm Ag sphere is not an efficient tool for TERS with visible excitation. Conversely, a spherical particle with an 80 nm diameter demonstrates a more pronounced enhancement, primarily centered around 440 nm in the visible spectrum. Notably, this resonance peak around 440 nm has been previously observed in experiments involving Ag particles of similar size (approximately 100 nm in diameter) fabricated at the apex of a SiO_2 tip [76]. However, it is essential to acknowledge that this enhancement is comparatively weaker than what is observed in Fig. 11.13i(c). Furthermore, the larger size of the particle disrupts the confinement of light and compromises spatial resolution, as evidenced by the cross-sectional line profiles of field intensities between the tips and substrates in the right panel of Fig. 11.13i(d).

As effort to verify the simulation result, the Kawata team also performed some experiments. As shown in Fig. 11.13ii, the Raman spectrum of a graphene sample was measured with (TERS) and without (confocal Raman) the Ag probe. The SEM image of the probe used in the experiment is shown in the inset. The spectrum in red shows a Raman spectrum when the probe tip was near the sample, while that in blue shows the Raman signal when the probe tip was far from the sample. With a connected multiple grain tip (Fig. 11.13ii(a)), the Raman signal intensities at G-band (1590 cm^{-1}) and 2D-band (2700 cm^{-1}) were increased with the enhancement factors of around 880 and 1100, respectively. The enhancement at G-band is weaker than that at 2D-band, which can be explained by the spectrally narrow resonance of the connected grain probe [43]. Figure 11.13ii(a) also shows the result of the same experiment but with a smooth probe. In this case, both G- and 2D-bands are only weakly enhanced with an enhancement factor of only 230. This experiment observation is consistent with the simulation result shown in Fig. 11.13i(b). Based on the results shown in Fig. 11.13ii, one can conclude that the roughness of metal is crucial for the enhancement factor in TERS. With a disconnected multiple grain tip (Fig. 11.13ii(b)), the Raman intensities of G- and 2D-bands were found to be increased by the enhancement factors of around 5500, which can be explained by the spectrally broad resonance of the disconnected grain probe [43].

In summary, the principal conclusion reached is that a collection of grains functions as an effective plasmonic antenna for both receiving and transmitting light at the nanoscale. Even a single particle can perform this function, but the efficiency significantly improves with the presence of multiple grains. The degree of enhancement is directly related to the number of grains present, indicating that there exists an optimal quantity for maximum efficiency. Moreover, it is anticipated that higher levels of enhancement can be achieved by employing a probe featuring grains of uniform size and identical antenna resonances. This can be achieved, for instance, by affixing monodisperse silver nanoparticles onto the surface of a SiO_2 cone through surface modification techniques. Additionally, creating a probe with the optimal number of grains represents a forthcoming technical challenge. The multi-grain tip structure under investigation holds substantial promise as an efficient TERS probe for molecular imaging and analysis at the nanoscale.

11.4 Applications of TERS

11.4.1 Chemical Reaction Monitoring

Chemical identification by optical means down to single-molecule sensitivity is a challenging task and usually requires large enhancements of the local fields acting on the molecule. TERS leverages the advanced imaging capabilities of sSNOM, extending beyond the limits of diffraction (for a comprehensive explanation, please refer to Sect. 11.2.1). This remarkable technology is a formidable analytical instrument, seamlessly fusing the exceptional spatial precision of imaging with the chemical selectivity of Raman spectroscopy.

TERS has solidified its position as an invaluable asset for scrutinizing nanoscale chemical reactions, endowing us with unparalleled insights into molecular mechanisms with exceptional precision [78–82].

Currently, heterogeneous catalysts play a pivotal role in the chemical industry, but acquiring molecular insights into functioning catalysts remains a significant challenge. Lantman et al. utilized TERS to inspect the catalytic processes at the nanoscale [77]. In this work, a monlayer of p-nitrothiophenol (pNTP) molecules was first assembled on gold nanoplates on glass coverslip (Fig. 11.14a). The silver-coated (by thermal evaporation) Si tips were used as TERS probes and 532 or 633 nm laser were used as excitation source. According to previous study, the pNTP molecules can be actived on a SERS silver substrate with 532 nm illumination [83] as a result of charge-transfer mechanism (see Sect. 11.2.2.2 in this Chapter). New peaks at 1390 and 1440 cm^{-1} bands can be attributed to either p-aminothiophenol (pATP) or p,p′-dimercaptoazobisbenzene (DMAB). From Fig. 11.14b, c, SERS measurements using 532 nm laser excitation clearly show a spectral change over time. After ~120s of laser illumination, the most prominent peak of the pNTP at 1335 cm^{-1} decreases, while DMAB related peaks at 1140, 1380 and 1,440 cm^{-1} increase in intensity, indicating the (partial) conversion of pNTP to DMAB. SERS measurements can yield valuable information on a "bulk" scale, but are limited in studies near single-molecule levels. With TERS, the observation area is well below the diffraction limit (see Sect. 11.2.2.1 in this Chapter). As the TERS activity is localized at the tip apex, which is also the catalytic particle in this arrangement, the system is flexible with respect to the material under study and position on the sample. To limit the effect of the green laser excitation and thus the extent of the reaction, only low power (0.7 mW) was used at 532 nm. To be able to link any observations on a molecular scale to the photoreaction, it is crucial to have a stable starting point. This means that a continuous monolayer is required. Thus, before starting the reaction, the TERS signal was monitored at 633 nm excitation for at least 1 min. Subsequently, the irradiation source was switched to 532 nm for a given time interval (for example, 30 s), then immediately switched back to 633 nm to follow the reaction as function of time. The results of this novel TERS mode are shown in Fig. 11.14d. Irradiating with green light changes the stable and complete pNTP monolayer (top) into a monolayer in which multiple spectral components are present. A plot of the relative peak areas of pNTP

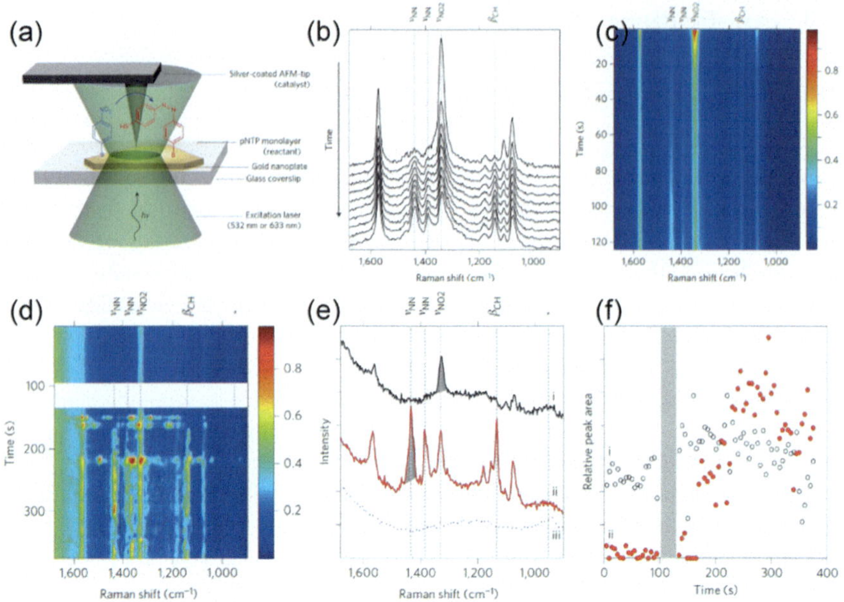

Fig. 11.14 Time-dependent TERS measurements on a gold nanoplate substrate [77]. **a** A monolayer of pNTP is assembled on a sample of gold nanoplates on glass. This sample is placed on a combination of an inverted Raman microscope and an AFM. Excitation and collection of the spectra occurs from below (bottom illumination).[20] **b** Time-dependent SERS measurements of the reduction of pNTP (top) to DMAB (bottom, spectrum mixed with that of pNTP). **c** Same series as in a, but shown as a colour-coded intensity map. **d** Time-dependent TERS spectra at 633 nm excitation shown before (top) and after (below white band) illumination. **e** Two spectra from **a** are shown. Asterisks in **a** and **b** indicate the location of the 950 cm^{-1} band of the SiO_2–glass signal of the glass substrate. **f** Peak areas as a function of time for the pNTP band at 1,335 cm^{-1} (i) and for the band at 1,440 cm^{-1} (ii), belonging to DMAB. Adapted with permission from author of [77]. Copyright (2012) (Nature nanotechnology)

and DMAB (Fig. 11.14c, d) confirms the catalytic activity by the appearance of the DMAB peaks after exposure to the green light. The simultaneously induced intensity fluctuations, show that the photoreaction upsets the monolayer stability. The change in monolayer structure can be explained directly by the reaction itself—either the change in functionality from NO_2 to NH_2 or even dimerization. Both the fluctuations and increase in DMAB content are reproducible observations. In this report [77], time-resolved TERS is proved to be able to monitor photocatalytic reactions at the nanoscale. The silver coated atomic force microscope tip to both enhance the Raman signal and to act as the catalyst.

While TERS has been employed extensively to investigate chemical adsorptions [77, 84] and related reactions [85–87] on molecular monolayers, it remains a formidable challenge to explore bond-breaking and bond-making processes, as

[20] See Fig. 11.10iii(a) and associated discussion.

well as the resulting structural changes within a well-defined single molecule at the single-bond level.

Additionally, there is an unresolved but crucial issue concerning the vertical resolving power of TERS and its ability to identify the chemical constituents at the base of an upright molecule. The primary limitation arises from the spatial extent of the AFM-TERS local plasmonic field, which typically spans only 5–10 nm (see Fig. 11.8). To overcome this limitation, the fusion of scanning tunneling microscopy with TERS, known as STM-TERS [80, 89, 90], allows for the acquisition of vibrational information with subnanometer spatial precision. STM-TERS can approach subatomic distances from molecules, achieving spatial resolutions at the angstrom scale, thereby facilitating the visualization of individual vibrational modes. This remarkable capability of STM-TERS is attributed to the significantly enhanced Raman signals within the plasmonic nanocavity [90, 91]. The primary contributor to this enhancement is the electromagnetic enhancement effect resulting from localized plasmon excitations near the metal nanostructures.

Wang et al. [88] utilized STM-TERS to inspect the chemical reaction, the bond breaking and forming of a up-standing melamine molecule absorbed on the Cu surface at the single-bond level. In this work, melamine molecules were first evaporated on the Cu(100) surface, followed by 6.2 K cooling. The electrochemically etched Ag tips were employed as TERS probes and a 532 nm laser was used for excitation. Figure 11.15a shows the schematic diagram of TERS measurements on a single upstanding melamine model molecule chemisorbed on Cu(100). As shown in Fig. 11.15c, d, the line-trace TERS over the melamine molecule were measured. From Fig. 11.15c, the symmetry vibration mode of N–H bonds (up in Fig. 11.15d) at 3453 cm^{-1} exhibits high intensity at the geometry center of the molecule, while the antisymmetric vibration mode (below in Fig. 11.15d) shows low intensity at the same position. Figure 11.15c reveals a spatial resolution down to around 2 Å at the single chemical-bond level. This phenomenon is expected for an up-standing structural model with two N–H bonds located at the top according to the interference effect. As the TERS measurement is performed under laser illumination, the photochemical reaction of the dehydrogenated molecule, i.e., the breakage of one N–H bond during the transformation into its tautomer by hydrogen transfer, can be tracked by the TERS spectral evolution. As is depicted in Fig. 11.15b, the disappearance of the symmetric (3453 cm^{-1}) and the antisymmetric (3572 cm^{-1}) vibration modes of N–H bonds indictase the breaking of single N–H bond. A red shift of the symmetric mode Raman peak to 3290 cm^{-1} demonstrates the weakened remaining N–H bonds. The newly emerged Raman peak at 3308 cm^{-1} indicates the formation of N–H bond in the cyanuric ring. The flat-lying melamine molecules on Cu surface cannot be observed as a result of the TERS selection rule. This work demonstrates the vertical resolving ability of subnanometer resolved TERS, which enables tracking in situ of the bond breaking and making as well as related chemical structural changes of an up-standing single melamine molecule chemically adsorbed on the Cu(100) surface. By monitoring the spectral evolution including peak positions and intensities as well as the emergence of new Raman peaks, not only can one determine the upstanding adsorption configuration of the molecule on the surface, but also unravel the bond breaking at both the top and bottom as well as the bond making in the cyanuric ring involved in a photon-induced tautomerization reaction.

While electromagnetic enhancement accounts for a significant portion of the enhancement observed in TERS, it does not fully elucidate the overall enhancement seen in experimental results. The remaining enhancement is ascribed to chemical enhancement, with a notable focus on charge transfer (CT) enhancement, which is discussed in greater detail in Sect. 11.2.2.2 in this Chapter. Regarding the primary EM enhancement, it has been observed that the components of the enhanced electric field parallel to the tip ($\boldsymbol{E}_z$) are more pronounced than those perpendicular to the tip ($\boldsymbol{E}_x$ and $\boldsymbol{E}_y$) (for more details, see Sect. 11.2.2.1 and Fig. 11.7). As a consequence,

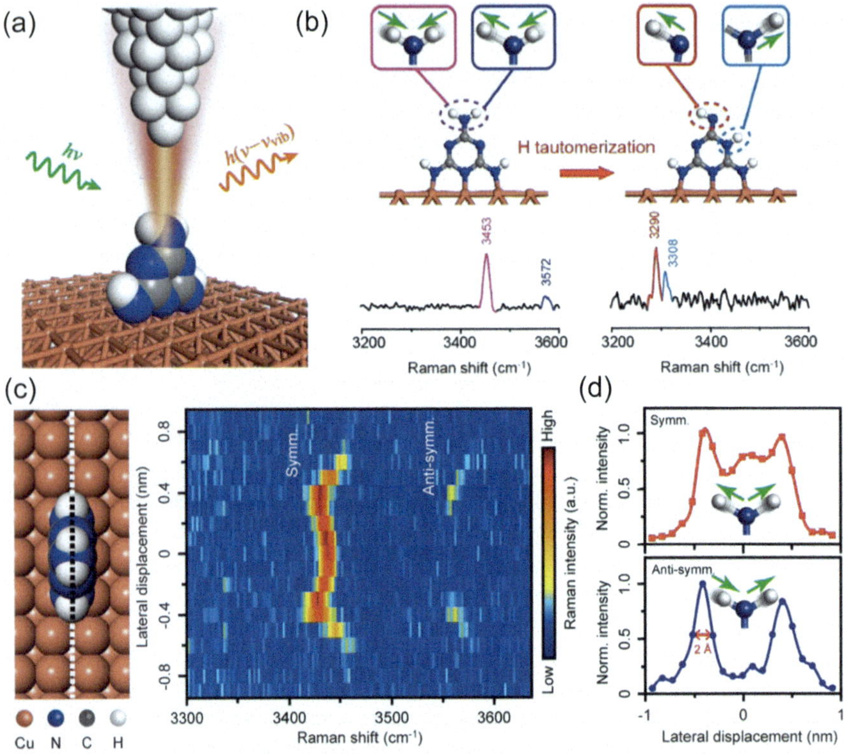

Fig. 11.15 TERS spectra of a single melamine molecule adsorbed on Cu(100) [88]. **a** Schematic diagram of TERS measurements. A strong and highly confined plasmonic field is generated at the Ag tip apex, which is responsible for the enhancement of the Raman signals from the single molecule. **b** Schematic illustration of TERS spectral variations upon the tautomerization of melamine. **c** Waterfall plot of 19 consecutive TERS spectra corresponding to the line-trace measurements indicated by the dotted line in the left panel. **d** Normalized Raman intensity profiles for the symmetric and antisymmetric stretching vibrations of N–H bonds at ~3434 cm^{-1} and ~3560 cm^{-1}, respectively. The spatial resolution of ~2 Å is labeled for the antisymmetric mode. Adapted with permission from author of [88]. Copyright (2021) (The Journal of Physical Chemistry Letters)

vibrational modes aligned parallel to the tip experience more substantial enhancement. This suggests that in order to harness the benefits of EM enhancement, either relatively small, planar molecules should adsorb in a tilted or perpendicular orientation to the substrate, or larger molecules should be explored, ensuring that their vibrational modes possess components perpendicular to the surface.

Jaculbia et al. [81] examined an isolated copper naphthalocyanine molecule adsorbed on a triple-layer NaCl substrate on Ag(111) using STM-TERS imaging. When the naphthalocyanine molecules lay flat on the NaCl surface, Raman images displayed distinct patterns depending on the vibrational mode's symmetry. The results underscore the significance of considering both perpendicular and parallel components of the electric field in relation to the STM tip.

Furthermore, it was observed that under resonance excitation conditions, the perpendicular electric field components can significantly contribute to the signal enhancement. This single-molecule study, conducted within a precisely defined environment, offers valuable insights into the Raman process within plasmonic nanocavities. Such insights have the potential to enhance the precision of nanoscale metrology for a wide range of molecular systems.

11.4.2 Low-Dimensional Materials

Low-dimensional materials, including single-walled carbon nanotubes (SWCNTs), graphene, transition metal dichalcogenide monolayers (TMDs) [92], and carbon nanomembranes (CNMs) [93], have emerged as some of the most extensively researched modern materials. They exhibit immense potential across a wide spectrum of applications, ranging from electronics and optoelectronics to photocatalysis and CO_2 reduction. These materials possess remarkable optical, electronic, and mechanical properties that distinguish them from their bulk counterparts. The distinctive physical properties of these materials are strongly influenced by local heterogeneities, such as edges, grain boundaries, defects, strain, doping, chemical bonding, and molecular orientation, owing to the strong dielectric screening effect. However, detecting this information can be challenging due to the weak signals emanating from very small areas on single or a few layers. As a result, there is a growing demand for techniques that can probe these heterogeneities and their impact on the physical properties of 2D semiconductors in a label-free and non-destructive manner. The introduction of TERS to low-dimensional materials not only facilitates a deeper understanding of the effects of local heterogeneities but also paves the way for novel quantum mechanical applications.[21]

[21] Due to the uniform properties and good stability, low-dimensional materials (especially graphene, SWCNT, TMDs) are ideal samples for evaluating the TERS enhancement factor ($|\boldsymbol{E}/\boldsymbol{E}_0|^4$) [40, 58, 87].

Two-dimensional (2D) transition metal dichalcogenides (TMDCs) have garnered significant attention as promising materials endowed with remarkable physicochemical properties. However, the fabrication process of these materials invariably introduces structural defects, albeit on a small scale, often at the nanometer level or even as singular atoms. Despite their diminutive size, these defects possess the capability to disrupt lattice symmetry, engender a quantum-confined environment, and alter the electronic band structures within TMDCs. Consequently, these defects play a pivotal role in imparting distinctive electronic properties to TMDCs, thereby influencing their electrical, optical, and chemical behaviors. In the work of Huang et al. [94] employed TERS, which can simultaneously obtain the morphologic and chemical fingerprint information with a nanometer spatial resolution and even single-molecule sensitivity assisted by the huge EM enhancement of the plasmonic Au or Ag tip, to probe the edge-related lattice structure and corresponding electronic properties of atomically thin MoS_2. The scheme to measure the TERS signals on edge defects in MoS_2 materials is shown in Fig. 11.16a. Monolayer (1 L) and bilayer (2 L) MoS_2 with different one-dimensional (1D) defects (1 L wrinkle, 1 L edge, 1–2 L step, and 2 L edge) transferred onto an atomically smooth Au substrate as the sample, and the AFM image of the sample was shown in Fig. 11.16b. With the observed distinctive spectral features related to the defects, the different Raman features at the edge and basal plane of MoS_2 were explored. Figure 11.16c shows the topographic height profile (left panel) and the corresponding color-coded intensity map of the line-trace TERS image (right panel) of the bilayer edge. The larger height of the bilayer MoS_2 (~1.8 nm) compared with the typical layer thickness (1.4–1.6 nm) is the result of a thin water layer existing between MoS_2 and the substrate, which can block the doping effect from the Au substrate. The line-trace TERS image of the edge matches well with the height profile. When the tip was positioned on the Au substrate, only the far-field Raman signal of the pristine bilayer MoS_2 was observed. To clearly identify the difference in the spectral features at the edge and basal plane, we present two representative Raman spectra in Fig. 11.16d. It is clear that distinctly different Raman features, including the two new Raman peaks at 220 and 396 cm^{-1}, show up at the edge and disappear on the basal plane, indicating a significant difference in physicochemical properties (such as the lattice structure and electronic property) between the edge and basal plane. As a result, transitional space forms between the edge and pristine MoS_2. The normalized intensities of two TERS peaks (396 and 406 cm^{-1}) as a function of tip position as the red and blue curves were plotted in Fig. 11.16e. The intensity becomes higher when the tip was moved closer to the edge, indicating a stronger double resonance Raman scattering (DRRS) involved band bending at the edge. The FWHM of the 396 cm^{-1} intensity profile is 7 nm. It should be pointed out that the measured TERS intensity profile is a convolution of the DRRS-involved band-bending region and the spatial resolution of TERS. Considering the typical spatial resolution of AFM-based TERS obtained in our lab of ~7 nm (also see Fig. 11.8), which is equal to the experimentally measured TERS intensity profile. Different from the 396 cm^{-1} peak, the A_{1g} peak (406 cm^{-1}) is sensitive to the electron density and its intensity becomes stronger when the tip is moved closer to the edge (Fig. 11.16e). The stronger signal at the edge is attributed to a lower electron density (Fig. 11.16f). Two factors may account for the different electron densities at the edge and basal plane. First, the Fermi level at the edge is higher than that on the basal plane, leading to a charge transfer from the edge to the basal plane. Second, the active edge site can be easily chemisorbed by O_2 or H_2O in air, which leads to the p-type doping induced by the oxygen species and a decreased electron density.

Robert et al. [95], combined TERS and photoinduced force microscopy (PiFM) synergistically and examined this instrumental approach on the 2D material molybdenum disulfide (MoS_2). The respective optical image, Raman, and TERS investigations of the MoS_2 monolayer flakes used throughout the experiments discussed here are shown in Fig. 11.17. As can be seen in Fig. 11.17i, the near field of the tightly focused (with NA = 1.25 objective) linearly polarized 532 nm laser spot was first mapped with PiFM. PiFM is a recent near-field optical technique based on the optical sample excitation and respective signal detection by sensing the resulting photoinduced force between tip and sample [96]. The magnitude of this force depends on the photoinduced molecular polarizability in the tip-sample junction (see Sect. 11.2.2.1

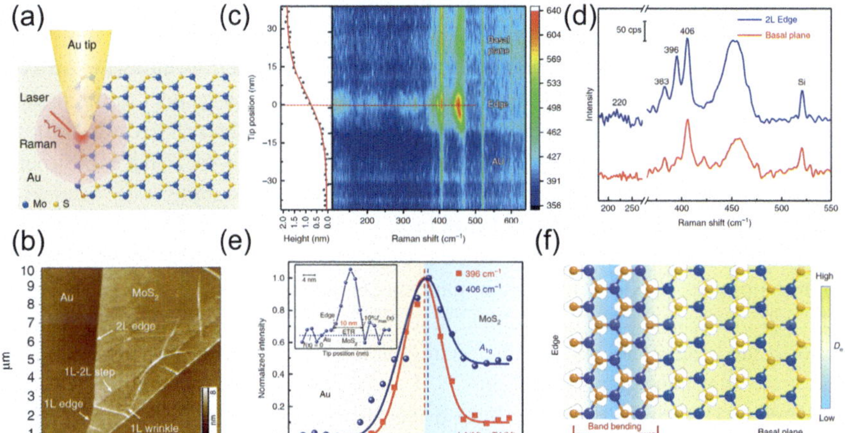

Fig. 11.16 TERS study of edge defects in the atomically thin MoS_2 [94]. **a** Schematic of a TERS configuration using an Au-coated AFM tip and an atomically smooth Au film with monolayer (1L) or bilayer (2L) MoS_2 on the surface. **b** AFM image of the mechanically exfoliated MoS2 with different types of 1D defect on an Au substrate. **c** TERS line image across the edge of a bilayer MoS_2. Left panel: topographic height profile of the edge of bilayer MoS_2. Right panel: a line-trace TERS image of the edge. Note that these are original TERS experimental spectra without any processing. **d** Two typical TERS spectra of a bilayer MoS_2 at the basal plane and the edge. **e** Plots of normalized intensities of two TERS peaks (396 and 406 cm^{-1}) with the tip position. The solid lines are the fitted results. Note that these spectra are the pure near-field signals and have been subtracted with the far-field signals. Inset is the enhanced TERS intensity profile of the ETR in the 2L edge after the deconvolution of the EM field the intensity. **f** Schematic diagram of band reconstruction and ETR of MoS_2 near the edge. De is the electron density. Adapted with permission from [94]. Copyright (2019) (Nature Communications)

in this Chapter). According to the theory [96], the overall measured force in PiFM is determined by the axial or longitude ($\boldsymbol{E}_z$) component of the electrical field, giving rise to the image contrast. In the case of tightly focused linearly polarized light, the $\boldsymbol{E}_z$ field manifest in the form double lobes which locate symmetrically by the optical axis, as shown in Fig. 11.17i. After finding position of the $\boldsymbol{E}_z$ field, the tip axis is fixed on it following by bringing the sample tip (in this case MoS_2) below the tip. In Fig. 11.17ii(a), (b) the optical microscopy and AFM image of a MoS_2 monolayer flake transferred on a glass coverslip is shown. Figure 11.17ii(c) presents a Raman spectrum of the flake. The spectrum reveals the two characteristic modes for a MoS_2 monolayer at 382 and 400 cm^{-1} (see Fig. 11.17ii(c)). The more enhanced peak in the TERS spectrum at 400 cm^{-1} refers to the A_{1g} out-of-plane vibration of the sulfur atoms in MoS_2, whereas the weaker peak at 382 cm^{-1} is assigned to the E_{2g} in-plane vibration of the two sulfur atoms with respect to the molybdenum atom in the sheet. Since the TERS tip is set into one of the two side lobes, z-polarization along the tip axis is obtained. Thus, the out-of-plane mode A_{1g} mode is strongly enhanced according to the TERS selection rules. Monolayer growth can be confirmed by analyzing the frequency difference of the two characteristic modes, being less than 20 cm^{-1} and equivalent with pure Raman measurements obtained for a MoS_2 monolayer by several groups. With respect to combined TERS and PiFM measurements, the A_{1g} out-of-plane vibrational mode (z-direction, along the tip axis) is perfectly suitable to probe the enhancement response with respect to the present electromagnetic fields in dependence of the focus position. Hence, the results indicate a significantly improved route toward the alignment efficiency of TERS experiments and establish MoS_2 as an outstandingly suitable TERS test sample.

The assembly of 2D hierarchical structures is a promising route for creating functional nanomaterials with nanoscale control and tunability. In this respect carbon nanomembranes (CNMs) molecular nanosheets with $\approx$1 nm thickness, deliver a universal platform for the hierarchical assembly of 2D materials. Marie et al. [54], demonstrated the synthesis and chemical characterization of hierarchically assembled molecular nanosheets based on CNM and covalently grafted, single-molecule layer cobalt(III) catalysts for the light driven hydrogen evolution reaction (HER). TERS was employed for the probing of both the transversal and longitudinal chemical information of the synthesized nanosheets with nanometer resolution. In this work, the inherent properties of CNMs, specifically their chemical and mechanical robustness as well as their ability for flexible surface functionalization, was used to assemble various functional molecular catalysts for HER on their surfaces. First, 4′-nitro-1,1′-biphenyl-4-thiol (NBPT) on a gold substrate was converted into short-range ordered NH_2-CNMs (Fig. 11.18i(a)) via low-energy electron irradiation induced cross-linking. Next, isonicotinic acid is covalently bound to the CNMs amino groups employing active esters for amide bond formation (iso-CNM, Fig. 11.18i(b)). Finally, these pyridine derivatives can act as axial ligands for Co^+Co^- double salt ions and exchange against their initial pyridine or chloride ligands (Co(III)-CNM, Fig. 11.18i(c)). Figure 11.18ii(a), (b) presents an example of a series of spectra measured on a NH_2-CNM membrane with a point-to-point distance of 25 nm, along with the associated waterfall plot for better clarity.

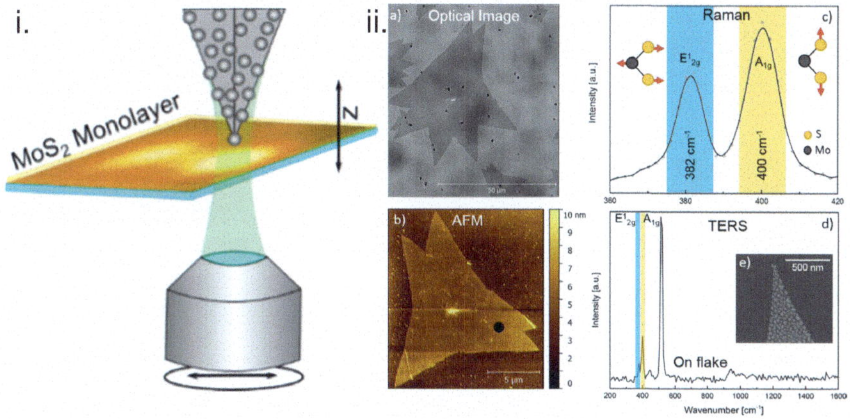

Fig. 11.17 Characterization of the MoS_2 monolayer flakes via TERS [95]. i. Schematic of the tip, sample, and illumination configuration. ii. Characterization of the MoS_2 monolayer flakes used. **a** Optical microscopy image. **b** AFM topography. **c** Raman spectra. **d** TERS spectra measured on a MoS_2 monolayer flake (black spot, the strong band at 520 cm^{-1} refers to Si of the cantilever). Inset **e** shows a representative SEM image of a TERS tip from the same batch used in all experiments. The frequency difference between E_{2g} and A_{1g} mode is less than 20 cm^{-1}, confirming monolayer growth [97]. Adapted with permission from author of [95]. Copyright (2019) (ACS Photonics)

The spectra primarily show the Si signal from the AFM tip at ≈520 and 1000 cm^{-1}. In addition, spurious signals occur which were not assigned due to the low signal to noise ratio. The low Raman scattering cross section of NH_2-CNM is a considerable advantage, as it allows to specifically distinguish the specific overlayers in the following experiments without interference. Figure 11.18ii(c), (d) shows the spectra and the respective waterfall plot measured on the iso-CNM sample. While the signal-to-noise ratio is low due to the low power used to avoid burning, the experiment clearly reveals the presence of all characteristic main bands of the isonicotine moiety at ≈1610, 1320, 950, 700, and 450 cm^{-1}. The ≈950 cm^{-1} signal attributed to isonicotinic acid is thus barely visible in some of the spectra due to the overlap. The absence of a strong C=O stretching signal at ≈1650–1700 cm^{-1} (carboxylic group) strongly indicates that the molecules are bound covalently via an amide bond on the surface and are not simply physisorbed. In TERS, vibrational modes with contributions of the scattering tensor along the tip axis are preferentially probed. Here, the amide group is expected to be perpendicular to the tip axis and thus, a relatively minor signal enhancement as shown in Fig. 11.3c is expected. The 400 × 400 nm^2 TERS mapping (with a 25 nm point-to-point distance) of the Co(III)-CNM is shown in Fig. 11.18ii(e), (f). The signal-to-noise ratio of the TERS signals is surprisingly high, especially considering the low laser power of 100 μW used for the first part of mapping, which suggests an unusually high scattering cross-section. The spectral positions of the major bands are consistent with the far-field Raman spectra and SERS spectra measured from the Co^+Co^- double complex salt drop casted on silver nanoparticles.

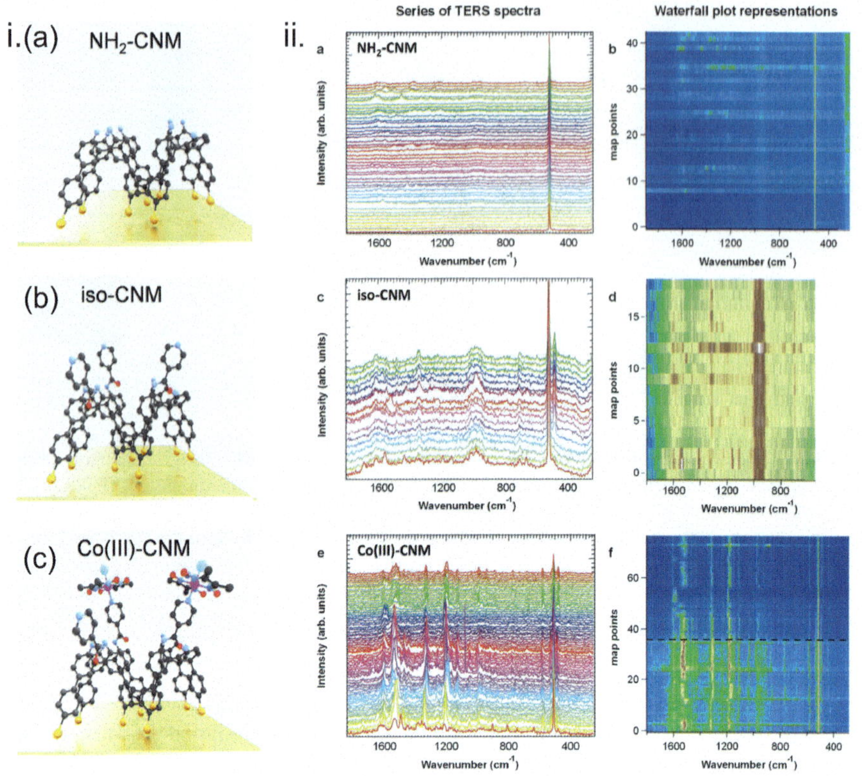

Fig. 11.18 **i. Schematic representation** [54]. **a** NH_2-CNM produced by the cross-linking of 4′-nitro-1,1′-biphenyl-4-thiol SAM via low-energy electron irradiation. **b** NH_2-CNM functionalized with isonicotinic acid through the formation of a covalent amide bond (iso-CNM), where in **c** the terminal pyridine groups allow axial ligand exchange with the complex salt $[Co(dmgH)_2(py)_2]^+[Co(dmgH)_2Cl_2]^-$ to immobilize Co(III) centers on the nanomembrane (Co(III)-CNM). Hydrogen atoms are not depicted in **c**–**e** for simplicity. ii. **a**–**f** TERS spectra and corresponding waterfall plots of a series of measurements on NH_2-CNM, iso-CNM, and Co(III)-CNM, respectively, with point-to-point distance of 25 nm. The dashed line in **f** indicates a position from which the laser power was reduced. Note that in (**d**), a different wavelength scaling was used to cut out the otherwise dominating Si-band of the tip, hence, improving the contrast for the actual iso-CNM. Adapted with permission from author of [54]. Copyright (2022) (Advanced Materials Interfaces)

Kumar et al. [98]. developed plasmonically-active TERS probes with a multilayer metal coating structure that can be successfully used within a liquid environment. Using these novel TERS probes, Kumar et al. have compared the plasmonic enhancement of TERS signals in air and water environments for both gap mode and non-gap mode configurations and show that in both cases the plasmonic enhancement decreases in water. TERS probes are typically prepared by coating commercial Si AFM cantilevers with a thin film (20–100 nm) of noble metals. However, the plasmonic metal coating of such probes is not stable when exposed to an aqueous environment. To improve the stability of the Ag coating in water, Kumar et al. first tried an adhesion layer of 3.5 nm Cr between the SiO_2 and the 100 nm Ag coating. The height profiles of SWCNTs confirm that the carbon nanotubes are indeed single-walled and are singly distributed over the sample surface. A high-resolution TERS image of the G-band intensity (at 1591 cm^{-1}) for a SWCNT measured in air with pixel resolution of 10 nm. The G-band intensity at each pixel was determined from the height of a Lorentzian fit to the TERS spectrum in the 1470–1671 cm^{-1} spectral range after linear background subtraction. To calculate the spatial resolution of the SWCNT TERS image, the procedure reported by Roy et al. [99], was adopted, which separates the near-field and far-field components in an intensity profile measured across the TERS image of a SWCNT through fitting it with two different Gaussian curves. The FWHM of the far-field Gaussian curve was set to 225 nm corresponding to a laser spot size of 450 nm. The near-field spatial resolution was estimated from the FWHM of the Gaussian curve to be $\approx$26 nm. The TERS imaging of a SWCNT was performed on the same sample immersed in water, using a 10 nm pixel size. Analysis of the corresponding line profile, yielded an estimated spatial resolution of approximately 28 nm. It's worth noting that both TERS images displayed intensity fluctuations along the length of the SWCNT. To obtain a more representative estimation of the spatial resolution in both the air and water environments for the two images displayed, the same fitting procedure was applied to three different line profiles across the TERS images. The result was an average spatial resolution of 28 $\pm$ 2 nm in air and 26 $\pm$ 2 nm in water. The spatial resolution obtained from these TERS images is consistent with the lateral size of the electric field enhancement, as calculated from numerical simulation maps in the non-gap mode for both air and water. Furthermore, a similar FWHM was observed from Gaussian fits to the line profiles in air (21 nm) and water. In summary, the TERS imaging of the SWCNT in both air and water environments revealed an estimated spatial resolution and the consistency of this resolution with the electric field enhancement and Gaussian fits underscores the reliability of the obtained results.

11.4.3 Biological and Life Sciences

In the realm of biological systems, the precise arrangement of biomolecules within cells, the molecular alignment of DNA/RNA, and the intricate secondary configurations of proteins and enzymes all wield significant influence. To capture these attributes, especially the molecular alignment, a high degree of sensitivity and spatial precision at the individual molecule scale is essential. TERS emerges as a powerful tool capable of not only discerning the distribution and alignment of biomolecules but also unraveling their structural intricacies. Remarkably, TERS achieves this without resorting to invasive measures that might disturb the natural state of the sample. It avoids any undesired interactions between the metal tip and the specimen, which could lead to alterations in the molecular structures.

Zhe et al. [100, 101], adopted gap-mode TERS for the detection of single-strand DNA. For gap-mode TERS, a silver tip is coupled with a gold substrate (see Sect. 11.2.2.1). The direct nucleic acid sequencing of a phage ssDNA (M13mp18) was performed with TERS. As is depicted in Fig. 11.19i(a), the DNA sample was deposited on the gold substrate via forced-air to align method to stretch the DNA for evading/minimizing DNA coiling. Mg^{2+} cations were added in the DNA buffer solution (pH 7.5) during the deposition to preferentially facilitate the adhesion of highly negatively charged DNA phosphate backbone to negatively charged gold surface. Compared to AFM, TERS provides a superior, subnanometer resolution of the DNA bundle edges. Shown by the red curve in Fig. 11.19i(c), the integrated intensities of the Raman signal peaks from 1630 to 1650 cm^{-1} can refer to DNA nucleobases cytosine (C), guanine (G), and thymine (T). The width of the DNA bundle is estimated to be 2.5 nm. The AFM curve exhibits a FWHM of 10 nm based on a Gaussian fitting. The TERS curve, which is the integrated intensity of the Raman signal from 1630 to 1650 cm^{-1}, exhibits a FWHM of 2.5 nm, far narrower than AFM. Similarly, in Fig. 11.19ii(a), the samples were imaged based on the same peak. The high contrast map highlights a single DNA strand (spot 1–24). Moreover, the distinguishable on-strand and off-strand signals in Fig. 11.19ii(b) prove that TERS was able to reach a spatial resolution of 0.5 nm. Identified DNA sequences are shown in Fig. 11.19ii(c). Because interferences of adjacent bases result in multicomponent signals, only the most possible bases below the bar charts were labeled. The multicomponent spectrum occurs due to the interference of adjacent bases. Mistakes may happen when the surrounding signals overwhelm the target base. Therefore, in Fig. 11.19ii(d), spot 7 and spot 11 were misread due to the strong signals of guanine and thymine. The accuracy shown in Fig. 11.19ii(c) is 91.7% (22/24). This works demonstrates that, as a straightforward optical sensing technique, TERS has the potential of becoming a next generation sequencing method for DNA/RNA as well as other important biological polymers such as polysaccharides, polypeptides, and even glyco-peptide conjugates.

In the wake of the COVID-19 pandemic, there has been a substantial increase in interest regarding viral detection. A cutting-edge technique that has captured significant attention is TERS [102, 103], which boasts the remarkable ability to analyze individual viruses. TERS not only facilitates the investigation of a virus's morphology but also provides valuable insights into its spectral characteristics. This method is especially advantageous when examining the surfaces of viruses, as it offers critical information about the proteins that play a crucial role in their structure and function. In recent experiments, Deckert et al. [102], employed TERS to investigate two distinct viruses: an enveloped influenza A virus H1N1 and a nonenveloped coxsackie virus B3 (CVB3) (shown in Fig. 11.20a). Figure 11.20a shows the schematic of the experiment. The building blocks of a virion are either DNA or RNA molecules surrounded by a protective scaffold of proteins, the so-called (nucleo)capsid (e.g., coxsasckievirus B3 [CVB3] virus in Fig. 11.20a). Moreover, there can be a lipid bilayer serving as an envelope. Protein spikes can protrude from the bilayer, forming a sea urchin-like structure (e.g., H1N1 virus in Fig. 11.20a). The swine H1N1 influenza virus particle is composed of a 2- to 6 nm lipid bilayer decorated with viral proteins and eight ribonucleoprotein (RNP) complexes. The RNPs consist of RNA strands,

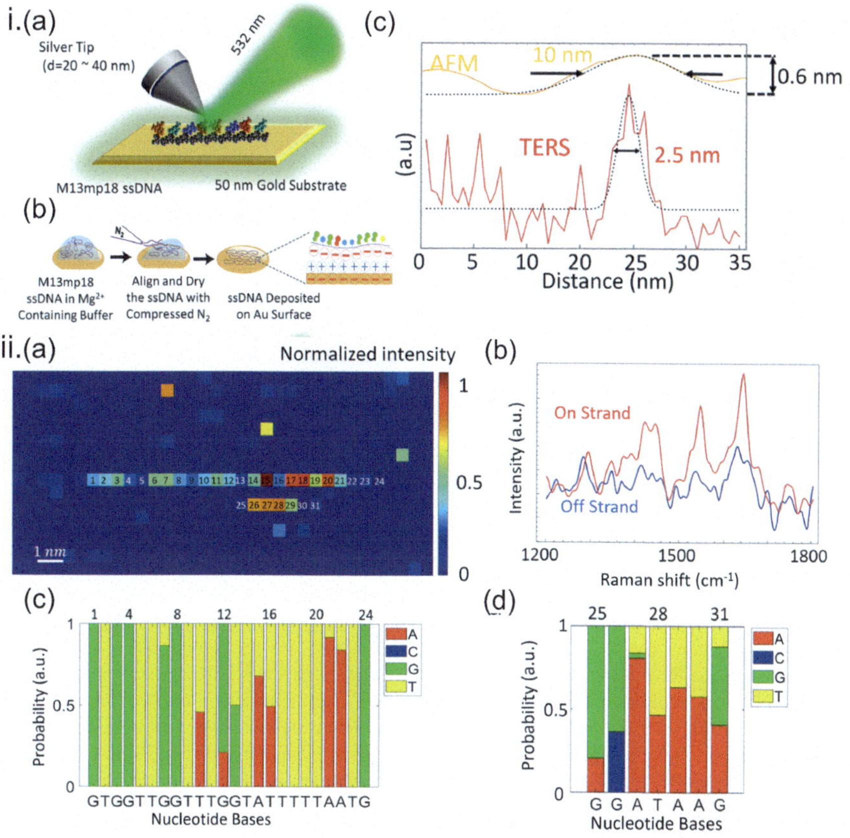

Fig. 11.19 **i. a** Schematic for TERS measurement of ssDNA molecules [100]. The diameter of the silver tip is 20–40 nm. Gap-mode TERS is formed by reducing tip-substrate distances to less than 1 nm. **b** The DNA deposition method. **c** AFM (yellow) and TERS (red) signals of ssDNA. **ii. DNA sequencing of sample 1 containing 1.0 pmol of M13mp18 DNA. a** TERS image of a single ssDNA segment with a step size of 0.5 nm. The plots show the integral intensities of the spectrum from 1630 to 1650 cm^{-1}. The numbers marking the map indicate the order of the sequences. **b** The "on strand" TERS spectrum of the pixel 7 and the "off strand" spectrum one step above the pixel 7 in panel (**a**). The substantial difference proves the 0.5 nm resolving capacity along the cross section. **c** A bar chart shows the probabilities Pi from the spot 1–24 labeled in panel a. The most probable bases are labeled at the bottom. Compared to the real DNA sequence GTGGTTCGTTCGGTATTTTTAATG, two errors are found at the spot 7 (G → C) and spot 11 (T → C). **d** The probabilities Pi from the spot 25–31. Two strands in panel a are 1 nm separated. The different sequences of spots 13–19 and spots 25–31 provides an evidence that TERS imaging can distinguish two parallel DNA molecules only separated by 1 nm. Adapted with permission from author of [100]. Copyright (2018) (Journal of the American Chemical Society)

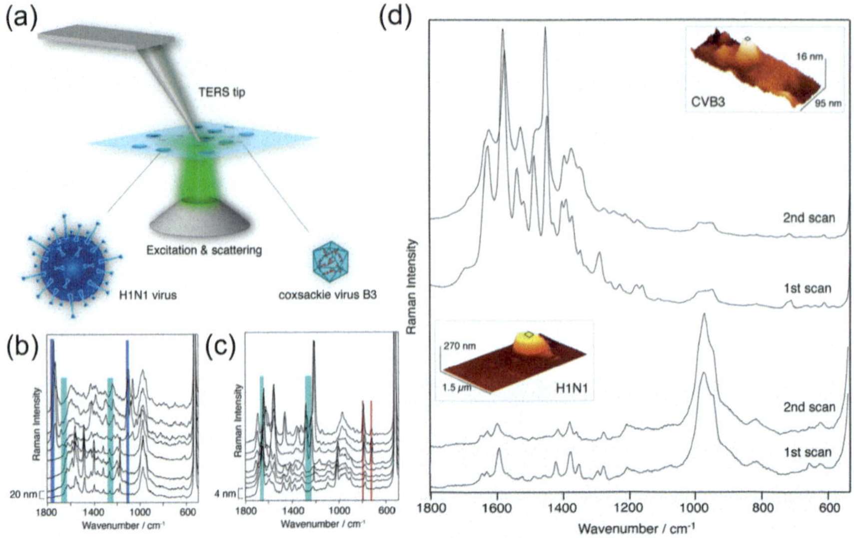

Fig. 11.20 Direct identification of a single virus with TERS [102]. **a** Schematics of the transmission TERS setup used in the experiments with the virus mixture spread on a slide and the laser illuminating the sample from below. In the model of an H1N1 virus protein spikes (cyan) protrude from the lipid bilayer (blue). In the model of a coxsackievirus B3 the protein lattice (cyan) is on the surface and the RNA strands (red) are inside the virus. **b** Selected TERS spectra recorded on a single H1N1 virus. **c** Selected TERS spectra recorded on a single coxsackievirus B3. In **b** and **c**, lipid marker bands (blue), protein marker bands (cyan), and RNA marker bands (red) are highlighted. **d** TERS spectra acquired during two repetitive topography scans on a single CVB3 (Top) and a single H1N1 virus (Bottom), respectively. The gray rectangles indicate the investigated areas on the particles. Adapted with permission from author of [102]. Copyright (2020) (Proceedings of the National Academy of Sciences)

which interact with numerous nucleoprotein molecules and some viral polymerase complexes. From AFM measurements (Fig. 11.20d) it is known that H1N1 shows pleomorphism that means size (150–400 nm) and shape (spherical or rodlike) vary. The dimension of the CVB3 virus (Fig. 11.20d) is 10 times smaller (20–30 nm), with an icosahedral structure (determined by crystallography), where the RNA is packed in a protein coat (capsid). Clearly, the different surface components of the viruses yield TERS spectra with information on proteins and lipids for H1N1 (Fig. 11.20b) and protein and RNA for CVB3 (Fig. 11.20c). Again, it is obvious that spectral variability on the virus surface occurs and a larger virus surface area must be considered. Nevertheless, a discrimination of both viruses solely based on the TERS spectra is quite straightforward. In Fig. 11.20d, TERS experiments acquired while scanning a defined area on a single H1N1 and CVB3 virus were presented. This way, all Raman signals during the scan were detected while the lateral resolution was intentionally decreased. In both samples the respective spectra acquired during subsequent topography scans indicate that the apparent spectral reproducibility is much higher compared to the single acquisitions in Fig. 11.20b, c, thus demonstrating the expected

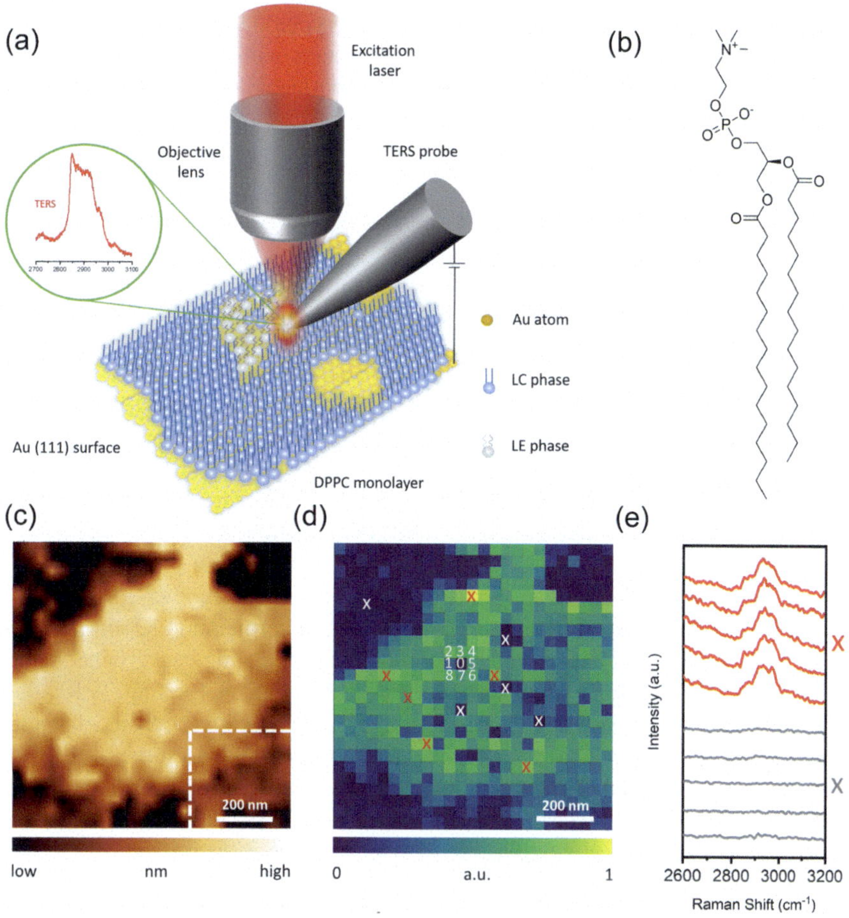

Fig. 11.21 STM-TERS imaging of lipid monolayers [104]. **a** Experimental set-up used for TERS imaging of supported DPPC monolayers. **b** Chemical structure of a DPPC molecule. **c** STM topography image of a DPPC monolayer region on Au (111). **d** TERS image of the C–H stretching (2800–3000 cm^{-1}) intensity measured in the area shown in (**c**). **e** TERS spectra measured at the locations of high signal intensity marked as "red x" and locations of low signal intensity marked as "white x" in (**d**). Adapted with permission from author of [104]. Copyright (2021) (Angewandte Chemie International Edition)

location averaging. This leads to a much faster overall data acquisition and much more representative spectral data. However, this comes at a cost that some information is consequently buried in an averaged dataset. In both samples, the respective spectra acquired during subsequent topography scans demonstrate significantly enhanced spectral reproducibility when compared to the individual acquisitions depicted in Fig. 11.20a. This underscores the successful implementation of location averaging as anticipated. Consequently, this approach leads to a notably expedited overall data acquisition process and yields more representative spectral data. However, this efficiency comes at the expense of burying certain details within an averaged dataset. For instance, in Fig. 11.20d, it's evident that the prominent lipid marker bands of the H1N1 virus, observed in the 'high-resolution' data acquisition, are not discernible.

11.5 TERS: A Promising Future Unveiled

In this Chapter, we provided a brief overview of Tip-Enhanced Raman Spectroscopy (TERS) technology, covering its historical origin, key concepts including near- and far-field effects, resolution limits, and field enhancement mechanisms. We also introduced very limited amount of its diverse applications. TERS has emerged as a robust analytical technique, poised for commercialization and broad adoption across various industries, promising significant advancements in materials science, electronics, pharmaceuticals, and beyond. Despite its numerous advantages, TERS remains a challenging technique, demanding meticulous optimization of experimental parameters and data analysis. The production of suitable TERS tips is still a complex issue, limiting its applicability in standard lab environments. Another open question requiring further research is data reproducibility and interpretation. Optimizing the tip-sample interaction, controlling laser power, and minimizing background signals are essential factors in TERS measurements. Moreover, interpreting TERS data necessitates the development of appropriate theoretical models that consider the intricate interaction between the tip, the sample, and the laser field. To successfully transition TERS from the laboratory to industrial applications, we must address several challenges, including standardization, instrumentation costs, data processing, interdisciplinary collaboration, and education. With ongoing research, innovation, and collaboration, TERS is set to play a pivotal role in shaping the future of nanoscience and technology.

In the realm of nanoscale characterization and analytical chemistry, TERS has emerged as a transformative technique in recent decades. This cutting-edge method empowers researchers to explore the molecular world with unparalleled precision, establishing itself as an invaluable tool across various scientific domains. Within the sphere of TERS applications, its potential in studying biological molecules and systems shines brightly. The use of ultraviolet (UV) excitation, coupled with high spatial resolution, has opened new frontiers in biological research, particularly in the context of cellular studies, thanks to the UV-resonance Raman effect. Furthermore, TERS' capabilities for nanoscale characterization of plasmon-enhanced reactions and molecular electronics are poised to deepen, driven by advancements in sub-nanoscale resolutions. These developments underscore the promising future of TERS, offering solutions to a wide range of challenges and expanding its horizons in ways that were once deemed unattainable.

Acknowledgements WW thanks for financial support through the DFG, CRC-TRR 234 "CataLight" Project C4. VD appreciates financial support via the DFG Collaborative Research Center SFB 1375 (NOA) Project CO_2.

References

1. C.V. Raman, K.S. Krishnan, Nature **121**, 501 (1928)
2. E.H. Hasdeo, A.R. Nugraha, M.S. Dresselhaus, R. Saito, Phys. Rev. B **94**, 075104 (2016)
3. T. Dieing, O. Hollricher, J. Toporski, *Confocal Raman Microscopy*, vol. 158 (Springer, 2011)
4. P.L. Stiles, J.A. Dieringer, N.C. Shah, R.P. Van Duyne, Annu. Rev. Anal. Chem. **1**, 601 (2008)
5. V. Markel, V. Shalaev, P. Zhang, W. Huynh, L. Tay, T. Haslett, M. Moskovits, Phys. Rev. B **59**, 10903 (1999)
6. S. Grésillon et al., Phys. Rev. Lett. **82**, 4520 (1999)
7. J.-X. Cheng and X.S. Xie, (ACS Publications, 2004), pp. 827
8. P. Verma, Chem. Rev. **117**, 6447 (2017)
9. T. Deckert-Gaudig, A. Taguchi, S. Kawata, V. Deckert, Chem. Soc. Rev. **46**, 4077 (2017)
10. L. Langelüddecke, P. Singh, V. Deckert, Appl. Spectrosc. **69**, 1357 (2015)
11. D.L. Jeanmaire, R.P. Van Duyne, J. Electroanal. Chem. Interfacial Electrochem. **84**, 1 (1977)
12. J. Wessel, JOSA B **2**, 1538 (1985)
13. E. Synge, London Edinburgh Dublin Philos. Mag. J. Sci. **6**, 356 (1928)
14. P. Bazylewski, S. Ezugwu, G. Fanchini, Appl. Sci. **7**, 973 (2017)
15. D.W. Pohl, W. Denk, M. Lanz, Appl. Phys. Lett. **44**, 651 (1984)
16. D.P. Tsai, A. Othonos, M. Moskovits, D. Uttamchandani, Appl. Phys. Lett. **64**, 1768 (1994)
17. D.A. Smith, S. Webster, M. Ayad, S.D. Evans, D. Fogherty, D. Batchelder, Ultramicroscopy **61**, 247 (1995)
18. C. Jahncke, M. Paesler, H. Hallen, Appl. Phys. Lett. **67**, 2483 (1995)
19. F. Zenhausern, M. O'boyle, H. Wickramasinghe, Appl. Phys. Lett. **65**, 1623 (1994)
20. G. Binnig, H. Rohrer, C. Gerber, E. Weibel, Appl. Phys. Lett. **40**, 178 (1982)
21. L. Novotny, B. Hecht, *Principles of Nano-Optics* (Cambridge university press, 2012)
22. R.M. Stöckle, Y.D. Suh, V. Deckert, R. Zenobi, Chem. Phys. Lett. **318**, 131 (2000)
23. M.S. Anderson, Appl. Phys. Lett. **76**, 3130 (2000)
24. N. Hayazawa, Y. Inouye, Z. Sekkat, S. Kawata, Opt. Commun. **183**, 333 (2000)
25. R.C. Rumpf, Electromagnetic analysis using finite difference time domain: metals and alternative grid schemes. EMPossible, pp. 12–13. https://empossible.thinkific.com/bundles/1d-2d-finite-difference-time-domain-with-matlab
26. S. Zhu, A. Yu, D. Hawley, R. Roy, Am. J. Phys. **54**, 601 (1986)
27. P. Busch, T. Heinonen, P. Lahti, Phys. Rep. **452**, 155 (2007)
28. F. H'dhili, R. Bachelot, A. Rumyantseva, G. Lerondel, P. Royer, J. Microsc. **209**, 214 (2003)
29. T. Treebupachatsakul, S. Shinnakerdchoke, S. Pechprasarn, Sensors **21**, 6164 (2021)
30. M. Esmann et al., Nat. Nanotechnol. **14**, 698 (2019)
31. K. Zhang et al., Anal. Chem. **93**, 7699 (2021)
32. S. Boroviks et al., Nat. Commun. **13**, 3105 (2022)
33. S. Berweger, J.M. Atkin, R.L. Olmon, M.B. Raschke, J. Phys. Chem. Lett. **1**, 3427 (2010)
34. W.L. Barnes, A. Dereux, T.W. Ebbesen, Nature **424**, 824 (2003)
35. M. Esmann, S.F. Becker, B.B. da Cunha, J.H. Brauer, R. Vogelgesang, P. Groß, C. Lienau, Beilstein J. Nanotechnol. **4**, 603 (2013)
36. M. Esmann, Probing single metal nanoparticles with a novel adiabatic-nanofocusing scanning near-field optical microscope (Carl von Ossietzky Universität Oldenburg, 2017)
37. B. Knoll, F. Keilmann, Opt. Commun. **182**, 321 (2000)
38. T. Wriedt, *The Mie theory: basics and Applications*, vol 53 (2012)

39. Jinxin Zhan. A new broadband, interferometric scanning near-field spectroscopy technique for probing the time-domain response of single nanoparticles (Carl von Ossietzky Universität Oldenburg, 2023)
40. Y.C. Martin, H.F. Hamann, H.K. Wickramasinghe, J. Appl. Phys. **89**, 5774 (2001)
41. B. Piglosiewicz, S. Schmidt, D.J. Park, J. Vogelsang, P. Groß, C. Manzoni, P. Farinello, G. Cerullo, C. Lienau, Nat. Photonics **8**, 37 (2014)
42. M. Kerker, D.-S. Wang, H. Chew, Appl. Opt. **19**, 3373 (1980)
43. A. Taguchi, J. Yu, P. Verma, S. Kawata, Nanoscale **7**, 17424 (2015)
44. A. Hartschuh, E.J. Sánchez, X.S. Xie, L. Novotny, Phys. Rev. Lett. **90**, 095503 (2003)
45. K. Kim, J.K. Yoon, J. Phys. Chem. B **109**, 20731 (2005)
46. K. Kitamura, K. Sakai, S. Noda, Opt. Express **18**, 4518 (2010)
47. B. Yang et al., Angew. Chem. Int. Ed. **62**, e202218799 (2023)
48. L. Xia, M. Chen, X. Zhao, Z. Zhang, J. Xia, H. Xu, M. Sun, J. Raman Spectrosc. **45**, 533 (2014)
49. M. Sun, S. Liu, M. Chen, H. Xu, J. Raman Spectrosc. **40**, 137 (2009)
50. M. Sun, S. Liu, Z. Li, J. Duan, M. Chen, H. Xu, J. Raman Spectrosc. **40**, 1172 (2009)
51. M. Sun, S. Wan, Y. Liu, Y. Jia, H. Xu, J. Raman Spectrosc. **39**, 402 (2008)
52. M. Sun, Y. Fang, Z. Yang, H. Xu, Phys. Chem. Chem. Phys. **11**, 9412 (2009)
53. M. Sun, J. Chem. Phys. **124** (2006)
54. M. Richard-Lacroix, M. Küllmer, A.L. Gaus, C. Neumann, C. Tontsch, M. von Delius, V. Deckert, A. Turchanin, Adv. Mater. Interfaces **9**, 2102389 (2022)
55. S.M. Mansfield, G. Kino, Appl. Phys. Lett. **57**, 2615 (1990)
56. B. Klarenaar, F. Brehmer, S. Welzel, H. Van Der Meiden, M. Van De Sanden, R. Engeln, Rev. Sci. Instrum. **86** (2015)
57. Q. Liu, S. Kim, X. Ma, N. Yu, Y. Zhu, S. Deng, R. Yan, H. Zhao, M. Liu, Nanoscale **11**, 7790 (2019)
58. R. Kato, K. Taguchi, R. Yadav, T. Umakoshi, P. Verma, Nanotechnology **31**, 335207 (2020)
59. Y.P. Huang et al., Angew. Chem. **130**, 7645 (2018)
60. A. Capaccio, A. Sasso, O. Tarallo, G. Rusciano, Nanoscale **12**, 24376 (2020)
61. T. Umakoshi, T.-A. Yano, Y. Saito, P. Verma, Appl. Phys. Express **5**, 052001 (2012)
62. X. Ma, Y. Zhu, N. Yu, S. Kim, Q. Liu, L. Apontti, D. Xu, R. Yan, M. Liu, Nano Lett. **19**, 100 (2018)
63. H. Furukawa, S. Kawata, Opt. Commun. **148**, 221 (1998)
64. E.J. Sánchez, L. Novotny, X.S. Xie, Phys. Rev. Lett. **82**, 4014 (1999)
65. B.-S. Yeo, W. Zhang, C. Vannier, R. Zenobi, Appl. Spectrosc. **60**, 1142 (2006)
66. S.A. Maier, P.G. Kik, H.A. Atwater, Appl. Phys. Lett. **81**, 1714 (2002)
67. M.L. Brongersma, J.W. Hartman, H.A. Atwater, Phys. Rev. B **62**, R16356 (2000)
68. S. Kawata, A. Ono, P. Verma, Nat. Photonics **2**, 438 (2008)
69. J.A. Fan et al., Science **328**, 1135 (2010)
70. S. Trautmann, M. Richard-Lacroix, A. Dathe, H. Schneidewind, J. Dellith, W. Fritzsche, V. Deckert, Nanoscale **10**, 9830 (2018)
71. A. Alabastri, X. Yang, A. Manjavacas, H.O. Everitt, P. Nordlander, ACS Nano **10**, 4835 (2016)
72. L. He, M. Rahaman, T.I. Madeira, D.R. Zahn, Nanomaterials **11**, 376 (2021)
73. L. Novotny, N. Van Hulst, Nat. Photonics **5**, 83 (2011)
74. T. Kalkbrenner, M. Ramstein, J. Mlynek, V. Sandoghdar, J. Microsc. **202**, 72 (2001)
75. P. Anger, P. Bharadwaj, L. Novotny, Phys. Rev. Lett. **96**, 113002 (2006)
76. I. Maouli, A. Taguchi, Y. Saito, S. Kawata, P. Verma, Appl. Phys. Express **8**, 032401 (2015)
77. E.M. van Schrojenstein Lantman, T. Deckert-Gaudig, A. J. Mank, V. Deckert, B.M. Weckhuysen, Nat. Nanotechnol. **7**, 583 (2012)
78. C.C. Neacsu, J. Dreyer, N. Behr, M.B. Raschke, Phys. Rev. B **73**, 193406 (2006)
79. J. Steidtner, B. Pettinger, Phys. Rev. Lett. **100**, 236101 (2008)
80. S. Jiang, Y. Zhang, R. Zhang, C. Hu, M. Liao, Y. Luo, J. Yang, Z. Dong, J. Hou, Nat. Nanotechnol. **10**, 865 (2015)
81. R.B. Jaculbia et al., Nat. Nanotechnol. **15**, 105 (2020)

82. E.C. Le Ru, P.G. Etchegoin, Annu. Rev. Phys. Chem. **63**, 65 (2012)
83. K. Kim, Y.M. Lee, H.B. Lee, Y. Park, T.Y. Bae, Y.M. Jung, C.H. Choi, K.S. Shin, J. Raman Spectrosc. **41**, 187 (2010)
84. J.-H. Zhong, X. Jin, L. Meng, X. Wang, H.-S. Su, Z.-L. Yang, C.T. Williams, B. Ren, Nat. Nanotechnol. **12**, 132 (2017)
85. N. Kumar, C.S. Wondergem, A.J. Wain, B.M. Weckhuysen, J. Phys. Chem. Lett. **10**, 1669 (2019)
86. K.-D. Park, E.A. Muller, V. Kravtsov, P.M. Sass, J. Dreyer, J.M. Atkin, M.B. Raschke, Nano Lett. **16**, 479 (2016)
87. M. Sun, Z. Zhang, H. Zheng, H. Xu, Sci. Rep. **2**, 647 (2012)
88. R.-P. Wang et al., J. Phys. Chem. Lett. **12**, 1961 (2021)
89. R. Zhang et al., Nature **498**, 82 (2013)
90. S.-Y. Ding, E.-M. You, Z.-Q. Tian, M. Moskovits, Chem. Soc. Rev. **46**, 4042 (2017)
91. F. Benz et al., Science **354**, 726 (2016)
92. S. Manzeli, D. Ovchinnikov, D. Pasquier, O.V. Yazyev, A. Kis, Nat. Rev. Mater. **2**, 1 (2017)
93. A. Turchanin, A. Gölzhäuser, Adv. Mater. **28**, 6075 (2016)
94. T.-X. Huang et al., Nat. Commun. **10**, 5544 (2019)
95. R. Meyer, S. Trautmann, K. Rezaei, A. George, A. Turchanin, V. Deckert, ACS Photonics **6**, 1191 (2019)
96. A.A. Sifat, J. Jahng, E.O. Potma, Chem. Soc. Rev. **51**, 4208 (2022)
97. C. Lee, H. Yan, L.E. Brus, T.F. Heinz, J. Hone, S. Ryu, ACS Nano **4**, 2695 (2010)
98. N. Kumar, W. Su, M. Veselý, B.M. Weckhuysen, A.J. Pollard, A.J. Wain, Nanoscale **10**, 1815 (2018)
99. D. Roy, J. Wang, C. Williams, J. Appl. Phys. **105** (2009)
100. Z. He et al., J. Am. Chem. Soc. **141**, 753 (2018)
101. Z. He, W. Qiu, M.E. Kizer, J. Wang, W. Chen, A.V. Sokolov, X. Wang, J. Hu, M.O. Scully, ACS Photonics **8**, 424 (2020)
102. V. Deckert, T. Deckert-Gaudig, D. Cialla-May, J. Popp, R. Zell, S. Deinhard-Emmer, A.V. Sokolov, Z. Yi, M.O. Scully, Proc. Natl. Acad. Sci. **117**, 27820 (2020)
103. E. Bailo, V. Deckert, Angew. Chem. Int. Ed. **47**, 1658 (2008)
104. Y. Pandey, N. Kumar, G. Goubert, R. Zenobi, Angew. Chem. Int. Ed. **60**, 19041 (2021)

Chapter 12
Hyper and Stimulated Raman Spectroscopy

Nilesh Kumar Rai

12.1 Introduction

Raman spectroscopy is a powerful technique for obtaining complete structural and dynamical information of the molecular systems. Presently, various Raman spectroscopic techniques are available based on either the principle of spontaneous (linear) Raman scattering or stimulated nonlinear scattering to account for many aspects of the molecular properties [1, 2]. Spontaneous or linear Raman scattering is a weak process due to low Raman scattering cross-section and the Raman spectroscopic techniques based on spontaneous Raman scattering are not found to be adequate to understand the molecular systems thoroughly. Recently developed stimulated Raman scattering is a coherent process that strongly enhances Raman scattering cross-section. The spectroscopic techniques based on the principle of stimulated scattering such as nonlinear Raman spectroscopic techniques are found to be suitable and useful to understand the molecular systems in a deeper manner. In view of the understanding of the principles behind nonlinear spectroscopic techniques including nonlinear Raman spectroscopy, many theories were developed and or advanced by various groups of scientists [3–8]. In both linear and nonlinear spectroscopy, the signal generation results from the polarization induced in the sample by the incident photons.

In a nonlinear process, two or more incident photons interact with the molecular system in order to generate a signal field. For input or incident field frequencies ω_1, $\omega_2 \ldots$, ω_n, the generated signal field will have a frequency $\omega_{sig} = \pm\omega_1 \pm\omega_2 \pm\omega_3 \pm \ldots \pm\omega_n$ where each of these combinations of the incident fields defines a unique nonlinear process. Also, the magnitudes of the incident fields ω_1, $\omega_2 \ldots$, ω_n, could be varied and hence lead to different nonlinear processes. If the number of incident fields is **n**, then, the corresponding nonlinear process is termed as nth order nonlinear

N. K. Rai (✉)
Department of Physics, C.M.P. Degree College, University of Allahabad, Prayagraj, Uttar Pradesh 211002, India
e-mail: nilesh.phy@cmpcollege.ac.in; nileshkrai162@gmail.com

D. K. Singh et al. (eds.), *Raman Spectroscopy*, Springer Series in Optical Sciences 248,
https://doi.org/10.1007/978-981-97-1703-3_12

process provided the molecular system being studied has non-zero *n*th order susceptibility. The most popular nonlinear Raman techniques are based on third order susceptibility which provides valuable structural and dynamical information of the molecular system. Such kind of third order nonlinear methods include stimulated Raman scattering (SRS) [9], coherent Stokes Raman scattering (CSRS) [10, 11], Femtosecond stimulated Raman scattering (FSRS) [12–14], coherent anti-Stokes Raman scattering (CARS) [15], Inverse Raman or Raman loss scattering [16, 17], and ultrafast Raman loss scattering (URLS) [18–20]. In particular, techniques based on SRS have been developed to carry out ultrafast time-resolved Raman spectroscopy [13].

Although techniques like stimulated Raman scattering (SRS) that involve multiple laser pulses or shifting of wavelengths fall under nonlinear Raman spectroscopy, however, the experimental methodologies such as sequence of pulses or spectral range of observations decide whether a given experiment is femtosecond stimulated Raman scattering (FSRS) or ultrafast Raman loss spectroscopy (URLS) or coherent Stokes Raman scattering (CSRS) etc. All of these techniques involve a Raman process leaving the molecule in the initial state or in a final, higher or lower vibrational state. Such Raman processes due to interaction of three fields provide opportunities to study either a higher or lower final state, which is normally termed as Stokes and anti-Stokes in linear Raman spectroscopy. In this chapter, theoretical concept SRS and experimental studies on excited state molecular dynamics using FSRS and URLS has been presented. Additionally, the experimental studies of Hyper-Raman spectroscopy which is an analogue to Raman spectroscopy and enables us to obtain information on molecular structure that cannot be obtained by Infrared absorption nor Raman spectroscopy has also been presented. In order to understand the three input field interactions with the molecular system and generation of the output signal field, a theoretical description of a third order nonlinear process has been given in the next section.

12.2 Theoretical Basis of Stimulated Raman Scattering

A material's response to an applied electromagnetic (EM) field, when a light beam travels through it, will either be linear or non-linear. The material act as linear media at low EM field intensities, with induced polarization, P, being linearly proportional to the applied EM field vector E ($P \alpha E$). However, the material response to the applied EM field becomes nonlinear at high field intensities. i.e., induced polarization in the medium remains no longer proportional to the applied electric field E but also depends on the higher powers of E. Thus, the total polarization induced in the system [21] can be written as in (12.1) and (12.2).

$$P \propto P^{(1)} + P^{(2)} + \cdots + P^{(n)} \tag{12.1}$$

$$P \propto \chi^{(1)}E + \chi^{(2)}E^{(2)} + \chi^{(3)}E^{(3)} + \cdots + \chi^{(n)}E^{(n)} \quad (12.2)$$

where $\chi^{(n)}$ is the nth order susceptibility of the material.

The induced polarization is linear and non-linear for $n = 1$ and $n \geq 2$ respectively and hence the corresponding susceptibilities $\chi^{(1)}$ and $\chi^{(n)}$, $n \geq 2$ are linear or nonlinear. Thus, the first term of (12.2) is the linear polarization and second, third and higher order terms give second order, third order and so on nonlinear polarizations. Second and third order polarizations involve two and three input fields interacting with the molecular system respectively. At low input field intensities, the higher order terms are negligible and only the linear term is retained. But for sufficiently high field intensities, the nonlinear terms begin to play a role. As discussed above, the *n*th order nonlinear process involves n-fields interacting with a molecular system and corresponding polarization can be given as.

$$P^{(n)} \propto \chi^{(n)}(\omega_{sig}; \pm\omega_1 \pm\omega_2 \pm\omega_3 \cdots \pm\omega_n)E_1(\pm\omega_1)E_2(\pm\omega_2)\cdots E_n(\pm\omega_n) \quad (12.3)$$

Second order nonlinear optical effects involve two fields interacting with the molecular system and are described by the second term in (12.2). If two input radiation fields of frequencies ω_1 and ω_2 are interacting with the medium, then the non-linear polarization induced by two input fields would radiate a signal field at a frequency given by any of the combinations, $\omega_{sig} = \pm\omega_1 \pm \omega_2$. Each of these combinations would correspond to different molecular processes. The combination $\omega_{sig} = \omega_1 + \omega_2$ or $\omega_{sig} = -\omega_1 - \omega_2$ is called sum frequency generation and represents the two photon absorption kind of process in the molecular system; the combination $\omega_{sig} = \omega_1 - \omega_2$ or $\omega_{sig} = -\omega_1 + \omega_2$ is called difference frequency generation and represents the inelastic Raman type of interaction. *i.e.* absorption at one frequency and scattering at another frequency; if $\omega_1 = \omega_2 = \omega$ then, the combination $\omega_{sig} = \omega_1 + \omega_2 = 2\omega$ or $\omega_{sig} = -\omega_1 - \omega_2 = -2\omega$ is called second harmonic generation whereas the combination $\omega_{sig} = \omega_1 - \omega_2 = 0$ or $\omega_{sig} = -\omega_1 + \omega_2 = 0$ is called optical rectification. The second order nonlinearity depends on the symmetry of the media. The second order nonlinear effects are observable for non-centrosymmetric materials and vanish for centrosymmetric materials.

Similar to second order nonlinear polarization as discussed above, third order nonlinear polarization also results in many terms corresponding to different third order nonlinear processes. All the third order nonlinear optical processes are circumscribed by a more general term called Four wave mixing in which interaction of three input fields with the molecular system results in the generation of the fourth signal field and can be described by the third term in (12.2). Third order optical nonlinearity is the lowest order nonlinearity observable in all media irrespective of their symmetry whether the media is Centro-symmetric or non-centrosymmetric.

If three input fields with frequencies ω_1, ω_2 and ω_3 with wave vectors $\overline{k}_1$, $\overline{k}_2$ and $\overline{k}_3$ respectively, are interacting with the molecular system, then the nonlinear polarization induced by the incident fields would radiate a signal field at frequency given by any of the combinations $\omega_{sig} = \pm\omega_1 \pm\omega_2 \pm\omega_3$ and the corresponding wave vector of the signal field would be given by any of the combinations $\overline{k}_{sig} = \pm\overline{k}_1 \pm\overline{k}_2 \pm\overline{k}_3$. The signal field $\overline{k}_{sig}$ may take a direction which is different from any of the input field wave vectors, $\overline{k}_1$ or $\overline{k}_2$or$\overline{k}_3$.

For efficient generation of signal, the phase of the induced polarization (for example, $\overline{k}_1 + \overline{k}_2 + \overline{k}_3$) should match with phase of the signal $\overline{k}_{sig}$ which is known as phase matching condition and is given by $\overline{k}_{sig} = \overline{k}_1 + \overline{k}_2 + \overline{k}_3$. The phase matching condition has to be fulfilled in any nonlinear optical processes [21–23] otherwise would result in the significant signal depletion depending on the extent of a vector called *phase mismatch vector* Δk which is given by $\Delta \mathrm{k} = \overline{k}_{sig} - \left(\overline{k}_1 + \overline{k}_2 + \overline{k}_3\right)$.

Based on the third order nonlinearity, nonlinear Raman spectroscopy is a highly versatile spectroscopic tool to investigate different material properties by the choice of any of the input frequencies $\omega_{sig} = \pm\omega_1 \pm\omega_2 \pm\omega_3$ and corresponding phase matching conditions, $\overline{k}_{sig} = \pm\overline{k}_1 \pm\overline{k}_2 \pm\overline{k}_3$. A typical example of one of the possible combinations is the sum frequency generation given by $\omega_{sig} = \omega_1 + \omega_2 + \omega_3$ which represent the three photon absorption kind of process. If all the input fields have the same frequency i.e. $\omega_1 = \omega_2 = \omega_3 = \omega$ then $\omega_{sig} = \omega_1 + \omega_2 + \omega_3 = 3\omega$ leads to the third harmonic generation. The combination like $\omega_{sig} = -\omega_1 - \omega_2 + \omega_3$ and $\omega_{sig} = \omega_1 + \omega_2 - \omega_3$ represents the optical rectification with, $\omega_3 = \omega_1 + \omega_2$. Similarly, the combination like,$\omega_{sig} = (\omega_1 - \omega_2) \pm \omega_3$ with $\omega_1 = \omega_3$ represents the stimulated Raman kind of process. In general, all the nonlinear Raman spectroscopic techniques are based on the stimulated Raman process. In Stimulated Raman process, the interaction of two input fields called Raman pump ($\omega_p = \omega_1$) and stoke field ($\omega_s = \omega_2$) with molecular system, excites a vibrational mode at frequency, $\omega_p - \omega_s = \omega_\upsilon$ where ω_υ is the frequency of the vibrational transition. The interaction of the third input field (called probe field) with the vibrational coherence prepared by the pump and Stoke field, result in the generation of a Raman signal.

The theoretical description of Stimulated Raman Scattering (SRS) can be given by diagrammatic perturbation theory using density matrix formalism [21]. In density matrix formalism, the equation of motion of density matrix $\rho(t)$ is given by $\frac{d}{dt}\rho_{nm}(t) = -\frac{\iota}{\hbar}[H, \rho(t)]_{nm}$

For simplicity, the subscripts have been dropped down.

$$\frac{d}{dt}\rho(t) = -\frac{\iota}{\hbar}[H, \rho(t)] \tag{12.4}$$

The Hamiltonian can be split into two parts,

$$\boldsymbol{H} = \boldsymbol{H}_0 + \widehat{V}(t) \tag{12.5}$$

where, $\boldsymbol{H}_0$ represents hamiltonian of free system or unperturbed hamiltonian and $\widehat{V}(t)$ represents the energy of interaction of the system with an external applied radiation field. We assume that this interaction energy is given adequately by electric dipole approximation as

$$\widehat{V}(t) = -\widehat{\mu} \cdot \underline{E}(t) \tag{12.6}$$

where $\widehat{\mu} = -e \cdot \widehat{r}$ denotes the electric dipole moment operator of the system and $\underline{E}(t)$ is the applied electromagnetic field and is given by

Thus, the density matrix $\rho(t)$ equation of motion is given by

$$\frac{d}{dt}\rho(t) = -\frac{\iota}{\hbar}[\widehat{H}_0, \widehat{\rho}(t)] - \frac{\iota}{\hbar}[-\widehat{\mu} \cdot \underline{E}(t), \widehat{\rho}(t)] \tag{12.7}$$

The (12.7) can be solved using perturbative methods. we therefore, seek a solution of density matrix $\rho(t)$ in the perturbation expansion as

$$\rho(t) = \rho^{(0)}(t) + \lambda\rho^{(1)}(t) + \lambda^{(2)}\rho^{(2)}(t) + \ldots \lambda^{(n)}\rho^{(n)}(t) \tag{12.8}$$

where $\lambda (= 0 \rightarrow 1)$ is a parameter that characterizes the strength of perturbation.

Now, the time-dependent polarization $\underline{P}(t)$ is given by

$$\underline{P}(t) = N\langle\widehat{\mu}(t)\rangle = NTr\big[\widehat{\mu} \cdot \widehat{\rho}(t)\big] \tag{12.9}$$

where N denotes the number density and $\langle\widehat{\mu}(t)\rangle$ denotes the expectation value of the induced dipole moment and is given by

$$\langle\widehat{\mu}(t)\rangle = Tr\big[\widehat{\mu} \cdot \widehat{\rho}(t)\big] \tag{12.10}$$

Substituting (12.5) into (12.4), and on solving the equation, the perturbative expansion of the polarization in powers of the electric field is given by

$$\underline{P}(t) = \underline{P}^{(1)}(t) + \underline{P}^{(2)}(t) + \underline{P}^{(3)}(t) + \ldots \underline{P}^{(n)}(t) \tag{12.11}$$

The nonlinear processes can be described by the nonlinear polarization. nth order polarization, $P^{(n)}(t)$, occurs in a system due to the interaction of the 'n' input fields and is given by

$$P^{(n)}(t) = NTr\left[\widehat{\mu} \cdot \widehat{\rho}^{(n)}(t)\right] \tag{12.12}$$

Particularly, the third order nonlinear polarization is given by

$$P^{(3)}(t) = NTr\left[\widehat{\mu} \cdot \widehat{\rho}^{(3)}(t)\right] \tag{12.13}$$

The nonlinear polarization can be described by the response function of the material. Response functions can, in general, be calculated in the time or in the frequency domain. The frequency domain response function is known as the susceptibility. The time domain response function can be described with the density matrix formalism where double sided Feynman diagrams can be used to represent the evolution of the ket and the bra states.

The SRS process can be explained with the help of the double sided Feynman diagrams and corresponding ladder diagrams. The Feynman diagrams of SRS process for the combination $\omega_{sig} = -\omega_p + \omega_{pr} + \omega_p$ of the incident field frequencies are given in Fig. 12.1a, b. The combination $\omega_{sig} = -\omega_p + \omega_{pr} + \omega_p$ of the incident field frequencies, there are eight Feynman diagrams which represent different SRS processes. Out of the eight Feynman diagrams for the SRS process, the two Feynman diagrams (ket/ket/ket and bra/bra/ket) are interpreted in terms of the density matrix evolution (Fig. 12.1a, b).

Let the system be initially at ground vibration level. The molecular vibrational states are represented by $|a\rangle$, $|c\rangle$ and $|b\rangle$.

Figure 12.1a represents the Feynman diagram (bra/bra/ket) that can be understood as follows: Let the system is initially in the lowest vibrational state where the density matrix $|a\rangle\langle a| = \rho_{aa} = 1$ represents the probability of the system being in the

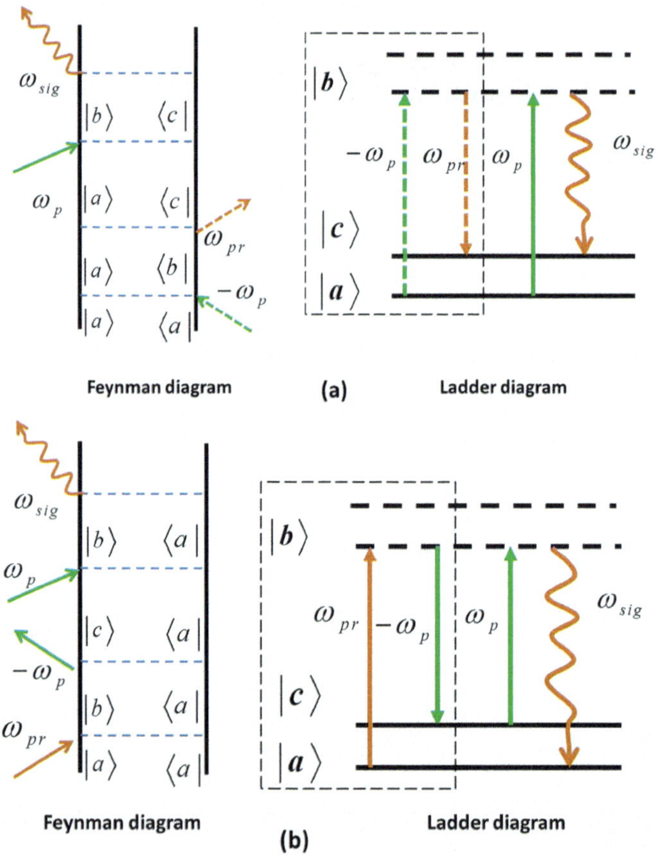

Fig. 12.1 **a** Feynman diagrams for SRS process: SRS gain and corresponding ladder diagram. **b** Feynman diagrams for SRS process: Inverse Raman and corresponding ladder diagram

state $|a\rangle$. After first interaction of the system with the light field in the bra side, the system evolves to a coherent state represented by $|a\rangle\langle b| = \rho_{ab}$ from the initial state $|a\rangle\langle a| = \rho_{aa}$. The second interaction of the system with the light field again in the bra side leads to a coherent state $|a\rangle\langle c| = \rho_{ac}$. The third interaction of the system with the light field results in the generation of Raman signal (conventionally on ket side) by leaving the system in the population state $|c\rangle\langle c| = \rho_{cc}$. The corresponding ladder diagram (Fig. 12.1a) also reveals the same information. The Raman pump and probe creates vibrational coherence in the system and interaction of third field (ω_p) with the system (lowest vibrational state) results in the generation of Raman signal. This kind of SRS process in which signal generation occurs as a gain signal on the stoke shifted frequency, is termed as the stoke shifted SRS gain process.

Similarly, Fig. 12.1b represents Feynman diagram (ket/ket/ket) and that can be understood as follows: Let the system is initially in the lowest vibrational state where the density matrix $|a\rangle\langle a| = \rho_{aa} = 1$ represents the probability of the system being in the state $|a\rangle$. After first interaction of the system with the light field either in the ket side, the system evolves to a coherent state represented by $|b\rangle\langle a| = \rho_{ba}$ from the initial state $|a\rangle\langle a| = \rho_{aa}$. The second interaction of the system with the light field again from the ket side leads to a coherent state $|c\rangle\langle a| = \rho_{ca}$. The third interaction of the system with the light field with ket side results in generation of Raman signal (conventionally on ket side) by leaving the system in the population state $|a\rangle\langle a| = \rho_{aa}$. The corresponding ladder diagram (Fig. 12.1b) also reveals the same information. The Raman pump and probe creates vibrational coherence in the system and interaction of third field (ω_p) with the system (Ist vibrational state) results in the generation of Raman signal. This kind of SRS process in which signal generation occurs as a loss signal on the anti-stoke shifted frequency, is termed as the Inverse Raman process.

12.3 Time Resolved Stimulated Raman Spectroscopy

In Stimulated Raman spectroscopy experiments, three optical fields interact with the material system to give the vibrational information. A simultaneous interaction of two optical fields called Raman pump and probe with the molecular system creates vibrational coherence and the interaction of the third field with it generates the Stimulated Raman signal. In femtosecond Stimulated Raman Spectroscopy, the Raman pump field is provided by a narrow bandwidth picosecond pulse and the Raman probe is provided by a broadband femtosecond pulse (white light continuum) which allows simultaneous detection of a large range of frequencies. In SRS, if a signal observed on the stoke side with respect to Raman pump pulse wavelength as the gain features on the probe pulse it is referred to as Femtosecond stimulated Raman spectroscopy (FSRS) whereas if a signal observed on the anti-stoke side with respect to Raman pump pulse wavelength as the loss features on the probe pulse it is referred to as Ultrafast Raman loss spectroscopy (URLS) provided the probe pulse contains all the frequencies (stoke as well as anti-stoke).

A schematic diagram of the time-resolved FSRS and URLS is shown in Fig. 12.2. The time-resolved FSRS and URLS can be used to probe the various photo physical processes such as vibrational relaxation, internal conversion, inter system crossing, intramolecular vibrational energy redistribution etc., and photochemical processes such as photo-isomerization, charge transfer, hydrogen abstraction etc.

The temporal sequence of Photo-pump ($\boldsymbol{\omega}_{\mathbf{pump}}$), Raman pump ($\boldsymbol{\omega}_{\mathbf{p}}$) and Raman probe ($\boldsymbol{\omega}_{\mathbf{pr}}$) pulses are shown in Fig. 12.2. The interaction of a photo-pump pulse with the molecular system, at any time t, sets up various photo physical and or photochemical processes by the photoexcitation to an excited electronic state at the time, t. The process is then probed by the spatially and temporally overlapped Raman pump-probe pair centered at time, t. The spectral resolution of the vibrational Raman spectra depends on the bandwidth of the Raman pump while the time resolution depends on the ability to resolve the delay time ($\Delta\tau$) between Photo-pump and the spatially and temporally overlapped Raman pump-probe pair. The nonlinear Raman signal generation has already been described above with the help of Feynman diagram and Ladder diagram.

The experimental setup for time resolved URLS described elsewhere [18–20] and a schematic diagram is shown in Fig. 12.3. A time-resolved URLS experiment requires three input pulsed laser fields as shown in Fig. 12.3. Figure 12.3 shows the schematic diagram for experimental set up for the time-resolved Raman experiments. The ground state molecular structure can be recorded with the spatially and temporally overlapped pair of picosecond Raman pump pulse and femtosecond Raman probe (white light continuum) pulse. The excited state structure and dynamics can be obtained by the Raman pump -probe pair along with the photo-pump which is described above.

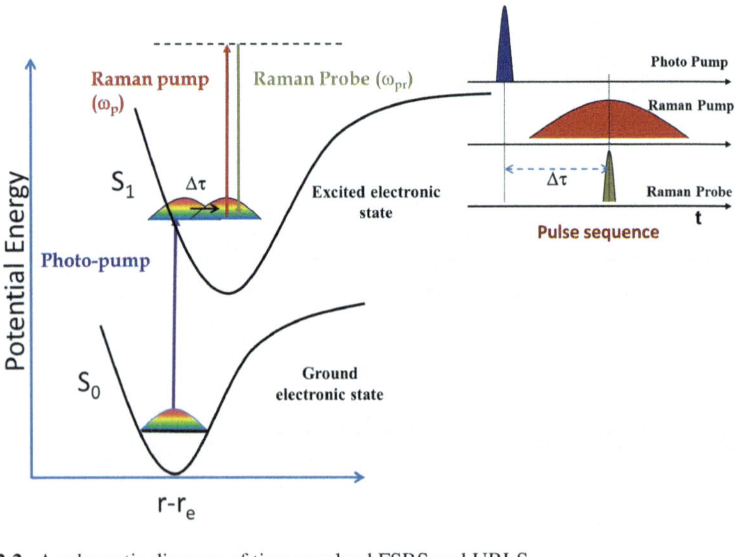

Fig. 12.2 A schematic diagram of time-resolved FSRS and URLS

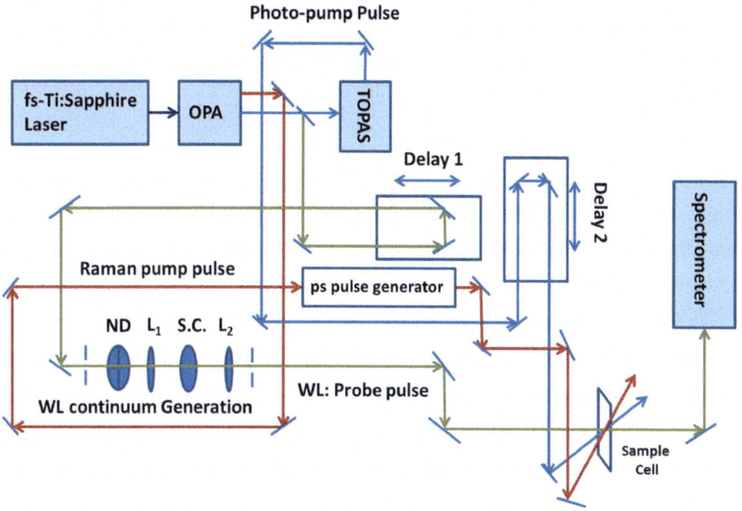

Fig. 12.3 Schematic diagram experimental setup of URLS

An amplified Ti-Sapphire laser (λ = 790 nm @ 2 mJ) system is used to perform the URLS measurements. One half part of the output of laser fundamental λ = 790 nm @ 1 mJ) is used to pump the OPA which provides visible to near IR region output to generate Raman pump. Out of the other half of the fundamental; a small fraction ~1 μJ is focused onto a 1 mm thick sapphire disk to generate the white light continuum (420–900 nm spectral windows) as probe pulse with average energy ~30 nJ and rest of the other half part of laser fundamental (λ = 790 nm @ ~1 mJ) is used as input to the TOPAS to generate the photo-excitation pump in the UV region. The delay 1 is the delay line to provide temporal overlap between Raman pump and Raman Probe while Delay 2 is the delay line which provides the various delays between photo pump and Raman pump-probe pair to perform the time resolved experiments. The photo pump, Raman pump and probe beams are mixed spatially and temporally in non-collinear geometry at the sample solution kept sample cell (1 mm thick path-length flow type curette) by setting the diameter of the photo pump, Raman pump and probe beams about ~500 μm, 300 μm and 50 μm, respectively. Finally, the probe (white light continuum, WL) pulse containing the URLS signal is collected and focused onto the slit of the spectrometer connected to a Liquid Nitrogen -cooled CCD. The URLS spectrum can be retrieved by subtracting the WL containing the signal from the WL without the signal.

12.3.1 Femtosecond Stimulated Raman Spectroscopy (FSRS)

Femtosecond stimulated Raman spectroscopy experiment [24] was attempted for first time in 1994 with a 300 fs time and 76 cm^{-1} frequency resolution and the recorded performance was observed below to that of a spontaneous Raman experiment. The first real Femtosecond stimulated Raman spectroscopy experiment on the photoisomerization of DCM dye with 250 fs time and 25 cm^{-1} frequency resolution [25] was reported in 1999 by the same group. In the following years, FSRS was used to report the first transient Raman spectra of photoisomerization of DCM dye with 250 fs time and 25 cm^{-1} frequency resolution [26] and vibrational relaxation of all *trans*-β-carotene with 300 fs time and 25 cm^{-1} frequency resolution [27, 28]. A sub-pico-second FSRS study on all-*trans*-neurosporene in hexane with 160 fs time and 32 cm^{-1} frequency resolution was reported by Rondonuwu et al. [29].

FSRS technique was advanced and popularized by the Mathies group. McCamant et. al. [30] demonstrated FSRS as a novel experimental technique that produces high-quality vibrational spectra with fast data acquisition time and giving a time-bandwidth product ~1.4 $pscm^{-1}$ whereas the same group reported the photo-isomerization of β-carotene [31] with a time-bandwidth product ~1.6 $pscm^{-1}$. Kukura et. al. [32] studied the vibrational structure and chemical reaction dynamics of the S_2 state of diphenyloctatetraene with the time-bandwidth product ~4.3 $pscm^{-1}$. A detailed description of the FSRS experiment was presented by McCamant et. al. [33] and the dependency of the stimulated Raman signal on experimental parameters like concentration, path length, and Raman pump power was studied and discussed with a benchmark time-bandwidth product ~1 $pscm^{-1}$ (100 fs time and 10 cm^{-1} frequency resolution). Kukura et al. [34] used FSRS to report on the structural observation of the primary isomerization in vision. Dasgupta et al. [35] utilized time-resolved FSRS technique to clarify the kinetics of the red-absorbing to the far-red-absorbing photoisomerization process and also to determine the molecular structure of the isomerization process. Fang et al. [36] have illustrated FSRS as a powerful approach to reveal the real-time nuclear dynamics that make up a multidimensional polyatomic reaction coordinate. In this study, they work out the time evolution of vibrational features of the electronically excited chromophore of Green Florescent Protein (GFP). The experimental results reveal the skeletal motions involved in the proton transfer that produces the fluorescent form of the protein. Kovalenko et al. [37] developed an experimental setup for FSRS with 60 fs and 10 cm^{-1} resolution. They reported the detailed description of experimental setup, signal processing, and data treatment by measuring FSRS spectra of stilbene in solution. Hoffman and Mathies [38] studied FSRS spectra of 4-nitro-40-dimethylamino-azobenzene (NDAB) to determine the structural transients in the trans-cis photoisomerization and also proposed the reaction mechanism. Weigel et al. [39] utilized FSRS technique to characterize the excited-state properties of riboflavin and Flavin adenine dinucleotide in polar solvents. In this study, Raman line shapes were observed to be affected by wave packet motion which resulted in the complex amplitude modulation of the stimulated

Raman spectra. Challa et al. [40] combined FSRS with a scanning multichannel technique (SMT) to averaged FSRS spectra with improved signal-to-noise ratios. Use of SMT with both CCDs and photodiode arrays detectors, has been observed to provide a simple method to extract weak signals in spectra dominated by systematic detector noise. In a FSRS study of pyranine in aqueous solution, the ultrafast conformational dynamics was observed by Liu et al. [41] during excited state proton transfer. The experimental results and analysis show an active participation of water molecule in excited state proton transfer in comparison to that of acetate molecule. The observed excited state conformational dynamics was found to be responsible for the photoacidity of pyranine in aqueous solution. Nakamura et al. [42] realized FSRS to study ultrafast structural dynamics in the electronic excited state of photoactive yellow protein. In this study, the hydrogen-bonding network surrounding the chromophore in the ground and excited states were investigated and hydrogen bonding dynamics was observed in the electronic excited state. Zhu et al. [43] developed a versatile femtosecond stimulated Raman spectroscopy setup with tunable pulses and reported the high spectral resolution ~12 cm^{-1}. Using this setup, fifteen times increment in the stimulated Raman gain signal of the 1627 cm^{-1} mode of photoacid pyranine in water was reported. Ferrante et al. [44] developed FSRS experimental setup with wavelength-tunable Raman pulses in the blue spectral region to study the temporal evolution of energy redistribution over different vibrational modes in myoglobin and proposed a mechanism through which initially stored vibrational energy in the highly excited Franck–Condon manifold is being transferred into low- and high-frequency modes with different timescales. Roy et al. [45] recorded the time resolved stimulated Raman spectra of TDPP-BBT in chlorobenzene at different pump-probe delay times which has been shown in Fig. 12.4. The ground state Raman spectrum is represented by black. The experimental result of the FSRS study reveals the formation of primary reactive coordinate by π-bridge torsion which drives the exciton relaxation-induced intramolecular charge transfer at ultrafast time scale in the donor-π-acceptor copolymer.

Recently, Kang et al. [46] have realized a FSRS setup with improved spectral resolution by employing, firstly, the 2nd-order diffraction in the double-pass grating filter technique. Using this FSRS experimental setup, they estimated the lowest spectral resolution, $\Delta\nu$, to be 2.5 cm^{-1}.

12.3.2 Ultrafast Raman Loss Spectroscopy (URLS)

Ultrafast Raman loss spectroscopy (URLS) is a coherent, sensitive and a new form of non-linear Raman spectroscopic technique. URLS is analogous to stimulated Raman spectroscopy (SRS) but is much more sensitive than SRS. URLS technique was developed by Umapathy's group [18–20] and the first URLS experiment was reported in 2009 [18]. URLS provides natural fluorescence rejection due to signal generation on the anti-stoke side. URLS is a self-phase matched process and does not require specific phase matching in the input beams for signal generation as in the

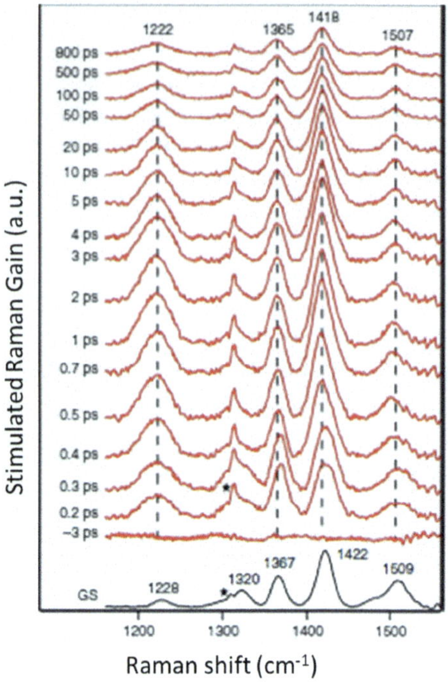

Fig. 12.4 Femtosecond stimulated Raman spectra of TDPP-BBT in chlorobenzene. Adapted from [45], Open Access, Springer Nature

case of coherent anti-stoke Raman Spectroscopy (CARS). Rai et al. [47] reported the basic principle of URLS technique developed in the laboratory. Roy et al. [48] used URLS to study the excited state dynamics of bis(phenylethynyl) benzene in different solvents. The experimental results reveal the mode-dependent kinetics and vibrational relaxation dynamics of high frequency vibrations at picosecond time scale. The time evolution of the transient Raman band intensities has also been reported that reveal the structural reorganization of Franck-Condon active non-planar rotamers to a planar conformation. In another study, Roy et al. [49] used URLS to study the excited state dynamics of bis(phenylethynyl)-benzene and reported the direct observation of the conformational relaxation dynamics of middle-ring and terminal ring twisted conformers. Kayal et al. [50] employed URLS to study the solvent dependent structural dynamics in entangled singlet and triplet states of the photo excited Thioxanthone. The experimental observations show that the intersystem crossing follows the vibrational cooling when Thioxanthone was photo excited in acetonitrile while it competes with the vibrational cooling when Thioxanthone was photo excited in methanol. Aslo, photo excitation of Thioxanthone in cyclohexane was observed to lead a rapid evolution towards the intersystem crossing. Jana et al. [51] employed URLS to unravel the photo excited state structural dynamics of 9, 10-bis(phenylethynyl)anthracene in n-hexane. Recently, Barak et al. [52] has implemented URLS to measure the distinct spectral

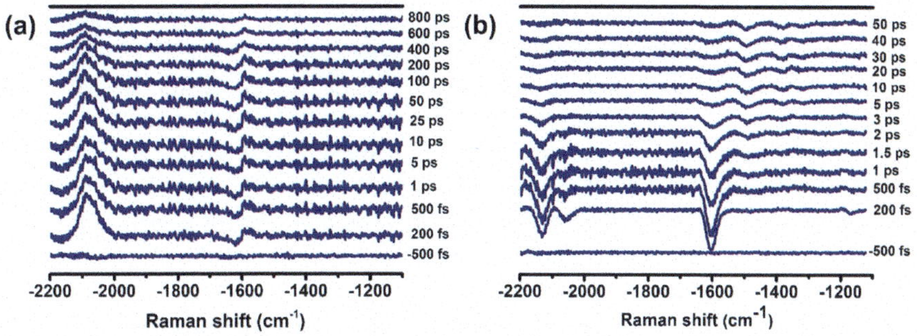

Fig. 12.5 Excited state dynamics: URLS spectra of DACN-DPA in **a** n-hexane, and **b** acetonitrile at different delay times. Adapted from [52], Open Access, Wiley

and temporal dynamics of different Raman bands associated with different vibrational modes of donor-π-bridge-acceptor system DACN-DPA (4-dimethylamino-4′-cyanodiphenylacetylene) in acetonitrile and n-hexane. The time-resolved URLS spectra of DACN-DPA in n-hexane and acetonitrile recorded at different delay times is shown in Fig. 12.5 [52]. The observation of a Raman band growth of at ~1492 cm^{-1} (Fig. 12.5b) on the sub-picosecond time scale was interpreted as structural modification associated with the intramolecular charge transfer. In another study, Barak et al. [53], has reported time resolved URLS spectra of 4-dimethylamino-4′-carbomethoxydiphenylacetylene (DACM-DPA) in different nonpolar and polar solvents. The result of this study reveals the solvent polarity dependent ultrafast structural dynamics of DACM-DPA and the formation of an intramolecular charge-separated state.

12.3.3 Hyper-Raman Spectroscopy

12.3.3.1 Introduction

Recent advances in laser technology have revolutionized the field of non-linear spectroscopy and as a result, based on the non-linear interaction of light with matter, wide variety of new spectroscopic techniques have been developed. These techniques provide information about systems not normally available from conventional types of spectroscopy. Hyper-Raman spectroscopy (HRS) is based on the principle of a kind of nonlinear scattering process that provides potential information about the molecules that is not generally available with techniques such as infrared (IR) or Raman spectroscopy. For example, the vibrational modes which are IR and Raman-inactive could give a hyper-Raman signal [54–56]. In principle, Hyper-Raman scattering is a three-photon process. HRS involves the absorption of two incident photons of frequency, ω, each followed by the emission of one photon of frequency ω′ (approximately twice the incident frequency ω), such that the difference (2ω–ω′) corresponds to the frequency (ω_{vib}) of a vibrational transition of the molecule.

Hyper-Raman scattering can be considered analogous to linear Raman scattering or coherent second harmonic generation. Spontaneous or linear Raman is based on inelastic processes where the frequency of the incident light is shifted by molecular vibrations whereas coherent second harmonic generation is a two photon process. The advantages of two photon processes is that due to different selection rules, a transition which is not allowed in one photon process but may be allowed in two photon processes. HRS has been recognized and realized as a useful experimental technique to probe the frequency of low energy vibrational transitions.

12.3.3.2 Theoretical Basis

When a laser light beam of frequency, ω and intensity, I_0 incident on a sample, the light-matter interaction produces a Stokes hyper-Raman band at frequency $(2\omega - \omega_a)$ and the band intensity [57–59] is given by

$$I_{hR,a}(2\omega - \omega_a) \propto \hbar I_0^2 (2\omega - \omega_a)^4 \left| \langle \psi_i | \beta(\omega, Q) | \psi_f \rangle \right|^2 \tag{12.14}$$

where, ω_a is the frequency of vibrational transition and $|\psi_i\rangle$ and $|\psi_f\rangle$ are the initial and final states of the system, respectively with population N. $\beta(\omega, Q)$ is the first hyperpolarizability at frequency ω and is a function of the nuclear displacements along the reduced normal coordinates, Q.

Under the Born–Oppenheimer approximation and neglecting the vibronic coupling, the electronic state of both the initial and final states can be taken to be the electronic ground state and then, can be integrated over the electronic degrees of freedom. Under this condition, the intensity can be written as

$$I_{hR,a}(2\omega - \omega_a) \propto \hbar I_0^2 (2\omega - \omega_a)^4 |\langle i | \underline{\beta}(\omega, Q) | f \rangle|^2 \tag{12.15}$$

where $|i\rangle$ and $|f\rangle$ are the initial and final vibrational states, respectively, and $\underline{\beta}$ is the electronic first hyperpolarizability of the ground state.

12.3.3.3 Energy Level Diagram

Hyper-Raman scattering is a three-photon process in which the two photon of frequency ω_1 and ω_2 are simultaneously annihilated and one photon of frequency $\omega_{sig} = \omega_1 + \omega_2 - \omega_a$ is created, where ω_a is the frequency of a vibrational transition of the material. If the molecule starts in a vibrationally excited state, the photon signal, $\omega_{Sig} = \omega_1 + \omega_2 + \omega_a$ will be generated/created at the anti-stoke side. In the experiment, ω_1 and ω_2 usually come from a single incident laser field ($\omega_1 = \omega_2 = \omega$) and hence identical photons.

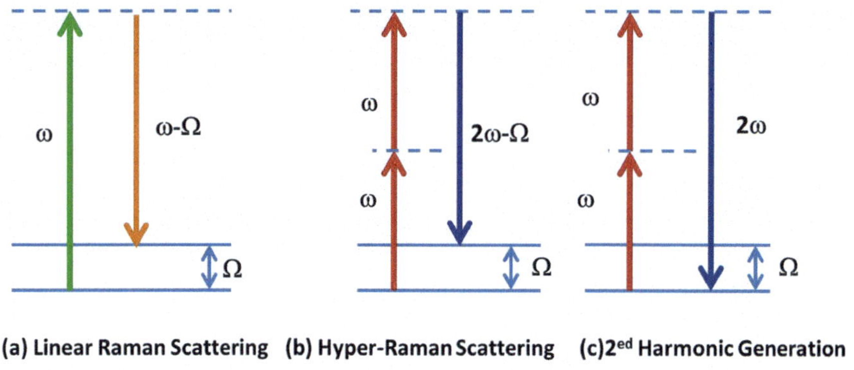

Fig. 12.6 **a** Energy level diagram of linear Raman, **b** hyper Raman scattering (HRS), **c** coherent second harmonic generation

Hyper Raman scattering can be considered as the two photon analog of linear Raman scattering Fig. 12.6a. The energy level diagram of hyper Raman scattering is shown in Fig. 12.6b. Hyper Raman scattering can also be considered as coherent second-harmonic generation (SHG) as shown in Fig. 12.6c.

12.3.3.4 Brief History and Recent Developments

Hyper-Raman scattering was first predicted in 1959 by Decius and Rauch [60] and experimentally observed in 1965 by Terhune and Savage [61] using a ruby laser and this was the first published report of hyper-Raman scattering. In this experiment, as a result of scattering of incident pulsed laser light, the scattered radiation was observed to be shifted slightly from the second harmonic of the laser, the scattered intensity was found to be proportional to the square of the laser intensity, and the frequency shifts was correspond to the known molecular vibrational frequencies. Long and Stanton [62] presented a theoretical description of Hyper Raman Scattering. In this report, the sum over molecular states formulation of hyper-Raman scattering has been presented which allows the effects of intermediate resonances to be explored. Andrews and Thirunamachandran [63] developed the theory of vibrational hyper Raman scattering using quantum electrodynamics. The theoretical result reveals the evolution of the hyperpolarizability tensor as a result of employing the electric dipole approximation. In this report, the time explicit expressions for scattering intensities for a wide variety of experimental configurations have also been presented.

Vogt and Presting [64] used an acousto-optically Q-switched Nd-YAG laser with a pulse repetition rate of 5 kHz and a pulse width of 150 ns to obtain the Hyper-Raman spectra of NaCl, KCl, KBr, KI, and RbI at room temperature and experimental result was compared and verified to that of Infrared Spectra. They also estimated the efficiency of absolute hyper-Raman scattering. Quinet et al. [59] reported the experimental investigation and theoretical interpretation of the Raman and hyper-Raman spectra of acetonitrile and its derivatives. Dobner and Fallnich [65] evaluated, demonstrated and discussed the capabilities of in-line interferometric femtosecond stimulated Raman scattering by obtaining the hyperspectral imaging of polymethyl methacrylate (PMMA) beads and lipid droplets in water. Cornaton et al. [66] used recursive implementation of response theory to work out the first computational treatment of the complete second-order vibrational perturbation treatment of hyper-Raman scattering spectroscopy. In view of getting better agreement with experimental spectra, anharmonic corrections were also done to the vibrational frequencies and the intensities and observed a larger effect on the frequencies than on the intensities after the inclusion of anharmonicity.

Recently, Okuno [67], has studied molecular liquids using Hyper-Raman spectroscopy. The molecular properties of four alcohols namely methanol, ethanol, 1-propanol, and 2-propanol have been investigated by recording the hyper Raman spectra using 532 nm excited. The work also reports the frequencies of the hydrogen bonded OH stretching mode in HR spectra, higher than that of Infrared (IR).

12.4 Conclusion and Future Perspective

In short, Stimulated Raman Spectroscopy such as URLS and FSRS are the excellent tool study various photo physical processes such as vibrational relaxation, internal conversion, inter system crossing, intramolecular vibrational energy redistribution etc., and photochemical processes such as photo-isomerization, charge transfer, hydrogen abstraction etc. which occur on an ultrafast time scale. Hyper-Raman spectroscopy has been recognized as a useful experimental tool to probe the frequency of low energy vibrational transitions which are difficult to probe with one photo process such as linear Raman scattering. Also HRS has proven the capability to probe those vibrational transitions which are IR and Raman inactive.

References

1. T. Tahara, H.O. Hamaguchi, Appl. Spectrosc. **47**, 391–398 (1993)
2. G. Eckhardt, R.W. Hellwarth, F.J. McClung, S.E. Schwarz, D. Weiner, E.J. Woodbury. Phys. Rev. Lett. **9**, 455–457 (1962)
3. N. Bloembergen, Am. J. Phys. **35**, 989–1023 (1967)
4. Y.R. Shen, N. Bloembergen, Phys. Rev. A **137**, 1787–1805 (1965)
5. N. Bloembergen, Science **216**, 1057–1064 (1982)
6. S. Mukamel *Principles of Nonlinear Optical Spectroscopy* (Oxford University Press, Oxford, 1995)
7. C.A. Marx, U. Harbola, S. Mukamel, Phys. Rev. A **77**, 022110-1-13 (2008)
8. S. Rahav, O. Roslyak, S. Mukamel, J. Chem. Phys. **131,** 194510-1-11 (2009)
9. R.W. Hellwarth, Phys. Rev. **130**, 1850–1852 (1963)
10. T. Hattori, A. Terasaki, T. Kobayashi, Phys. Rev. A **35**, 715–724 (1987)
11. M.J. Stimson, D.J. Ulness, A.C. Albrecht, Chem. Phys. Lett. **263**, 185–190 (1996)
12. M. Yoshizawa, M. Kurosawa. Phys. Rev. A. **61**, 013808-1-6 (1999)
13. P. Kukura, D.W. McCamant, R.A. Mathies, J. Phys. Chem. A **108**, 5921–5925 (2004)
14. P. Kukura, S. Yoon, R.A. Mathies, Anal. Chem. **78**, 5952–5959 (2006)
15. H. Kano, H.O. Hamaguchi, Anal. Chem. **79**, 8967–8973 (2007)
16. W.J. Jones, B.P. Stoicheff, Phys. Rev. Lett. **13**, 657–659 (1964)
17. R.R. Frontiera, S. Shim, R.A. Mathies. J. Chem. Phys. **129**, 064507-1-6 (2008)
18. S. Umapathy, A. Lakshmanna, B. Mallick, J. Raman Spectrosc. **40**, 235–237 (2009)
19. A. Lakshmanna, B. Mallick, S. Umapathy, Curr. Sci. **97**, 210–216 (2009)
20. B. Mallick, A. Lakshmanna, S. Umapathy, J. Raman Spectrosc. **42**, 1883–1890 (2011)
21. R.W. Boyd, *Nonlinear Optics* (Academic Press, San Diego, 1992)
22. Y.R. Shen, *The Principles of Nonlinear Optics* (Wiley, New York, 1984)
23. N. Bloembergen, *Nonlinear Optics* (Benjamin, New York, 1965)
24. M. Yoshizawa, Y. Hattori, T. Kobayashi, Phys. Rev. B **49**, 13259–13262 (1994)
25. M. Yoshizawa, M. Kurosawa, Phys. Rev. A **61**, 013808-1-6 (1999)
26. M. Yoshizawa, M. Kubo, M. Kurosawa, J. Lumin. **87**, 739–741 (2000)
27. M. Yoshizawa, H. Aoki, H. Hashimoto, Phys. Rev. B **63**, 180301-1-4 (2001)
28. M. Yoshizawa, H. Aoki, H. Hashimoto, Bull. Chem. Soc. Jpn. **75**, 949–955 (2002)
29. F.S. Rondonuwu, Y. Watanabe, J.-P. Zhang, K. Furuichi, Y. Koyama, Chem. Phys. Lett. **357**, 376–384 (2002)
30. D.W. McCamant, P. Kukura, R.A. Mathies, Appl. Spectrosc. **57**, 1317–1323 (2003)
31. D.W. McCamant, P. Kukura, R.A. Mathies, J. Phys. Chem. A **107**, 8208–8214 (2003)
32. P. Kukura, D.W. McCamant, P.H. Davis, R.A. Mathies, Chem. Phys. Lett. **382**, 81–86 (2003)
33. D.W. McCamant, P. Kukura, S. Yoon, R.A. Mathies, Rev. Sci. Instrum. **75**, 4971–4980 (2004)
34. P. Kukura, D.W. McCamant, S. Yoon, D.B. Wandschneider, R.A. Mathies, Science **310**, 1006–1009 (2005)
35. J. Dasgupta, R.R. Frontiera, K.C. Taylor, J.C. Lagarias, R.A Mathies, PNAS, USA **106**, 1784–1789 (2009)
36. C. Fang, R.R. Frontiera, R. Tran, R.A. Mathies, Nature **462**, 200–2004 (2009)
37. S.A. Kovalenko, A.L. Dobryakov, N.P. Ernsting, Rev. Sci. Instrum. **82**, 063102-1-9 (2011)
38. D.P. Hoffman, R.A. Mathies, Phys. Chem. Chem. Phys. **14**, 6298–6306 (2012)
39. A. Weigel, A. Dobryakov, B. Klaumunzer, M. Sajadi, P. Saalfrank, N.P. Ernsting, J. Phys. Chem. B **115**, 3656–3680 (2011)
40. J.R. Challa, Y. Du, D.W. McCamant, Appl. Spectrosc. **66**, 227–229 (2012)
41. W. Liu, F. Han, C. Smith, C. Fang, J. Phys. Chem. B **116**, 10535–10550 (2012)
42. R. Nakamura, N. Hamada, K. Abe, M. Yoshizawa, J. Phys. Chem. B **116**, 14768–14775 (2012)
43. L. Zhu, W. Liu, C. Fang, Appl. Phy. Lett. **105**, 041106-1-5 (2014)
44. C. Ferrante, E. Pontecorvo, G. Cerullo, M.H. Vos, T. Scopigno, Nat. Chem. **8**, 1137–1143 (2016)

45. P. Roy, A. Jha, V.B. Yasarapudi, T. Ram, B. Puttaraju, S. Patil, J. Dasgupta, Nat. Commun. **8**, 1716-1-10 (2017)
46. D.G. Kang, K.C. Woo, D.H. Kang, C. Park, S.K. Kim, Sci. Rep. **11**, 3361-1-7 (2021)
47. N.K. Rai, A.Y. Lakshmamma, V.V. Namboodiri, S. Umapathy, J. Chem. Sci. **124**, 177–186 (2012)
48. K. Roy, S. Kayal, F. Ariese, A. Beeby, S. Umapathy, J. Chem. Phys. **146**, 064303-1-9 (2017)
49. K. Roy, S. Kayal, V.R. Kumar, A. Beeby, F. Ariese, S. Umapathy, J. Phys. Chem. A **121**, 6538–6546 (2017)
50. S. Kayal, K. Roy, Y.A. Lakshmanna, S. Umapathy, J. Phys. Chem. A **122**, 6048–6054 (2018)
51. S. Jana, Y.A. Lakshmanna, S. Umapathy, Phys. Chem. Chem. Phys. **21**, 14341–14349 (2019)
52. A. Barak, N. Dhiman, F. Sturm, F. Rauch, Y.A. Lakshmanna, K.S. Findlay, A. Beeby, T.B. Marder, S. Umapathy, ChemPhotoChem **6**, e202200146-1-10 (2022)
53. A. Barak, N. Dhiman, F. Sturm, F. Rauch, Y.A. Lakshmanna, K.S. Findlay, A. Beeby, T.B. Marder, S. Umapathy, J. Phys. Chem. C **127**, 5855–5865 (2023)
54. A.M. Kelley, Annu. Rev. Phys. Chem. **61**, 41–61 (2010)
55. W.P. Acker, D.H. Leach, R.K. Chang, Chem. Phys. Lett. **155**, 491–499 (1989)
56. A. Mohammed, H. Ågren, M. Ringholm, A.J. Thorvaldsen, K. Ruud, Mol. Phys. **110**, 2315–2320 (2012)
57. O. Quinet, B. Champagne, J. Chem. Phys. **117**, 2481–2488 (2003)
58. O. Quinet, B. Champagne, V. Rodriguez, J. Chem. Phys. **121**, 4705–4710 (2004)
59. O. Quinet, B. Champagne, V. Rodriguez, J. Chem. Phys. **124**, 244312-1-12 (2006)
60. J.C. Decius, J.E. Rauch, *At the Ohio State Symposium on Molecular Structure and Spectroscopy*, paper 48 (1959)
61. R.W. Terhune, P.D. Maker, C.M. Savage, Phys. Rev. Lett. **14**, 681–684 (1965)
62. D.A. Long, L. Stanton, Proc. R. Soc. Lond. A **318**, 441–457 (1970)
63. D.L. Andrews, T. Thirunamachandran, J. Chem. Phys. **68**, 2941–2951 (1978)
64. H. Vogt, H. Presting, Phy. Rev. B **31**, 6731–6738 (1985)
65. S. Dobner, C. Fallnich, J. Chem. Phys. **140**, 084201-1-6 (2014)
66. Y. Cornaton, M. Ringholm, K. Ruud, Phys. Chem. Chem. Phys. **18**, 22331–22342 (2016)
67. M. Okuno, J. Raman Spectrosc. **52**, 849–856 (2021)

Chapter 13
Coherent Anti-Stokes Raman Spectroscopy (CARS)

Francesco Mazza, Leonardo Castellanos, Dmitrii Kliukin, and Alexis Bohlin

13.1 Introduction

Optical diagnostics are the main experimental tools for the quantitative measurements in chemically reacting flows, with unparalleled performance in terms of precision and accuracy, spatio-temporal resolution, and response time [1, 2]. The use of various forms of light-matter interaction to extract information on the velocity field, chemical composition and local thermodynamic properties allows for in situ measurements to be performed in harsh measurement conditions that physical probes could not withstand. Secondly, while the introduction of physical probes inevitably alters the flowfield we set out to measure [3], optical diagnostics can realize truly non-perturbative measurements. The coherent nature of the light provided by laser sources can be used to focus the light with lenses which translates into an excellent spatial resolution unachievable with alternative measurement techniques. Furthermore, the availability of pulsed laser sources with repetition rates in the kHz range or faster results in the "freeze of the flow field" with high sampling rates that are thus capable of measuring highly dynamic and transients phenomena in reacting flows [2]. In view of this, many specialized laser-based techniques have been developed to measure key physical and chemical properties in gas-phase reacting flows. Chief among these is laser spectroscopy, which allows for quantitative measurements of important scalar quantities in harsh environments such as flames, plasmas and high-speed/high-enthalpy flows [4, 5]. Raman spectroscopy in particular provides non-perturbative, in situ thermometry in high-temperature flames, as well as concentration measurements of major

F. Mazza · L. Castellanos · D. Kliukin · A. Bohlin (✉)
Department of Aerospace Engineering, Delft University of Technology, Kluyverweg 1, 2629 HS Delft, The Netherlands
e-mail: alexis.bohlin@gknaerospace.com

A. Bohlin
Space Campus, Luleå University of Technology, Bengt Hultqvists Väg 1, 98128 Kiruna, Sweden

Inspection Technologies, GKN Aerospace Engine Systems, Trollhättan, Sweden

D. K. Singh et al. (eds.), *Raman Spectroscopy*, Springer Series in Optical Sciences 248,
https://doi.org/10.1007/978-981-97-1703-3_13

species, and Raman-based techniques are the most commonly employed for scalar determination in combustion environments [1].

Raman scattering is nevertheless a rather inefficient optical process, the typical Raman cross-section being in the order of 10^{-29} cm^2/molecule/steradian in gas-phase media [6]. In addition, spontaneous Raman scattering is an incoherent process, leading to the isotropic emission of the signal: single-shot Raman spectroscopy thus requires large collection optics and suffer from the often presence of luminous background due to e.g. flame chemiluminescence or fluorescence in plasma reactors. The intensity of the scattered radiation can be enhanced by several orders of magnitude if the molecules can be made to re-emit the incident light coherently. In coherent anti-Stokes Raman spectroscopy (CARS) we implement this idea by employing additional input laser fields to excite the Raman modes to be probed in the scattering process. The resulting CARS signal is spatially and temporally coherent, such that it can be detected remotely from the measurement location and easily be discriminated from the incoherent background in luminous harsh environments.

The potential of CARS for scalar determination in gas-phase media [7, 8] was realized soon after the first observation of this phenomenon by Maker and Terhune [9], and continuous efforts over the last five decades have made this technique into the gold standard for gas-phase thermometry and concentration measurements [10]. In this context, two diverging trends can be identified in the development of gas-phase CARS diagnostics. The first is the strive for diagnostics completeness, developing new diagnostics schemes to extend the number of scalar fields observed simultaneously on the basis of a single-shot measurement. This kind of research efforts lead to e.g. multiplex CARS for the simultaneous interrogation of multiple Raman-active species [11, 12], and to the extension of CARS to measure velocity, as well as density and pressure [13, 14]. On the other hand, the relevant use of CARS as a diagnostic tool for industrial applications pushed the technique in the opposite direction, aiming to the development of portable, field-deployable systems, capable of robust operations in practical measurement scenarios [15–17].

The present chapters describes some of the most recent advancements in gas-phase CARS with particular focus on two-beam femtosecond/picosecond (fs/ps) CARS [18] and its development in terms of correlated imaging measurements and wideband multiplex detection capabilities on the one hand, and simplified system architecture on the other. This recent progress establishes two-beam fs/ps CARS as the state-of-the-art for multi-fields scalar measurements in chemically reacting flows and provide a convincing pathway for the prospect use in a wide range of industrial applications, with unparalleled measurements fidelity and spatio-temporal resolution.

13.2 Fundamentals of Coherent Anti-Stokes Raman Scattering

Contrary to linear spontaneous Raman scattering, CARS is a third-order non-linear optical process, consisting of the four-wave mixing (FWM) of three input laser fields—called the "pump" (oscillating at optical frequency ω_{pump}), the "Stokes" (ω_{Stokes}), and the "probe" (ω_{probe})—to produce the CARS polarization field [19]:

$$P^{(3)}_{\text{CARS}} = \frac{3i\omega_{\text{CARS}}}{2nc}\chi^{(3)}_{\text{CARS}}E_{\text{pump}}E^{*}_{\text{Stokes}}E_{\text{probe}} \tag{13.1}$$

where ω_{CARS} is the frequency of the CARS signal, $\chi^{(3)}_{\text{CARS}}$ is the third-order non-linear optical susceptibility of the optical medium, l is the interaction length, and nc is the speed of light in the medium.

As illustrated in Fig. 13.1, the frequency of the CARS field is dictated by the conservation of energy in the light-matter interaction, such that:

$$\omega_{\text{pump}} - \omega_{\text{Stokes}} + \omega_{\text{probe}} - \omega_{\text{CARS}} = 0 \tag{13.2}$$

where the negative sign in front of the Stokes and CARS frequency indicate that these field are emitted from the medium as a result of, respectively, stimulated and spontaneous Raman scattering of the incident pump and Stokes fields.

Two different kind of CARS processes are illustrated in Fig. 13.1a, b: respectively, a ro-vibrationally resonant and a non-resonant (NR) CARS process. The former is realized when the frequency difference between the pump and Stokes fields is resonant to a rotational-vibrational mode of the optical medium, such that the Raman mode is coherently excited. The establishment of rotational-vibrational coherent in the molecular medium results in the coherent scattering of the incident probe field, such that the scattered radiation is spatially coherent. The propagation direction of the CARS signal satisfies the so-called "phase matching" for the non-linear optical process, conserving linear momentum in the process:

$$\Delta\boldsymbol{k} = \boldsymbol{k}_{\text{pump}} - \boldsymbol{k}_{\text{Stokes}} + \boldsymbol{k}_{\text{probe}} - \boldsymbol{k}_{\text{CARS}} = 0 \tag{13.3}$$

where $\boldsymbol{\Delta k}$ is the phase mismatch, and $\boldsymbol{k}_i$ is the wave vector of the ith field involved in the FWM process. The phase matching of the CARS process in a non-collinear BOXCARS [20] geometry is sketched in Fig. 13.1c.

The resonant CARS process is subject to the same selection rules that apply to spontaneous Raman scattering. In the simplest case of CARS on simple homonuclear diatomic molecules such as N_2, the selection rules for pure-rotational and ro-vibrational CARS are respectively:

$$\begin{aligned} \Delta v &= 0 \\ \Delta J &= \pm 2 \end{aligned} \tag{13.4}$$

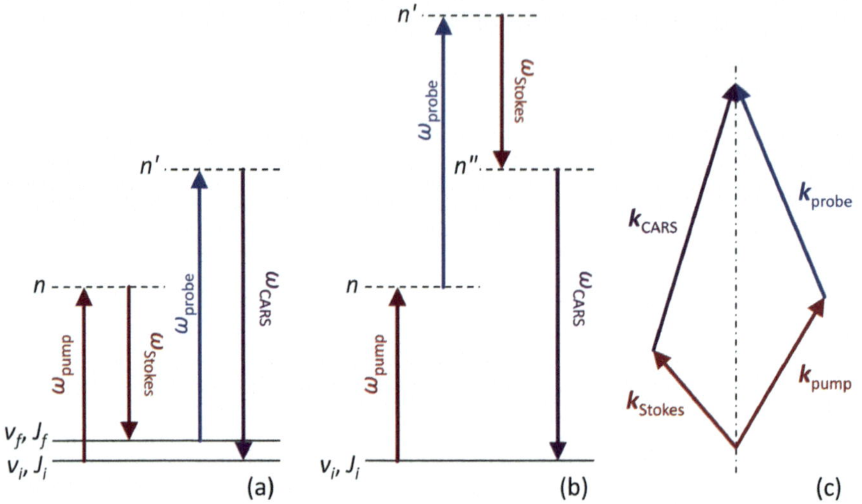

Fig. 13.1 Energy and momentum conservation in the CARS process. Energy diagram for the resonant **a** and non-resonant. **b** CARS processes. In the former, the frequency difference between the pump and Stokes fields is resonant to the molecular transition between the initial (v_i, J_i) and final rotational-vibrational (v_f, J_f) energy states. n, n', n'': virtual electronic states. **c** Phase-matching of the CARS process. The planar BOXCARS phase-marching sketched here allows for a perfectly phase-matched CARS process (i.e. $\Delta \boldsymbol{k} = \boldsymbol{0}$) with crossed-beam geometry

and

$$\begin{aligned} \Delta v &= 1 \\ \Delta J &= 0, \pm 2 \end{aligned} \tag{13.5}$$

where v is the vibrational quantum number and J is the total angular momentum (or rotational) quantum number, and the spectral branches corresponding to the rotational selection rules $\Delta J = -2, 0, 2$ are indicated according to the usual notation as O-, Q- and S-branch, respectively [21]. Through the third-order non-linear optical susceptibility in (13.1) the CARS signal carries information about the statistical population distribution of the Raman-active molecules over their rotational and vibrational energy manifolds. For a thermalized gas-phase medium, this corresponds to the rotational and vibrational Boltzmann distribution: CARS can thus be employed to perform high-fidelity thermometry in gas-phase media.

The NR CARS process in Fig. 13.1b does not involve any rotational or vibrational resonance in the optical medium: upon the interaction with the pump, Stokes and probe fields, the interacting molecule is found in a "virtual" electronic state, for which an instantaneous dephasing can be assumed. The spectrum of the NR CARS signal is then a continuum which carries no information on the internal energy distribution of the scatterers. Nevertheless, this signal plays an important role in the CARS experimental routine, as we will discuss in the following.

13.2.1 Frequency-Resolved CARS

Since its first demonstration in gas-phase media [8], CARS has been developed in the most powerful laser diagnostics to measure temperature and major species concentrations in combustion environments [4]. Of all the many research efforts spent in the development of gas-phase CARS: of particular relevance to the content of present chapter is the development of dual-broadband CARS, first introduced by Eckbreth and Anderson for multi-species spectroscopy [11], and subsequently demonstrated in Aldén et al. for pure-rotational CARS spectroscopy [12].

The main idea behind pure-rotational dual-broadband CARS is illustrated in Fig. 13.2a: by employing broadband pump and Stokes pulses, the whole manifold of pure-rotational modes can be simultaneously excited, allowing for the single-shot detection of all the different rotationally Raman-active species present in the probe volume. Since the first demonstration by Aldén and co-workers, pure-rotational dual-broadband has been vastly employed in combustion diagnostics, both in laboratory flames [22, 23] and in industry-relevant applications [24, 25].

The use of nanosecond (ns) laser sources, typically Nd:YAG lasers, provides high spectral resolution in CARS measurements, allowing for resolving e.g. closely-spaced rotational lines in the ro-vibrational spectrum of N_2 [26], or assigning the spectral lines due to different species in the pure-rotational Raman spectrum [27]. Nevertheless, there are a number of factors that ultimately limit the accuracy and precision of ns CARS techniques. The most significant is certainly the impact of molecular collisions, which determine the dephasing of the rotational-vibrational Raman coherence on a timescale of typically some hundreds ps in atmospheric gas-phase media. The ns pulses employed to realize frequency-domain CARS are much longer than that, hence the collisional Raman linewidths are paramount to accurate CARS measurements [28]. As the collisional environment is typically unknown—the concentration of major collisional species being indeed an objective of CARS measurements–, molecular collisions represent a significant source of uncertainty for ns CARS. Similarly the interference with NR CARS signal, whose intensity depends on the local composition of the optical medium in the probe volume, adds to the uncertainty in the evaluation of the CARS spectra especially at high pressure [24]. Furthermore, the multi-mode broadband dye lasers employed to realize multiplex CARS are affected by mode noise [29], which limits the inherent single-shot precision of CARS thermometry to ~3%.

These limitations have pushed further the development of novel experimental approaches to gas-phase CARS diagnostics, leading to the introduction of time-resolved CARS techniques, which solve most of the major limitations of ns CARS.

13.2.2 Time-Resolved CARS

The development of time-resolved CARS techniques has been propelled by the recent commercial availability of regenerative fs laser amplifiers, providing mJ-level energies with pulse duration <100 fs.

Employing short fs laser pulses, the rotational-vibrational coherence of the target molecules can be impulsively excited and maximized, thus enhancing the intensity of the CARS signal significantly [30]. As seen in Fig. 13.2, the molecular coherence, as encoded in the third-order non-linear optical susceptibility $\chi^{(3)}_{\text{CARS}}$ in (13.1), shows quantum revivals with a period of some ps for the simple diatomic molecules investigated by gas-phase CARS. Hence, the use of ultrashort (ps or fs) probe pulses results in high temporal resolution, such that the evolution of the Raman coherence can be measured in a time-resolved fashion. The main dephasing mechanism for the Raman coherence in gas-phase atmospheric media is due to molecular collisions: time-resolved CARS measurements can be performed below the typical collisional timescale (~100 ps), resulting in near collision-independent measurements [31]. The benefit is two-fold: on the one hand, this reduces the measurement uncertainty due to the unknown collisional environment [32] and, on the other, the collisional Raman linewidths can be directly measured by the time-resolved dephasing of the CARS signal [33, 34]. Furthermore, the high peak power of fs laser pulses, combined with the most efficient impulsive excitation of the Raman coherence, allows for extending the CARS capability beyond point measurements to realize imaging spectroscopy [18, 35–37], as we will detail in the following section.

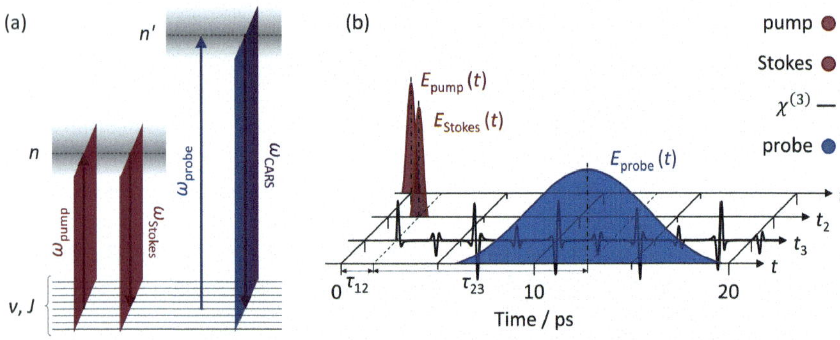

Fig. 13.2 Dual-broadband pure-rotational CARS. **a** Energy diagram for the dual-broadband CARS process. By employing broadband pump and Stokes pulses multiple Raman modes can be simultaneously excited: the excitation bandwidth corresponds to the spectral cross-correlation of the two pulses $E_{\text{pump}}(\omega) \star E_{\text{Stokes}}(\omega)$. **b** Schematic of the hybrid fs/ps CARS process in the time-domain. The combined interaction of the fs pump and Stokes laser pulses excites the non-linear optical response $\chi^{(3)}$ of the gas-phase medium (N_2 is simulated here), which evolves in time through quantum revivals. The longer ps probe pulse "reads" multiple revival peaks, which enhance the encoding of the underlying Raman modes in the frequency-resolved CARS spectrum

A detailed overview of the many time-resolved CARS techniques developed over the past two decades is beyond the scope of the present chapter, and the reader is referred to the specialized literature (see e.g. [2, 10]). In the following, referencing is only made to hybrid fs/ps CARS, which combines broadband fs and relatively narrowband ps laser pulses to perform simultaneously time- and frequency-resolved measurements [38]. As shown in Fig. 13.2b, fs pump and Stokes pulses are used to impulsively excite the Raman coherence in the gas-phase medium, while a ps laser pulse serves as the probe, providing sufficient spectral resolution for analyzing the CARS spectrum. The possibility of delaying the probe pulse with respect to the fs pump and Stokes pulses furthermore allows for time-gating the generation of the NR CARS background. Due to the instantaneous dephasing of the electronic coherence in the virtual states of Fig. 13.1b, this signal is only generated when the pump, Stokes and probe fields are temporally overlapped. In the past ten years, hybrid fs/ps CARS has become established as the state-of-the-art for scalar measurements in gas-phase media, allowing for high-fidelity, in situ thermometry and species concentration measurements in combustion [2, 39] and plasma environments [40, 41].

13.3 Two-Beam fs/ps CARS Instrument

One of the most significant advancements in hybrid fs/ps CARS for gas-phase diagnostics was the development of two-beam fs/ps CARS. This technique is based on a novel phase-matching scheme introduced by Bohlin and Kliewer [18], and first demonstrated by the authors by performing pure-rotational fs/ps CARS over a one-dimensional field-of-view (FOV). Figure 13.3 illustrates the principle of two-beam fs/ps CARS in the time- and frequency-domain. Contrary to the conventional implementation of hybrid fs/ps CARS—which employs two fs laser pulses at distinct carrier frequencies to provide the pump and Stokes fields and excite the target Raman modes–, the two-beam phase-matching scheme makes use of a single broadband fs pump/Stokes laser pulses to excite the Raman coherence. As a laser pulse with duration <100 fs has a large bandwidth spanning some tens of nm (for a carrier wavelength of 800 nm), constructive pump/Stokes frequency pairs can be found across the entire bandwidth of a single fs laser pulse. For example a 45 fs laser pulse, as employed in [18], thus has sufficient bandwidth to effectively excite Raman modes in the pure-rotational region of the spectrum up to ~500 cm^{-1}.

This allows for a great simplification of the hybrid fs/ps CARS instrument as a single regenerative fs laser amplifier provides the laser pulse needed to excite the pure-rotational coherence of most species of interest in gas-phase media. The authors reported significantly improved signal levels, owing to the pulse energy available to the pump and Stokes fields provided by the single fs laser pulse, and to the ease of alignment of the two laser beams. The inherent spatial and temporal overlap of the pump and Stokes fields guaranteed by this two-beam scheme was furthermore proposed to improve the robustness of hybrid fs/ps CARS against beam steering in turbulent measurement environments.

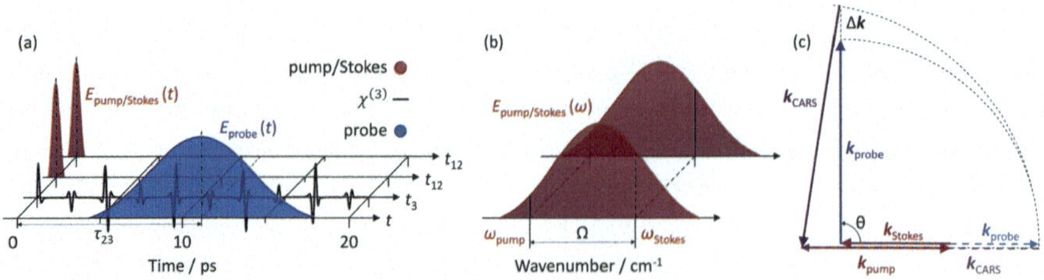

Fig. 13.3 The two-beam CARS phase-matching scheme. **a** Schematic of two-beam fs/ps CARS in the time domain. $E_{\text{pump/Stokes}}(t)$, and $E_{\text{probe}}(t)$ represent the temporal envelops of the combined pump/Stokes and probe pulses, respectively. $\chi^{(3)}$ is the nonlinear optical susceptibility of the gas-phase medium due to the coherent motion of its molecular components. t_{12} and t_3 are the coherence timescales during the interaction with the input laser pulses, while τ_{23} is the intra-pulse delay. **b** Spectral excitation in two-beam fs/ps CARS. Constructive pump/Stokes frequency-pairs at Raman shifts $\Omega = |\omega_{\text{pump}} - \omega_{\text{Stokes}}|$ are found across the bandwidth of a single broadband fs excitation pulse. Since the pulse has finite bandwidth, the excitation efficiency decreases monotonically with increasing Raman shift. **c** Phase-matching of two-beam fs/ps CARS. The two-beam fs/ps CARS process sketched here can only be perfectly phase-matched (i.e. $\Delta \boldsymbol{k} = 0$) if all the laser beams are collinear. Adopting a crossed-beam geometry (i.e. $\theta \neq 0°$, with crossing angle θ), necessarily results in a phase-mismatch that increases with larger Raman shifts and crossing angle [18]

A significant advantage of the two-beam scheme over alternative phase-matching geometries, such as BOXCARS [20] and collinear single-beam CARS [42], is the high spatial resolution available and its straightforward tuneability, achieved by adjusting the incident crossing angle between the pump/Stokes and probe beams. Bohlin and Kliewer demonstrated a longitudinal resolution of their two-beam fs/ps CARS instrument, as dictated by the interaction length between the two laser beams, of ~50 μm for a crossing angle of 90°, an improvement of almost two orders of magnitude with respect of typical BOXCARS configurations [36]. It should be noted nonetheless that the improvement in the spatial resolution comes at the expense of a perfect phase-matching geometry, introducing a phase-mismatch that grows with the crossing angle and Raman shift, ultimately limiting the efficiency of the CARS signal generation for high-energy Raman modes.

The most consequential development of gas-phase CARS introduced with the two-beam phase-matching scheme, is undoubtedly the possibility of performing coherent imaging spectroscopy over linear and planar FOVs. Employing cylindrical focusing optics to form the laser beams into sheets, crossed over a ~6 mm line at the measurement location, and coherent imaging optics to perform spatial imaging spectroscopy, the authors demonstrated the possibility of retrieving spatially-correlated spectral information over this 1D FOV with an ~60 μm line-spread function (LSF). Following work from the same research group demonstrated the use of this technique to investigate the flame-wall interaction in a head-on quenching (HOQ) burner, resolving the temperature gradient evolution through the impingement of the CH_4/air flame onto a metal wall of the HOQ burner and its subsequent quenching [43]. More recently, Escofet-Martin and co-authors have employed a similar two-beam fs/ps

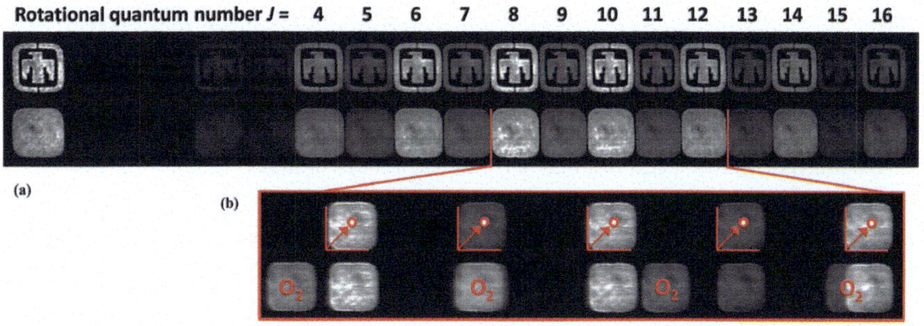

Fig. 13.4 Two-beam fs/ps planar CARS imaging (2D-CARS). **a** Single-shot hyperspectral N_2 CARS image recorded at room temperature. In the first row, a mask is placed on the probe beam path to illustrate the imaging quality of the CARS instrument. **b** High-spectral resolution 2D CARS images acquired in N_2 (first row) and O_2 (second row), each square image corresponding to an isolated line in the pure-rotational Raman spectrum. Each image consists of ~125 × 120 pixels, corresponding to ~15000 spatially correlated spectra recorded with a single-shot. Adapted with permission from Bohlin and Kliewer [37]. © 2013 AIP Publishing LLC

CARS instrument to study the transient heat transfer in the boundary layer at the flame-wall interface [44].

Furthermore, this two-beam fs/ps CARS phase matching scheme was later extended to perform hyperspectral imaging, as shown in Fig. 13.4. Crossing the pump/Stokes pulse laser formed into a laser sheet with an unfocused collimated probe beam, Bohlin and Kliewer showed that the CARS signal can be generated and recorded over the resulting ~2 × 20 mm planar FOV, thus realizing for the first time 2D CARS in gas-phase media [37]. The great diagnostic potential of this 2D CARS instrument was then demonstrated by the same authors, by simultaneously mapping the temperature and species concentration fields over planar slices in a laminar premixed CH_4/air flame [45] and in a turbulent H_2 jet flame [46].

13.3.1 State-of-the-Art Capabilities

As a continuation on these works we have recently introduced several advancements to the pure-rotational two-beam fs/ps CARS technique. For example, the possibility of employing a single ultrafast regenerative amplifier laser system to build a two-beam fs/ps CARS instrument with single-shot 1D imaging capabilities at temperatures higher than 2000 K was first demonstrated in Castellanos et al. [47]. This system employs an ultrafast regenerative amplifier system (7.5 mJ, 1 kHz, Astrella Coherent) as the single laser source to generate both the broadband fs pump/Stokes pulse and the relatively narrowband probe pulse, as shown in Fig. 13.5. The laser amplifier provides two repetition-wise synchronized output beams. The first one is a 4.5 mJ compressed near transform-limited (TL) 35 fs duration laser pulse centered at a wavelength of 800 nm, which undergoes second-harmonic bandwidth compression

(SHBC, Light Conversion) to generate a narrowband probe pulse. In the SHBC unit the 35 fs TL pulse is equally split in two pulses, to which a conjugate linear chirp is applied before recombining them in a $\chi^{(2)}$ non-linear crystal [48]. The sum-frequency generation (SFG) of the phase-conjugated chirped pulses results in an ~4 ps probe pulse centered at 400 nm, obtained with a ~30% conversion efficiency. Third-order chirp terms arising in the SHBC result in the detrimental presence of spectral side-bands in the spectrum of the ps probe pulse [49], which are here suppressed by introducing a tunable pulse shaper, consisting of two high-resolution transmission gratings (3039 lines/mm, >90% transmission efficiency, Ibsen Photonics), two cylindrical lenses with a 300 mm focal length, and a mechanical slit, all mounted in a 4f arrangement. By tuning the slit aperture in the Fourier plane of the pulse shaper, the probe duration can be tuned in the range ~5–15 ps, with a corresponding pulse energy of 1.2–0.3 mJ. The second, uncompressed, output of the ultrafast regenerative amplifier is sent to an external grating compressor to generate the 35 fs TL pump/Stokes pulse: the use of an external compressor unit allows for a flexible compensation (pre-chirping) of dispersion terms along the optical path, maximizing the excitation bandwidth at the measurement location. The linear polarization states of the pump/Stokes and probe pulses are independently controlled by half-wave plates for 800 nm and 400 nm, respectively. An optical delay line controls the relative arrival of the pump/Stokes and probe pulses at the measurement location.

The same two-beam phase matching geometry of [18] is adopted with cylindrical lenses (CL_h, f: = 500 and 300 mm, respectively) forming the pump/Stokes and probe beams into two laser sheets, and is crossed with an incident angle of ~3°. The FOV is set to ~1.3 mm by employing an astigmatic focusing configuration, with cylindrical

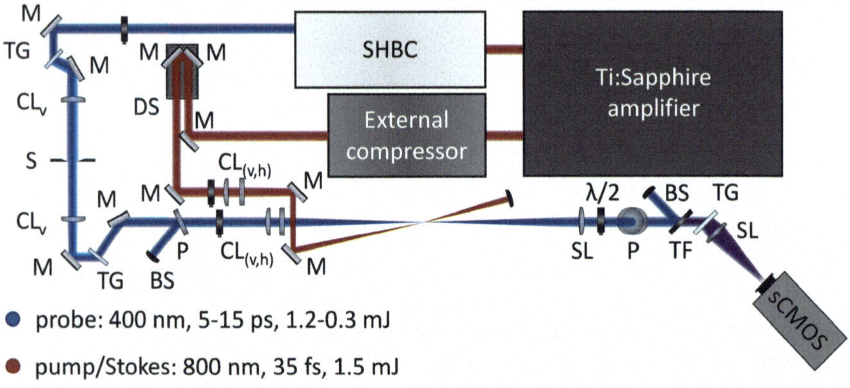

Fig. 13.5 Schematic drawing of the two-beam fs/ps 1D-CARS imaging instrument developed in [47]. A single Ti:sapphire ultrafast regenerative amplifier system is employed to generate the 35 fs pump/Stokes and the ~5–15 ps probe pulses. Cylindrical lenses, mounted in an astigmatic focusing configuration, form the laser beams into two sheets at the measurement location, where they cross to form a ~1.3 mm measurement line. TG, transmission grating; S, slit; CL, cylindrical lens with vertical (v) and horizontal (h) symmetry axes; λ/2, half-wave plate; B, burner; P, polarizer; SL, spherical lens; M, mirror

lenses with vertical symmetry axes (CL_V, f: = 1000 and 300 mm) in combination with horizontal axes compressing the height of the pump/Stokes and probe laser sheets. The CARS signal is collected in a wide-field slit-less coherent imaging spectrometer and imaged 1:1 onto a sCMOS detector (Zyla, Andor), allowing for 200 × 2048 pixel image acquisition at 1 kHz frame rate, thus matching the repetition rate of the laser system. This is the first use of a sCMOS camera in the context of gas-phase fs/ps CARS proving that—despite the limited depth of the quantum well, as compared to conventionally employed back-illuminated water-cooled CCD devices—such detectors can be effectively employed to perform CARS imaging at high repetition rates in combustion environments. The combination of high-resolution transmission grating and 400 mm imaging lens allows for spectral images to be recorded over the whole pure-rotational region of the Raman spectrum 0–525 cm^{-1}, with spectral resolution of 0.25 cm^{-1}/pixel, allowing for the simultaneous detection of major combustion species (for example: N_2, O_2 and H_2).

The 1D-CARS system capability as a powerful combustion diagnostics tool is thus demonstrated by performing simultaneous thermometry and relative O_2/N_2 concentration measurements across a laminar premixed CH_4/air flame front, as shown in Fig. 13.6. The gradients in the two scalar fields are simultaneously resolved along the 1D FOV and tracked in time with a repetition rate of 1 kHz, much faster than the 13 Hz oscillations induced by buoyancy effects in the flame, and well-resolved in the temporal traces of Fig. 13.6. These flame measurements furthermore provide a benchmark for the performance of the CARS instrument in terms of thermometric precision and accuracy, which are estimated to be 0.7% and 2.5% respectively, at a temperature of ~1700 K.

As compared to previous implementations of two-beam fs/ps CARS imaging instruments, such as in [18, 43, 44], the use of a single laser source to provide both the fs pump/Stokes and the ps probe pulses guarantees the inherent synchronization of the pulses at the measurement location, with no temporal jitter. Moreover, the use of a 400 nm probe, as generated by the SHBC from the 800 nm fs pulses, when compared to conventional external ps lasers with a carrier wavelength of 532 nm [18, 43, 44], is beneficial in terms of signal generation and spatial resolution. The intensity of the CARS signal scales with the squared frequency of the signal itself, such that reducing the probe wavelength from 532 to 400 nm results in a gain in the signal intensity by a factor ~2. In addition, employing a shorter wavelength improves the focusing properties of the probe pulse and the diffraction-limited resolution of the imaging optics in the CARS spectrometer. In [47] the line-spread function of the imaging spectrometer is measured to be ~20 μm, a significant improvement over the best reported value of ~40 μm in [50]. The temperature and O_2 concentration gradients in Fig. 13.6 are thus smoothly sampled by ~25 independent measurement points. Its superior spatial resolution makes this CARS imaging instrument ideal for the detailed investigation of the physics of premixed H_2 flame propagation, which is characterized by a steep temperature gradient whose spatial undersampling can give a large measurement uncertainty.

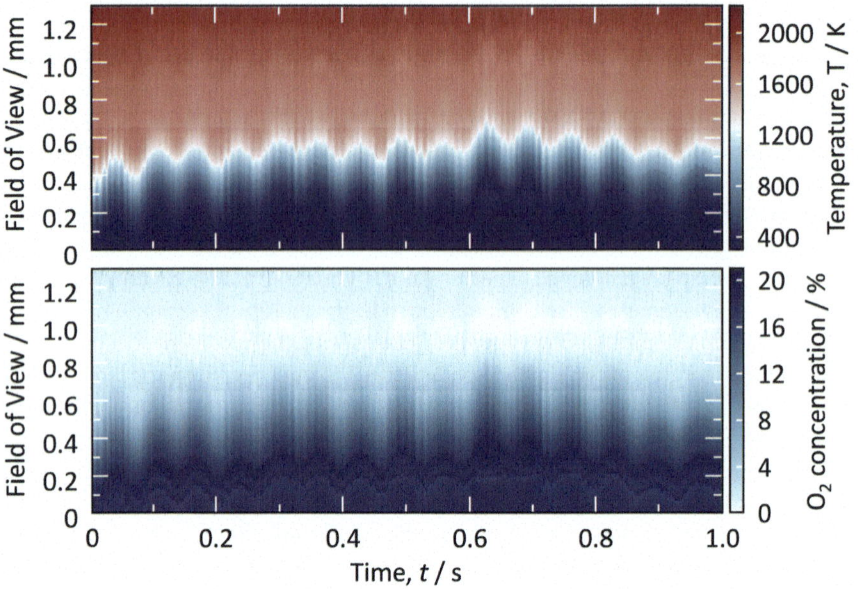

Fig. 13.6 Simultaneous thermometry and relative O_2/N_2 concentration measurements by two-beam fs/ps 1D-CARS imaging across a laminar premixed CH_4/air flame front. **a** Temperature gradient across the flame front mapped by single-shot fs/ps 1D-CARS at 1 kHz. The inflection point in the temperature profile ($T \approx 1200$ K at this flame condition) is used to define the location of the flame front at each single-shot spectral image. **b** Relative O_2/N_2 concentration measured by pure-rotational fs/ps CARS across the laminar flame front. As expected the O_2 concentration gradients is inversely correlated to the local temperature profile

Finally, the space-time correlated resolution obtained with the CARS imaging measurements allows for deconvolving fluctuations in the experimental flame platform, due e.g. to buoyancy forces (as observed in Fig. 13.6), Helmholtz resonances in the burner, or fluctuation in the mass flow-controllers. Apart from the importance on deducing the flame physics more accurately, the capability in space-time correlated resolution is also beneficial at the assessment of the performance of the fs/ps CARS instrument itself. If the uncertainty in the boundary conditions is not mapped properly, the presence of large fluctuations built-in at the benchmarking experimental platform can largely shadow the inherent stochastic error, which is solely dependent of the laser diagnostic instrument. Barros and co-workers, for example, have recently reported a thermometric precision better than 0.5% in a laminar premixed CH_4/air flame when removing the dynamic contribution due to flame oscillations [51]. In [47] as well as in Mazza et al. [52], analogous flame oscillations are resolved in the temporal evolution of the temperature gradient, measured over the 1D FOV. This space-time correlated resolution thus allows for defining the flame front location at each single-shot spectral image—identified as the position of the inflection point in the temperature gradient–, and tracking its oscillations as shown in Fig. 13.7. Since these oscillations are only mechanical in nature, the measurement statistics

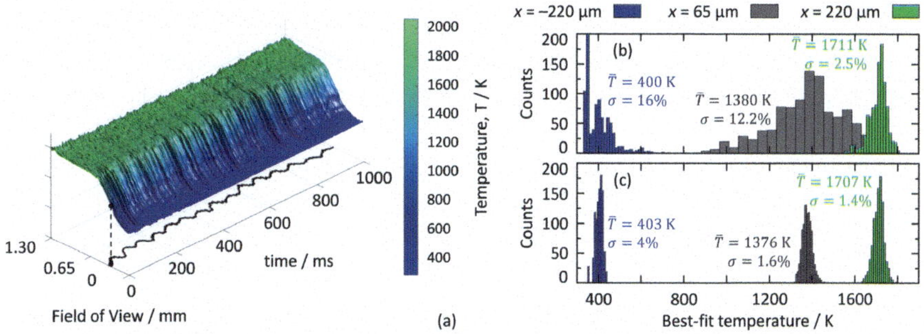

Fig. 13.7 Conditional mapping of the flame front by fs/ps 1D-CARS imaging. **a** Temperature gradient across the laminar premixed CH_4/air flame front, as mapped by single-shot fs/ps 1D-CARS imaging over the ~1.3 mm FOV performed at 1 kHz. The location of the flame front is subject to low-frequency oscillations at ~13 Hz, which affect the evaluation of the single-shot precision of the CARS instrument. **b** Histogram of the temperature measured by pure-rotational fs/ps CARS at three distinct locations along the flame front. In the reactants (blue) 220 μm before the temperature inflection point ($x = 0$ mm); inside the reaction layer (gray) at $x = 65$ μm; and in the combustion products (green) at $x = 220$ μm after the temperature inflection point. **c** Histogram of the temperature measured at three locations conditional to the position of the flame front in each single-shot spectral image. The conditional mapping of the flame front position removes the measurement uncertainty arising from the physical oscillations in the flame platform. The inherent single-shot precision of the fs/ps CARS instrument can thus be assessed

can thus be collected conditional to the distance from the flame front, rather than at a fixed point over the FOV. Figure 13.7 illustrates the substantial improvement in the resulting standard deviation of the temperature statistics measured over 1000 single-shot spectra at three locations in the flame. Upon deconvolving the physical fluctuations of the flame platform used to benchmark the fs/ps CARS instrument, the thermometric precision is consistently found to be ~0.7% in [47, 52].

13.3.2 Limitations

The complex chemistry often encountered in chemically reacting flows in industrial applications pose rather demanding requirements on the fidelity of the experimental data available to validate predictive engineering models. Temperature, in particular, is the key scalar controlling the chemical reaction rates and thus determining the chemical pathways for the reactions. An instructive example of this being the impact of the flame temperature on the formation of NO_x in methane combustion at the pressure and temperature found in a typical aircraft turbo-engines (inlet conditions: 30 atm, 900 K). A measurement uncertainty of ~1% in the measured temperature results in a change of ~20% in the amount of thermal NO_x predicted by Arrhenius-type reaction rates [53]. The high spectral stability of presently commercially-available fs laser systems, and the development of time-resolved CARS techniques –minimizing the impact of the unknown collisional environment on the CARS spectra, and the relating experimental uncertainty, even at pressures as high as 70 bar! [49]—presently make hybrid fs/ps CARS the gold standard for spatially resolved thermometry. The single-shot precision and the accuracy of hybrid fs/ps CARS at flame temperature are estimated to be ~1–2% and ~2–3%, respectively [50, 54, 55]. Furthermore, the possibility of employing a single ultrafast regenerative amplifier system to generate all the laser pulses has recently pushed the precision of hybrid fs/ps CARS thermometry consistently below the 1% limit [47, 51, 52, 54, 56].

An outstanding source of uncertainty in two-beam fs/ps CARS measurements is the excitation efficiency of the various Raman modes within the bandwidth of the fs pump/Stokes pulse [57–59]. As a single broadband fs laser pulse is employed to impulsively excite the rotational Raman coherence, the number of constructive pump/Stokes frequency pairs available to excite a specific Raman mode is determined by the spectral autocorrelation of the fs pulse itself. This can be measured by the spectrum of the NR CARS signal generated in a medium with no active Raman modes, e.g. in Argon (Ar). As no rotational-vibrational coherence is generated in the medium, the NR CARS signal is instantaneously dephased and its spectrum is a continuum profile that maps the spectral excitation efficiency. The conventional fs/ps CARS experimental routine thus entails the measurement of the NR CARS spectrum, to subdivide the resonant CARS spectra (spectral referencing) as shown in Fig. 13.8, such that quantitative information about the local temperature and composition of the gas-phase medium can be retrieved. The NR CARS signal is then measured ex situ in a non-resonant Ar, sequentially to the CARS experiment. The effective excitation efficiency profile available at the measurement location (in situ) can nonetheless be affected by the shot-to-shot stability of the laser source employed [29, 57, 60, 61], as well as by the propagation of the laser pulses to the measurement location.

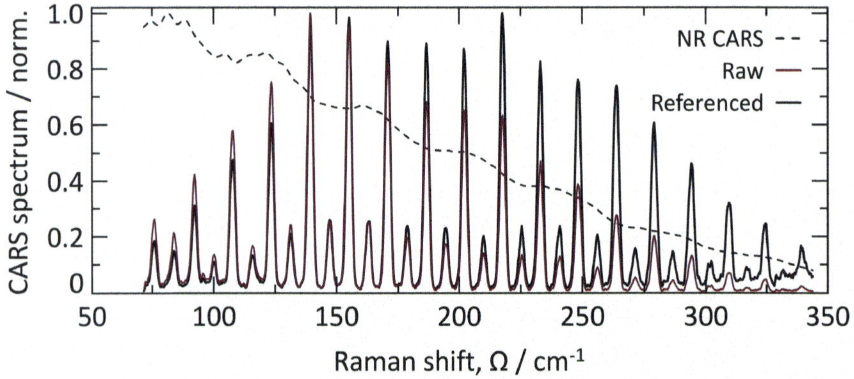

Fig. 13.8 Spectral referencing for pure-rotational CARS. The single-shot pure-rotational N_2 CARS spectrum (solid red line), acquired in a laminar premixed CH_4/air flame, is subdivided by the NR CARS spectrum acquired ex situ in room-temperature Ar (dashed line). The solid black line is the resulting referenced spectrum that can meaningfully be compared to the rotational Boltzmann population distribution of the N_2 molecules

Different schemes have been demonstrated for the concurrent ex situ generation of the NR CARS signal top monitor the shot-to-shot fluctuations in the spectral excitation efficiency [16, 58, 59]. Nevertheless, these cannot account for the temporal chirp introduced to short fs pump/Stokes pulses as they propagate through different optical materials towards the measurement location. Hence, the impact of the spectral referencing protocol on the accuracy of fs/ps CARS thermometry is still largely unexplored.

Besides there is an increasing interest in the development of fs/ps CARS to realize complementary spectroscopy, by simultaneously investigating the pure-rotational and ro-vibrational regions of the Raman spectrum [62]. In the context of gas phase diagnostics, there are several motivations to extend the applicability of two-beam fs/ps CARS beyond the pure-rotational region:

- Some molecules that play a major role in combustion environments and other chemically reacting flows are difficult to detect and quantify by means of pure-rotational CARS. For example, major combustion products or pollutant species (e.g. H_2O [63] and CO [64]) have extremely low rotational Raman cross-sections, such that they give rise to weak CARS signals in high temperature environments. The example of water vapor is rather illustrative: H_2O is a major product of combustion processes and CARS techniques have been applied to its ro-vibrational (ν_1 mode) spectrum since the 1980s to measure temperature and concentrations in flames [65–68]. On the other hand, the application of pure-rotational CARS to water vapor lead Nordström and co-workers to the conclusion that it would not be possible to use this technique to measure temperature and H_2O concentrations in combustion environments [63]. Even worse is the case of CH_4, a spherical-top polyatomic molecule with no rotational degree of freedom in its ground vibrational state, and thus rotational Raman-inactive.
- The H_2 molecule is rotationally Raman-active however, since it is light weight, has a relatively low moment of inertia spinning around its centro-symmetric axis, and has an uncommonly large rotational constant ($B \approx 60.8$ cm^{-1} [69]), such that its pure-rotational spectrum spans more than 1500 cm^{-1} at flame temperature [70]. The application of fs/ps CARS to investigate the H_2 spectrum in the pure-rotational region is thus limited to the first two spectral lines at 354 and 587 cm^{-1}, as shown in Fig. 13.9. While these are typically sufficient to measure the relative H_2/N_2 concentration at low temperature [71, 72], the shift in the rotational Boltzmann population distribution with the increasing temperature ultimately limits the detection of H_2 by pure-rotational fs/ps CARS [73].
- Alongside its use in combustion diagnostics, gas-phase CARS has found vast application to perform thermometry in harsh plasma environments. The use of laser spectroscopy in this context is particularly useful as it can provide complementary information on the vibrational and rotational Boltzmann population distributions where significant thermal non-equilibrium conditions exist (e.g. dielectric-barrier discharges [74] and arc discharges [75]). The ability of simultaneously performing pure-rotational and ro-vibrational spectroscopy on the same Raman-active molecule, e.g. by dual-broadband dual-pump fs/ps CARS as in

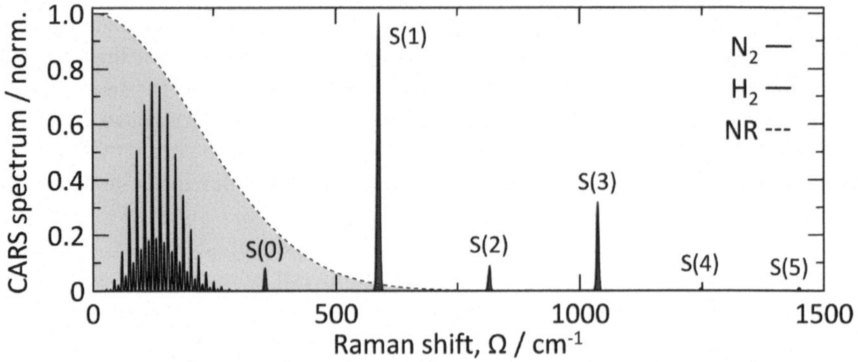

Fig. 13.9 Pure-rotational two-beam fs/ps CARS for simultaneous thermometry and relative H_2/N_2 concentration measurements. The simulated excitation profile provided by a 35 fs TL pump/Stokes pulse is represented by the shaded red area: only the first two lines in the pure-rotational Raman spectrum of H_2 can be coherently excited by the spectral breadth of this pump/Stokes pulse

[40] makes fs/ps CARS the ideal tool to investigate rotational-vibrational non-equilibrium. On these grounds ultrabroadband CARS has been developed, which employs supercontinuum laser pulses—with bandwidth of many tens or hundreds of nm—to simultaneously excite the rotational and vibrational coherence of the Raman-active molecules over a large portion of the Raman spectrum. A first example of this approach is found in Tedder et al., who employed a broadband ns dye laser ($\sim$18 nm FWHM) to perform dual-pump CARS, simultaneously probing N_2, O_2, H_2, C_2H_4, CO and CO_2 in the spectral region ~1000–2500 cm^{-1} [76].

13.4 Polarization-Sensitive CARS Imaging Spectrometer

The measurement uncertainty introduced by the spectral referencing protocol in pure-rotational fs/ps CARS thermometry is addressed in Mazza et al. by introducing a novel experimental protocol for the concurrent in situ generation of the NR CARS signal [52]. This protocol is implemented by a polarization approach [62, 77], whereby the linear polarization states of the pump/Stokes and probe pulses is determined such that the resonant and NR CARS signal are cross-polarized. Introducing a polarization gate in the wide-field coherent imaging spectrometer is thus possible to discriminate the two signal components and record them simultaneously in distinct detection channels.

To this effect, a novel polarization-sensitive coherent imaging spectrometer is developed in [52], as illustrated in Fig. 13.10. Half-wave plates for 800 and 400 mm control the linear polarization of the pump/Stokes and probe pulses respectively, and the relative polarization angle between the two is set to be $\tan^{-1}\sqrt{6} \approx 67.8°$. This "magical" angle is demonstrated to generated the NR CARS signal with linear polarization orthogonal to that of the resonant CARS signal for completely depolarized

Raman transitions, as is the case for pure-rotational spectra [73]. The two signal components are nearly co-propagating with the probe beam, for the two-beam phase matching geometry and shallow crossing angle employed. These are collimated by a 400 mm spherical lens and their polarization is rotated by a 400 nm half-wave plate, to align it to the P- and S-polarization direction of a thin-film polarizer. After the half-wave plate the resonant signal is P-polarized with respect to and transmitted through the polarizer, while the NR signal is reflected by it and collected in a second detection channel. The latter consists of a mirror and a second 400 nm half-wave plate, which turns the polarization of the NR signal by 90°: the signal components have the same polarization, such that the diffraction efficiency of the transmission grating (Ibsen Photonics) is >90% in both detection channels. An angle-tunable band-pass filter (bandwidth FWHM: 20 at 400 nm, Semrock) suppresses the probe beam before the transmission grating. Both CARS signals are imaged 1:1 onto two 200 × 2048 pixels regions of the sCMOS detector by the 400 mm spherical lens, so that high-resolution CARS spectra can be acquired in the pure-rotational region up to ~500 cm^{-1}. This spectrometer design allows for multidimensional fs/ps CARS, as the spatio-temporal and spectral resolutions are complemented by the sensitivity to the linear polarization angle of the CARS signal. An innovative design combining the polarization-sensitive dual-channel configuration with 2D hyperspectral imaging capabilities is further proposed in [78]: this extends the CARS measurement capability over 5 dimensions (x–y in space, in spectrum, in time, and in polarization), while avoiding the convolution of space and frequency for relatively broadband spectral features.

The use of the coherent imaging spectrometer to address the measurement uncertainty inherent to the spectral referencing protocol [52] is addressed in the next paragraph. In Kliukin et al. the polarization sensitivity of the spectrometer is exploited to

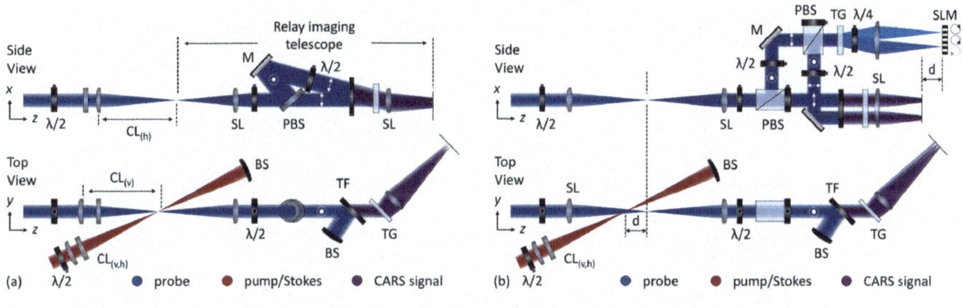

Fig. 13.10 Schematic of the polarization-sensitive coherent imaging spectrometer. Two separate detection channels for P- and S-polarized light, with orientation determined with respect to the polarization beam splitter (PBS), are relay-imaged onto the sCMOS sensor. **a** 1D CARS imaging configuration with astigmatic cylindrical focusing of the pump/Stokes and probe beams. Adapted with permission from Mazza et al. [52]. © 2020 Elsevier Inc. **b** 2D hyperspectral imaging. A 4f pulse shaper, consisting of transmission grating (TG), spherical lens (SL) and spatial light modulator (SLM), allows for the application of the in situ referencing protocol to the 2D CARS spectral images. Adapted with permission from Kliukin and Bohlin [78] M, mirror; λ/2 half-wave plate; λ/4, quarter-wave plate; BS, beam stop; TF, tunable band-pass filter; CL, cylindrical lens

investigate the origin of a high-order cascaded CARS (CCARS) signal and proposed as a novel diagnostic tool to measure local fluctuations in the flow composition with unparalleled sensitivity [79]. This work is illustrated in Sect. 13.4.2. Finally, the polarization approach developed for the discrimination of the resonant and NR CARS signals can be extended to realize dual-probe fs/ps CARS [80–82] and measure the effect of the collisional environment on the pure-rotational Raman coherence. As discussed in Sect. 13.4.3 this approach can be employed to determine the concentration of water vapor in H_2/air flames [72], which cannot be attained by conventional frequency-resolved CARS in the pure-rotational region.

13.4.1 In Situ Referencing of the Spectral Excitation Efficiency

The polarization-sensitive coherent imaging spectrometer described in the previous paragraph is employed in Mazza et al. to perform coherent Raman imaging thermometry with in situ referencing of the spectral excitation efficiency [52]. The resonant and NR CARS signals are simultaneously generated over the 1D FOV placed orthogonally across a laminar premixed CH_4/air flame front, as shown in Fig. 13.11.

The gradient in the temperature and O_2 concentration along the reaction layer of the flame is evidently manifested in the resonant CARS image detected in

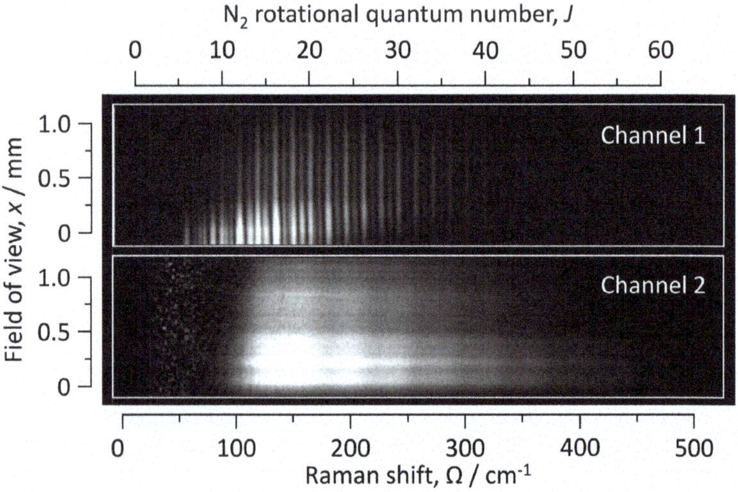

Fig. 13.11 Single-shot fs/ps CARS imaging in the pure-rotational region of the Raman spectrum, with simultaneous detection of the resonant and NR CARS signals. The measurements are performed across a laminar premixed CH4/air flame front ($\phi = 0.95$). The resonant CARS signal is detected "Channel 1" and correspond to the pure-rotational CARS spectrum of N_2 and O_2. The broadband NR CARS signal is detected in "Channel 2". Adapted with permission from Mazza et al. [52]. © 2020 Elsevier Inc.

"Channel 1". In the lower portion of the FOV (i.e. from $x = 0$ mm) the cold reactants are probed and the pure-rotational CARS spectra present the spectral signatures of N_2 and O_2, the main component of ambient air. In addition, only lower rotational energy states are thermally populated at room temperature according to Boltzmann statistics. Moving inside the thin chemical reaction layer of the flame front, O_2 is consumed in the combustion reaction and the heat released by it increases the local temperature, as measured by the pure-rotational Raman spectrum on N_2, which spreads to higher rotational energy states. The temperature change is also evident in the reducing intensity of the NR CARS signal along FOV in "Channel 2", in agreement with the quadratic scaling of the CARS signal intensity with the number density of the gas-phase medium. Each NR spectrum along the FOV of Fig. 13.11 maps the excitation efficiency of the pure-rotational Raman modes in the corresponding resonant CARS spectrum at the same location along the measurement line.

In order to understand the impact of the spectral referencing protocol on the accuracy of pure-rotational fs/ps CARS thermometry, the NR CARS spectra recorded in situ simultaneously to the flame spectra are compared to the NR spectra acquired sequentially in a room-temperature Ar flow (ex situ) in Fig. 13.12. The difference between the two spectral referencing protocol is clarified by comparing the NR CARS spectra measured ex situ and in situ at two different locations along the 1D FOV of the fs/ps CARS instrument. At $x = 0.4$ mm, corresponding to the average location of the temperature inflection point in the flame front measurements, the two average NR CARS spectra perfectly coincide. Adopting either the ex situ or in situ referencing protocols results in the same normalization of the resonant CARS spectra: the conventional ex situ referencing protocol does not introduce any systematic bias in pure-rotational fs/ps CARS thermometry at this location. The situation is quite different when considering the NR CARS spectrum generated in the hot product gases of the flame at location $x = 1$ mm, and comparing it to the NR CARS spectrum measured in Ar at the same location. The effective excitation efficiency available at this measurement location, as mapped by the in situ NR CARS spectrum, is larger than assessed by the NR spectrum measured ex situ. The conventional experimental protocol is thus predicted to introduce a systematic overestimation of the local temperature at this measurement location in the flame.

The mismatch between the spectral excitation efficiency profiles measured ex- and in situ is explained in [52] by the fs pump/Stokes pulse propagation to the measurement location and the impact of the temporal chirp imparted by different optical materials on the effective spectral excitation efficiency. In particular, the temperature gradient across the FOV entails a corresponding gradient in the effective refractive index of the optical medium. Hence, different portions of the pump/Stokes beams experience a different group velocity dispersion (GVD) while propagating through the flame to the measurement location. The relatively cold gas-phase medium in the reactants and pre-heat zone of the flame introduces a larger GVD than the rarefied hot product gases, resulting in the observed variation of the spectral excitation efficiency over the FOV of the fs/ps CARS instrument. The conventional ex situ referencing protocol is unable to account for the effect of the local temperature and composition on the fs pump/Stokes pulse propagation, and thus introduce a systematic bias in

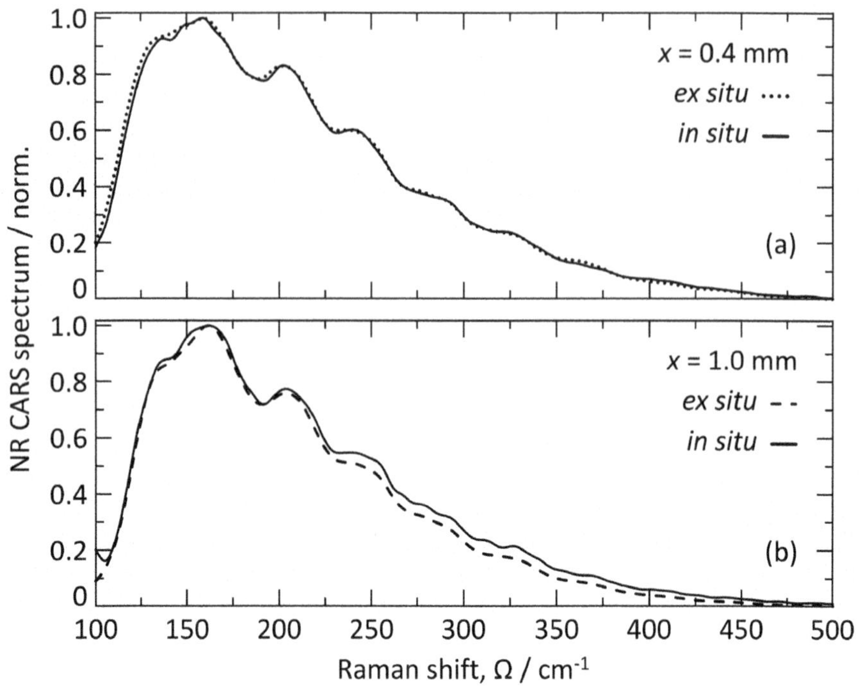

Fig. 13.12 Ex situ and in situ referencing for 1D pure-rotational fs/ps CARS imaging. **a** Comparison between the ex- (dotted line) and in situ (solid line) spectral excitation efficiency profiles at $x = 0.4$ mm. **b** Comparison of the excitation profiles measured ex- (dashed line) and in situ (solid line) in the products of the flame at location $x = 1$ mm

the fs/ps CARS thermometry. This is estimated in the flame experiment of [52] to amount to ~1.3% of the evaluated temperature.

These results show how the spectral referencing protocol can introduce an additional uncertainty in pure-rotational fs/ps CARS thermometry, which is most relevant when short fs laser pulses (<50 fs) are employed to impulsively excite the rotational Raman coherence. This uncertainty should be taken into account when pursuing exact CARS diagnostics, to cope with the demanding fidelity requirements for experimental measurements in chemically reacting flows. The novel experimental protocol for the in situ referencing of the spectral excitation efficiency is thus proposed to further advance state-of-the-art pure-rotational fs/ps CARS, pushing the measurement uncertainty (both stochastic and systematic) consistently below the "dream-limit" of 1%.

13.4.2 CCARS

The large peak power offered by fs laser pulses not only allows for the efficient excitation of the Raman coherence and the possibility of extending the measurement capability to fs/ps CARS imaging, but can also results in higher-order non-linear optical effects accompanying the CARS process [83].

One such higher-order process is the cascading effect of the CARS signal depicted in Fig. 13.13. When the amplitude of the CARS field is sufficiently high, it can be significantly scattered by the Raman-active molecules, effectively acting as a probe field in a second-order coherent scattering process. This cascading effect can proceed to higher orders, as reported e.g. by Liu et al., who observed CCARS up to the 58th order in gaseous CO_2 [84]. As the spectrum of the CARS signal in Fig. 13.13d is discrete in nature—each spectral line corresponding to a pure-rotational Raman transition of N_2–, the resulting CCARS spectrum is the superposition of the pure-rotational Raman spectra originated by its individual lines. This explains the ~4:3 intensity ratio for even and odd lines in the CCARS spectrum, while the well-known 4:1 ratio in the CARS spectrum is explained by the nuclear spin multiplicity of the N_2 molecule. Furthermore, the pure-rotational O- and S-branch CCARS spectra are shifted by $6B$ (with rotational constant $B \approx 2\ \text{cm}^{-1}$ for N_2) on both sides of each spectral line of the CARS spectrum. As a result, the CCARS lines appear exactly between the lines of the CARS spectrum and are well resolved by the coherent imaging spectrometer discussed above. The theory developed to describe the polarization dependence of the CARS signal on the polarization states of the pump/Stokes and probe pulses can be easily extended to describe the CCARS signal, as sketched in Fig. 13.13d. By employing a relative polarization angle of 70.5° between the pump/Stokes and probe pulses, the CARS and CCARS signals are cross-polarized and can be independently recorded in the two detection channels of the polarization-sensitive coherent imaging spectrometer.

The simultaneous single-shot detection of the CARS and CCARS signals in gas-phase media offers a unique opportunity to develop a highly non-linear optical diagnostic tool with unparalleled sensitivity to local changes in species concentrations. This is because of the CCARS process being a $\chi^{(5)}$ non-linear optical process, characterized by the following scaling law:

$$I_{\text{CCARS}} \propto \left|\chi^{(3)}_{\text{CARS}}\right|^4 \left(\frac{\omega_{\text{CARS}}\omega_{\text{CCARS}}}{c^2}\right)^2 l^4 I^4_{\text{pump/Stokes}} I_{\text{probe}} \tag{13.6}$$

where $I_{\text{pump/Stokes}}$ and I_{probe} are the intensities of the pump/Stokes and probe laser pulses, and ω_{CCARS} is the frequency of the CCARS field. Through the third-order non-linear optical susceptibility of the optical medium $\chi^{(3)}_{\text{CARS}}$ the CCARS signal scales with the fourth power of the number density of molecular scatterers, $I_{\text{CCARS}} \sim N^4$. This scaling relationship is verified in [79] by varying the relative concentration of N_2 in binary mixture with Ar from 30 to 80%. The CARS and CCARS signal are simultaneously recorded in the two detection channels of the polarization-sensitive

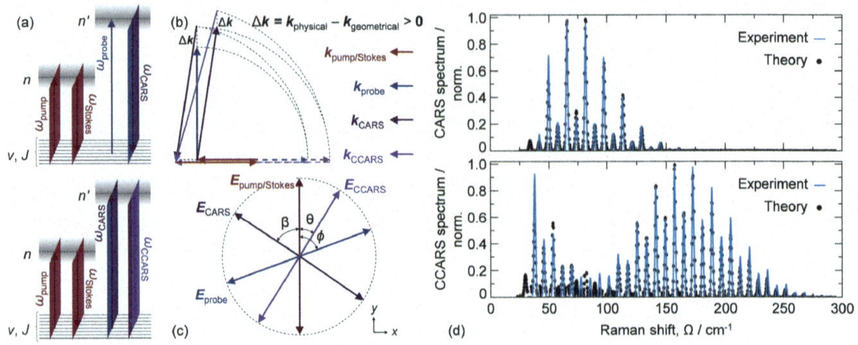

Fig. 13.13 Cascaded fs/ps CARS. **a** Energy diagram of two-beam pure-rotational fs/ps CARS and CCARS. **b** Two-beam quasi phase-matching diagram for the CARS and CCARS processes with (i) all-collinear beams and (ii) 90° crossing angle between the pump/Stokes and probe beams. The CCARS signal is generated by the CARS field acting as a probe beam. **c** Polarization settings to generate the CARS and CCARS signals with orthogonal polarization states. **d** Single-shot fs/ps CARS (top) and CCARS (bottom) spectra recorded in room-temperature N_2. Adapted with permission from Kliukin et al. [79]. © 2021 John Wiley & Sons Ltd.

coherent imaging spectrometer and subdivided to measure the absolute concentration of N_2 molecules in the probed volume, according to $I_{\text{CCARS}}/I_{\text{CARS}} \sim N^2$.

The simultaneous detection of the CARS and CCARS signals can thus be employed as a calibration-free diagnostic for absolute concentration and density measurements. In particular, owing to exponential-order higher sensitivity of the CCARS signal to the number density of the molecular scatterers, this technique could find application in the quantification of local fluctuations in bulk mixture composition, due e.g. to differential diffusion effects. A prospect application is the measurement of the H_2 transport in turbulent flows, where the large mass diffusivity of the lightweight H_2 molecule is known to play a significant role, and experimental data are much needed for the validation of numerical models [85].

13.4.3 CARS Probing the Invisible!

The low rotational Raman cross-section of the H_2O molecule and the complexity of its pure-rotational Raman spectrum [63] present a formidable obstacle to the detection and quantification of water vapor in high-temperature environments via pure-rotational fs/ps CARS. On the other hand, water vapor is one of the major products of the combustion reaction, and the only one such in the case of H_2/air flames: its quantification is thus critical to be able to monitor the progress of the chemical reaction in situ. While a pure-rotational CARS signal generated by the water molecules in these environments, this is effectively "invisible" to the fs/ps CARS instrument, as it is largely overshadowed by the orders-of-magnitude stronger pure-rotational CARS signal from N_2 molecules [63].

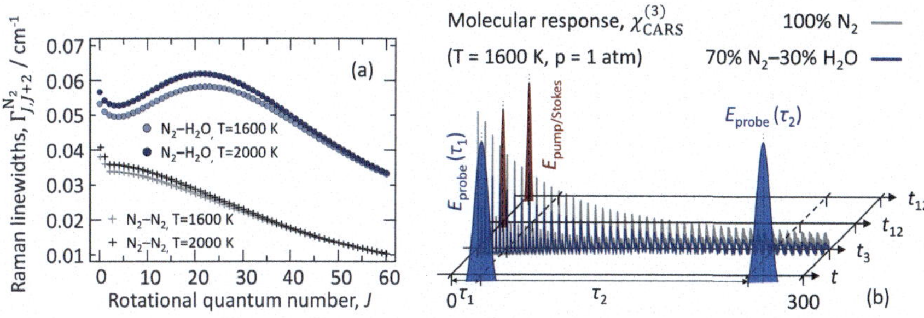

Fig. 13.14 Detection scheme for water vapor via time-resolved measurements of the collisional dephasing of the pure-rotational Raman coherence of N_2. **a** Collisional Raman linewidths for N_2 (radiator) perturbed by N_2 and H_2O collisional partners at combustion-relevant temperatures. **b** Dual-probe time-resolved fs/ps CARS measurements of the collisional dephasing of the pure-rotational Raman coherence of the N_2 molecules

Castellanos et al. present an alternative approach to the measurement of water vapor concentrations in H_2 flames by the time-resolved collisional dephasing of the pure-rotational N_2 CARS signal [72]. While the signature of the H_2O molecules can hardly be detected in the pure-rotational Raman spectrum, these play a significant impact on the evolution of the pure-rotational coherence of the more easily-detected N_2 molecules. Under atmospheric pressure conditions, the evolution of the rotational coherence of a thermal ensemble of N_2 molecules is determined by inelastic collisions between the coherently-rotating N_2 molecules (radiators) and other collisional partners (perturbers) [86]. The rotational energy transfer (RET) determined by inelastic collisions is thus responsible for the dephasing for the dephasing of the pure-rotational CARS signal. The RET rates, quantified by the collisional Raman linewidths $\Gamma^{N_2}_{J,J+2}$, depend on: (i) the rotational state of the radiator, (ii) the temperature of the gas-phase medium, and (iii) the nature of the perturber molecules as seen in Fig. 13.14. In particular, H_2O perturbers are responsible for a larger dephasing of the N_2 CARS signal than N_2 perturbers and the N_2-H_2O Raman linewidths present a largely non-monotonic dependence on the rotational quantum number J [87].

In H_2/air flames, water vapor is the only major species that cannot be directly detected in the frequency domain, while N_2, O_2 and H_2 all present strong signatures in the pure-rotational region of the Raman spectrum. Temperature, density and relative O_2/N_2 and H_2/N_2 concentrations can be simultaneously measured by use of the pure-rotational fs/ps CARS instrument described in the previous section. Hence, the collisional environment and resulting dephasing of the pure-rotational N_2 CARS signal can be specified up to one unknown: the H_2O concentration. The diagnostic strategy presented in [72] exploits the simultaneous spectral and temporal resolution of this pure-rotational fs/ps CARS instrument to measure the collisional dephasing of the N_2 CARS signal (resulting in the collisional Raman linewidths $\Gamma^{N_2}_{J\rightarrow J+2}$) and deduce the H_2O mole fraction according to

$$X_{H_2O} = \frac{1}{\Gamma_{J\to J+2}(N_2 \ - H_2O)}\left[\Gamma^{N_2}_{J\to J+2} - \sum_k X_k \cdot \Gamma_{J\to J+2}(N_2 \ - M_k)\right] \quad (13.7)$$

where $\Gamma_{J\to J+2}$ are the Raman linewidths due to inelastic collisions with the kth molecular perturber (N_2, O_2 or H_2) present in the probe volume with mole fraction X_k. Near collisional-independent measurements of the temperature and relative O_2/N_2 and H_2/N_2 concentrations are performed thus at short probe delay ($\tau_1 \approx 20$ ps), while the H_2O concentration is estimated by the collisional dephasing of the N_2 CARS signal at longer probe delays ($\tau_2 > 100$ ps), as shown in Fig. 13.15. The whole set of relevant scalar fields are thus measured across a laminar H_2/air diffusion flame front in [72]: temperature is measured simultaneously to all the major combustion species, yielding absolute concentration measurements of N_2, O_2, H_2, and H_2O.

As the temperature and concentrations are assumed as constant in a laminar flow, sequential measurements of the collisional dephasing of the pure-rotational N_2 CARS signal are performed, by delaying the probe pulse up to 360 ps. The extension of this diagnostic technique to turbulent flows, where the local temperature and composition fields are highly dynamic, requires the collisional dephasing to be measured simultaneously to temperature and relative O_2/N_2 and H_2/N_2 concentrations. This is achieved in [72] by employing a dual-probe CARS scheme [80–82], whereby the probe beam is split in two pulses: the first probe pulse is delayed by 20 ps with respect to the pump/Stokes pulse, the second one by 250 ps. A novel polarization approach is furthermore demonstrated in Castellanos et al., such that the two probe beams are collinear, and their linear polarization is respectively orthogonal and perpendicular to the linear polarization of the pump/Stokes beam. The resulting CARS signals are thus generated, at time $t = 20$ ps and 250 ps, with orthogonal relative polarization and can be independently recorded in the two detection channels of the polarization-sensitive coherent imaging spectrometer. Figure 13.15 shows two single-shot pure-rotational fs/ps CARS imaging spectra acquired simultaneously at 20 ps and 250 ps at the center of the turbulent canonical "H3" flame provided on the TU Darmstadt/DLR Stuttgart burner [88]. This is a turbulent jet flame, with a center fuel stream consisting of a mixture of 50% N_2–50% H_2. The first two spectral lines of H_2 are detected up to 600 cm^{-1}, allowing for simultaneous thermometry and H_2/N_2 concentration measurements in this region of the flame. The whole set of major scalar fields are thus simultaneously mapped on a single laser-shot over the 1D FOV of the fs/ps CARS instrument. An example is given in Fig. 13.16, where the temperature and absolute concentrations of O_2 and H_2O are simultaneously measured over a sample of 300 single-shot pure-rotational CARS spectra, acquired in the reaction zone of the turbulent jet flame (6.75 mm from its centerline).

The J-dependence of the collisional Raman linewidths shown in Fig. 13.17 offers an additional degree of freedom that could be exploited to extend this diagnostic approach to more complex collisional environments. For example, the time-resolved dephasing of the spectrally-integrated CARS signal could be employed to measure

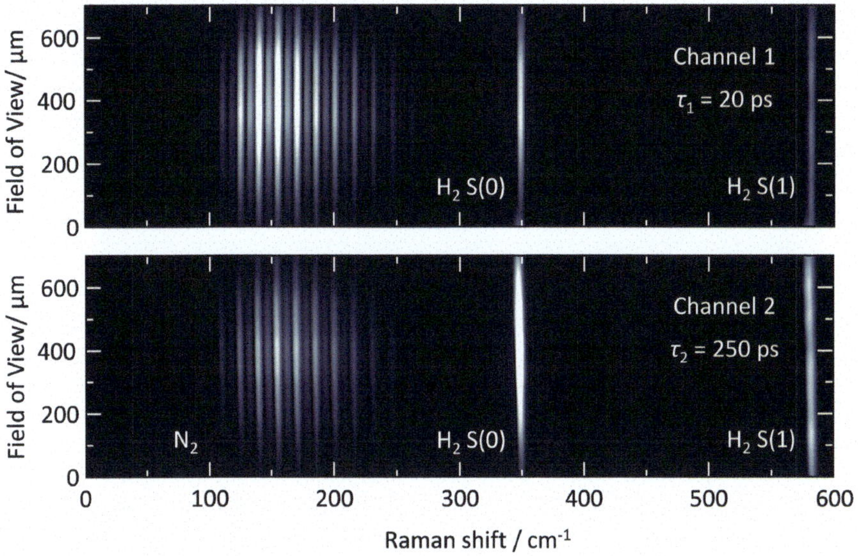

Fig. 13.15 Single-shot dual-probe fs/ps CARS imaging measurements in the turbulent canonical "H3" flame provided on the TU Darmstadt/DLR Stuttgart burner [88]. Time- and frequency-resolved hybrid fs/ps CARS allows for the simultaneous mapping of the whole set of major scalar fields in the flame: the temperature and relative O_2/N_2 and H_2/N_2 concentrations are measured in the frequency domain at $\tau_1 = 20$ ps, while the H_2O concentrations are deduced from the time-resolved collisional dephasing of the N_2 CARS signal at $\tau_2 = 250$ ps

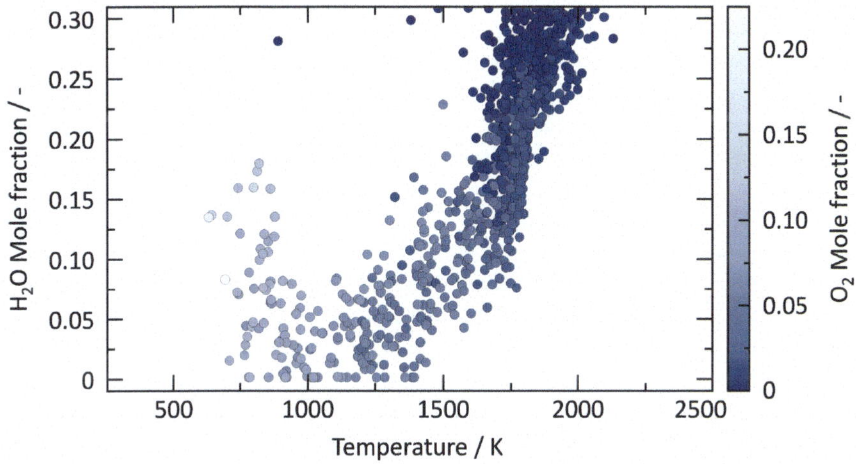

Fig. 13.16 Complete scalar characterization in a turbulent hydrogen jet flame. Each point in the scatter plot represents the local temperature and absolute O_2 and H_2O mole fractions measured 6.75 mm from the centerline of the turbulent jet flame. 300 single-shot fs/ps CARS spectra are acquired in the sample. As expected, strong positive and negative correlations are found between temperature and H_2O and O_2 mole fractions, respectively. Adapted with permission from Castellanos et al. [72]. © 2022 Elsevier Inc.

the local pressure, while the specific dephasing of well-resolved lines in the pure-rotational N_2 spectrum would be sensitive to the H_2O mole fraction. The species-specific dependence of the collisional Raman linewidths on the rotational quantum number could also allow for extending this detection scheme to another species invisible in the pure-rotational Raman spectrum, such as CH_4.

13.5 Ultrabroadband fs/ps CARS with In Situ Filamentation

Over the past ten years, the pursuit of two-beam fs/ps CARS as the ultimate laser spectroscopy for complete scalar determination in chemically reacting flows has led to the development of ultrabroadband fs/ps CARS.

This technique makes use of a compressed supercontinuum laser pulse, a single pump/Stokes pulse, to impulsively excite the pure-rotational and ro-vibrational coherence of the target molecules. Compressing the temporal duration of the pump/Stokes pulse to below ~20 fs the spectral excitation window can be extended beyond the pure-rotational region to low-frequency vibrational modes in the vibrational fingerprint region 500–1800 cm^{-1} (e.g. bending modes of hydrocarbon molecules [50, 89]). Even shorter fs laser pulses can excite the vibrations of light diatomic molecules up to the ro-vibrational Raman spectrum of H_2 at 4200 cm^{-1}, which has a vibrational period of ~8 fs [90]. Ultrabroadband fs/ps CRS can thus be used to perform simultaneous thermometry and absolute concentration measurements of all the major molecular species present in the probed volume, as well as multi-modal thermometry in non-equilibrium environments.

The first demonstration of two-beam ultrabroadband fs/ps CARS imaging for gas-phase detection was presented by Bohlin and Kliewer, who employed a hollow-core fiber and a set of chirped mirrors to generate a 7 fs compressed supercontinuum [60]. This was employed as the ultrabroadband pump/Stokes pulse in the two-beam 1D phase-matching scheme of [18] to perform wideband spectroscopy from the pure-rotational up to the C–H stretch region (~3200 cm^{-1}) of the Raman spectrum. The same authors demonstrated the full diagnostic potential of ultrabroadband fs/ps CARS imaging by achieving wideband chemical detection across a laminar premixed C_2H_4/air flame front. The simultaneous excitation of the whole Raman spectrum up to 4200 cm^{-1} and the ability of performing single-shot spatially correlated measurements across the 1D FOV allowed for the mapping across the chemical reaction layer of the flame [50]. Ultrabroadband fs/ps CARS was later employed as a combustion diagnostic tool to concurrently map the temperature and major species at the flame-wall interaction region on a side-wall-quenching burner [91]. The application of pure-rotational fs/ps CARS to perform direct thermometry on H_2 was then investigated in Courtney et al., where a 7 fs supercontinuum pump/Stokes pulse was employed to excite 8 spectral lines of the pure-rotational H_2 Raman spectrum up to ~1820 cm^{-1}.

The current state-of-the-art in ultrabroadband fs/ps CARS for gas-phase diagnostics is undoubtedly represented by the use of a compressed supercontinuum ro realize 2D hyperspectral imaging of the whole Raman spectrum up to 4200 cm^{-1} [46]. Bohlin and Kliewer demonstrated single-shot hyperspectral CARS imaging to measure the 2D temperature field over planar sections of a H_2 jet flame, as well as to simultaneously map the reactants (H_2 and O_2) distribution across the flame front.

A recent example of the use ultrabroadband fs/ps CARS for the study of industry-relevant chemical processes is found in Ran et al., where this technique was employed for the in situ investigation of carbonaceous feedstock gasification in a CO_2 atmosphere [92]. The authors employed an optical parametric chirped-pulse amplifier (OPCPA) to generate a 7 fs supercontinuum pump/Stokes pulse to simultaneously excite the ro-vibrational Raman spectra of CO_2 and CO. Simultaneous CO_2 CARS thermometry and absolute CO concentration measurements were performed in a gas oven, providing experimental insight in the carbon gasification process. The possibility of performing multiplex spectroscopy on the whole set of Raman-active species present in the probe volume, as well as accurate multi-mode thermometry, makes ultrabroadband fs/ps CARS an appealing to for a vast range of industrial applications relating to the propulsion and power generation [25, 76, 93].

Nevertheless, a number of technical difficulties need to be addressed to make ultrabroadband fs/ps CARS a mature and robust diagnostic tool for extensive use in industrial applications. The first question concerns the shot-to-shot stability of the supercontinuum light source employed to generate the ultrabroadband pump/Stokes pulse. As pointed out by Bohlin and Kliewer, the 7 fs supercontinuum output of the hollow-core fiber was affected by spectral fluctuations, which negatively

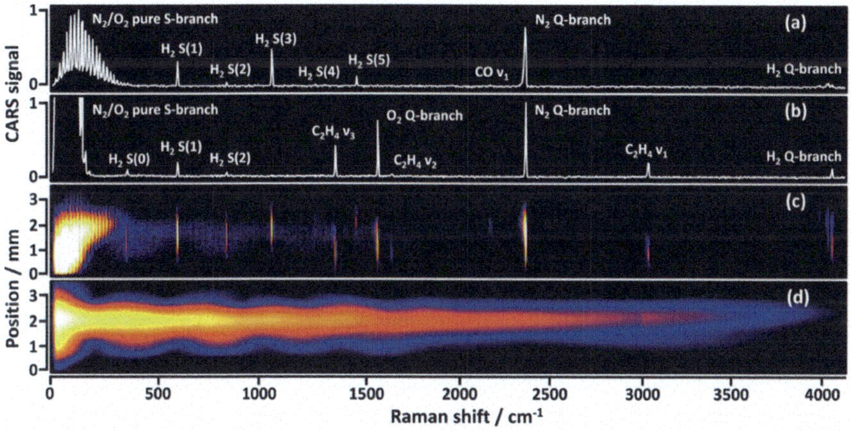

Fig. 13.17 Ultrabroadband fs/ps 1D-CARS imaging across a laminar premixed C_2H_4/air flame front. The progression of the chemical reaction along the flame front is mapped by single-shot spectral imaging along the 3 mm FOV, allowing for simultaneous thermometry and detection of all the major combustion species. Adapted with permission from Bohlin and Kliewer [50]. © 2015 American Chemical Society

affected the reproducibility of ultrabroadband fs/ps CARS thermometry resulting in a single-shot precision of ~5%, as compared to the ~1% precision of the same CARS instrument employing a 45 fs pump/Stokes pulse [50]. The main obstacle to the application of ultrabroadband fs/ps CARS in the vast majority of representative full-scale experimental platforms and industrial applications is the need to transmit the compressed supercontinuum to the measurement location. This is a non-trivial engineering problem as the large GVD introduced by the optical material of thick optical windows, in optically-accessible platforms, can reduce the effective excitation bandwidth provided by the supercontinuum pump/Stokes pulse. An example of the impact of the optical window on the effective excitation bandwidth is found in Richardson et al. where 1D pure-rotational fs/ps CARS imaging was employed to map the temperature profile in the post-detonation fireball of a commercial detonator [94]. The significant non-linear chirp introduced by the 13 mm optical window of the explosion containment housing resulted in a limitation of the excitation bandwidth to ~200 cm^{-1}. Moreover, supercontinuum fs laser pulses, having a much larger bandwidth, are even more sensitive to dispersion of their spectral components. This problem is even more significant when the harsh experimental conditions of e.g. high-pressure rocket and aircraft combustors are considered: the large vibrational, thermal and pressure loads on the optical windows can change the local GVD of the pulse to an extent that is difficult to foresee and compensate beforehand. Furthermore, even after its transmission through the optical material of the window, the supercontinuum pulse can be significantly affected by the local temperature and composition of the gas-phase medium along its path to the probe volume. As pointed out in the previous section, the unknown effective spectral excitation profile is a source of uncertainty in fs/ps CARS measurements when the ex situ referencing protocol is employed [52]. The magnitude of the GVD introduced along the optical path of the fs pump/Stokes pulse is of course larger the shorter the duration of the compressed supercontinuum pulse is. Hence, the conventional ex situ referencing protocol can be expected to introduced even larger an uncertainty to ultrabroadband fs/ps CARS measurements.

13.5.1 Soft Pulse Compression Behind Thick Optical Windows

A novel strategy for the realization of ultrabroadband fs/ps CARS measurements behind thick optical windows is introduced in Mazza et al. by employing fs laser-induced filamentation [95]. This phenomenon can be described as the non-linear propagation of high-peak irradiance fs laser pulses in transparent optical media [96]: the combination of optical Kerr effect and laser induced ionization results in cycles of self-focusing and plasma defocusing. Hence, while propagating within the resulting plasma filament, which acts a waveguide, the laser pulse experiences self-compression to a few- or even single-cycle pulse. The filamentation of fs laser pulses has been demonstrated to produce a compressed supercontinuum, which can excite the whole set of pure-rotational and ro-vibrational Raman modes up to 4200 cm^{-1} [90, 97].

Fs laser-induced filamentation can thus be employed to achieve a "soft" pulse compression –to < 20 fs– directly behind thick optical windows, according to a strategy called "in situ generation/in situ use" of the compressed supercontinuum. In Fig. 13.18a a relatively long 35 fs pulse (1.5 mJ/pulse) is transmitted "pre-chirped" through a 22 mm-thick BK7 glass window, placed after a 500 mm spherical focusing lens, and generates a ~13 mm long filament in air. The compressed supercontinuum output by the filament acts as the ultrabroadband pump/Stokes pulse coherently exciting ro-vibrational Raman modes up to ~2000 cm^{-1} [89]. Adopting the two-beam phase-matching scheme described above, spatially-resolved measurements can be performed with the probe beam crossing the supercontinuum pump/Stokes beam 3–5 mm after the end of the filament [95].

Two remarkable observations make in situ fs laser-induced filamentation well-suited to realize ultrabroadband fs/ps CARS. On the one hand, the filament is experimentally confirmed to propagate through steep temperature and composition gradients as encountered, e.g. through the flame front of the laminar premixed CH_4/air flame in Fig. 13.18b. This is explained by the change in the effective refractive index affecting the optical Kerr effect and the plasma density to a similar extent, such that the dynamic equilibrium between self-focusing and plasma defocusing is maintained as the light propagates in this filament through the flame front. On the other hand, for increasing energy in the fs laser pulse, the optical Kerr effect determines a non-linear shift of the focus towards the focusing lens, such that the filament grows backwards and the exit position from the filament remains the same. This is crucial to guarantee the non-intrusiveness of ultrabroadband fs/ps CARS with in situ filamentation as gas-phase diagnostics tool. By maintaining the probe volume at a fixed distance

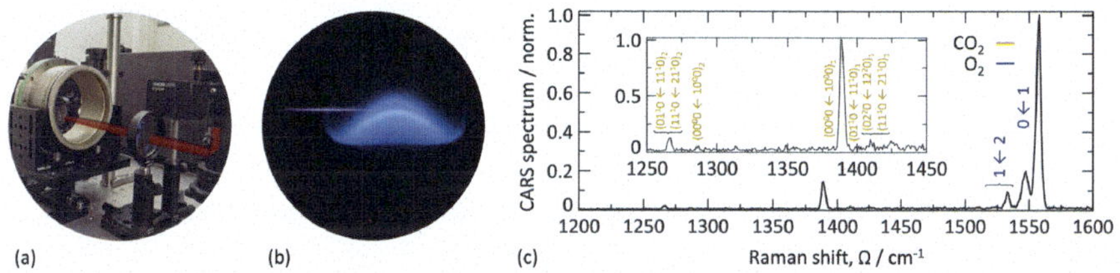

Fig. 13.18 Ultrabroadband fs/ps CARS behind thick optical windows via in situ fs laser-induced filamentation. **a** Transmission of the "pre-chirped" 35 fs pump/Stokes pulse through a 22 mm BK7 window placed ~20 cm after the 500 mm spherical focusing lens. **b** In situ generation/in situ use of the compressed supercontinuum via fs laser-induced filamentation in a laminar premixed CH_4/air flame. The CARS probe volume is located ~4 mm after the end of the filament to avoid ionization of the gas-phase medium in the probe volume and to ensure the use of this light source for non-intrusive laser diagnostics. **c** Single-shot ro-vibrational CO_2–O_2 CARS spectrum acquired in the hot products of the flame behind the optical window. At the flame temperature (T ~ 1620 K) the first vibrational hot band of in the ro-vibrational Raman spectrum of O_2 is detected, the ro-vibrational Raman spectrum of polyatomic CO_2 is much more complex and multiple hot bands are detected in the CO_2 Fermi dyad [95]

from the end of the filament, we can guarantee that no ionization is present in the probe volume [95].

The use of in situ laser-induced filamentation to perform ultrabroadband fs/ps CARS thermometry is first demonstrated in [95], by simultaneously investigating the ro-vibrational Raman spectra of O_2 and CO_2 in a laminar premixed CH_4/air flame. The performance of ro-vibrational CO_2 CARS thermometry is found to be comparable to state-of-the-art pure-rotational fs/ps CARS at temperatures higher than 2000 K. In particular, the complex ro-vibrational Raman spectrum of CO_2, consisting of the Fermi dyad shown in Fig. 13.18c, has a strong temperature dependence that makes it a suitable candidate for CARS thermometry. The precision of CO_2 CARS thermometry, assessed over a sample of 40-shot-averaged ultrabroadband CARS spectra, is better than 1% at the highest temperature measured (T ~ 2130 K). More recently, Tian and co-workers have demonstrated the use of in situ filamentation to perform single-shot ro-vibrational N_2 CARS thermometry at 1 kHz in a flat CH_4/air flame [98].

The first proof-of-principle demonstration of the in situ generation/in situ use strategy to achieve ultrabroadband fs/ps CARS behind a thick optical window is given in Mazza et al., where direct pure-rotational fs/ps CARS is employed on H_2 to measure the temperature profile across a laminar H_2/air diffusion flame [73]. As mentioned above, a 22 mm BK7 glass window is placed in the pump/Stokes beam path ~20 cm after the spherical focusing lens, such that the pulse is compressed directly behind it via in situ filamentation. The window introduces a linear chirp of 424 fs^2, as measured by attenuating the pump/Stokes to ~50 μJ/pulse to avoid the onset of filamentation and recording the NR CARS spectrum in a flow of Ar. This linear chirp is easily compensated by the external compressor unit shown in Fig. 13.5, which can pre-chirp the fs pulse introducing a linear dispersion well in excess of ± 4000 fs^2. The external compressor is thus employed to control the phase of the fs pulse prior its filamentation behind the window: a negative chirp of -492 fs^2 is experimentally found to optimize the filamentation process in the H_2/air flame [73], as shown in Fig. 13.19. The resulting supercontinuum has a duration of 17 fs, as measured by the ultrabroadband NR CARS signal generated in Ar. This compressed supercontinuum has a bandwidth sufficient to impulsively excite the pure-rotational coherence of not only diatomic molecules such as O_2 and N_2, but also of H_2 whose Raman spectrum spans more than 1500 cm^{-1} at flame temperatures.

Unfortunately the ultrabroadband NR CARS spectrum in Fig. 13.19b, as recorded ex situ in a room-temperature Ar flow, is not fully representative of the ultrabroadband excitation efficiency achieved by in situ filamentation in the flame environment. As pointed out in the previous section, the NR CARS spectrum acquired ex situ cannot account for the impact of the local temperature and composition fields along the beam path to the measurement location. Furthermore, in the case of in situ generation/in situ use of the compressed supercontinuum the magnitude of the pulse self-compression via fs laser-induced filamentation strongly depends on the local temperature and composition of the gas-phase medium, as seen in Fig. 13.20. As different gases have a different electronic contribution to the third-order optical susceptibility $\chi^{(3)}_{NR}$, so

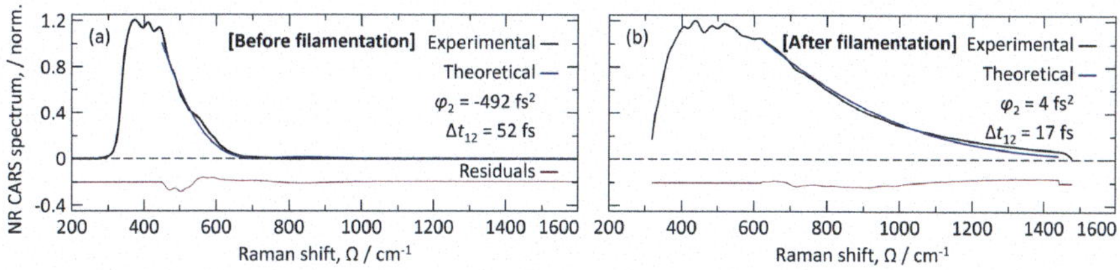

Fig. 13.19 Pump/Stokes pulse self-compression behind the optical window via in situ fs laser-induced filamentation. **a** 1000 shot-averaged NR CARS spectrum of the chirped pump/Stokes pulse, optimised for pulse compression via in situ filamentation in a laminar H_2/air diffusion flame. **b** 1000 shot-averaged ultrabroadband NR CARS spectrum generated by in situ filamentation in an Ar flow. Adapted with permission from Mazza et al. [73]. © 2022 Optica Publishing Group

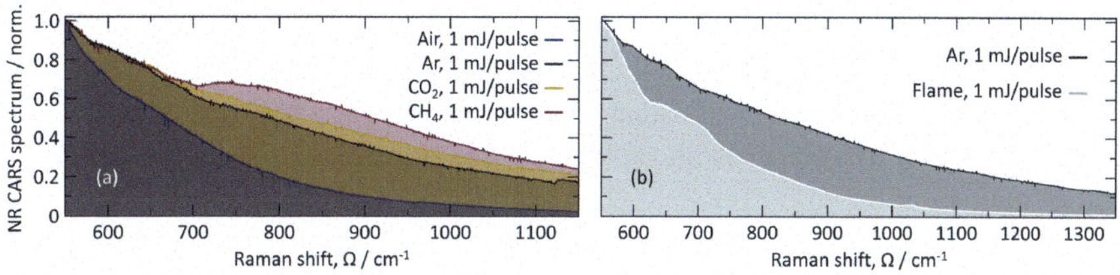

Fig. 13.20 Impact of the local boundary conditions to fs laser-induced filamentation on the in situ generation of the compressed supercontinuum. **a** Single-shot ultrabroadband NR CARS spectra generated by in situ filamentation of a 1 mJ, 35 fs TL pulse in room-temperature flows of: air (blue), Ar (black), CO_2 (yellow), and CH_4 (red). **b** Single-shot ultrabroadband NR CARS spectra generated by in situ filamentation of a 1 mJ, 35 fs TL pulse in a room-temperature flow of Ar (black) and in the hot products of a laminar H_2/air diffusion flame (gray)

does the optical Kerr effect responsible for the onset of fs laser-induced filamentation. In particular, the pulse self-compression attained by in situ filamentation in Ar is significantly larger than in air. In addition, the local temperature of the gas-phase medium has a significant impact on the non-linear refractive index of the medium, thus affecting both the magnitude of the optical Kerr effect and the plasma density inside the filament. The bandwidth of the compressed supercontinuum generated in the H_2/air flame is thus much smaller than measured ex situ in Ar: the ex situ referencing protocol is then expected to introduce a significant uncertainty in ultrabroadband fs/ps CARS with in situ filamentation.

13.5.2 Ultrabroadband fs/ps CARS with In Situ Referencing

In situ fs laser-induced filamentation is thus demonstrated to be a viable avenue to ultrabroadband fs/ps CARS behind thick optical windows. Nevertheless, the uncertainty due to the effective excitation bandwidth provided by the compressed supercontinuum pulse at the measurement location still needs to be addressed.

This is done in Mazza et al. by combining the in situ generation/in situ use of the compressed supercontinuum with the in situ referencing discussed in the previous section [73]. The polarization-sensitive coherent imaging CARS spectrometer described above is adapted to perform ultrabroadband fs/ps CARS in the vibrational fingerprint region of the Raman spectrum ~300–1500 cm^{-1}. This is done first of all by ensuring that the fs pump/Stokes pulse has linear polarization state, and that no significant ellipticity is introduced upon propagation through the optical window. A linear polarization is required to avoid the possibility of polarization ellipse rotation during the filamentation process: the compressed supercontinuum pump/Stokes pulse thus maintains the polarization of the input fs laser pulse. By choosing the appropriate relative polarization angle of the probe pulse, the ultrabroadband NR and pure-rotational H_2 CARS signals are cross-polarized. A Glan-laser polarizer (10^5 extinction ratio, Thorlabs) is employed in the polarization-sensitive spectrometer to separate the two signal components, and a broadband Fresnel rhomb rotates the linear polarization of the ultrabroadband NR CARS signal. A 100 mm imaging lens focuses the two signals onto the sCMOS detector with a dispersion of 0.93 cm^{-1}/pixel, resulting in a detection bandwidth of 1900 cm^{-1}, ultimately limited to 1200 cm^{-1} by the bandpass filter employed to suppress the co-propagating probe pulse.

The use of ultrabroadband fs/ps CARS to achieve accurate pure-rotational H2 thermometry behind the thick optical window is thus demonstrated in [73] by performing spatially resolved measurements across a laminar H_2/air diffusion flame front. The probe volume is placed ~0.5 mm above the rim of a Bunsen burner and moved at 17 radial locations from the center of the fuel stream (50% H_2–50% H_2) towards the oxidizer stream (room air). Single-shot measurements of the ultrabroadband NR and pure-rotational H_2 CARs signals are performed at each measurement location, as shown in Fig. 13.21. The NR CARS spectra acquired in situ at each location in the flame are compared to the NR CARS spectrum acquired ex situ in Ar: a dramatic change in the spectral excitation bandwidth achieved by in situ filamentation is demonstrated. The dependence of the pulse self-compression upon the local boundary conditions to the filamentation process in the flame is furthermore confirmed by the progressive reduction of the excitation bandwidth in the progression from the center of the burner towards the high-temperature reaction zone of the flame front.

Hence, the NR CARS spectrum acquired ex situ cannot faithfully map the effective excitation profile provided by the compressed supercontinuum generated via in situ filamentation in the flame. The conventional ex situ referencing protocol therefore results in a severe distortion of the normalized H_2 CARS spectra leading to a dramatic underestimation of the temperature measured by pure-rotational H_2 CARS thermometry, as shown in Fig. 13.22. Pure-rotational N_2 CARS measurements are also performed at each measurement location, providing a reference to validate the performance of pure-rotational H_2 CARS thermometry, and compare the two referencing protocols. As expected, the ex situ referencing protocol introduces a systematic bias in the measured temperature, which is larger the more the pulse compression via in situ filamentation is affected by the temperature increase in the flame. This systematic error varies from 10 to 20% at relatively low temperature in the fuel stream, up to 80% in the reaction zone of the flame. The in situ

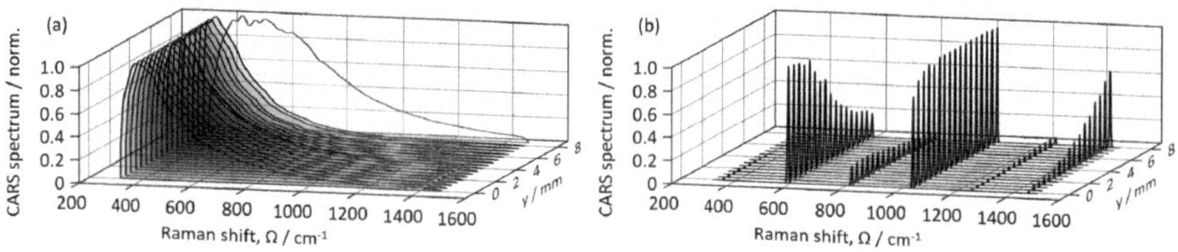

Fig. 13.21 Pure-rotational fs/ps CARS on H_2 with in situ referencing of the ultrabroadband spectral excitation. **a** 1200 shot-averaged ultrabroadband NR CARS spectra acquired in situ (black) at 17 locations from the center of the burner ($y = 0$ mm) to the flame front ($y = 8$ mm), and ex situ in room-temperature Ar (red). **b** Single-shot pure-rotational H_2 CARS spectra acquired across the flame front. Up to six pure-rotational lines are detected simultaneously over the ~300–1500 cm^{-1} spectral range. Adapted with permission from Mazza et al. [73]. © 2022 Optica Publishing Group

referencing protocol, on the contrary, allows for a correct spectral referencing of the pure-rotational H2 CARS spectra, such that the resulting accuracy is better than 1% at most measurement locations. The precision of pure-rotational H_2 CARS with in situ referencing is comparable to that of pure-rotational N_2 CARS, with a standard deviation over the 1200 single-shot spectra in each sample estimated in the range 1–3% for measurements up to location $y = 6.5$ mm. At the last three measurement locations, the reduction in the SNR due to the higher temperature and the almost complete consumption of the H_2 fuel negatively impacts the measurement precision, increasing the standard deviation up to 9% at location $y = 8$ mm.

Ultrabroadband fs/ps CARS not only allows for direct high-fidelity measurements on H_2 behind thick optical windows, but also provides new insights in the molecular transport processes taking place in the flame environment. In particular, the direct experimental observation of differential and preferential diffusion of H_2 is reported in [73]. These effects relate to the high mass diffusivity of the lightweight H_2 molecule: the former refers to its mass diffusivity coefficient being larger than other species in the combustion environment, the latter is due to the mass diffusivity being larger than the thermal diffusivity. Both effects play a critical role in the propagation of H_2 flames and their experimental quantification is critical to validate numerical modes. Differential diffusion of H_2 in the fuel stream of the laminar H_2/air diffusion flame of [73] is evident in the temperature and relative H_2/N_2 concentration profiles measured ~1 mm above the burner rim, as shown in Fig. 13.23. Within the first ~3.5 mm from the center of the burner the H_2 mole fraction reduces from 50 to 40%, while the temperature only slightly increases by ~57 K. This is due to the differential diffusion of H_2, which is transported ahead of N_2 towards the reaction zone, where is consumed in the combustion reaction.

In conclusion to the present section, it is worth noting that the in situ referencing protocol could be applied to ultrabroadband fs/ps CARS for a broader range of molecules besides H_2. The polarization approach developed for the implementation of in situ referencing protocol is applicable to any strongly depolarized Raman transition, with depolarization ratio $\rho > 0.5$. Moreover the same relative polarization angle between the pump/Stokes and probe pulses employed in the case of pure-rotational fs/ps CARS in [52, 73] (i.e. 67.8°) can be used for the in situ referencing of any

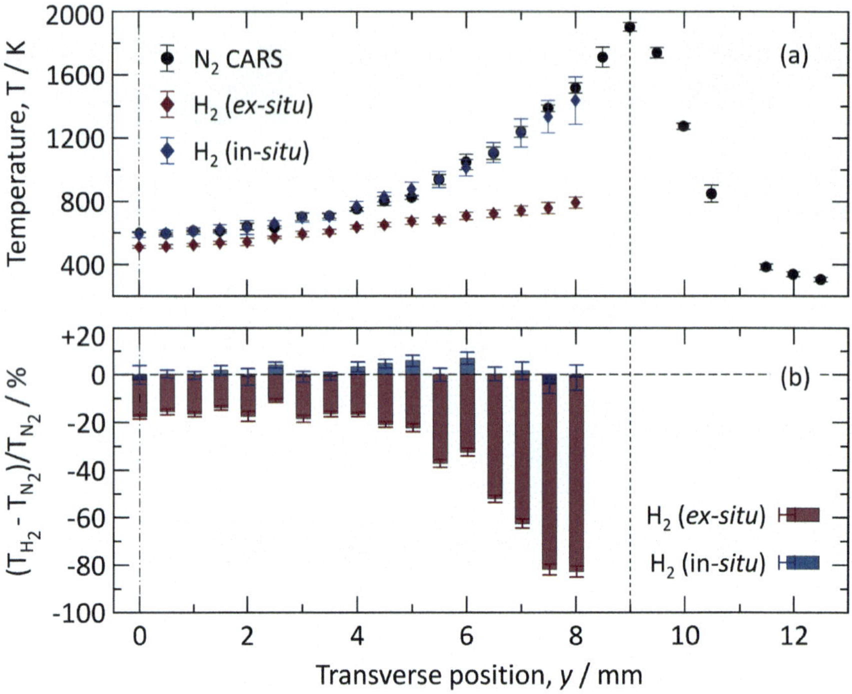

Fig. 13.22 Pure-rotational H_2 fs/ps CARS thermometry. **a** Comparison between the temperature profile across the laminar H_2/air diffusion flame front, a measured by pure-rotational N_2 and H_2 CARS thermometry. **b** Accuracy and precision of pure-rotational H_2 CARS thermometry, assuming the temperature measured by pure-rotational N_2 CARS as a reference. The conventional ex situ referencing protocol leads to a significant underestimate of the temperature measured by pure-rotational H_2 CARS. Adapted with permission from Mazza et al. [73]. © 2022 Optica Publishing Group

completely depolarized ($\rho = 0.75$) Raman spectrum. This has important implications for the possible use of ultrabroadband fs/ps CARS as a diagnostic tool for industrial chemical processes involving larger polyatomic molecules, whose least-symmetric vibrational Raman modes are often completely depolarized. An example of this is the recent use of ultrabroadband fs/ps CARS to investigate the ro-vibrational Raman spectrum associated with the ν_2 (bending) mode of CH_4 [89].

Although long-overlooked in favor of the C–H stretch (ν_1) mode, the ν_2 mode Raman spectrum offers an interesting window for the in situ monitoring of chemical and molecular transport processes in combustion environments and chemical reactors. As seen in Fig. 13.24, this Raman spectrum sits in the middle of the vibrational fingerprint region and can be simultaneously detected with other species of interest, such as O_2, CO_2 and, most importantly, H_2 [89]. The possibility of detecting CH_4 and H_2 in the same spectral window and apply the in situ referencing protocol to accurately measure their concentrations could make fs/ps CARS a powerful tool to

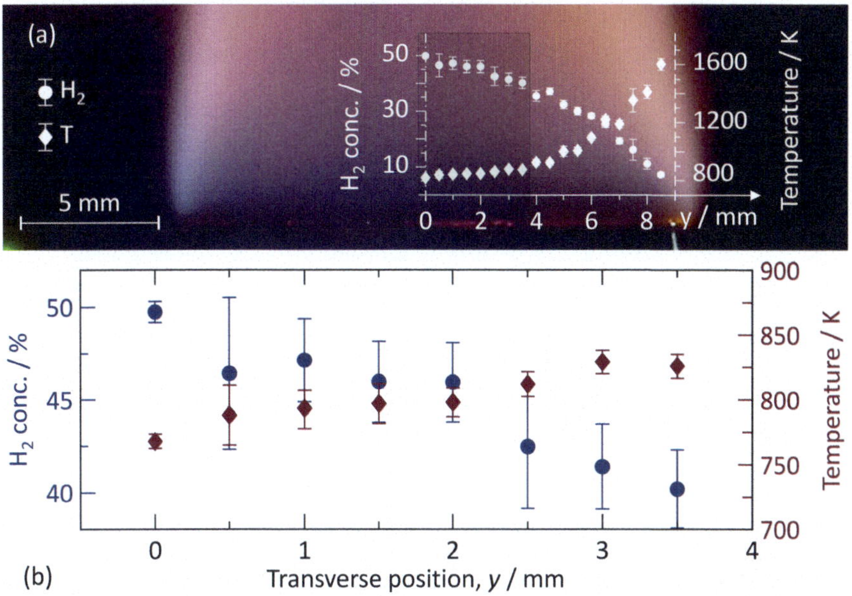

Fig. 13.23 Direct experimental observation of H_2 preferential and differential diffusion in a laminar H_2/air diffusion flame. **a** Simultaneous CARS measurement of the temperature and relative H_2/N_2 concentration across the flame front. The black box represents the area of interest for H_2 preferential diffusion. **b** While the temperature profile is almost flat in the region $y = 0$–3.5 mm the H_2 mole fraction shows a relative variation of 20%, owing to the differential diffusion of H_2 in the fuel stream. Adapted with permission from Mazza et al. [73]. © 2022 Optica Publishing Group

study the chemical reforming of CH_4 for carbon-neutral H_2 production in catalytic or plasma reactors.

13.6 Summary and Conclusions

We have illustrated some of the recent advancements in two-beam fs/ps CARS and discussed their potential towards the development of a single field-deployable diagnostic instrument for complete scalar determination in industry-relevant applications.

Two competing approaches to the extension of the two-beam fs/ps CARS instrument to simultaneous multi-fields measurements have been discussed:

- Pure-rotational fs/ps CARS imaging can achieve space-time correlated resolution, which is important for diagnostics completeness, i.e. the instrument is capable of measuring temperature, density, pressure and concentrations of major species simultaneously in gas-phase applications. In theory all the rotationally Raman-active species present in the probe volume could be detected by this technique, but in practice some of them are "invisible" in the CARS spectrum, due to their

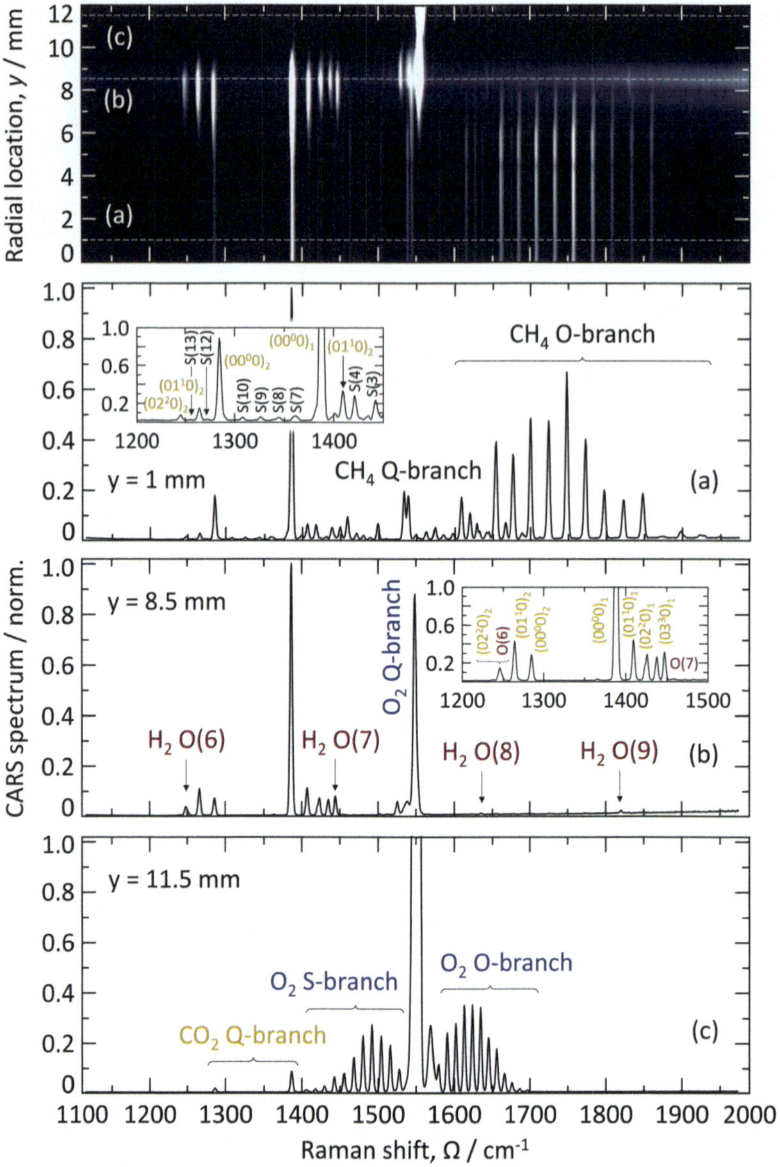

Fig. 13.24 Spatially-resolved ultrabroadband CARS in the vibrational fingerprint region of the Raman spectrum across the laminar CH_4/air diffusion flame. **a** Low-temperature ro-vibrational spectra of CO_2 and CH_4 in the fuel stream. **b** At temperatures >1300 K CH_4 pyrolysis produces H_2. Four lines of the pure-rotational H_2 CARS spectrum are detected in the spectral region ~1100–2000 cm^{-1}. **c** The oxidizer is ambient air so that the CARS spectrum is dominated by the ro-vibrational O_2 Q-branch at ~1553 cm^{-1}; a small amount of CO_2 diffuses into the oxidizer stream. Adapted with permission from Mazza et al. [89]. © 2013 AIP Publishing LLC

low rotational Raman cross-section and complex rotational structure. The use of time-resolved CARS allows for measuring concentrations of these species, via their collisional dephasing effect on the CARS signal. Furthermore, by introducing polarization control strategies for the CARS signal generation, one more "dimension" is available to CARS spectroscopy. This polarization-sensitivity can be employed to develop new diagnostic strategies for e.g. the concurrent, in situ measurements of the spectral excitation efficiency profile, or for high-sensitivity absolute concentration measurements via pure-rotational CCARS.

- Alternatively, a compressed supercontinuum pulse can simultaneously excite all the Raman-active modes to perform ultrabroadband fs/ps CARS on the pure-rotational and ro-vibrational regions of the Raman spectrum. Moreover, multi-modal CARS thermometry can be performed in non-equilibrium environments. This approach is most powerful when combined with one- or two-dimensional imaging arrangements, such that space-frequency-correlated measurements can map the in situ progression of chemical reactions and physical transport processes in rapid reacting flows. The need to transport the compressed supercontinuum through thick optical windows in industry-relevant measurement scenarios (e.g. high-pressure combustion chambers), while preserving its wide excitation bandwidth, is a long-standing limitation to the practical use of ultrabroadband fs/ps CARS. The use of fs laser-induced filamentation as an ultrabroadband coherent light source is proposed as a straightforward solution to this issue. The in situ generation/in situ use of the compressed supercontinuum is thus demonstrated to perform single-shot fs/ps CARS in the vibrational fingerprint region in laboratory flames behind a thick optical window. The impact of the local boundary conditions to the supercontinuum generation in the reacting flow is effectively quantified by the adoption of the in situ spectral referencing protocol.

Both the above-mentioned approaches can be implemented with minor modification of the same fs/ps CARS instrument, employing a single ultrafast regenerative amplifier laser system. This fs laser system design is cost-effective and greatly simplifies the operations of the CARS instrument: the need for complex beam alignment, often mentioned as a downside to CARS diagnostics, is largely relaxed with the adoption of the two-beam phase matching scheme and inherent synchronization of the two pulses. Furthermore, the introduction of an external compressor unit to pre-chirp the pump/Stokes pulses grants a large flexibility to the system in terms of its applicability in optically-accessible applications.

In conclusion, two-beam fs/ps CARS is the most-powerful diagnostic technique for high-fidelity multi-field measurements in harsh gas-phase environments and chemically reacting flows. The possibility of single-shot complete scalar determination with unparalleled spatial-temporal resolution and its robust, cost-effective implementation with a single amplified femtosecond laser system make this technique suitable for a wide range of applications with particular regard to the propulsion and power industry.

References

1. K. Kohse-Höinghaus, R.S. Barlow, M. Aldén, J. Wolfrum, Proc. Combust. Inst. **30**, 89 (2005)
2. A. Farooq, A.B.S. Alquaity, M. Raza, E.F. Nasir, S. Yao, W. Ren, Prog. Energy Combust. Sci. **91**, 100997 (2022)
3. R.L. Farrow, P.L. Mattern, L.A. Rahn, Appl. Opt. **21**, 3119 (1982)
4. A.C. Eckbreth, *Laser Diagnostics for Combustion Temperature and Species* (CRC Press, London, 1996)
5. W.R. Lempert, I.V. Adamovich, J. Phys. D Appl. Phys. **47**, 433001 (2014)
6. C.M. Penney, R.L. St. Peters, M. Lapp, J. Opt. Soc. Am. **64**, 712 (1974)
7. P.R. Régnier, J.-P.E. Taran, Appl. Phys. Lett. **23**, 240 (1973)
8. F. Moya, S.A.J. Druet, J.-P.E. Taran, Opt. Commun. **13**, 169 (1975)
9. P.D. Maker, R.W. Terhune, Phys. Rev. **137**, 801 (1965)
10. S. Roy, J.R. Gord, A.K. Patnaik, Prog. Energy Combust. Sci. **36**, 280 (2010)
11. A.C. Eckbreth, T.J. Anderson, Appl. Opt. **24**, 2731 (1985)
12. M. Aldén, P.-E. Bengtsson, H. Edner, Appl. Opt. **25**, 4493 (1986)
13. E.K. Gustafson, J.C. McDaniel, R.L. Byer, IEEE J. Quantum Electron. **QE-17**, 2258 (1981)
14. R.E. Foglesong, S.M. Green, R.P. Lucht, J.C. Dutton, AIAA J. **36**, 234 (1998)
15. A.C. Eckbreth, Combust. Flame **39**, 133 (1980)
16. F. Grisch, P. Bouchardy, W. Clauss, Aerosp. Sci. Technol. **7**, 317 (2003)
17. M. Scherman, R. Santagata, E. Lin, P. Nicolas, J.-P. Faleni, A. Vincent-Randonnier, P. Cherubini, F. Guichard, A. Mohamed, D. Gaffie, B. Attal-tretout, A. Bresson, J. Raman Spectrosc. **52**, 1643 (2021)
18. A. Bohlin, B.D. Patterson, C.J. Kliewer, J. Chem. Phys. **138**, 081102 (2013)
19. R.W. Boyd, *Nonlinear Optics*, 3rd edn. (Academic Press, 2008)
20. A.C. Eckbreth, Appl. Phys. Lett. **32**, 421 (1978)
21. D.A. Long, *The Raman Effect: a Unified Treatment of the Theory of Raman Scattering by Molecules*, 1st ed. (Wiley, 2002)
22. M. Aldén, P.-E. Bengtsson, H. Edner, S. Kröll, D. Nilsson, Appl. Opt. **28**, 3206 (1989)
23. S. Roy, T.R. Meyer, R.P. Lucht, M. Afzelius, P.-E. Bengtsson, J.R. Gord, Opt. Lett. **29**, 1843 (2004)
24. C. Brackmann, J. Bood, M. Afzelius, P.-E. Bengtsson, Meas. Sci. Technol. **15**, 13 (2004)
25. T.R. Meyer, S. Roy, R.P. Lucht, J.R. Gord, Combust. Flame **142**, 52 (2005)
26. R.J. Hall, Combust. Flame **35**, 47 (1979)
27. I.R. Beattie, T.R. Gilson, D.A. Greenhalgh, Nature **276**, 378 (1978)
28. A. Bohlin, F. Vestin, P. Joubert, P.-E. Bengtsson, J. Raman Spectrosc. **40**, 788 (2009)
29. S. Kröll, M. Aldén, T. Berglind, R.J. Hall, Appl. Opt. **26**, 1068 (1987)
30. R.P. Lucht, P.J. Kinnius, S. Roy, J.R. Gord, J. Chem. Phys. **127**, 044316 (2007)
31. J.D. Miller, C.E. Dedic, S. Roy, J.R. Gord, T.R. Meyer, Opt. Express **20**, 5003 (2012)
32. M. Afzelius, P.-E. Bengtsson, J. Bood, J. Bonamy, F. Chaussard, H. Berger, T. Dreier, Appl. Phys. B **75**, 771 (2002)
33. C.J. Kliewer, A. Bohlin, E. Nordström, B.D. Patterson, P.E. Bengtsson, T.B. Settersten, Appl. Phys. B **108**, 419 (2012)
34. C. Meißner, J.I. Hölzer, T. Seeger, Appl. Opt. **58**, 47 (2019)
35. W.D. Kulatilaka, H.U. Stauffer, J.R. Gord, S. Roy, Opt. Lett. **36**, 4182 (2011)
36. C.J. Kliewer, Opt. Lett. **37**, 229 (2012)
37. A. Bohlin, C.J. Kliewer, J. Chem. Phys. **138**, 221101 (2013)
38. B.D. Prince, A. Chakraborty, B.M. Prince, H.U. Stauffer, J. Chem. Phys. **125**, 044502 (2006)
39. S. Legros, C. Brunet, P. Domingo-Alvarez, P. Malbois, E. Salaun, G. Godard, M. Caceres, B. Barviau, G. Cabot, B. Renou, G. Lartigue, V. Moureau, S. Puggelli, S. Richard, M. Abdelkrim, F. Grisch, Combust. Flame **224**, 273 (2021)
40. C.E. Dedic, T.R. Meyer, J.B. Michael, Optica **4**, 563 (2017)
41. S.A. Steinmetz, T.Y. Chen, B.M. Goldberg, C.M. Limbach, C.J. Kliewer, Opt. Lett. **47**, 5429 (2022)

42. S. Roy, P. Wrzesinski, D. Pestov, T. Gunaratne, M. Dantus, J.R. Gord, Appl. Phys. Lett. **95**, 074102 (2009)
43. A. Bohlin, M. Mann, B.D. Patterson, A. Dreizler, C.J. Kliewer, Proc. Combust. Inst. **35**, 3723 (2015)
44. D. Escofet-Martin, A.O. Ojo, N.T. Mecker, M.A. Linne, B. Peterson, Proc. Combust. Inst. **38**, 1579 (2021)
45. A. Bohlin, C.J. Kliewer, J. Phys. Chem. Lett. **5**, 1243 (2014)
46. A. Bohlin, C.J. Kliewer, Appl. Phys. Lett. **105**, 161111 (2014)
47. L. Castellanos, F. Mazza, D. Kliukin, A. Bohlin, Opt. Lett. **45**, 4662 (2020)
48. S.P. Kearney, D.J. Scoglietti, Opt. Lett. **38**, 833 (2013)
49. T.L. Courtney, N.T. Mecker, B.D. Patterson, M. Linne, C.J. Kliewer, Appl. Phys. Lett. **114**, 101107 (2019)
50. A. Bohlin, C.J. Kliewer, J. Phys. Chem. Lett. **6**, 643 (2015)
51. J. Barros, M. Scherman, E. Lin, N. Fdida, R. Santagata, B. Attal-Tretout, A. Bresson, Opt. Express **28**, 34656 (2020)
52. F. Mazza, L. Castellanos, D. Kliukin, A. Bohlin, Proc. Combust. Inst. **38**, 1895 (2021)
53. S.M. Correa, Combust. Sci. Technol. **87**, 329 (1993)
54. S.P. Kearney, Combust. Flame **162**, 1748 (2015)
55. D.R. Richardson, H.U. Stauffer, S. Roy, J.R. Gord, Appl. Opt. **56**, 37 (2017)
56. Z. Tian, H. Zhao, H. Wei, Y. Tan, Y. Li, Appl. Opt. **61**, 4500 (2022)
57. M. Pealat, P. Bouchardy, M. Lefebvre, J.-P.E. Taran, Appl. Opt. **24**, 1012 (1985)
58. E.H. van Veen, D.J.E.M. Roekaerts, Appl. Opt. **44**, 6995 (2005)
59. D.K. Lauriola, P.S. Hsu, N. Jiang, M.N. Slipchenko, T.R. Meyer, S. Roy, Opt. Lett. **46**, 5489 (2021)
60. A. Bohlin, C.J. Kliewer, Appl. Phys. Lett. **104**, 031107 (2014)
61. J.E. Retter, D.R. Richardson, S.P. Kearney, Appl. Phys. B **126**, 1 (2020)
62. R.P. Lucht, R.E. Palmer, M.A. Maris, Opt. Lett. **12**, 386 (1987)
63. E. Nordström, A. Bohlin, P.-E. Bengtsson, J. Raman Spectrosc. **44**, 1322 (2013)
64. C. Brackmann, J. Bood, P.-E. Bengtsson, T. Seeger, M. Schenk, A. Leipertz, Appl. Opt. **41**, 564 (2002)
65. R.J. Hall, J.A. Shirley, A.C. Eckbreth, Opt. Lett. **4**, 87 (1979)
66. R.J. Hall, J.A. Shirley, Appl. Opt. **37**, 196 (1983)
67. D.A. Greenhalgh, R.J. Hall, F.M. Porter, W.A. England, J. Raman Spectrosc. **15**, 71 (1984)
68. F.M. Porter, D.R. Williams, Appl. Phys. B **54**, 103 (1992)
69. G. Herzberg, *Molecular Spectra and Molecular Structure. I. Spectra of Diatomic Molecules*, 2nd edn. (Van Nostrand Reinhold Company, Inc., 1950)
70. T.L. Courtney, A. Bohlin, B.D. Patterson, C.J. Kliewer, J. Chem. Phys. **146**, 224202 (2017)
71. J.E. Retter, G.S. Elliott, Appl. Opt. **58**, 2557 (2019)
72. L. Castellanos, F. Mazza, A. Bohlin, Proc. Combust. Inst. **39**, 1279 (2022)
73. F. Mazza, A. Stutvoet, L. Castellanos, D. Kliukin, A. Bohlin, Opt. Express **30**, 35232 (2022)
74. A. Montello, Z. Yin, D. Burnette, I.V. Adamovich, W.R. Lempert, J. Phys. D Appl. Phys. **46**, 464002 (2013)
75. J. Zhu, A. Ehn, J. Gao, C. Kong, M. Aldén, M. Salewski, F. Leipold, Y. Kusano, Z. Li, Opt. Express **25**, 20243 (2017)
76. S.A. Tedder, J.L. Wheeler, A.D. Cutler, P.M. Danehy, Appl. Opt. **49**, 1305 (2010)
77. F. Vestin, M. Afzelius, P.-E. Bengtsson, Proc. Combust. Inst. **31**, 833 (2007)
78. D. Kliukin, A. Bohlin, Patent P126758NL00/*Simultaneous Space-Time Resolution Broadband Coherent Raman Microscope with In-Situ Spectral Referencing* (2021)
79. D. Kliukin, F. Mazza, L. Castellanos, A. Bohlin, J. Raman Spectrosc. **52**, 1589 (2021)
80. B.D. Patterson, Y. Gao, T. Seeger, C.J. Kliewer, Opt. Lett. **38**, 4566 (2013)
81. S.P. Kearney, P.M. Danehy, Opt. Lett. **40**, 4082 (2015)
82. D. Escofet-Martin, A.O. Ojo, J. Collins, N.T. Mecker, M.A. Linne, B. Peterson, Opt. Lett. **45**, 4758 (2020)
83. L. Gong, W. Zheng, Y. Ma, Z. Huang, Nat. Photonics **14**, 115 (2020)

84. Z. Liu, J. Yao, H. Zhang, B. Xu, J. Chen, F. Zhang, Z. Zhang, Y. Wan, W. Chu, Z. Wang, Y. Cheng, Phys. Rev. A **101**, 043404 (2020)
85. J. Jiang, X. Jiang, M. Zhu, Int. J. Hydrog. Energy **40**, 15709 (2015)
86. N. Owschimikow, F. Königsmann, J. Maurer, P. Giese, A. Ott, B. Schmidt, N. Schwentner, J. Chem. Phys. **133**, 044311 (2010)
87. J. Bonamy, D. Robert, J.M. Hartmann, M.L. Gonze, R. Saint-Loup, H. Berger, J. Chem. Phys. **91**, 5916 (1989)
88. W. Meier, S. Prucker, M.-H. Cao, W. Stricker, Combust. Sci. Technol. **118**, 293 (1996)
89. F. Mazza, O. Thornquist, L. Castellanos, T. Butterworth, C. Richard, V. Boudon, A. Bohlin, J. Chem. Phys. **158**, 094201 (2023)
90. J.H. Odhner, D.A. Romanov, R.J. Levis, Phys. Rev. Lett. **103**, 075005 (2009)
91. A. Bohlin, C. Jainski, B.D. Patterson, A. Dreizler, C.J. Kliewer, Proc. Combust. Inst. **36**, 4557 (2017)
92. Y. Ran, A. Boden, F. Küster, F. An, A. Richter, S. Guhl, S. Nolte, R. Ackermann, Appl. Phys. Lett. **119**, 243905 (2021)
93. A.D. Cutler, L.M.L. Cantu, E.C.A. Gallo, R. Baurle, P.M. Danehy, R. Rockwell, C. Goyne, J. McDaniel, AIAA J. **53**, 2762 (2015)
94. D.R. Richardson, S.P. Kearney, D.R. Guildenbecher, Proc. Combust. Inst. **38**, 1657 (2021)
95. F. Mazza, N. Griffioen, L. Castellanos, D. Kliukin, A. Bohlin, Combust. Flame **237**, 111738 (2022)
96. A. Couairon, A. Mysyrowicz, Phys. Rep. **441**, 47 (2007)
97. F. Calegari, C. Vozzi, S. De Silvestri, S. Stagira, Opt. Lett. **33**, 2922 (2008)
98. Z. Tian, H. Zhao, Y. Gao, H. Wei, Y. Tan, Y. Li, Appl. Phys. Lett. **121**, 081102 (2022)

Chapter 14
Imaging Based Raman Spectroscopy

Jiro Karlo, Ashish Kumar Dhillon, Syed S. Razi, Soumik Siddhanta, and S. P. Singh

14.1 Introduction

Biological systems are made of dynamic, diverse, and complex biomolecules. The heterogeneity of this system increases from a single cell level to a tissue or a whole organism. Raman imaging is an analytical technique combining microscopy and Raman spectroscopy for exploring the dynamic and multiplex monitoring of the biomolecules [1]. In this chapter, we have aimed to summarise how Raman imaging can serve as a holistic approach in identifying and mapping biomolecular gradients in biological systems. Along with the working principles, we have also discussed its application in disease diagnosis, microbiology, food, and agriculture technology [1, 2]. Raman spectroscopy is based on the Raman effect which is the inelastic scattering of light based on the molecular vibration of the chemical bonds. Raman effect is a weak phenomenon i.e., 1 in 10^6 photons undergo inelastic scattering [1–4]. This inelastically scattered light provides both qualitative and quasi-quantitative information. Raman spectroscopy has very low sensitivity towards water which aids in its application in biochemical analysis. Overall, it is known as a stain-free, non-destructive, and non-invasive technique with minimal sample preparation [1, 5, 6]. Raman imaging generates an image with pseudo colour from the focused sample. It is made up pixels and each pixel has a single complete Raman spectrum. This provides information about spatial distribution of different biomolecules as per the location of pixels. Therefore, the pseudo colour of the Raman image corresponds

J. Karlo · S. P. Singh (✉)
Department of Biosciences and Bioengineering, Indian Institute of Technology Dharwad, Dharwad, India
e-mail: ssingh@iitdh.ac.in

A. K. Dhillon · S. Siddhanta
Department of Chemistry, Indian Institute of Technology Delhi, Delhi, India

S. S. Razi
Department of Chemistry, Gaya College, Bihar, India

D. K. Singh et al. (eds.), *Raman Spectroscopy*, Springer Series in Optical Sciences 248,
https://doi.org/10.1007/978-981-97-1703-3_14

to one value obtained from the Raman spectrum with respect to each pixel position [2, 5]. For acquiring the Raman image, a Raman microscope is used with motorised axes (x–y- and/or z-axis). First, the sample is focused and the area on the sample is chosen with a defined axis length. Now within this defined area, Raman spectra are acquired from every single spot where the sample beneath is moved in a point-by-point manner covering the whole targeted area. The objective lens, focus, step size, targeted area, acquisition time and laser power need to be optimized for better Raman image acquisition sensitivity.

The first Raman microscope was successfully designed and implemented in the year 1974 by Delhaye and Dhamelincourt [3, 7, 8]. In terms of conventional usage, point Raman spectroscopy has been used and explored extensively. However, due to the small laser spot size, it is not feasible to cover a large sample area. With the advancement in the spectrometer, chemometric-based processing algorithm, microscope features and laser, Raman imaging was made possible and is constructively evolving. Raman imaging technique has emerged as a highly sensitive technique to explore and acquire one-dimensional (wavenumber or Raman shift), two-spatial dimensional (x and y) and three-dimensional (hypercube) spatial information such as chemical composition and gradient, properties, purity, and homogeneity of the sample. In contrast to the gold standard fluorescence imaging, Raman imaging is label-free. Vibrational tags or Raman tags such as stable isotope labelled atoms such as deuterium, carbon-13, and nitrogen-15 are often used. Its effect on the function and chemical properties of the molecule or the sample is nearly negligible up to a larger extent. In biological samples such as cells, the Raman image comes with the merit of multiplexing where different cellular components as well as biomolecules can be simultaneously mapped [1, 8–10]. Detailed and high-resolution images of tissues have been generated through raster scanning coupled Raman mapping. Raster scanning refers to the rapid scanning of pixel points along the horizontal line which runs parallel to each other covering the entire sample area of interest. However, it is a slow imaging method with lengthy data acquisition times, making it unsuitable for intra-operative or online applications [11, 12].

Raman imaging modes can be widely classified into—spontaneous Raman imaging, surface-enhanced Raman scattering (SERS) imaging, tip-enhanced Raman spectroscopy (TERS) imaging and coherent anti-stokes (CARS) Raman imaging. Spontaneous Raman imaging is based on the spontaneous Raman scattering of monochromatic light. However, the spontaneous Raman scattering is weak, [2] and SERS has been proved to enhance the Raman signal by nanoparticles which is based on two principles—firstly, excitation of surface plasmons and secondly, the sample substrate charge transfer complex [13, 14]. Spontaneous and surface-enhanced Raman spectroscopy imaging has drawbacks when it comes to diffraction limit and spatial resolution, whereas TERS can be utilized for chemical characterization in tens of nanometres [15, 16]. CARS allow for video-rate image acquisition in a non-invasive way [17]. Raman imaging is not without limitations, such as it can be relatively slow and affected by fluorescence interference; however, it is a non-destructive and label-free approach with high spatial resolution. It is compatible with multiple sample types and multiplex monitoring. In the following sections a brief description of above-mentioned Raman imaging methods have been provided with suitable examples.

14.2 Spontaneous Raman Imaging and Biological Applications

Spontaneous Raman imaging is based on the inelastic scattering of photons when an incident light interacts with a sample. They are broadly divided into point scanning, line scanning and wild field or area scanning Raman imaging [1, 2]. In point scanning, the Raman spectrum is acquired using a point scan laser at each spatial location of the sample. It provides confocally high spectral resolution with full spectral coverage. However, for a heterogenous sample such as biofluid or tissue scanning, a small area of a few microns in diameter is not a suitable approach. Point by point mapping along a line and scanning an area with a motorized axis can aid to some extent. However, the drawback of point mapping is that it results in photobleaching of samples due to prolonged exposure to laser and slow scanning time [3, 18–21]. Line scanning uses a line-shaped scanning laser beam that is ordered into a line perpendicular to the sample, allowing Raman spectra acquisition from multiple points at a given time. The acquired Raman spectra are detected using a two-dimensional array detector and creating a false colour image. This method is an expansion of point mapping with an advantage of faster acquisition of Raman spectra over a larger sample area, and with higher laser power and reduced exposure giving better optical resolution. Furthermore, due to the higher cost of the line-scanning system and complex chemometrics, it is still underused as compared to the point-scanning method. Another drawback of this system is that the intensity is highest at the centre of the line and fades towards the edges of line, and uneven illumination along the line. However, this unevenness has been corrected using the Powell lens [3, 21–24]. In wide-field Raman imaging, the large planar sample area of interest is illuminated by the incident laser light with bigger laser spot size, a wide-aperture optical filter, and a two-dimensional CCD. With the development of the tuneable laser and various filters, wide-field Raman imaging has evolved very rapidly. It is highly advantageous from the real time Raman imaging and permits a rapid wavelength tuning for analysis [3, 21, 25, 26].

Spontaneous Raman imaging has been studied extensively in many areas such as microbiology, animal cells, cancer cells, plant cells, and others. This imaging approach has been efficient in imaging single live cells, time-lapse imaging of cytokinesis in live cells, distribution of mitochondria in cells, and drug-induced changes in live cells [2, 11, 27, 28]. The combination of stimulated Raman scattering and spontaneous Raman microscopy enables imaging of plant cell walls and enhances our understanding of the intricate structures and functions in a label-free manner [29] A study reported that tracking the biological processes such as cell mitosis, apoptosis, and cell differentiation complex in biological systems is feasible through high-speed hyperspectral Raman imaging [30–34]. Raman imaging has been used for diagnosing different cancer types such as colon cancer, liver cancer, pancreatic cancer, skin cancer, breast cancer, and prostate cancer [35–40]. Recently glycosylated and non-glycosylated proteins were distinguished in normal and cancerous tissue using Raman imaging, revealing deregulated proteins, lipids, and glycans metabolism in breast and brain tumours [41]. The uptake and distribution of lycopene in prostate cancer cells, revealing subcellular localization and accumulation in lipid membranes of organelles was studied using Raman imaging [42]. Raman imaging revealed distinct differences in spectral patterns in protein and lipid bands between prostate cancer cells and those cultured in 3D bone mimetic scaffolds, offering insights into sub-cellular responses and the identification of specific cellular constituents [43]. *Cucumis sativus* plant tissues reveal spatial variations in cellulose and lignin contributions within cell walls, aiding in the characterization of tissue substructures and providing insights into tissue morphology for future plant imaging studies [44]. IDH1 (Isocitrate Dehydrogenase) mutations in various cancers produce D-2-hydroxyglutarate, promoting tumorigenesis and which alters cellular metabolism. Treatment of IDH1-mutated cells with FK866, an NAD + (Nicotinamide Adenine Dinucleotide) salvage pathway inhibitor, decreases viability and leads to changes in protein and lipid levels, as detected by Raman imaging is shown in Fig. 14.1 [45] A recent study has applied Raman imaging for visualizing and quantifying intricate cellular components namely cholesterol, phospholipids (PLPs), triacylglycerides (TAGs), Nucleus, Cytoplasm as shown in Fig. 14.2 [46]. Although there is a need for better spatial resolution, sensitivity, and imaging speed to obtain better spontaneous Raman images, these have been widely explored in studying the biological sample.

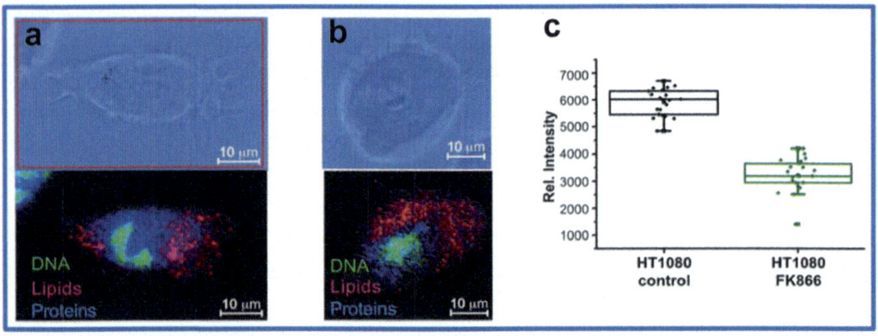

Fig. 14.1 Metabolic changes detection using Raman imgaing microscopy in live cells **a** untreated HT1080 cells and **b** FK866 treated cells were examined to observe the distribution of DNA (green), lipids (red), and proteins (blue); **c** Box and whisker plots were generated to analyse the lipid peak intensity (2845 cm^{-1}) from the cells. The regions of the cells analysed excluded the nucleus, allowing assessment of data variability and the reduction in lipid content after FK866 treatment. Adapted with permission from [45]. Copyright © 2018 Biosensor (Basel)

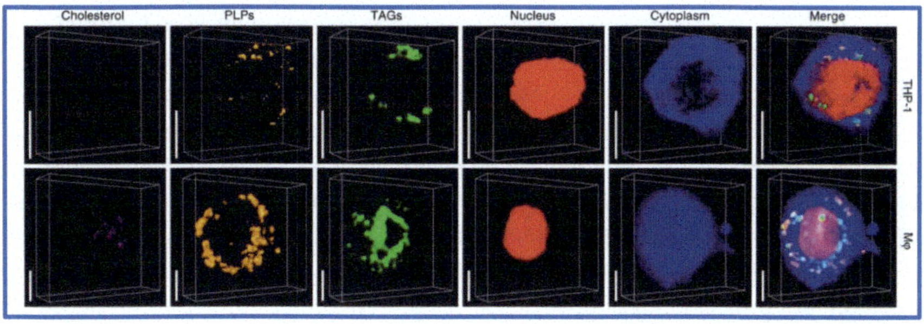

Fig. 14.2 Quantitative volumetric Raman image of differentiated macrophage showing important subcellular components of THP-1 cells and THP-1 differentiated macrophages (M_{φ}). Adapted with permission from [46]. Copyright © 2017 Nature Communication

14.3 Imaging with Surface-Enhanced Raman Spectroscopy

Raman microscopy presents an opportunity to investigate biological systems and their operations not only in a label-free manner but also by utilizing vibrational tags, which offer innovative optical imaging techniques for biological and medical research [47]. Raman spectroscopy (RS) stands as a promising and competitive technique in the realm of molecular imaging, owing to its capability to provide molecular and chemical information regarding the tissue of interest [48]. Nevertheless, Raman scattering exhibits intrinsically weak strength, with only one inelastically scattered photon for every 10^7 elastically scattered photons, significantly limiting its potential for biomedical applications [49]. The inherent feebleness of Raman signals can be attributed to their molecular origin, referred to as the Raman cross-section. Surface-enhanced Raman spectroscopy (SERS) is the most commonly employed form of RS in the field of medicine and biology. SERS leverages a combination of electromagnetic and chemical enhancement mechanisms, resulting in a substantial amplification of Raman signal intensity, as depicted in Fig. 14.1a. Electromagnetic enhancement (EME) arises due to the concentration of the electromagnetic field on the surface of nanoparticles (NPs). While direct chemical contact between molecules and the plasmonic substrate is not mandatory for this enhancement, maintaining an appropriate proximity between a molecule and the NP within a range of 1–10 nm is crucial. The spatially confined regions with robust electric fields, known as hotspots, play a pivotal role in augmenting EME [50]. Chemical Enhancement (CE) is achieved through the formation of NP-molecule complexes necessitating a direct interaction between a molecule and NPs. CE predominantly relies on chemical bonds and other electronic components [51]. For example, extended system molecules possessing easily polarizable electrons, such as benzene, naphthalene, and anthracene, exhibit pronounced Raman scattering, while small molecules lacking electron-rich atoms, such as H_2 and CO, exhibit weak Raman scattering due to their diminutive size. SERS retains the advantages of conventional Raman spectroscopy, including the provision of structural details with minimal interference from water.

As biomedical science advances, researchers have increasingly focused on detecting and characterizing analytes present in minute quantities in complex biological environments. SERS emerges as the optimal approach for detecting and characterizing low analyte concentrations within such contexts [52]. Given that different functional groups possess distinct vibrational frequencies, each SERS-active chemical yields a unique enhanced spectrum. Several active compounds exhibit robust SERS signals even at extremely low concentrations, making them valuable as Raman reporters for imaging or detecting biomarkers in biological samples [53]. The critical step in conducting a SERS experiment lies in the selection and utilization of the SERS substrate. It has been established that only substrates featuring nanoscale "roughness," such as rough electrodes and nanoparticles, are capable of enhancing SERS [54]. Raman scattering in SERS predominantly occurs within noble metallic nanostructures characterized by high local optical fields. Notably, the shape, composition, size, and interparticle spacing of metallic nanostructures significantly influence the surface plasmon resonance (SPR) and electromagnetic fields. Nanoparticles are well-suited for targeted imaging and detection in biological samples due to their ease of dispersion in solution-phase systems [55]. In contrast, planar substrates with patterned metallic surfaces, immobilized on a platform, exhibit greater chemical stability and resistance to aggregation compared to nanoparticles. These substrates have found applications in in vivo and in vitro SERS biosensing [56, 57]. Raman-based biosensing techniques can be broadly categorized into two classes: labeled and label-free modalities. Numerous molecules, such as 4-mercaptobenzoic acid (4-MBA), 4-aminothiophenol (4-ATP), etc. [58] been used to detect various biological and chemical substances [59], including bacteria [60], viruses [61], DNA [62], cancer biomarkers [63], glucose [64], explosives [65], volatile organic chemicals [66] and transition metals [67]. There are groups that demonstrated the label free Raman imaging for examination of protein distribution in apoptotic [68], mitotic cells, and DNA [68]. In the case of label-free imaging, super-resolution optical imaging has emerged as a powerful tool for surpassing the diffraction limit and gaining insights into the characteristics of hot spots that facilitate maximum electromagnetic field enhancement. Super-resolution SERS imaging, even with wide-field illumination from a narrow-line-width laser source, can achieve rapid acquisitions on the order of microseconds to milliseconds with nanometer spatial resolution. Single fluorescent dyes are used in biological systems to label and precisely pinpoint spectroscopically silent entities, such as proteins or DNA, with a resolution of about 5 nm [69]. This technique has been applied to examine specific areas such as macrophage cells, bacteria, and DNA fragments in super-resolution SERS imaging of biological samples [70]. The majority of studies have employed nanoparticle-based SERS labels or probes to generate super-resolution images on structured surfaces.

In vivo imaging represents a robust methodology for investigating the localized effects of medications and biological processes within living tissues or organs. Traditional Raman scattering from cellular components is typically weak due to the low concentrations of native constituents within cells and the impact of cellular degradation processes (cell degeneration) [53]. To overcome these limitations, SERS has emerged as a powerful tool for swiftly acquiring informative signals from intracellular environments. Recent advancements in SERS technology have led to the development of techniques for localized intracellular measurements, primarily achieved by introducing nanoparticles into cellular organelles. It is crucial to note that the distribution of bare nanoparticles within cells lacks control [71]. Consequently, many studies focused on in vivo SERS imaging utilize SERS tags, which are either injected into animals and accumulate within tumors via the Enhanced Permeation and Retention (EPR) effect or are directly targeted to specific cells or tumors through injection or biorecognition surface ligands as shown in Fig. 14.3b [72]. The creation of targeted SERS nanoprobe systems necessitates the conjugation of functional biomolecules to the SERS substrate. The fundamental concept of SERS enhancement revolves around the adsorption or close proximity of analytes to metal surfaces. However, it is essential to ensure that functionalized nanoparticles do not excessively cover the metal surface, as this can diminish the efficacy of SERS detection when altered reagents significantly affect the Raman signal [73]. In addition to intracellular measurements, the outer cell walls can also serve as a source of Raman information. Pioneering work by Kneipp et al. demonstrated the study of single living cells using Au nanoparticles, which were internalized through fluid phase uptake [74]. The Raman bands acquired primarily originated from native cellular constituents such as DNA, RNA, phenylalanine, and tyrosine. This challenging yet crucial approach enables the detection of single-celled organisms like bacteria and yeast cells, as exemplified by Sujith et al. work with living yeast cells [75]. Furthermore, Kahraman and colleagues reported similar detection methodologies involving bacterial cells coated with Au and Ag nanoparticles, expanding the applicability of SERS in biological research [76].

Recent innovations have seen the utilization of SERS nanoparticles in detecting circulating tumor cells (CTCs) in human peripheral blood, particularly in patients with head and neck squamous cell cancer. These SERS nanoparticles incorporate environment-sensitive Raman reporters that respond to physicochemical changes within live cells and tissues [77]. Examples of such reporters include indocyanine green (ICG) for thermosensitivity, MBA, 2-aminothiophenol, and 4-mercaptopyridine for pH sensitivity [78]. To create SERS-encoded amphiphilic nanoparticles, Song et al. tagged a pH-sensitive hydrophobic brush, a hydrophilic polymer brush, and a Raman reporter on the surface of gold nanoparticles (Fig. 14.3b). Innovative approaches have also emerged, such as the development of SERS-encoded amphiphilic nanoparticles that self-assemble into vesicles containing anticancer medications [79]. These vesicles exhibit strong SERS signals when tagged with cancer-targeting ligands on their outer surfaces, facilitating targeted cancer cell interaction. In other groups, SERS sensing probes were created by Jaebum Choo et al. to target and image particular cancer markers in living cells [80]. In order to image the local distribution of CD24 and CD44 tags co-expressed on MDA-MB231 breast cancer cells and afterward identify the signature of a particular molecular interaction, anti-CD24 Rubpy/RuITC DMNPs and anti-CD44 MGITC/FITC DMNPs were created. The functionalization of nucleic acids and proteins does not significantly disturb the abundance of structural information provided by SERS data, which serves as a targeting probing method for nucleus detection in living cells, as shown in Fig. 14.3c [71]. Further, Jaebum Choo et al. effectively used a unique metal nanoprobe

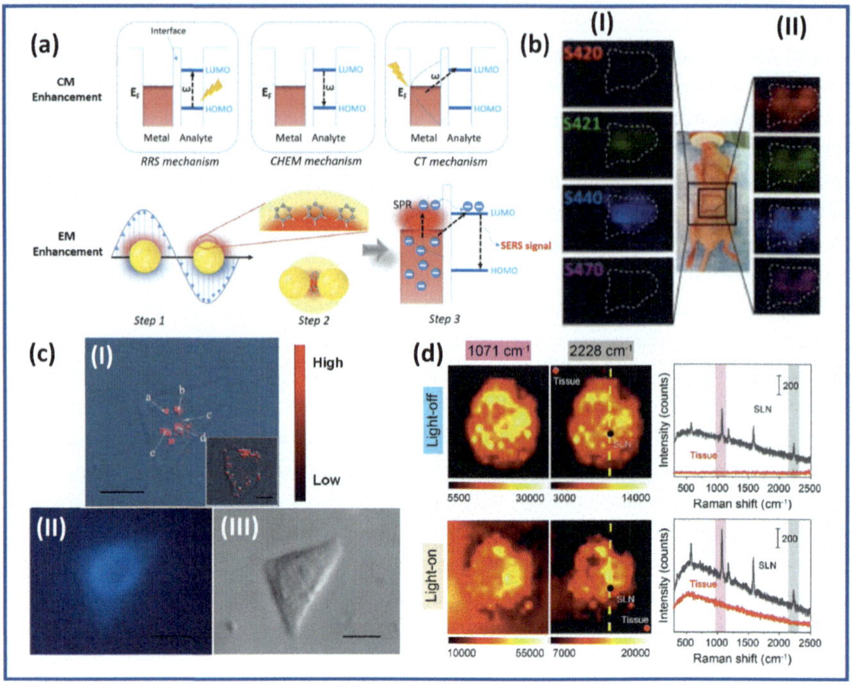

Fig. 14.3 **a** SERS enhancement modes: chemical enhancement (CM) and electromagnetic enhancement (EM). Therein, CM enhancement includes three contributions: RRS mechanism, CHEM mechanism, and CT mechanism; EM enhancement is described through a three-step process. Adapted with permission from [54]. Copyright © 2022 American Chemical Society. **b** In vivo SERS imaging, SARI system to image SERS tags: two tags are seen at 1 h post injection (I), whereas all four probes are evident at 2 h post injection (II). Adapted with permission from [78]. Copyright © 2019 American Chemical Society (C)(I) Merged SERS image made up of a SERS map by the intensity from 600 to 1700 cm^{-1} and an optical transmission image. The insert is the cell incubated with NLS mutant functionalized nanoparticles as the control experiment. (ii) Fluorescent image of the cell after nuclear staining by Hoechst 33258. (iii) Differential interference contrast micrograph of the cell. Adapted with permission from [77]. Copyright © 2009 American Chemical Society. (D) (Left) Images of the same SLN after injection of 8 h at 1071 and 2228 cm^{-1} with halogen light on and off, (Right side) Raman spectra at the SLN in situ (black) and adjacent tissues (red). Adapted with permission from [103]. Copyright © 2022 American Chemical Society

they had made that was conjugated with particular antibodies for the highly sensitive biological imaging of HEK293 cells that were expressing PLC-1 cancer markers [81]. It has been discovered that cancerous and non-cancerous cells can be distinguished with great clarity using SERS imaging and antibody-conjugated metal nanoprobes.

Wang et al. have presented a novel methodology for the growth of gold nanostars with strong near-infrared (NIR) absorbance capabilities. These nanostars have been employed for the visualization and treatment of cancer cells through photothermal treatment (PTT) [82]. They utilized SiO_2@Au nanoparticles (NPs) for in vitro imaging using the HCT 116 cell line, demonstrating high sensitivity at the single-particle level under 785 nm NIR laser excitation. In vivo imaging using NIR light has received a lot of interest as a solution to these concerns as it has excellent tissue penetration. One of the benefits of in vivo imaging probes is their multiplexing capacity, which allows for the simultaneous observation of the effects of many medications and their various tumour-targeting capabilities. Given the limited bandwidth (<2 nm) of the Raman signal, SERS utilizing NIR-active NPs have been employed for multiplex imaging. The SERS signal may also be recorded in swine tissue as deep as 7 mm, thanks to NIR light and detection benefits [83]. As tissues are denser than cells, their Raman spectra can be acquired without augmentation. SERS, however, can reduce signal acquisition time and quench the luminous background. Also, preparing a tissue sample is easier than one for a cell experiment. Either tissue fragments or tissue slices can be combined with the SERS substrates. Since tissue samples are typically collected for clinical or medical purposes, visual diagnosis is the main focus of SERS research on tissues [83, 84]. Prior to the SERS detection, frozen tissue samples obtained from hospitals are crushed and combined with Ag nanoparticles. The extent of SERS enhancement is subject to analyte-specific factors, with certain SERS-active compounds, such as Cresyl violet, MBA, and rhodamine, capable of generating high-intensity SERS signals at extremely low concentrations. Due to their high SERS intensity and spectrum specificity, the SERS active molecular reporter can distinguish the target analytes. A common method of detecting chemicals that are unique to antibodies is based on SERS tags. SERS may simultaneously detect several nucleic acids. For instance, Mirkin's team utilized dye-labeled gold nanoparticles to perform multiplex viral DNA or RNA detection [85], employing different reporters to label various oligonucleotides for distinguishing complementary sequences via hybridization processes. However, the non-homogeneous composition of cellular components may render some nanoparticles ineffective for identifying metallic nanoparticle distribution within cells. To address this limitation, Nithipatikom et al. utilized modified Ag nanoparticles to identify enzymes within cells [86]. Prior to SERS detection in their work, antibodies were employed to bind desired enzymes or receptors with Raman markers. By altering the functional residues, SERS nanotags can be created to operate as chemical sensors. Based on Ag nanoparticles modified with 4-mercaptobenzoic acid, Talley and colleagues created an intracellular pH sensor that produces distinct Raman signals at various pH levels [87]. The intracellular pH sensing system developed by Kneipp et al. using 4-mercaptobenzoic acid and Au nano aggregates was exhibited [88]. Ag nanoparticles functionalized with 2-aminobenzenethiol have been used by Wang et al. to detect living cells using SERS at various pH levels. The pH-dependent SERS spectra can be used to measure environmental pH values between 3.0 and 8.0. By combining nanoparticles with organic molecules to form composite organic–inorganic nanoparticles (COINs), Su and colleagues have shown how to create SERS spectra by incorporating an organic Raman reporter into the COINs [89]. They used Raman reporters for various proteins to show that COINs can detect multiple detections. Raman reporter-tagged gold nanoparticles and SERS were used by Sun et al. to develop a method for visualizing DNA micro-patterns [90]. SERS tags attached to antibodies can be used to detect cancer in tissues, cultured cells, and even live animals. According to Sha and colleagues, blood sample analysis revealed the presence of breast cancer cells [91]. They coupled epithelial cell-specific antibody functionalized magnetic nanoparticles with anti-her2 antibody-linked SERS tags. It is difficult to identify cancer in living animals using an optical approach. Jaydeep et al. employed gold nanoflowers for SERS-based sensing of foreign compounds at higher consistency and picomolar concentrations [92]. Also, the malignant cells were imaged using gold nanoflowers. The cytotoxicity of these nanostructures was also investigated using the MTT assay, utilizing the human lung cancer cell line A549, which exhibits good biocompatibility. Another noteworthy class of SERS tags includes metal carbonyls, which have been utilized as cellular imaging probes. Despite weak Raman scattering in metal carbonyls due to their low concentration, 3D Raman imaging demonstrated their localization in the nuclear membrane and nucleolus. For instance, the triosmium carbonyl cluster $Os_3(u\text{-}H)_2(CO)_{10}$, conjugated to gold nanoparticles, served as a pioneering example of using metal carbonyls as Raman reporters in SERS tags for live cell imaging [93]. These conjugates exhibited aqueous dispersibility, stability, non-toxicity, and ease of biofunctionalization for targeted delivery, resulting in ultrasensitive detection [94]. Additionally, a functional chip was developed by depositing the triosmium carbonyl cluster $Os_3(CO)_{10}(u\text{-}H)(u\text{-}SCH_2CH_2COOH)$ onto BMFON, followed by the attachment of A1AT antibodies to the carboxylic acid group. This chip demonstrated promise as a biomarker for label-free detection of alpha-1 antitrypsin (A1AT) in blood cancer [95]. Urine samples showed distinct differences in A1AT concentrations between healthy individuals and bladder cancer patients, indicating its potential for clinical use.

In recent developments, SERS probes have also been employed for drug delivery and detection, including ablation and photothermal therapies. This approach holds promise for theranostics due to the biodegradable and non-toxic nature of nanoprobes, which selectively accumulate in diseased tissues while sparing healthy ones [96]. These nanoprobes can report biochemical and morphological properties, deliver therapeutic drugs, and provide high-contrast, low-interference Raman imaging in natural light nanoprobes, as shown in Fig. 14.1d [97]. Potential applications in humans include therapy, drug delivery, image-guided tumor margin delineation, labeled or label-free identification of malignant tissues, and non-invasive, label-free assessment of metastatic potential in neoplasms.

14.4 Imaging with Tip-Enhanced Raman Scattering

To gain deeper insights into the fundamental molecular mechanisms governing cellular processes and the onset of diseases, there is a concerted effort to elucidate the precise molecular structures of biomolecules and their interactions with exceptional spatial precision. In pursuit of these objectives, tip-enhanced Raman scattering (TERS) spectroscopy emerges as an immensely promising technique [98]. TERS instruments amalgamate cutting-edge technologies to facilitate reliable nanoscale structural and chemical imaging of biomolecules and biomaterials, even within unconventional or poorly regulated environments [99]. Surface-enhanced Raman scattering (SERS) spectroscopy and scanning probe microscopy (SPM), such as atomic force microscopy (AFM) and scanning tunnelling microscopy (STM), are combined in TERS [100, 101]. A metal SPM probe, often referred to as the TERS tip, serves as the active plasmonic nanostructure in TERS. The TERS tip has plasmonic characteristics and acts as a "lightning rod" and a "nano-antenna." With non-resonant light excitation, a tapered tip can act as a "lightning rod," accumulating surface charges at the apex [102]. Under resonant light irradiation, it operates as a "nano-antenna," facilitating the interconversion between local and propagative electromagnetic fields. The amplified Raman signal is detected by a TERS tip's plasmonic, "lightning-rod," and "nano-antenna" effects. This signal derives from the nanoscale "hotspot" formed below the tip at the sample surface [100]. Due to the limited size of the probed region in TERS, a single-point measurement does not adequately represent the composition and structure of large heterogeneous samples. In response, the "hotspot" is systematically moved across the sample in nanoscale increments, generating a cartography of its chemical composition and structural intricacies [101]. TERS has made it possible to identify distinctive spectral fingerprint regions that distinguish nucleic acids, proteins, and lipids [102]. The efficient characterization of molecules is the foundation for a detailed examination of complex biosystems, including cells and viruses. The primary goal of TERS imaging would be to explore molecular mechanisms of biological processes, such as the interaction of proteins and cell membranes with nanoscale SOR, in aqueous or even physiologic conditions [103]. This necessitates the identification of target lipids within membrane compositions and the nanoscale scrutiny of lipid microdomains, including potential glycolipoprotein lipid rafts, within membranes. A comprehensive exploration of the nexus between amyloid fibrils and biological membranes, particularly in the context of neurodegenerative disorders, promises insights into the role of fibril polymorphism in amyloidogenesis. The vast majority of TERS investigations have been carried out in dry environments, which can be suitable for characterizing individual biomolecules in the early stages but is insufficient to investigate their biological activity [104]. For use in biology and catalysis, numerous research teams worldwide have made and are trying to adapt TERS to the liquid environment. This means that TERS tips can be manufactured to scan several samples in liquid without the tip metal layer delaminating. According to several recent findings, commercial AFM probes with noble metal coatings (added adhesion or capping layers) can create practical and reliable TERS tips for AFM-based TERS systems [105]. These most recent developments are positive for advancing the method for biological applications.

To describe biomolecules, four optical configurations of TERS devices are used shown in Fig. 14.4a [104]. The TERS tip's lighting is primarily where they diverge. Bottom illumination is the most widely used arrangement, which enables the effective collection of scattered photons and offers a high signal-to-noise ratio (SNR) for TERS [106]. A TIR objective with a high NA is used in this geometry to produce a sizable sample area exposed to evanescent field illumination. Due to this, the measured far-field signal is reduced compared to the conventional bottom lighting setup. Top- and side-illumination setups are frequently used to analyze optically opaque or not transparent enough samples. The state-of-the-art TERS technology exhibits remarkable adaptability and ultrahigh sensitivity, as demonstrated through extensive investigations into various biomolecules such as nucleic acids, proteins, peptides, lipid membranes, and more complex systems like cells and viruses. This nanoscale analysis of the secondary structure of amyloid fibrils, protofibrils, and oligomers implicated in neurodegenerative diseases has proven instrumental in elucidating the toxicity of these species and amyloid formation pathways, as well as their interactions with phospholipids [52]. Comparing the amounts of protein and fat in individual viral particles allowed researchers to identify various viral strains from one another.

An innovative approach has been devised to enhance the sensitivity of TERS bio-imaging, involving the amalgamation of the plasmonic attributes of gold-coated probes with additional amplification due to the interaction between metallized probes and gold substrates (referred to as gap-mode TERS). Gold substrates characterized by

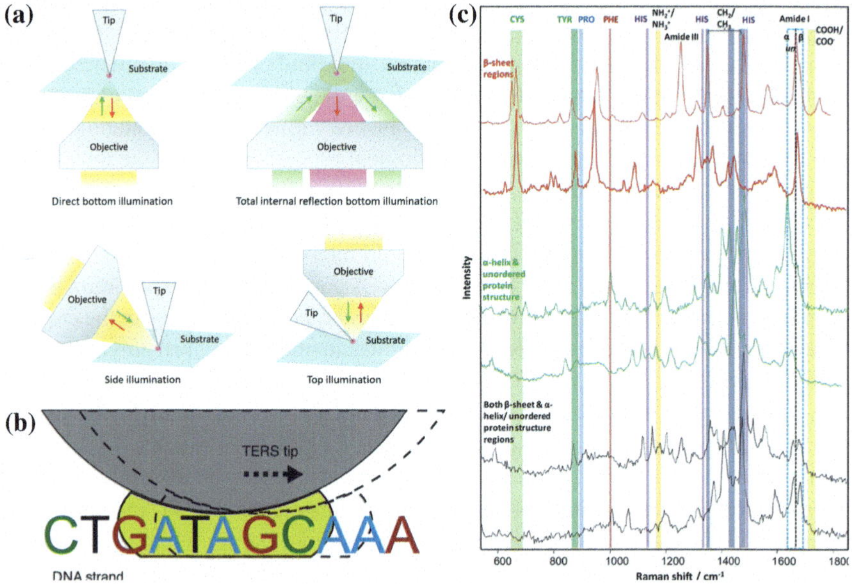

Fig. 14.4 **a** Direct bottom (I), total internal reflection bottom (II), side (III), and top (IV) illumination configurations in TERS instruments. Green and red arrows indicate the excitation light and the scattered light, respectively. Adapted with permission from [104]. Copyright © 2022 Royal Society of Chemistry. **b** Direct Base Sequencing: Sequence information can be obtained by laterally shifting the probe in intervals of one base-to-base distance. The yellow area refers to the enhancing region at the first tip position. Adapted with permission from [114]. Copyright © 2013 American Chemical Society. **c** Insulin fibril TERS spectra are assigned to one of three groups according to the dominant protein secondary structure revealed by amide I. β-sheet clusters (red), α-helix/unordered regions (green) and mixed regions with significant contributions from all secondary structures (black) are shown. Adapted with permission from [116]. Copyright © 2012 American Chemical Society

atomically flat surfaces have been tailored for TERS measurements using the template stripping technique [107]. The quality of TERS probes plays a pivotal role in TERS imaging, impacting the enhancement factor, nanometric spatial resolution, and signal stability. Current developments in single nucleobase resolution TERS examinations of DNA/RNA indicate the method's appealing application for fluorescent labelling and direct nucleic acid sequencing without amplification as shown in Fig. 14.4b [108]. This achievement is realized through the use of gold substrates coated with a layer of DNA molecules electrostatically assembled on them in conjunction with titanium/gold-coated Atomic Force Microscopy (AFM) probes. These findings have established DNA plasmids as a reference standard for nano-spectroscopy investigations into DNA structure. This improved methodology lays a robust foundation for further exploration into DNA damage, its repair mechanisms, and molecular alterations resulting from interactions between DNA and other macromolecules, a task made more complex by the fact that interactions within DNA or RNA molecules can engender variations in the anticipated vibrational spectra. Numerous variables can impact the vibrational spectra in a TERS experiment [108]. Different selection principles may be in effect due to system perturbation from the strong electromagnetic near field at the tip apex. It has also been demonstrated that variations in the vibrational spectra can result from the physical force applied by the TERS probe. Shifts in the TER spectra can happen due to variations in the immediate environment because very little analyte is detected in TERS. TERS peak positions can move due to variations in the distance between the probe and the molecule, further complicating data processing.

TERS has been used to analyze various biological materials, including viral particles, microbial cells, and macromolecules like lipids, proteins, and nucleic acids [109]. The TERS spectra of insulin fibrils can be categorized into three groups based on the predominant protein secondary structures, as elucidated by the amide I bands, specifically β-sheet clusters (red), β-helix/unordered regions (green), and mixed regions with substantial contributions from all secondary structures (black), as exemplified in Fig. 14.4c [110]. However, many variables, including optical and tip design and properties, operating environment and conditions, and far-field scattering background, must be considered to ensure the reproducibility and reliability of TERS measurements. Some practical issues in TERS are sample degradation, tip contamination, and thermal drift. When exposed to heat, biological macromolecules frequently lose their shape due to the breakdown of intermolecular interactions. The damage threshold varies from molecule to molecule, although temperatures beyond 50–60 °C may be harmful. TERS experiments may also encounter various forms of tip contamination. Contaminants adhering to tip metal coatings, which serve as Surface-Enhanced Raman Scattering (SERS) substrates, can generate spectral features resembling those of the sample spectrum. Moreover, thermal drift, closely tied to the stability of the Scanning Probe Microscopy (SPM) system, constitutes a significant instrumental challenge in TERS investigations [111].

Nonetheless, executing TERS studies, particularly TERS imaging, is still challenging in liquid media. Optimizing photon excitation and collection in top- and side-illumination configurations, currently the most common liquid applications, is a significant challenge. This makes the experiments incompatible, uncertain, and expensive. TERS on biological samples in their natural context would face additional challenges because of the bio-sample heterogeneity and complexity, further creating difficulties in spectral interpretation.

14.5 Imaging with Carbon Nanotubes

Since their first discovery, carbon nanotubes (CNTs), particularly single- and multi-walled carbon nanotubes (SWNTs), have gained prominence as materials. Many groups have reported a wide range of CNT-based bio-sensors with various sensing methodologies for detecting various biological substances [112]. Another highly explored area of research is the utilization of CNTs as imaging probes for bio-imaging applications. SWNTs, in particular, possess unique physical properties that make them attractive for this purpose [113]. SWNTs are highly effective in Raman spectroscopy due to their strong resonant Raman scattering and an exceptionally broad scattering cross-section [114]. Surface-enhanced Raman scattering (SERS) has been employed to enhance Raman signals further by incorporating noble metal nanoparticles onto SWNTs. This enhancement enables significantly faster Raman imaging of biological specimens, as illustrated in Fig. 14.5a, b [115]. SWNTs, when appropriately surface-functionalized, have also found applications in in vivo multimodal tracking through Raman imaging and stem cell labeling. The distinctive Raman peaks associated with SWNTs, namely the radial breathing mode (RBM, 100 cm^{-1}–300 cm^{-1}) and the tangential G band (1580 cm^{-1}), correspond to vibrational modes of carbon atoms in the radial and tangential directions, respectively. These peaks are typically present in SWNTs. In a pioneering study by Heller et al. in 2005, SWNTs' intrinsic Raman characteristics were exploited for live cell imaging [116]. They utilized SWNTs wrapped in DNA oligonucleotides as live cell tracking markers. Upon stimulation with a 785 nm laser, the RBM peak of SWNTs was detected within living 3T3 cells without the need for additional fluorescent markers. SWNTs exhibited robust Raman scattering signals without significant quenching or photobleaching, making them suitable for long-term tracking in biological samples, unlike organic dyes and quantum dots which tend to degrade over time [117]. High SWNT Raman signals were seen during Raman imaging of those cells, providing conclusive evidence of nanotube cellular penetration without additional fluorescent markers. Due to SWNTs' inert chemical structure and excellent photo-stability, SWNTs can be tracked and imaged over an extended period in biological systems [118]. This was further demonstrated by Liu et al. in a subsequent study that followed the biodistribution of SWNT in mice over three months using ex vivo Raman spectroscopy measurements [117]. Before Raman imaging, the SWNT-RGD compound was intravenously delivered into U87MG tumour-bearing mice. The temporal information of SWNT levels in tumours was disclosed by Raman imaging of tumours on living mice. High SWNT Raman signals were seen on the tumours of mice injected with SWNT-RGD, as was predicted and in agreement with an earlier work employing positron emission tomography (PET) to scan radio-labelled SWNTs in mice [119]. Lamprecht et al. succeeded in mapping and tracking the intracellular distribution of carbon nanotubes following targeted delivery to cancer cells, utilizing the inherent Raman G band of carbon nanotubes [120]. In a recent study, Kang et al. investigated the cellular uptake of SWNTs using high-speed confocal Raman imaging. They captured videos of two cell-intrinsic and nine nanotube-derived Raman signals in RAW 264 macrophages to distinguish SWNTs with different indices and aggregation states within the cells [121]. The benefits of Raman spectroscopy for molecular imaging of living cells were underlined in this study. Targeting ligand conjugated SWNTs have also been used to demonstrate in vivo Raman imaging of mouse cancers [122].

Liu et al. achieved the first multiplexed Raman imaging of live cells by conjugating three types of SWNTs with distinct ^{13}C doping ratios and Raman G-band peaks to various targeting ligands for cell line identification [112]. Multiplexed Raman imaging was accomplished simultaneously using spectrum unmixing. This advancement holds promise for multiplexed in vitro cell imaging and ex vivo tissue slice labeling. Enhancing the Raman signals of these tags is crucial for reducing imaging time, as Raman imaging acquisition time is primarily influenced by the Raman signals of the tags. One approach to achieving this enhancement involves attaching gold nanoparticles to covalently functionalized SWNTs, creating noble metal-coated pristine SWNTs for use as SERS nanoprobes in Raman imaging of biological samples. In conventional Raman spectroscopy, single-walled carbon nanotubes (SWNTs) typically exhibit diminished Raman signals due to structural damage incurred during functionalization, such as oxidation processes (e.g., oxidization) [123]. In that study, pure SWNTs that preserved high Raman scattering were functionalized using single-strand DNA. These nanotubes were first coated with negatively charged gold nanoseeds, which caused the formation of positively or negatively charged gold or silver nano-shells on the surface of SWNTs. In the solution phase, the noble metal coated SWNTs demonstrated excellent SERS effects, with a maximum enhancement factor of over 20 [124]. The relatively strong Raman signals of SWNTs, which do not quench or bleach, allow for long-term monitoring and imaging. Multiplexed Raman imaging is made possible by adjusting the Raman shifts by altering the isotope composition of SWNTs. Heller et al. published the first study using Raman imaging of SWNTs in living cells in 2005 [125]. DNA was employed to functionalize Hipco SWNTs before combining them with 3T3 fibroblast and myoblast stem cells.

SWNTs outperform organic molecules regarding chemical stability, inertness, and powerful Raman signals. Long-term tracking and imaging are made possible by this [116]. Another work that proved the multiplexed cell identification/imaging using SWNTs with various isotope compositions was feasible for simultaneously probing and imaging several biological species [125]. The three-colour SWNT mixture was incubated with three different cell types, including BT474 (human breast cancer, Her2+), U87MG (human glioblastoma, Integrin Rv3+), and MDA-MB-468 (human breast cancer, Her1+). After that, the cells were imaged using confocal Raman spectroscopy. Three different cell kinds were visible in the Raman images, and a unique Raman colour characterized each. SWNTs with regulated diameters also display

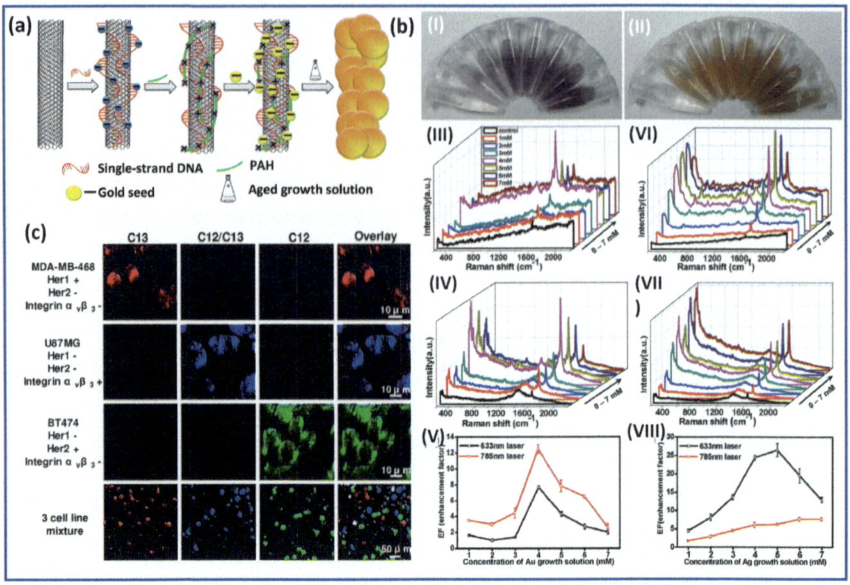

Fig. 14.5 **a** Illustration of the synthetic procedure of SWNTs-metal nanocomposite. **b** Studies on the SERS effects. (I and II) Photos of SWNT-Au-PEG (I) and SWNT-Ag-PEG (II) solutions prepared by adding different concentrations of aged growth solutions. From left to right were samples added with increasing concentrations of growth solutions (0–7 mM). (III and IV) Raman spectra of SWNT-Au-PEG prepared by adding different concentrations of growth solutions taken under 633 nm (III) and 785 nm (IV) laser excitation (VI and VII). Raman spectra of SWNT-Ag-PEG samples prepared by adding different concentrations of silver growth solutions taken under 633 nm (VI) and 785 nm (VII) laser excitation. (V and VIII) SERS EFs of SWNT-Au-PEG (V) and SWNT-Ag-PEG (VIII) samples with increasing concentrations of growth solutions. (**A**) and (**B**) are adapted with permission from [115]. Copyright © 2012 American Chemical Society. **c** Multicolour Raman imaging with SWNTs. Deconvoluted confocal Raman spectroscopy images of three different cell lines after incubation with a mixture of the three-color SWNTs (top 3 rows, red, blue, green colors are Raman intensities of C12, C12/C13, and C13 SWNTs, respectively). In the bottom row, a mixture of three cell lines was incubated with the three-color SWNT mixture. Those images clearly show a mixture of cells with differentiated Raman labeling by three types of SWNTs. Adapted with permission from [120]. Copyright © 2008 American Chemical Society

different Raman peaks in their radial-breathing modes, which can also be employed for multicolor Raman imaging and isotope-dependent Raman colours as shown in Fig. 14.5c [112] SWNTs are potential new Raman tags for biological system multiplexed detection and imaging when used collectively.

Further, targeted Raman imaging of cancer cells marked with folic acid (FA) conjugated SWNT-Au nanocomposite (SWNT-Au-PEG-FA) is made possible by utilizing a near-infrared (NIR) laser as the excitation source [75]. The SWNTs-Au-PEG-FA nanocomposite delivers noticeably better photothermal cancer cell-killing efficacy due to the gold shell's significant surface plasmon resonance absorption. This demonstrates the method for creating water-soluble noble metal coated SWNTs with a strong SERS effect ideal for labelling and quick Raman spectroscopic imaging of biological samples. Hence, the SWNT-Au-PEG nanocomposite made here may be a fascinating optical theranostics probe for treating and imaging cancer.

Raman spectroscopy displays non-invasive deep-tissue molecular pictures in a living individual [80]. An exemplary imaging system can be created by utilizing surface-enhanced Raman scattering nanoparticles and single-wall carbon nanotubes that can be used for small-animal Raman imaging, whole-body Raman imaging, nanoparticle pharmacokinetics, multiplexing, and in vivo tumour targeting. Future research with different targeted Raman nanoparticles will be built on the capacity of Raman spectroscopy to pinpoint targeted SWNTs in a tumour model noninvasively [122]. The non-invasive, extremely sensitive Raman imaging technology to evaluate the effectiveness of novel diagnostic approaches and therapeutics in small animal models ultimately leads to advancements in the treatment of cancer patients.

14.6 Imaging with CARS

Over the last decade, researchers have utilized CARS (Coherent Anti-Stokes Raman Spectroscopy) to investigate the metabolic processes involving lipids, proteins, nucleic acids, and various metabolites within living cells and simple model organisms. New applications for tissue biology and histopathology have been made possible by CARS imaging method. A few illustrations of the numerous recent successes of CARS imaging have been covered in this section. CARS involves employing two synchronized pulse trains on a laser scanning microscope. Since CARS uses the same laser source and microscopy setup, however, make use of various signal-detection tools. Furthermore, single-pulse CARS microscopy imaging has been performed by using intra-pulse excitation of broadband laser rather than two beams to create pump and Stokes pulses [126]. A wide-field CARS microscope with a dark-field objective lens has been created in addition to laser scanning.

Coherent Raman imaging (CRI) has become a promising toolkit that holds the potential for transition from laboratory settings to clinical and surgical applications. This is due to its capability for rapid video-rate imaging, non-invasive and non-destructive nature, and its ability to provide chemically-specific images of tissues without relying on external contrast agents [127–129]. By fine-tuning the disparity in frequency between the excitation sources, namely the pump and Stokes lasers, to align with the vibrational resonance of the specific target molecule, it becomes possible to generate coherent and selective Raman active vibrations. This phenomenon occurs through the nonlinear interaction of two or more ultrafast laser sources [127–129]. Pulsed lasers, operating at picosecond or femtosecond durations and characterized by short duty cycles, are employed to achieve nonlinear excitation at low average power levels. This enables the essential nonlinear mixing process to take place. Images can be produced by quickly scanning the tissue with the combined beams. CARS imaging techniques are commonly preferred due to their practicality and inherent capability to discern Raman features, facilitating the differentiation and identification of various species. To achieve high photon densities, researchers utilize high numerical aperture

(NA) microscope objectives to tightly focus the light, thereby promoting nonlinear interactions between the co-propagating laser sources. The direction and specific modalities of CARS imaging can be finely tuned, making it particularly versatile for research involving cells in controlled laboratory settings or thin sections of tissue samples removed from the body.

Single-cell imaging has made extensive use of CARS microscopy with robust signal, particularly in the C–H stretching area, offering hitherto unheard-of insight into C–H rich subcellular constituents like lipids and proteins. As an example, Nan et al. employed CH_2 stretching vibrations to perform CARS imaging of neutral lipid droplets within live fibroblast cells. Furthermore, their study delved into the behavior of lipid droplets during the differentiation process of 3T3-L1 cells [130–132]. Paar et al. employed time-lapse CARS imaging to examine the metabolism of intracellular lipid droplets during the process of lipolysis [131–133]. Following forskolin stimulation, Fig. 14.6 depicts the gradual reduction of neutral lipid content within 3T3-L1 adipocytes (Panel A). CARS was utilized to examine the in vitro and in vivo behavior of cancer cells as well as the impact of a high-fat diet on cancer metastasis. Furthermore, CARS has been applied to the analysis of circulating prostate tumor cells, revealing a seven-fold increase in C–H intensity when compared to leukocytes [134]. The label-free chemical mapping of cellular components has yielded promising outcomes in the recent advancements of hyperspectral CARS imaging. For instance, in a study by Di Napoli et al. human adipose-derived stem cells were fed with various ratios of saturated palmitic acid and unsaturated linolenic acid [135]. They showed images of the resonance ratio of the peaks at 2,930 cm^{-1} (from the protein CH_3 group) and 2,855 cm^{-1}, as well as the hyperspectral CARS images of cellular components following this treatment and the related CARS spectra (from the lipid CH_2 group). This method served to emphasize the variation in lipid content in these circumstances and the fingerprint region has also undergone a comparable study. The recent developments in the hyperspectral CARS imaging have shown intriguing results for the label-free chemical mapping of cellular components [135]. Studies of lipid metabolism and fat accumulation using CARS microscopy have used Caenorhabditis elegans as a live test subject [133–135]. By observing the intricate architecture of the organism and the geometrical arrangement of lipid droplets, Hellerer et al. employed CARS microscopy to investigate the relationship between genetic variants and metabolic routes for storing lipid in living C. elegans [133] In a comparison of label-free CARS microscopy and a conventional dye-labeled method for imaging C. elegans, Yen et al. showed that CARS offers direct and non-invasive quantitative measurement of lipids which overcomes several drawbacks of the conventional method [134]. CARS was also used to investigate how genetic alteration affects how neutral lipids are stored by C. elegans and how changes to the amount of lipid unsaturation [135].

Cholesterol crystals have also been found using CARS microscopy in the atherosclerotic tissue of sick mice and humans. Broadband CARS was recently used by Camp et al. to demonstrate 3D tissue imaging, which allowed for the differentiation of collagen, DNA, proteins, and lipids based on their characteristic spectra (Fig. 14.6,

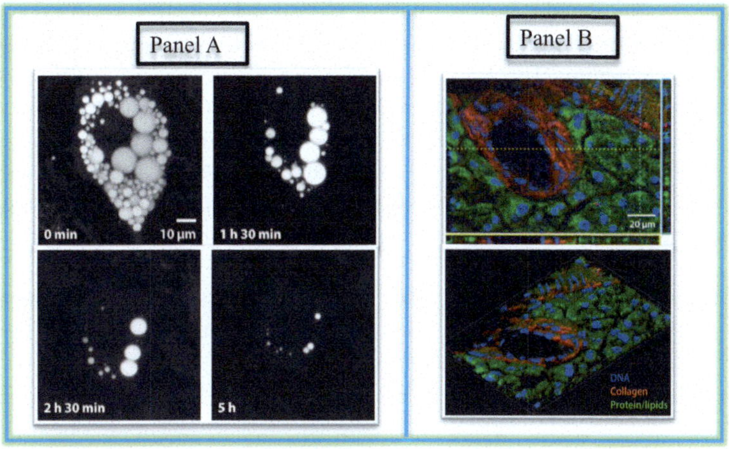

Fig. 14.6 Coherent anti-Stokes Raman scattering (CARS) images. (Panel A) Cellular lipid depletion in 3T3-L1 adipocytes images with different time zone. Permission adapted from Reference 4a. (Panel B) A murine pancreatic duct's 3D broadband CARS imaging reveals the distributions of DNA, collagen, protein, and lipids. Adapted with permission from [136]. Copyright © 2014 Nature Photonics

panel B) [136]. In their research, hyperspectral CARS with exceptional spectral resolution (10 cm^{-1}) spanned a wide Raman spectral window of 500–3,500 cm^{-1}. High-quality CARS imaging of both the C–H stretching region and the fingerprint region was made possible by the heterodyne amplification of weak Raman transitions by the non-resonant background [136].

For objects smaller than a micron, CRI approaches offer a signal strength that is 10,000 times greater than spontaneous Raman, cutting down the hours-long acquisition durations of spontaneous Raman imaging to fractions of a second [137]. This remarkable speed advantage has been achieved not only for tissue mapping but also for various devices and applications utilized in tissue analysis within or close to the operating room. The advantages of optical imaging based on Coherent based optical imaging encompass the elimination of time-consuming staining and sample preparation, the reduction in the quantity of biopsy tissue required for analysis, and the confinement of specific tissue characteristics to those for which staining methods are firmly established. Despite the considerable technical and translational obstacles, König and his team initially documented their early efforts to adapt multiphoton tomographic devices for conducting CARS measurements in human skin [138]. To generate synchronized pumps and Stokes pulses, the first device combined an optical parametric oscillator (OPO) with a tunable NIR femtosecond laser oscillator [138]. All of these techniques make use of CARS imaging, but they are all limited by the background signals that are non-resonant and interfere with CARS imaging as well as the quadratic relationship between the CARS signal and the concentration of Raman scatterers. In contrast, background-free, linearly concentration-dependent signals can be seen in stimulated Raman scattering. Hirose and colleagues utilized a single-mode fiber to transmit laser light into a rigid endoscope unit equipped with integrated galvanometric scanning. This setup involved a pair of relay lenses positioned within a 12 mm diameter, 270 mm long tube. Their aim was to explore the potential of robot-assisted surgery for visualizing nerves intraoperatively [139]. CARS imaging has a huge field of view (around 650 m in diameter) with great homogeneity and minimum distortion because to the utilization of free-space optical components. Long imaging periods were necessary despite this performance, but according to the authors, performance might be enhanced by managing chromatic aberrations, laser spot sizes, and focused spot sizes. For these technologies to have an influence on clinical medicine and surgery in the future, it will be essential to be able to optimize system designs to satisfy application-specific needs.

14.7 Conclusion and Future Outlook

Raman imaging is a constructively evolving analytical technique that has been widely applied in different fields of biological science and bioengineering. Due to its non-invasive, non-destructive, and label-free approach giving the chemical composition and gradient information about the sample. It has drawn the attention of researchers from multiple disciplines who are exploring its potential for better advancements and development. In this chapter, we have summarized the different aspects of Raman imaging, including its principles, working setup, and recent applications. In the near future, there is a need to develop translational Spontaneous Raman imaging, SERS imaging, TERS imaging, and CARS imaging approaches to address it as a routine analytical testing technique. The scanning speed of Raman imaging is one of the key targets for improving the efficiency of Raman imaging. In the future, with the development of advanced detectors and image processing, the time-consuming Raman imaging data acquisition and image processing should be addressed. CARS imaging has been seen with better video rate spatial resolution and needs to be further explored in the imaging of the biological sample. Nowadays, many other techniques such as scanning electron microscopy, atomic force microscopy, and fluorescence microscopy are integrated with the Raman microspectroscopy, resulting in the development of a multimodal imaging systems that can provide a better understanding of the composition, gradient, structural, surface topology, and other complementary information. In addition to this, a portable Raman imaging system with easy-to-use translational features rather than conventional laboratory Raman set up for practical use in the fields of the pharmaceutical, bioprocess and food industry, medical operative field, forensic investigation, threat detection, and environmental monitoring would expand its applicability and improve its accessibility for non-expert users. However, while developing such approaches, sensitivity should not be compromised, and it should be cost effective. However, molecular specificity and minimal sample preparation aid Raman's imaging efficiency. Raman imaging can be further upgraded and enhanced by achieving a higher spatial resolution. This goal can be reached by detailed sample probing, which can be achieved with better nanofabrication. The heterogeneous composition of the sample may result in overlapping Raman bands, so the development of novel Raman probes targeting the molecule of interest and enhancing the Raman signal should be prioritized. This approach can also help detect biomolecules in low abundance. With the ongoing advancement in developing better detectors, lasers, nanofabrication which will improve sensitivity, spatial resolution, and imaging speed, the versatility of Raman imaging will expand to various fields in biosciences.

References

1. Y. Shen, F. Hu, W. Min, Raman imaging of small biomolecules. Annu. Rev. Biophys. **48**, 347 (2019)
2. B. Durrant, M. Trappett, D. Shipp, I. Notingher, Recent developments in spontaneous raman imaging of living biological cells. Curr. Opin. Chem. Biol. **51**, 138 (2019)
3. C. Krafft, I.W. Schie, T. Meyer, M. Schmitt, J. Popp, Developments in spontaneous and coherent Raman scattering microscopic imaging for biomedical applications. Chem. Soc. Rev. **45**, 1819 (2016)
4. C.V. Raman, K.S. Krishnan, A new type of secondary radiation. Nature **121**, 501 (1928)
5. L. Ashton, K.A. Hollywood, R. Goodacre, Making colourful sense of Raman images of single cells. Analyst **140**, 1852 (2015)
6. J. Karlo, A.K. Dhillon, S. Siddhanta, S.P. Singh, Monitoring of microbial proteome dynamics using Raman stable isotope probing. J. Biophotonics **16**, (2023)
7. J. Barbillat, P. Dhamelincourt, M. Delhaye, E. Da Silva, Raman confocal microprobing, imaging and fibre-optic remote sensing: a further step in molecular analysis. J. Raman Spectrosc. **25**, 3 (1994)
8. M. Delhaye, P. Dhamelincourt, Raman microprobe and microscope with laser excitation. J. Raman Spectrosc. **3**, 33 (1975)
9. S. Ai (ed.), *Pharmaceutical Applications of Raman Spectroscopy* (Wiley, Hoboken, NJ, USA, 2007)
10. I.J. Pence, C.L. Evans, Translational biophotonics with Raman imaging: clinical applications and beyond. Analyst **146**, 6379 (2021)
11. R. Smith, K.L. Wright, L. Ashton, Raman spectroscopy: an evolving technique for live cell studies. Analyst **141**, 3590 (2016)
12. M.G. Lizio, R. Boitor, I. Notingher, Selective-sampling Raman imaging techniques for ex vivo assessment of surgical margins in cancer surgery. Analyst **146**, 3799 (2021)
13. S. Laing, L.E. Jamieson, K. Faulds, D. Graham, Surface-enhanced Raman spectroscopy for in vivo biosensing. Nat. Rev. Chem. **1**, 0060 (2017)
14. A.I. Pérez-Jiménez, D. Lyu, Z. Lu, G. Liu, B. Ren, Surface-enhanced Raman spectroscopy: benefits, trade-offs and future developments. Chem. Sci. **11**, 4563 (2020)
15. R. Böhme, D. Cialla, M. Richter, P. Rösch, J. Popp, V. Deckert, Biochemical imaging below the diffraction limit—Probing cellular membrane related structures by tip-enhanced Raman spectroscopy (TERS). J. Biophotonics **3**, 455 (2010)
16. A. Belu, C. Mahoney, K. Wormuth, Chemical imaging of drug eluting coatings: combining surface analysis and confocal Raman microscopy. J. Control. Release **126**, 111 (2008)
17. J.-X. Cheng, X.S. Xie, Coherent anti-stokes Raman scattering microscopy: instrumentation, theory, and applications. J. Phys. Chem. B **108**, 827 (2004)
18. Z. Liao, F. Sinjab, H.M. Elsheikha, I. Notingher, Optical sectioning in multifoci Raman hyperspectral imaging. J. Raman Spectrosc. **49**, 1660 (2018)

19. A.A. Abalymov, R.A. Verkhovskii, M.V. Novoselova, B.V. Parakhonskiy, D.A. Gorin, A.M. Yashchenok, G.B. Sukhorukov, Live-cell imaging by confocal Raman and fluorescence microscopy recognizes the crystal structure of calcium carbonate particles in HeLa cells. Biotechnol. J. **13**, 1800071 (2018)
20. S. Managò, N. Migliaccio, M. Terracciano, M. Napolitano, N.M. Martucci, L. De Stefano, I. Rendina, A.C. De Luca, A. Lamberti, I. Rea, Internalization kinetics and cytoplasmic localization of functionalized diatomite nanoparticles in cancer cells by Raman imaging. J. Biophotonics **11**, e201700207 (2018)
21. S. Schlücker, M.D. Schaeberle, S.W. Huffman, I.W. Levin, Raman microspectroscopy: a comparison of point, line, and wide-field imaging methodologies. Anal. Chem. **75**, 4312 (2003)
22. J. Karlo, R. Prasad, S.P. Singh, Biophotonics in food technology: quo Vadis? J. Agric. Food Res. **11**, 100482 (2023)
23. J. Qi, W.-C. Shih, Performance of line-scan Raman microscopy for high-throughput chemical imaging of cell population. Appl. Opt. **53**, 2881 (2014)
24. M. Bowden, D.J. Gardiner, G. Rice, D.L. Gerrard, Line-scanned micro Raman spectroscopy using a cooled CCD imaging detector. J. Raman Spectrosc. **21**, 37 (1990)
25. A. Masic, J.C. Weaver, Large area sub-micron chemical imaging of magnesium in sea urchin teeth. J. Struct. Biol. **189**, 269 (2015)
26. S. Yang, B. Li, A. Akkus, O. Akkus, L. Lang, Wide-field Raman imaging of dental lesions. Analyst **139**, 3107 (2014)
27. A.F. Palonpon, M. Sodeoka, K. Fujita, Molecular imaging of live cells by Raman microscopy. Curr. Opin. Chem. Biol. **17**, 708 (2013)
28. A.F. Palonpon, J. Ando, H. Yamakoshi, K. Dodo, M. Sodeoka, S. Kawata, K. Fujita, Raman and SERS microscopy for molecular imaging of live cells. Nat. Protoc. **8**, 677 (2013)
29. N. Zhu, Y. Yang, M. Ji, D. Wu, K. Chen, Label-free visualization of lignin deposition in loquats using complementary stimulated and spontaneous Raman microscopy. Hortic. Res. **6**, 72 (2019)
30. J. Ando, A.F. Palonpon, M. Sodeoka, K. Fujita, High-speed Raman imaging of cellular processes. Curr. Opin. Chem. Biol. **33**, 16 (2016)
31. K. Hamada, K. Fujita, N.I. Smith, M. Kobayashi, Y. Inouye, S. Kawata, Raman microscopy for dynamic molecular imaging of living cells. J. Biomed. Opt. **13**, 1 (2008)
32. M. Okada, N.I. Smith, A.F. Palonpon, H. Endo, S. Kawata, M. Sodeoka, K. Fujita, Label-free Raman observation of cytochrome C dynamics during apoptosis. Proc. Natl. Acad. Sci. **109**, 28 (2012)
33. F. C. Pascut, S. Kalra, V. George, N. Welch, C. Denning, I. Notingher, Non-invasive label-free monitoring the cardiac differentiation of human embryonic stem cells in-vitro by Raman spectroscopy. Biochimica et Biophysica Acta (BBA)—General Subjects **1830**, 3517 (2013)
34. S. Takanezawa, S. Morita, Y. Ozaki, Y. Sako, Raman spectral dynamics of single cells in the early stages of growth factor stimulation. Biophys. J. **108**, 2148 (2015)
35. B. Brozek-Pluska, J. Musial, R. Kordek, H. Abramczyk, Analysis of human colon by Raman spectroscopy and imaging-elucidation of biochemical changes in carcinogenesis. Int. J. Mol. Sci. **20**, 3398 (2019)
36. L. Huang et al., Rapid, label-free histopathological diagnosis of liver cancer based on Raman spectroscopy and deep learning. Nat. Commun. **14**, 48 (2023)
37. Z. Li et al., Detection of pancreatic cancer by convolutional-neural-network-assisted spontaneous Raman spectroscopy with critical feature visualization. Neural Netw. **144**, 455 (2021)
38. A.N. Ramya, J.S. Arya, M. Madhukrishnan, S. Shamjith, M.S. Vidyalekshmi, K.K. Maiti, Raman imaging: an impending approach towards cancer diagnosis. Chem. Asian J. **16**, 409 (2021)
39. H. Abramczyk, B. Brozek-Pluska, Raman imaging in biochemical and biomedical applications. Diagnosis and treatment of breast cancer. Chem. Rev. **113**, 5766 (2013)

40. F. Gaba, W.J. Tipping, M. Salji, K. Faulds, D. Graham, H.Y. Leung, Raman spectroscopy in prostate cancer: techniques, applications and advancements. Cancers (Basel) **14**, 1535 (2022)
41. M. Kopec, A. Imiela, H. Abramczyk, Monitoring glycosylation metabolism in brain and breast cancer by Raman imaging. Sci. Rep. **9**, 166 (2019)
42. B.T. Scarpitti, C. Chitchumroonchokchai, S.K. Clinton, Z.D. Schultz, In vitro imaging of lycopene delivery to prostate cancer cells. Anal. Chem. **94**, 5106 (2022)
43. S. Kar, S.V. Jaswandkar, K.S. Katti, J.W. Kang, P.T.C. So, R. Paulmurugan, D. Liepmann, R. Venkatesan, D.R. Katti, Label-free discrimination of tumorigenesis stages using in vitro prostate cancer bone metastasis model by Raman imaging. Sci. Rep. **12**, 8050 (2022)
44. I. Zeise, Z. Heiner, S. Holz, M. Joester, C. Büttner, J. Kneipp, Raman imaging of plant cell walls in sections of cucumis sativus. Plants **7**, 7 (2018)
45. M. Larion, T. Dowdy, V. Ruiz-Rodado, M. Meyer, H. Song, W. Zhang, D. Davis, M. Gilbert, A. Lita, Detection of metabolic changes induced via drug treatments in live cancer cells and tissue using Raman imaging microscopy. Biosensors (Basel) **9**, 5 (2018)
46. C. Kallepitis, M.S. Bergholt, M.M. Mazo, V. Leonardo, S.C. Skaalure, S.A. Maynard, M.M. Stevens, Quantitative volumetric Raman imaging of three dimensional cell cultures. Nat. Commun. **8**, 14843 (2017)
47. Y.H. Ong, M. Lim, Q. Liu, Comparison of principal component analysis and biochemical component analysis in Raman spectroscopy for the discrimination of apoptosis and necrosis in K562 Leukemia cells. Opt. Express **20**, 22158 (2012)
48. I. Pence, A. Mahadevan-Jansen, Clinical instrumentation and applications of Raman spectroscopy. Chem. Soc. Rev. **45**, 1958 (2016)
49. A. Chaichi, A. Prasad, M. Gartia, Raman spectroscopy and microscopy applications in cardiovascular diseases: from molecules to organs. Biosensors (Basel) **8**, 107 (2018)
50. Q.D. Mai, H.A. Nguyen, T.L.H. Phung, N. Xuan Dinh, Q.H. Tran, T.Q. Doan, A.-T. Le, Silver nanoparticles-based SERS platform towards detecting chloramphenicol and amoxicillin: an experimental insight into the role of HOMO–LUMO energy levels of the analyte in the SERS signal and charge transfer process. J. Phys. Chem. C **126**, 7778 (2022)
51. R. Pilot, R. Signorini, C. Durante, L. Orian, M. Bhamidipati, L. Fabris, A review on surface-enhanced Raman scattering. Biosensors (Basel) **9**, 57 (2019)
52. R.L. McCreery, *Raman Spectroscopy for Chemical Analysis* (Wiley, Hoboken, NJ, USA, 2000)
53. W. Xie, P. Qiu, C. Mao, Bio-imaging, detection and analysis by using nanostructures as SERS substrates. J. Mater. Chem. **21**, 5190 (2011)
54. C.L. Haynes, R.P. Van Duyne, Nanosphere lithography: a versatile nanofabrication tool for studies of size-dependent nanoparticle optics. J. Phys. Chem. B **105**, 5599 (2001)
55. Y.E. Ejorh, W.H. Ilsley, B.G. Ooi, Elucidating the chemisorption phenomena in SERS studies via computational modeling. Opt. Photonics J. **08**, 212 (2018)
56. J.M. Yuen, N.C. Shah, J.T. Walsh, M.R. Glucksberg, R.P. Van Duyne, Transcutaneous glucose sensing by surface-enhanced spatially offset Raman spectroscopy in a rat model. Anal. Chem. **82**, 8382 (2010)
57. D.A. Stuart, C.R. Yonzon, X. Zhang, O. Lyandres, N.C. Shah, M.R. Glucksberg, J.T. Walsh, R.P. Van Duyne, Glucose sensing using near-infrared surface-enhanced Raman spectroscopy: gold surfaces, 10-day stability, and improved accuracy. Anal. Chem. **77**, 4013 (2005)
58. M.K. Gregas, F. Yan, J. Scaffidi, H.-N. Wang, T. Vo-Dinh, Characterization of nanoprobe uptake in single cells: spatial and temporal tracking via SERS labeling and modulation of surface charge. Nanomedicine **7**, 115 (2011)
59. R.A. Tripp, R.A. Dluhy, Y. Zhao, Novel nanostructures for SERS biosensing. Nano Today **3**, 31 (2008)
60. W.R. Premasiri, D.T. Moir, M.S. Klempner, N. Krieger, G. Jones, L.D. Ziegler, Characterization of the surface enhanced Raman scattering (SERS) of bacteria. J. Phys. Chem. B **109**, 312 (2005)
61. S. Shanmukh, L. Jones, J. Driskell, Y. Zhao, R. Dluhy, R.A. Tripp, Rapid and sensitive detection of respiratory virus molecular signatures using a silver nanorod array SERS substrate. Nano Lett. **6**, 2630 (2006)

62. Y. Wen, W. Wang, Z. Zhang, L. Xu, H. Du, X. Zhang, Y. Song, Controllable and reproducible construction of a SERS substrate and its sensing applications. Nanoscale **5**, 523 (2013)
63. L. Guerrini, E. Pazos, C. Penas, M.E. Vázquez, J.L. Mascareñas, R.A. Alvarez-Puebla, Highly sensitive SERS quantification of the oncogenic protein C-Jun in cellular extracts. J. Am. Chem. Soc. **135**, 10314 (2013)
64. K.E. Shafer-Peltier, C.L. Haynes, M.R. Glucksberg, R.P. Van Duyne, Toward a glucose biosensor based on surface-enhanced Raman scattering. J. Am. Chem. Soc. **125**, 588 (2003)
65. D.A. Stuart, K.B. Biggs, R.P. Van Duyne, Surface-enhanced Raman spectroscopy of half-mustard agent. Analyst **131**, 568 (2006)
66. K. Kim, J.W. Lee, D. Shin, J.-Y. Choi, K.S. Shin, Organic isocyanide-adsorbed gold nanostructure: a SERS sensory device for indirect peak-shift detection of volatile organic compounds. Analyst **137**, 1930 (2012)
67. K. Kim, J.W. Lee, K.S. Shin, Cyanide SERS as a platform for detection of volatile organic compounds and hazardous transition metal ions. Analyst **138**, 2988 (2013)
68. C. Matthäus, S. Boydston-White, M. Miljković, M. Romeo, M. Diem, Raman and infrared microspectral imaging of mitotic cells. Appl. Spectrosc. **60**, 1 (2006)
69. N. Uzunbajakava, A. Lenferink, Y. Kraan, E. Volokhina, G. Vrensen, J. Greve, C. Otto, Nonresonant confocal Raman imaging of DNA and protein distribution in apoptotic cells. Biophys. J. **84**, 3968 (2003)
70. E.J. Titus, M.L. Weber, S.M. Stranahan, K.A. Willets, Super-resolution SERS imaging beyond the single-molecule limit: an isotope-edited approach. Nano Lett. **12**, 5103 (2012)
71. W. Xie, L. Wang, Y. Zhang, L. Su, A. Shen, J. Tan, J. Hu, Nuclear targeted nanoprobe for single living cell detection by surface-enhanced Raman scattering. Bioconjug. Chem. **20**, 768 (2009)
72. E. Lenzi, D. Jimenez de Aberasturi, L.M. Liz-Marzán, Surface-enhanced Raman scattering tags for three-dimensional bioimaging and biomarker detection. ACS Sens **4**, 1126 (2019)
73. A.G. Tkachenko, H. Xie, D. Coleman, W. Glomm, J. Ryan, M.F. Anderson, S. Franzen, D.L. Feldheim, Multifunctional gold nanoparticle−Peptide complexes for nuclear targeting. J. Am. Chem. Soc. **125**, 4700 (2003)
74. K. Kneipp, A.S. Haka, H. Kneipp, K. Badizadegan, N. Yoshizawa, C. Boone, K.E. Shafer-Peltier, J.T. Motz, R.R. Dasari, M.S. Feld, Surface-enhanced Raman spectroscopy in single living cells using gold nanoparticles. Appl. Spectrosc. **56**, 150 (2002)
75. A. Sujith, T. Itoh, H. Abe, K. Yoshida, M.S. Kiran, V. Biju, M. Ishikawa, Imaging the cell wall of living single yeast cells using surface-enhanced Raman spectroscopy. Anal. Bioanal. Chem. **394**, 1803 (2009)
76. M. Kahraman, A.I. Zamaleeva, R.F. Fakhrullin, M. Culha, Layer-by-layer coating of bacteria with noble metal nanoparticles for surface-enhanced Raman scattering. Anal. Bioanal. Chem. **395**, 2559 (2009)
77. R. Singh, H.S. Nalwa, Medical applications of nanoparticles in biological imaging, cell labeling, antimicrobial agents, and anticancer nanodrugs. J. Biomed. Nanotechnol. **7**, 489 (2011)
78. Y. Hang, J. Boryczka, N. Wu, Visible-light and near-infrared fluorescence and surface-enhanced Raman scattering point-of-care sensing and bio-imaging: a review. Chem. Soc. Rev. **51**, 329 (2022)
79. H. Zheng, G. Chen, L.A. DeLouise, Z. Lou, Detection of the cancer marker CD146 expression in melanoma cells with semiconductor quantum dot label. J. Biomed. Nanotechnol. **6**, 303 (2010)
80. S. Lee, H. Chon, S.-Y. Yoon, E.K. Lee, S.-I. Chang, D.W. Lim, J. Choo, Fabrication of SERS-fluorescence dual modal nanoprobes and application to multiplex cancer cell imaging. Nanoscale **4**, 124 (2012)
81. S. Lee, S. Kim, J. Choo, S.Y. Shin, Y.H. Lee, H.Y. Choi, S. Ha, K. Kang, C.H. Oh, Biological imaging of HEK293 cells expressing PLCγ1 using surface-enhanced Raman microscopy. Anal. Chem. **79**, 916 (2007)

82. S. Bock et al., Highly sensitive near-infrared SERS nanoprobes for in vivo imaging using gold-assembled silica nanoparticles with controllable nanogaps. J. Nanobiotechnol. **20**, 130 (2022)
83. Ö. Aydin, M. Altaş, M. Kahraman, Ö.F. Bayrak, M. Çulha, Differentiation of healthy brain tissue and tumors using surface-enhanced Raman scattering. Appl. Spectrosc. **63**, 1095 (2009)
84. Y.C. Cao, R. Jin, C.A. Mirkin, Nanoparticles with Raman spectroscopic fingerprints for DNA and RNA detection. Science (1979) **297**, 1536 (2002)
85. X. Wu, H. Liu, J. Liu, K.N. Haley, J.A. Treadway, J.P. Larson, N. Ge, F. Peale, M.P. Bruchez, Immunofluorescent labeling of cancer marker Her2 and other cellular targets with semiconductor quantum dots. Nat. Biotechnol. **21**, 41 (2003)
86. K. Nithipatikom, M.J. McCoy, S.R. Hawi, K. Nakamoto, F. Adar, W.B. Campbell, Characterization and application of Raman labels for confocal Raman microspectroscopic detection of cellular proteins in single cells. Anal. Biochem. **322**, 198 (2003)
87. C.E. Talley, L. Jusinski, C.W. Hollars, S.M. Lane, T. Huser, Intracellular PH sensors based on surface-enhanced Raman scattering. Anal. Chem. **76**, 7064 (2004)
88. J. Kneipp, H. Kneipp, B. Wittig, K. Kneipp, One- and two-photon excited optical PH probing for cells using surface-enhanced Raman and hyper-Raman nanosensors. Nano Lett. **7**, 2819 (2007)
89. X. Su, J. Zhang, L. Sun, T.-W. Koo, S. Chan, N. Sundararajan, M. Yamakawa, A.A. Berlin, Composite organic–inorganic nanoparticles (COINs) with chemically encoded optical signatures. Nano Lett. **5**, 49 (2005)
90. S. Sun, D. Thompson, U. Schmidt, D. Graham, G.J. Leggett, Micro-/nano-patterning of DNA and rapid readout with SERS tags. Chem. Commun. **46**, 5292 (2010)
91. M.Y. Sha, H. Xu, M.J. Natan, R. Cromer, Surface-enhanced Raman scattering tags for rapid and homogeneous detection of circulating tumor cells in the presence of human whole blood. J. Am. Chem. Soc. **130**, 17214 (2008)
92. A.S. Patel, S. Juneja, P.K. Kanaujia, V. Maurya, G.V. Prakash, A. Chakraborti, J. Bhattacharya, Gold nanoflowers as efficient hosts for SERS based sensing and bio-imaging. Nano-Struct. Nano-Objects **16**, 329 (2018)
93. K.V. Kong, Z. Lam, W.D. Goh, W.K. Leong, M. Olivo, Metal carbonyl-gold nanoparticle conjugates for live-cell SERS imaging. Angew. Chem. Int. Ed. **51**, 9796 (2012)
94. Z. Lam, K.V. Kong, M. Olivo, W.K. Leong, Vibrational spectroscopy of metal carbonyls for bio-imaging and -sensing. Analyst **141**, 1569 (2016)
95. K.V. Kong, W.K. Leong, Z. Lam, T. Gong, D. Goh, W.K.O. Lau, M. Olivo, A rapid and label-free SERS detection method for biomarkers in clinical biofluids. Small n/a (2014)
96. S.K. Paidi, A. Rizwan, C. Zheng, M. Cheng, K. Glunde, I. Barman, Label-free Raman spectroscopy detects stromal adaptations in premetastatic lungs primed by breast cancer. Cancer Res. **77**, 247 (2017)
97. S. Zhu, B. Deng, F. Liu, J. Li, L. Lin, J. Ye, Surface-enhanced Raman scattering bioimaging with an ultrahigh signal-to-background ratio under ambient light. ACS Appl. Mater. Interfaces **14**, 8876 (2022)
98. S. Bonhommeau, G.S. Cooney, Y. Huang, Nanoscale chemical characterization of biomolecules using tip-enhanced Raman spectroscopy. Chem. Soc. Rev. **51**, 2416 (2022)
99. R. Zhang, X. Zhang, H. Wang, Y. Zhang, S. Jiang, C. Hu, Y. Zhang, Y. Luo, Z. Dong, Distinguishing individual DNA bases in a network by non-resonant tip-enhanced Raman scattering. Angew. Chem. Int. Ed. **56**, 5561 (2017)
100. S. Bonhommeau, S. Lecomte, Tip-enhanced Raman spectroscopy: a tool for nanoscale chemical and structural characterization of biomolecules. ChemPhysChem **19**, 8 (2018)
101. T. Schmid, L. Opilik, C. Blum, R. Zenobi, Nanoscale chemical imaging using tip-enhanced Raman spectroscopy: a critical review. Angew. Chem. Int. Ed. **52**, 5940 (2013)
102. J. Stadler, T. Schmid, R. Zenobi, Developments in and practical guidelines for tip-enhanced Raman spectroscopy. Nanoscale **4**, 1856 (2012)
103. E. Lipiec, J. Kaderli, J. Kobierski, R. Riek, K. Skirlińska-Nosek, K. Sofińska, M. Szymoński, R. Zenobi, Nanoscale hyperspectral imaging of amyloid secondary structures in liquid. Angew. Chem. Int. Ed. **60**, 4545 (2021)

104. Y.-P. Huang et al., Shell-isolated tip-enhanced Raman and fluorescence spectroscopy. Angew. Chem. Int. Ed. **57**, 7523 (2018)
105. N. Kumar, B.M. Weckhuysen, A.J. Wain, A.J. Pollard, Nanoscale chemical imaging using tip-enhanced Raman spectroscopy. Nat. Protoc. **14**, 1169 (2019)
106. D. Talaga, A. Bremner, T. Buffeteau, R.A.L. Vallée, S. Lecomte, S. Bonhommeau, Total internal reflection tip-enhanced Raman spectroscopy of cytochrome C. J. Phys. Chem. Lett. **11**, 3835 (2020)
107. S. Seweryn, K. Skirlińska-Nosek, K. Sofińska, K. Szajna, J. Kobierski, K. Awsiuk, M. Szymoński, E. Lipiec, Optimization of tip-enhanced Raman spectroscopy for probing the chemical structure of DNA. Spectrochim. Acta A Mol. Biomol. Spectrosc. **281**, 121595 (2022)
108. E.A. Pozzi, M.D. Sonntag, N. Jiang, J.M. Klingsporn, M.C. Hersam, R.P. Van Duyne, Tip-enhanced Raman imaging: an emergent tool for probing biology at the nanoscale. ACS Nano **7**, 885 (2013)
109. Z. He et al., Tip-enhanced Raman imaging of single-stranded DNA with single base resolution. J. Am. Chem. Soc. **141**, 753 (2019)
110. D. Kurouski, T. Deckert-Gaudig, V. Deckert, I.K. Lednev, Structure and composition of insulin fibril surfaces probed by TERS. J. Am. Chem. Soc. **134**, 13323 (2012)
111. D. Wang, P. He, Z. Wang, G. Li, N. Majed, A.Z. Gu, Advances in single cell raman spectroscopy technologies for biological and environmental applications. Curr. Opin. Biotechnol. **64**, 218 (2020)
112. K. Besteman, J.-O. Lee, F.G.M. Wiertz, H.A. Heering, C. Dekker, Enzyme-coated carbon nanotubes as single-molecule biosensors. Nano Lett. **3**, 727 (2003)
113. L. Wu et al., A green synthesis of carbon nanoparticles from honey and their use in real-time photoacoustic imaging. Nano Res. **6**, 312 (2013)
114. Z. Liu, X. Li, S.M. Tabakman, K. Jiang, S. Fan, H. Dai, Multiplexed multicolor Raman imaging of live cells with isotopically modified single walled carbon nanotubes. J. Am. Chem. Soc. **130**, 13540 (2008)
115. X. Wang, C. Wang, L. Cheng, S.-T. Lee, Z. Liu, Noble metal coated single-walled carbon nanotubes for applications in surface enhanced Raman scattering imaging and photothermal therapy. J. Am. Chem. Soc. **134**, 7414 (2012)
116. D.A. Heller, S. Baik, T.E. Eurell, M.S. Strano, Single-walled carbon nanotube spectroscopy in live cells: towards long-term labels and optical sensors. Adv. Mater. **17**, 2793 (2005)
117. Z. Liu, C. Davis, W. Cai, L. He, X. Chen, H. Dai, Circulation and long-term fate of functionalized, biocompatible single-walled carbon nanotubes in mice probed by Raman spectroscopy. Proc. Natl. Acad. Sci. **105**, 1410 (2008)
118. Z. Liu, K. Yang, S.-T. Lee, Single-walled carbon nanotubes in biomedical imaging. J. Mater. Chem. **21**, 586 (2011)
119. C. Lamprecht, N. Gierlinger, E. Heister, B. Unterauer, B. Plochberger, M. Brameshuber, P. Hinterdorfer, S. Hild, A. Ebner, Mapping the intracellular distribution of carbon nanotubes after targeted delivery to carcinoma cells using confocal Raman imaging as a label-free technique. J. Phys.: Condens. Matter **24**, 164206 (2012)
120. J.W. Kang, F.T. Nguyen, N. Lue, R.R. Dasari, D.A. Heller, Measuring uptake dynamics of multiple identifiable carbon nanotube species via high-speed confocal Raman imaging of live cells. Nano Lett. **12**, 6170 (2012)
121. C. Zavaleta, A. de la Zerda, Z. Liu, S. Keren, Z. Cheng, M. Schipper, X. Chen, H. Dai, S.S. Gambhir, Noninvasive Raman spectroscopy in living mice for evaluation of tumor targeting with carbon nanotubes. Nano Lett. **8**, 2800 (2008)
122. L. Beqa, Z. Fan, A.K. Singh, D. Senapati, P.C. Ray, Gold nano-popcorn attached SWCNT hybrid nanomaterial for targeted diagnosis and photothermal therapy of human breast cancer cells. ACS Appl. Mater. Interfaces **3**, 3316 (2011)
123. H. Gong, R. Peng, Z. Liu, Carbon nanotubes for biomedical imaging: the recent advances. Adv. Drug Deliv. Rev. **65**, 1951 (2013)
124. C.L. Zavaleta, B.R. Smith, I. Walton, W. Doering, G. Davis, B. Shojaei, M.J. Natan, S.S. Gambhir, Multiplexed imaging of surface enhanced Raman scattering nanotags in living mice using noninvasive Raman spectroscopy. Proc. Natl. Acad. Sci. **106**, 13511 (2009)

125. S. Keren, C. Zavaleta, Z. Cheng, A. de la Zerda, O. Gheysens, S.S. Gambhir, Noninvasive molecular imaging of small living subjects using Raman spectroscopy. Proc. Natl. Acad. Sci. **105**, 5844 (2008)
126. C.W. Freudiger, W. Yang, G.R. Holtom, N. Peyghambarian, X.S. Xie, K.Q. Kieu, Stimulated Raman scattering microscopy with a robust fibre laser source. Nat. Photonics **8**, 153 (2014)
127. C.L. Evans, E.O. Potma, M. Puoris'haag, D. Côté, C.P. Lin, X.S. Xie, Chemical imaging of tissue in vivo with video-rate coherent anti-stokes Raman scattering microscopy. Proc. Natnl Acad. Sci. **102**, 16807 (2005)
128. J.-X. Cheng, X.S. Xie, Vibrational spectroscopic imaging of living systems: an emerging platform for biology and medicine. Science (1979) **350**, aaa8870 (2015)
129. M. Ji et al., Rapid, label-free detection of brain tumors with stimulated Raman scattering microscopy. Sci. Transl. Med. **5** (2013)
130. X. Nan, J.-X. Cheng, X.S. Xie, Vibrational imaging of lipid droplets in live fibroblast cells with coherent anti-stokes Raman scattering microscopy. J. Lipid Res. **44**, 2202 (2003)
131. H.A. Rinia, K.N.J. Burger, M. Bonn, M. Müller, Quantitative label-free imaging of lipid composition and packing of individual cellular lipid droplets using multiplex CARS microscopy. Biophys. J. **95**, 4908 (2008)
132. A. Fussell, E. Garbacik, H. Offerhaus, P. Kleinebudde, C. Strachan, In situ dissolution analysis using coherent anti-stokes Raman scattering (CARS) and hyperspectral CARS microscopy. Eur. J. Pharm. Biopharm. **85**, 1141 (2013)
133. M. Paar et al., Remodeling of lipid droplets during lipolysis and growth in adipocytes. J. Biol. Chem. **287**, 11164 (2012)
134. R. Mitra, O. Chao, Y. Urasaki, O.B. Goodman, T.T. Le, Detection of lipid-rich prostate circulating tumour cells with coherent anti-stokes Raman scattering microscopy. BMC Cancer **12**, 540 (2012)
135. C. Di Napoli, I. Pope, F. Masia, P. Watson, W. Langbein, P. Borri, Hyperspectral and differential CARS microscopy for quantitative chemical imaging in human adipocytes. Biomed. Opt. Express **5**, 1378 (2014)
136. C.H. Camp Jr., Y.J. Lee, J.M. Heddleston, C.M. Hartshorn, A.R.H. Walker, J.N. Rich, J.D. Lathia, M.T. Cicerone, High-speed coherent Raman fingerprint imaging of biological tissues. Nat. Photonics **8**, 627 (2014)
137. M. Lee, C.S. Herrington, M. Ravindra, K. Sepp, A. Davies, A.N. Hulme, V.G. Brunton, Recent advances in the use of stimulated Raman scattering in histopathology. Analyst **146**, 789 (2021)
138. K. König, H.G. Breunig, R. Bückle, M. Kellner-Höfer, M. Weinigel, E. Büttner, W. Sterry, J. Lademann, Optical skin biopsies by clinical CARS and multiphoton fluorescence/SHG tomography. Laser Phys. Lett. **8**, 465 (2011)
139. K. Hirose, T. Aoki, T. Furukawa, S. Fukushima, H. Niioka, S. Deguchi, M. Hashimoto, Coherent anti-stokes Raman scattering rigid endoscope toward robot-assisted surgery. Biomed. Opt. Express **9**, 387 (2018)

SPRINGER NATURE

GPSR Compliance

The European Union's (EU) General Product Safety Regulation (GPSR) is a set of rules that requires consumer products to be safe and our obligations to ensure this.

If you have any concerns about our products, you can contact us on ProductSafety@springernature.com

In case Publisher is established outside the EU, the EU authorized representative is:

Springer Nature Customer Service Center GmbH
Europaplatz 3
69115 Heidelberg, Germany

The manufacturer's authorised representative in the EU is Springer Nature Customer Service Centre GmbH, Europaplatz 3, 69115 Heidelberg, Germany. If you have any concerns regarding our products, please contact ProductSafety@springernature.com

Printed and bound by CPI Group (UK) Ltd, Croydon, CR0 4YY
01/10/2026
02225935-0011